EOSINOPHIL ULTRASTRUCTURE

EOSINOPHIL ULTRASTRUCTURE

Atlas of Eosinophil Cell Biology and Pathology

Rossana C. N. Melo

Ann M. Dvorak

Peter F. Weller

ELSEVIER

ACADEMIC PRESS
An imprint of Elsevier

Academic Press is an imprint of Elsevier
125 London Wall, London EC2Y 5AS, United Kingdom
525 B Street, Suite 1650, San Diego, CA 92101, United States
50 Hampshire Street, 5th Floor, Cambridge, MA 02139, United States
The Boulevard, Langford Lane, Kidlington, Oxford OX5 1GB, United Kingdom

Notices
Knowledge and best practice in this field are constantly changing. As new research and experience broaden our understanding, changes in research methods, professional practices, or medical treatment may become necessary.

Practitioners and researchers must always rely on their own experience and knowledge in evaluating and using any information, methods, compounds, or experiments described herein. In using such information or methods they should be mindful of their own safety and the safety of others, including parties for whom they have a professional responsibility.

To the fullest extent of the law, neither the Publisher nor the authors, contributors, or editors, assume any liability for any injury and/or damage to persons or property as a matter of products liability, negligence or otherwise, or from any use or operation of any methods, products, instructions, or ideas contained in the material herein.

Library of Congress Cataloging-in-Publication Data
A catalog record for this book is available from the Library of Congress

British Library Cataloguing-in-Publication Data
A catalogue record for this book is available from the British Library

ISBN 978-0-323-99413-2

For information on all Academic Press publications
visit our website at https://www.elsevier.com/books-and-journals

Publisher: Andre Gerhard Wolff
Acquisitions Editor: Linda Versteeg-buschman
Editorial Project Manager: Timothy Bennett
Production Project Manager: Swapna Srinivasan
Cover Designer: Kennedy Bonjour (front) and Miles Hitchen (back)

Typeset by STRAIVE, India

Contents

About the authors

Rossana C. N. Melo, MS, PhD
Professor of Cell Biology
Federal University of Juiz de Fora
Juiz de Fora, Minas Gerais, Brazil
Visiting Scientist from 2002 to 2019
Beth Israel Deaconess Medical Center
Harvard University
Boston, Massachusetts, United States
https://orcid.org/0000-0003-1736-0806

Ann M. Dvorak, MD
Professor Emerita of Pathology
Beth Israel Deaconess Medical Center
Harvard Medical School
Boston, Massachusetts, United States

Peter F. Weller, MD
William B. Castle Professor of Medicine
Harvard Medical School
Beth Israel Deaconess Medical Center
Professor of Immunology and Infectious Diseases
Harvard Medical School
Harvard T. H. Chan School of Public Health
Boston, Massachusetts, United States
https://orcid.org/0000-0001-9580-560X

The authors have decades of experience focused on studies of eosinophils in health and disease. The collaborative interactions of the authors include a pioneering pathologist—Dr. Ann Dvorak—who has singularly advanced our structural knowledge through her elegant career-long electron microscopy of eosinophils, and a physician investigator—Dr. Peter Weller—who has studied eosinophil functioning and the roles of eosinophils in diseases. A distinguished cell biologist—Dr. Rossana Melo—joining with the other authors, has applied novel electron microscopic techniques that have advanced our insights into the structural bases of eosinophil biology. Drawing on their studies together, the authors provide unique expertise and experience related to human eosinophils, and the atlas of eosinophil ultrastructure is a product of their longstanding studies.

Preface

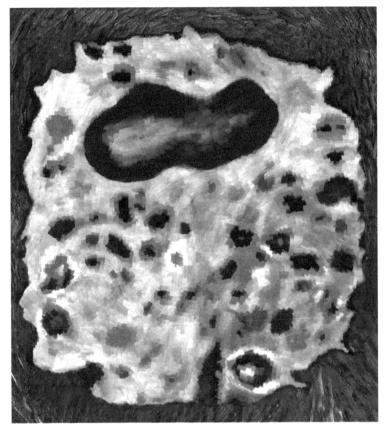

A human eosinophil seen under the transmission electron microscope and pseudocolored in Van Gogh's style.

Eosin dye was one of the most brilliant hues of Vincent van Gogh's palette applied to create a variety of nuances in his astonishing paintings more than 150 years ago. Around the same time, the eosinophil, a cell of the immune system, was brought to the attention when Paul Ehrlich captured the bright pink/orange/red nuances of its granules on eosin staining under the light microscope. Since then, this *eosin lover* has been considered one of the most beautiful and enigmatic cells. As important elements of disease processes and maintenance of homeostasis, eosinophils have a vast history of research with remarkable scientific advances, but there is still much to learn regarding their life.

Under the electron microscope, eosinophils were revealed with an additional striking visual—their granules bear an electron-dense crystalline core placidly enmeshed in an electron-lucent matrix—which is not found in any other cells, thus giving eosinophils an elegant ultrastructural signature and making electron microscopy (EM) a powerful tool to investigate them.

Eosinophil granules are complex and contain a multitude of proteins. Release of the granule contents is a critical event of the eosinophil function that only EM can uncover in detail. In this atlas, we explore the ultrastructure of eosinophils, illustrating the moods and the contrasts of their life and death. What emerges is the awesome eosinophil subcellular world that only EM can provide.

The atlas is divided into three main sections. The first section—*The Cell Biology of Human Eosinophils*—includes the general ultrastructure of eosinophils found in the bone marrow, peripheral blood, and tissues. We provide a core of essential knowledge to identify both mature and immature eosinophils and to understand eosinophil biological processes, such as activation and degranulation, observed mostly in experimental conditions. Eosinophils are shown under a blend of conventional and advanced electron microscopic techniques such as immunonanogold EM to uncover specific subcellular localization of cytokines and other proteins and electron tomography to show tridimensional aspects of eosinophil organelles.

The second section—*Eosinophil in Human Diseases*—explores the ultrastructure of eosinophils seen in body fluids and biopsy specimens from patients with eosinophilic disorders. The cellular features and structural events associated with eosinophil functioning, as well as its microenvironment, are depicted in varied human diseases, thus providing a link between basic science and clinical aspects.

The third section—*The Cell Biology of Mouse Eosinophils*—is focused on the ultrastructure of mature and immature eosinophils of mouse models, which are extensively used in studies aiming to understand the mechanisms underlying eosinophil-associated diseases. We highlight the ultrastructural differences between human and mouse eosinophils and the responses of mouse eosinophils to activation conditions and in the context of diseases.

This atlas is a unique and comprehensive reference; it is a guide for scientists who use EM as a tool to study eosinophil structural biology, cellular immunology, innate and adaptive immunity, immune responses to pathogens, inflammation, and pathology. By uncovering the eosinophil universe, the atlas also serves as a source to cell biologists pursuing to understand the remarkable structure-function relationships during a cell journey in a dynamic environment.

As a portrait, every image captured by EM reflects at high resolution a moment of the eosinophil lifetime experience and invites the viewer to engage with and to learn from this multifaceted cell. In this atlas, the eosinophil is under the spotlight in time moments and scenes of its enigmatic life, which have been motivating our scientific careers and fascinating us for decades.

We hope you will enjoy the fantastic eosinophil world!

Rossana C. N. Melo
Ann M. Dvorak
Peter F. Weller

Acknowledgments

In our compilation of this *Eosinophil Ultrastructure Atlas*, we have many to thank and recognize for their efforts, encouragement, and support.

For the decades-long studies of eosinophils and their structure and function, we acknowledge with gratitude:

Family support: RCNM: Sarah M. Salles, my daughter, who has made all the difference to me. AMD: Harold Dvorak, MD, Emeritus Professor of Pathology at Harvard Medical School and former Chair of the Department of Pathology at Beth Israel Deaconess Medical Center. PFW: Anne Nicholson-Weller, MD, Professor of Medicine at Harvard Medical School.

Technical support: We thank all the technicians at the Electron Microscopy Unit, Department of Pathology, BIDMC, for the outstanding technical assistance, especially Rita Monahan-Earley, Ellen Morgan, Tracey Sciutto, and Kit Pyne. The time and dedication that you have put into our studies during decades were just amazing. We are also grateful to the staff at Harvard Medical School EM Facility (in particular to Maria Ericsson), Centro de Microscopia Eletrônica (UFMG, Brazil), and CAPI (ICB, UFMG, Brazil) for kind assistance in more recent years.

We extend our sincere thanks to the members of our research groups at BIDMC/Boston (USA) and Laboratory of Cellular Biology (UFJF, Brazil), past and present, for their collective efforts and enthusiasm. In particular, we thank Thiago P. Silva and Vitor H. Neves for the tremendous assistance with the figures of this atlas and Kennedy Bonjour for the beautiful cover design.

Funding support: We thank the research agencies NIH (USA) and FAPEMIG and CNPq (Brazil). AMD and PFW have been supported by many NIH grants, including AI05645, AI022571, and AI020241. For decades, AI 020241, first as R01s and then as an R37, along with a Brazilian collaborative funding supplement, supported research on eosinophil form and function. RCNM has been supported by CNPq and FAPEMIG grants, including a fellowship of research productivity granted by CNPq.

We also acknowledge the contributions of those with eosinophilic disorders who provided valuable samples and generated support through the MWV Leukocyte Research Fund to advance our studies.

PFW acknowledges with much appreciation the efforts of his colleague, Rossana Melo. Not only does the compilation and editing of this atlas constitute an enormous effort on Rossana's part, but also her technical expertise as it has been applied to the study of eosinophils has progressively advanced our insights. Electron microscopy has technically advanced from the early studies of eosinophils. With refined methodological approaches, including better preservation of intracellular membranous structures and applications of pre-embedding immunonanogold immunolabeling, Rossana's contributions have immensely advanced our understandings of the cellular functioning of eosinophils.

RCNM: My deepest gratitude to PFW and AMD. From my initial meeting with Peter in Brazil in 2001 and invitation to join his lab at Harvard, my great appreciation is for continuously inspiring me as an exemplary scientist and human being. And to Ann, a brilliant scientist with a passion for electron microscopy, I could not have wished for a better colleague.

Finally, we express our gratitude to the very capable Elsevier team who helped with this project in many ways. Special thanks to the Senior Acquisitions Editor, Linda Versteeg-Buschman, the Editorial Project Manager, Timothy Bennett, and the Production Project Manager Swapna Srinivasan for their kind coordination and a great deal of support.

We dedicate this volume to the late Michael W. Verronneau, who, with his family and friends, was an active advocate for continued research on eosinophils and their associated diseases.

Abbreviations

AA	arachidonic acid
[³H]-AA	tritiated arachidonic acid
ABPA	allergic bronchopulmonary aspergillosis
ADRP (also known as adipophilin or PLIN-2)	adipose differentiation-related protein
ANCA	antineutrophil cytoplasmic antibody
APO-1	apoptosis antigen 1
Bcl-2	B-cell lymphoma 2 protein
BFA	brefeldin-A
CASP	caspase
CCL5 [also known as RANTES (regulated on activation, normal T cell expressed and secreted)]	chemokine C-C motif ligand 5
CCL11 (also known as eotaxin-1)	chemokine C-C motif ligand 11
CCR3	C-C chemokine receptor type 3
CD9	cluster of differentiation 9
CD63	cluster of differentiation 63
CD34	cluster of differentiation 34
CD95	cluster of differentiation 95
CysLTR	cysteinyl leukotriene receptor
CLC	Charcot-Leyden crystal
CLC-P	Charcot-Leyden crystal protein
COX	cyclooxygenase
cPLA2	cytosolic phospholipase A_2
DKO	double knockout
3D	three-dimension
EADs	eosinophil-associated diseases
EARs	eosinophil-associated RNases
ECM	extracellular matrix
ECP (also known as RNase 3)	eosinophil cationic protein
ECRS	eosinophilic chronic rhinosinusitis
EDN (also known as RNase2)	eosinophil-derived neurotoxin
EGIDs	eosinophilic gastrointestinal disorders
EGPA	eosinophilic granulomatosis with polyangiitis
EoE	eosinophilic esophagitis
EoSVs	eosinophil sombrero vesicles
EoPs	eosinophil progenitors
EM	electron microscopy
EPO (also known as EPX)	eosinophil peroxidase
EPX (also known as EPO)	eosinophil peroxidase
ER	endoplasmic reticulum
ETosis	extracellular trap cell death
EETosis	eosinophil extracellular trap cell death
ETs	extracellular traps
EETs	eosinophil extracellular traps
EVs	extracellular vesicles
FEGs	free extracellular granules
cFEGs	clusters of FEGs

Gal-10	galectin-10
GM-CSF	granulocyte-macrophage colony-stimulating factor
hCMPs	human common myeloid progenitors
HE	hematoxylin and eosin
HES	hypereosinophilic syndrome
HIV	human immunodeficiency virus
IBD	inflammatory bowel disease
IFN-γ	interferon-gamma
IFNGR1	interferon-gamma receptor 1, also termed interferon-gamma receptor alpha chain
IL-3	interleukin-3
IL-4	interleukin-4
IL-4Rα	interleukin-4 receptor alpha chain
IL-5	interleukin-5
IL-5Rα	interleukin-5 receptor alpha chain
IL-6	interleukin-6
IL-12	interleukin-12
IL-13	interleukin 13
IL-31	interleukin 31
IL-33	interleukin-33
Immuno-EM	immuno-electron microscopy
IAV	influenza A virus
LB	lipid body
LD	lipid droplet
LOX	lipoxygenase
LTC4	leukotriene C4
5-LO	5-lipoxygenase
MBP-1 (also known as MBP and PRG2)	major basic protein 1
MMP9	matrix metalloproteinase-9
MVB	multivesicular body
MVs	microvesicles
NADPH-oxidase	nicotinamide adenine dinucleotide phosphate oxidase
NETs	neutrophil extracellular traps
NETosis	neutrophil extracellular trap cell death
NFAT	nuclear factor of activated T cells
OSCC	oral squamous cell carcinoma
OVA	ovalbumin
PAF	platelet-activating factor
PDGF	platelet-derived growth factor
PDGFRA	platelet-derived growth factor receptor alpha
PDI	protein disulfide isomerase
PGD2	prostaglandin D2
PLIN (also known as PLIN-1)	perilipin
PLIN-1	perilipin 1
PLIN-2	perilipin 2
PLIN-3 (also known as tail-interacting protein of 47 kDa)	perilipin 3
PMA	phorbol 12-myristate 13-acetate
PMD	piecemeal degranulation
PRR	pattern-recognition receptor
RER	rough endoplasmic reticulum
ROS	reactive oxygen species
RSV	respiratory syncytial virus
SARS-CoV-2	severe acute respiratory syndrome coronavirus-type 2
SCF	stem cell factor

SEM	scanning electron microscopy
S.E.M.	standard error of the mean
SNARE	soluble *N*-ethylmaleimide-sensitive factor attachment protein receptor
STX17	syntaxin 17
TEM	transmission electron microscopy
Tg	transgenic
TGF-α	transforming growth factor-alpha
Th1	T helper 1
Th2	T helper 2
TLR	toll-like receptor
TNF-α	tumor necrosis factor-alpha
TNFR	TNF-α receptor
TRAIL-R	TNF-related apoptosis-inducing ligand receptor
VAMP2	vesicle-associated membrane protein-2
VEGF	vascular endothelial growth factor
VPF	vascular permeability factor
WT	wild type

The cell biology of human eosinophils

1

Introduction

By examining inflammatory exudates from humans and other species during the first half of the 19th century, early investigators noticed the presence of so-called "granule blood cells" or "compound inflammatory globules" (reviewed in[1]). These distinct granular cells were certainly the first observations of eosinophils. However, the term "eosinophil" was introduced only in 1879 by Paul Ehrlich.[2] By applying new staining methods to the study of blood films, Ehrlich coined the name "eosinophil" to define cells with a large number of coarse cytoplasmic granules having a high affinity for eosin and other acid dyes, which he termed "alpha-granules."[1] Based mainly on the presence of this population of brilliant orange/red-stained granules, Ehrlich described the occurrence and distribution of eosinophils in different species, tissues, and conditions, thus providing cardinal early contributions to the fields of leukocyte biology, immunology, and pathology.[1] In fact, eosinophils, terminally differentiated, bone-marrow-derived cells of the immune system, have a broad distribution in tissues and varied functions related to both immune homeostasis and immunity.[3] Eosinophils are critical components of mechanisms related not only to the innate host defense but also to the tissue homeostasis through immune, remodeling/repair, and metabolic activities, both locally and systemically.[3–6]

The "alpha-granules," later termed specific granules, constitute the central morphologic feature of mature eosinophils and, as correctly speculated by Ehrlich, are storage sites for secretory products; for this reason, they are also referred to as secretory granules.[7,8] The acidophilic nature of eosinophil specific granules is due to the massive amount of four cationic (basic) proteins stored within them: major basic protein 1 (MBP-1) (also known as MBP and PRG2), eosinophil cationic protein (ECP) (also known as RNase3), eosinophil-derived neurotoxin (EDN) (also known as RNase2), and eosinophil peroxidase (EPX) (also known as EPO).[3] It is now recognized that the collection of proteins stored mainly as preformed products within the specific granules of human eosinophils also includes many cytokines, chemokines, and growth factors.[3,8,9] Based on their capacity to release these products, eosinophils have been recognized as multifunctional cells involved not only in host responses such as the responses to helminth infection and allergic diseases but also in diverse "noneffector" immunoregulatory situations in health and disease.[3,10]

In addition to their stunning and typical staining under light microscopy (Fig. 1.1), an intriguing and unique feature of eosinophil secretory granules was unveiled with the advent of transmission electron microscopy (TEM) and ultrathin sectioning circa 70 years after Ehrlich's discoveries. At that time, pioneering works conducted on peripheral blood and blood marrow of humans and other species considered that "the high resolving power of the electron microscope promises to reveal details previously unsuspected, as well as to extend and clarify existing knowledge concerning the cytology of blood cells, both normal and pathologic."[11] In fact, at high resolution, eosinophil secretory granules were described during the 1950s as biconvex discs bounded by a membrane and usually containing in their equatorial region inclusions of a dense, crystalloid material embedded in a less dense matrix[12–15] (Fig. 1.1).

Hence, early applications of electron microscopy (EM) showed that eosinophils in both humans and other species contained an ultrastructural signature, perhaps more specific than Ehrlich's eosin staining, which enables unambiguous identification of these cells. This means that eosinophils can be promptly recognized by conventional TEM without any particular labeling. The simple finding of eosinophil crystalline granules canonically defines the eosinophil lineage in multiple species.[9] Moreover, conventional TEM is a powerful tool not only to identify intact eosinophils in human fluids and tissues but also to detect infiltration of these cells even after their cell disruption and death. TEM enables the recognition of core-containing intact eosinophil granules when these are deposited extracellularly through cytolytic degranulation, a common situation reported in tissue sites of eosinophil-associated diseases (EADs).[16,17]

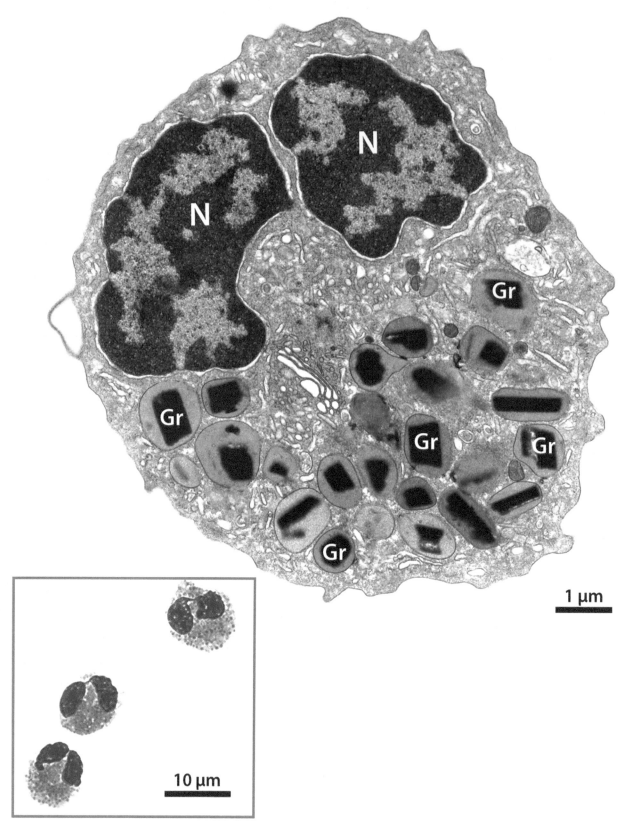

FIG. 1.1   Mature human eosinophils isolated from the peripheral blood are observed under light microscopy (LM) and transmission electron microscopy (TEM). Secretory granules (Gr) appear as highly acidophilic structures at LM after hematoxylin and eosin (HE) staining and as bicompartmental organelles with a central electron-dense core surrounded by a less dense matrix at TEM. Note the bilobed nucleus (N). The electron micrograph is also shown on the book cover.

TEM is very informative in demonstrating the complex and diverse secretory activities of eosinophils.[18] Our group has a special interest in the study of secretion mechanisms underlying eosinophil responses and has been using ultrastructural analyses to understand how eosinophil immune mediators are mobilized and released.[19–25] Since TEM is the only technique with resolution sufficient to clearly identify and distinguish between different modes of cell secretion, distinct structural scenarios can be observed within eosinophils in response to in vitro stimulation and in human diseases.[18,26] In fact, by applying TEM after optimal fixation, we found remarkable ultrastructural differences between resting and activated human eosinophils. Morphologic patterns of eosinophil degranulation, including piecemeal degranulation, classical/compound exocytosis, and cytolysis, will be discussed in detail in Chapter 3.

Progress in the comprehension of eosinophil functional activities through the application of advanced TEM showed us that the secretory granules of human eosinophils are not merely storage containers.[20] Electron tomography, a technique that allows three-dimensional (3D) visualization in high resolution,[27] brought a new view of specific granules as organelles with membranous subcompartments organized as an aggregate of flattened tubular networks and tubules with interconnections in some planes as well as connections with the granule limiting membrane.[20] Electron tomography in conjunction with conventional TEM also enabled the identification of a distinct vesicular system with typical morphology—large tubular carriers termed by us eosinophil sombrero vesicles (EoSVs) because of their appearance as a "Mexican hat"—actively involved in intracellular trafficking of specific proteins from secretory granules.[19,21]

Initial conventional and immunogold EM studies have focused on granules within eosinophils and their compartmentalized content of varied eosinophil proteins, and localized MBP-1 to the granule core and other cationic proteins such as ECP and EPO to the matrix of the granules.[28,29] Subsequent advances in EM methodologies, including better preservation of cytosolic vesicular compartments (that was not feasible and hence was lacking from most early EM studies), more sensitive immuno-EM labeling, and detection strategies have revealed detailed subcellular localization of cytokines and other specific proteins within human eosinophils as shown in Chapter 5. The Atlas highlights, for example, the subcellular localization of the tetraspanins CD63 and CD9 found as robust constitutive pools in the cytoplasm and at the cell surface of human eosinophils, respectively,[23,30] and the presence of cytokines such as interleukin-4 (IL-4)[31] and interferon-gamma (IFN-γ)[25] being carried in association with the membranes of vesicular carriers (EoSVs) additionally to their localization as preformed pools in specific granules. Remarkably, MBP-1 was also consistently localized in EoSVs, thus disclosing, for the first time, that vesicular traffic is an important pathway to release this cationic protein.[22] By applying immunonanogold EM, we have also demonstrated intracellular traffic of cytokine receptors, such as IL-4R alpha[31] and Notch signaling molecules.[32] Our ultrastructural immunolabeling studies also unveiled intracellular sites within human eosinophils in which galectin-10 (Gal-10), a carbohydrate-binding protein associated with eosinophilic inflammatory diseases and considered a potential marker of eosinophil inflammation,[33–36] is stored.[37] Because Gal-10 characteristically forms bipyramidal, hexagonal, or amorphous crystals known as Charcot-Leyden crystals (CLCs),[38] frequently found in tissues from patients with eosinophilic disorders,[39–41] we also provided insights from these ultrastructural studies into how this protein is released from eosinophils and crystallized to form extracellular CLCs in connection with cytolysis.

Other ultrastructural aspects of the human eosinophil's life and death have been studied by us and are explored in this Atlas. These include the ultrastructural architecture of eosinophils participating in a wide variety of human diseases (Chapter 8), based largely on studies of diagnostic ultrastructural pathology, and under in vitro activation with a plethora of agonists after cell isolation from the peripheral blood.[42–44] As EM enables imaging of the object of interest at high resolution in its structural context,[45] ultrastructural studies have been providing critical insights into the responses of tissue eosinophils to their microenvironment including interactions with other inflammatory cells.

We also emphasize ultrastructural views of extracellular trap cell death (ETosis), a nicotinamide adenine dinucleotide phosphate (NADPH) oxidase-dependent pathway identified in human eosinophils, which culminates in the nuclear envelope and plasma membrane disintegration, deposition of free extracellular granules (FEGs), and development of filamentous chromatin structures termed extracellular traps—all morphological aspects clearly identified by EM.[46,47] These ultrastructural features characterize eosinophil ETosis, a process involved in inflammation and/or efficient elimination of pathogens detailed in Chapter 6.[46–49]

Our ultrastructural qualitative in vitro and in vivo studies are integrated with quantitative evaluations that have been fundamental to understanding how the eosinophil increases its resources of organelles/structures in response to different stimuli and inflammatory conditions. We address quantitative ultrastructural evaluation of structures and organelles, which are particularly formed in response to cell activation/inflammation, such as lipid bodies, EoSVs, CLCs, and extracellular vesicles. Applications of ultrastructural cytochemical, autoradiographic, and immunocytochemical techniques, all highlighted in this Atlas, have contributed to a better understanding of the eosinophil functional capabilities and to distinguish organelles and structures regularly present in mature eosinophils that are also altered in number and appearance in certain situations and diseases.

A. The cell biology of human eosinophils

Finally, the Atlas brings a comprehensive view of human immature eosinophils (Chapter 7) from ultrastructural studies conducted on cells collected from the bone marrow or eosinophils arising in culture systems either completely or partially supportive for the differentiation, proliferation, and maturation of these cells.

As noted, applications of both conventional and advanced EM techniques have provided substantial insights into the structural organization, the cellular content of biomolecules, and functional capabilities of human eosinophils, but these applications require critical attention with the samples and many technical challenges. Our ability to obtain interpretable material for both diagnostic purposes and research depends greatly on how the material is handled from the beginning. Key elements in this initial stage include the appropriate collection of the cells/tissues, which means special care in specimen manipulation and precise fixation. Inappropriate handling can easily activate human eosinophils and harm these cells. For example, the steps of eosinophil isolation should be done with critical care while fixation of these cells is promptly done while the cells are still in suspension and not after pelleting, which may lead to structural changes. Cells fixed in suspension are then pelleted in agar so that uniformly distributed cells can be processed as blocks of tissues. Other methodological strategies for accurate handling of eosinophils have been used in our studies. When eosinophils are kept adhered on slides to evaluate events such as cytolysis, cell migration, or formation of extracellular DNA traps,[46,47] all procedures for TEM are directly done on the slide surface, thus enabling visualization of these events exactly at the sites in which they occurred, without any interference with the cell morphology. The application of immunonanogold EM to eosinophils requires specific procedures to provide sensitive antigen detection alongside detailed information on the cell structure as discussed in our protocol developed primarily to capture immune mediators and other molecules in these cells.[50] For example, successful localization of cytokines within human eosinophils requires immunolabeling before the specimens are processed for EM (pre-embedding immuno-EM), which enables improved antigen preservation.[50]

In summary, in this Atlas, we review the ultrastructural studies of human eosinophils that we have done for over 30 years, which allowed the identification of new structural aspects of these cells and the advancement of knowledge regarding their mechanisms of functioning.

A. The cell biology of human eosinophils

2

Mature eosinophils: General morphology

2.1 Overview

The purpose of this chapter is to describe the ultrastructural architecture of human eosinophils as observed in normal conditions. These cells are ~ 12–15 μm in diameter[51] and, like other granulocytes, are characterized by a segmented nucleus usually seen as a bilobed organelle, large cytoplasmic granules, and an irregular surface with short cell projections (Fig. 2.1).[9,40,52–54] These surface protrusions can be observed by routine TEM (Fig. 2.1), cytochemical ultrastructural demonstration of nonspecific esterase activity (α-naphthyl acetate esterase), which label this ectoenzyme at the cell surface[55] (Fig. 2.2) and notably by scanning electron microscopy (SEM), which shows the cell surface in 3D (Fig. 2.3). As noted, a population of cytoplasmic secretory granules with unique morphology (crystalline granules) readily identifies eosinophils (Figs. 2.1 and 2.2), but these cells are also characterized by the presence of other distinctive organelles and structures: very osmiophilic lipid bodies (Fig. 2.2) and vesiculotubular structures termed eosinophil sombrero vesicles (EoSVs).[9] Additional cytoplasmic organelles common to other cells include mitochondria, Golgi apparatus, variable amounts of smooth tubules and small vesicles, free ribosomes, minimal amounts of the peripheral rough endoplasmic reticulum (RER), endosomes, and glycogen particles/aggregates.[40] These and other ultrastructural features of mature human eosinophils are detailed below.

2.2 Nucleus

As granulocytes, mature human eosinophils observed in the peripheral blood (Figs. 2.4–2.7) and tissues (Figs. 2.8 and 2.9) exhibit a segmented nucleus, that is, a nucleus subdivided into a varying number of lobes (polylobed).[9,56] Thus, when these cells are visualized by TEM, they can appear, depending on the plane of section, with up to four nuclear lobes (Figs. 2.4–2.9). However, a bilobed nucleus is the most frequent appearance for human eosinophils in thin sections (Figs. 2.1, 2.2, 2.4, 2.5, and 2.8), likely because two nuclear lobes are more prominent and well defined for the majority of the cells, thus explaining why eosinophils are routinely considered a bilobed cell. In an earlier study, in which eosinophils were evaluated by light microscopy in blood smears of normal donors, the "lobe index" (mean number of nuclear lobes per eosinophil) was 2.06 with 70%–90% of eosinophils classified as bilobed and the remaining as unsegmented or three-lobed, while four-lobed eosinophils were rarely noticed.[57] By TEM, the presence of four lobes in thin sections of human eosinophils (Fig. 2.7) is also identified just occasionally, whereas cells with three lobes (Figs. 2.4, 2.6, and 2.9) are more commonly found. The nuclear lobes of human eosinophils and other granulocytes are connected by a thin bridge of chromatin, which can be occasionally observed in areas of deep constrictions of the nucleus when the section plane passes through these areas (Fig. 2.10). Of note, an increase in the frequency of blood eosinophils with three-four nuclear lobes or with hypersegmentation (more than four lobes) was reported in association with some pathological conditions.[57–60] It is also important to consider that, because TEM images are obtained from sectioned samples, human eosinophils can show just one of the lobes, a small fraction of the nucleus or even the nucleus may not be visible in the section plane.

Another feature of the eosinophil nucleus is a condensed marginal chromatin (heterochromatin) with a clear distinction between this and the remaining electron-lucent chromatin (euchromatin) (Figs. 2.10 and 2.11). This characteristic can be lost when eosinophil nuclei undergo chromatin decondensation during different processes of cell death, as described in Chapter 6. The nucleolus is not evident in the eosinophil nucleus but can be sporadically seen in electron micrographs (Fig. 2.9).

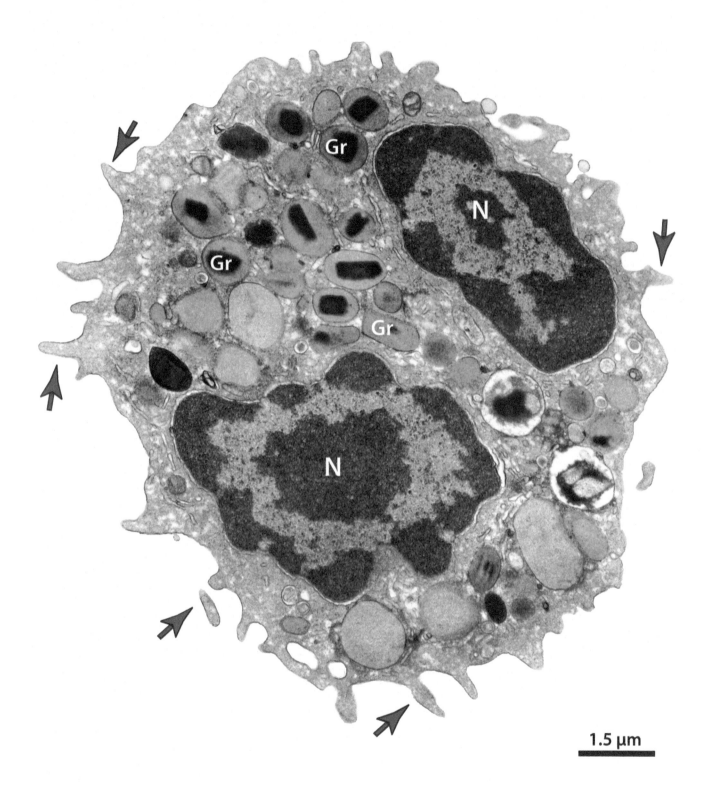

FIG. 2.1 Electron micrograph from a human eosinophil isolated from the peripheral blood. Note the bilobed nucleus (N), specific granules (Gr) occupying most of the cytoplasm, and the irregular cell surface with short cell projections *(arrows)*.

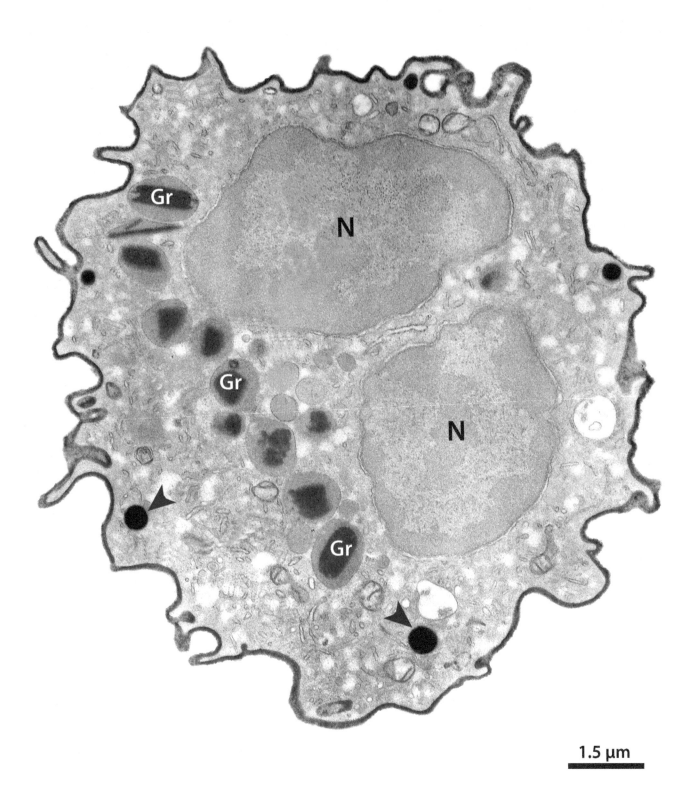

FIG. 2.2 The surface of a human eosinophil is highlighted after application of a cytochemical technique for ultrastructural demonstration of nonspecific esterase activity. *Gr*, secretory granules; *N*, nucleus. Lipid bodies are indicated by *arrowheads*.

A. The cell biology of human eosinophils

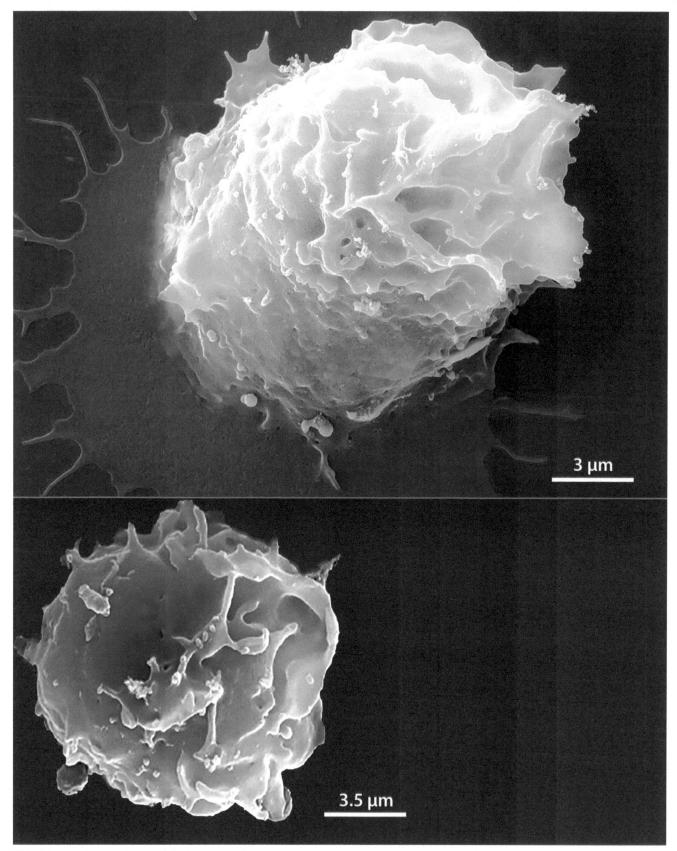

FIG. 2.3   Blood human eosinophils observed under scanning electron microscopy (SEM). As a feature of granulocytes and other leukocytes, the cell surface is characterized by numerous cellular projections. The high membrane surface area facilitates migration and phagocytosis events. *Top panel* shows a pseudocolored cell.

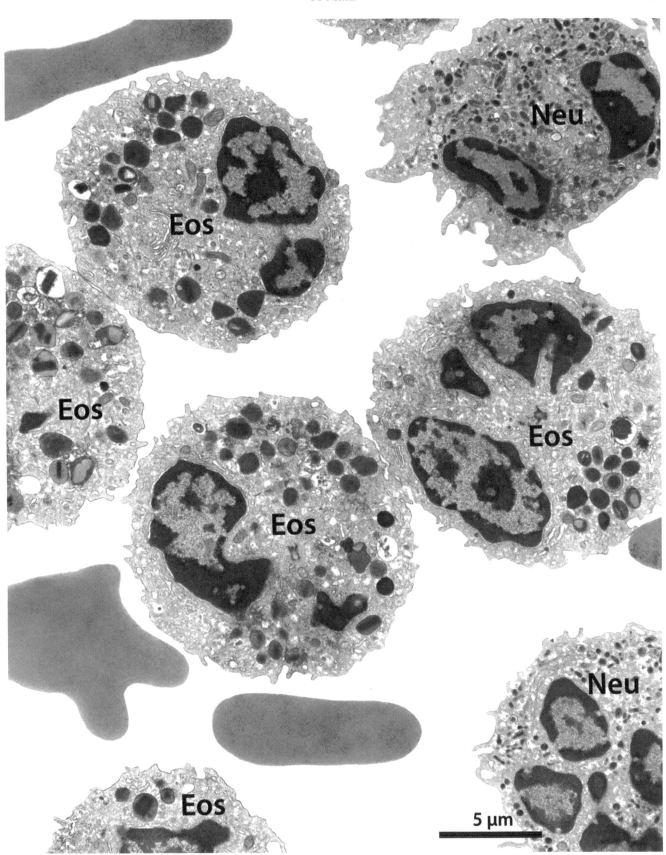

FIG. 2.4 Electron micrograph showing human granulocytes in the peripheral blood. Eosinophils (Eos), neutrophils (Neu), and red blood cells *(colored in red)* with typical ultrastructure are observed. The polylobed nucleus *(colored in purple)* of these leukocytes is noted at different section planes.

A. The cell biology of human eosinophils

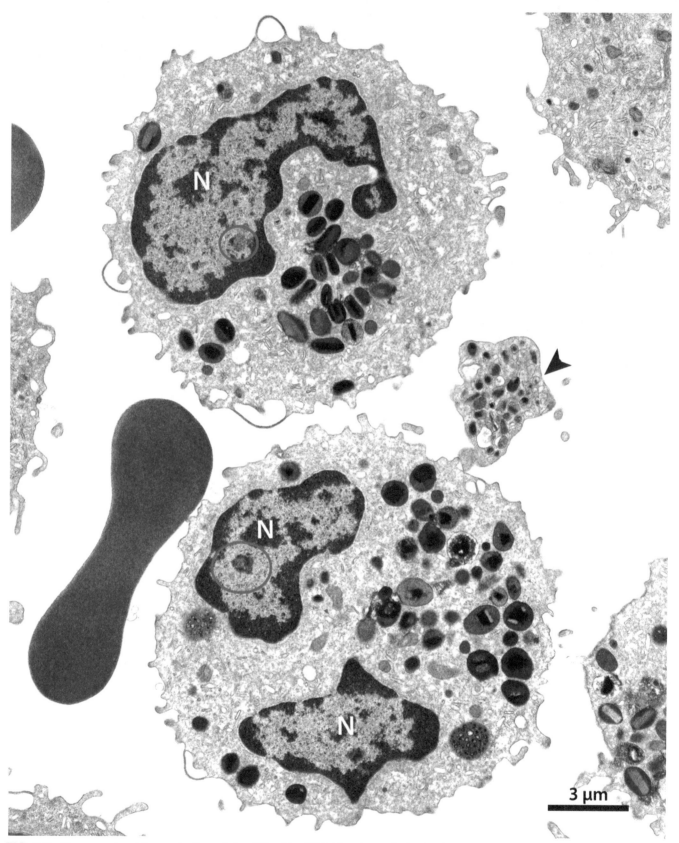

FIG. 2.5   Human eosinophils seen in the peripheral blood by TEM. Note the nucleolus *(circles)* within the lobes of the nucleus (N). Specific granules are distributed in the cytoplasm. A platelet is indicated *(arrowhead)*.

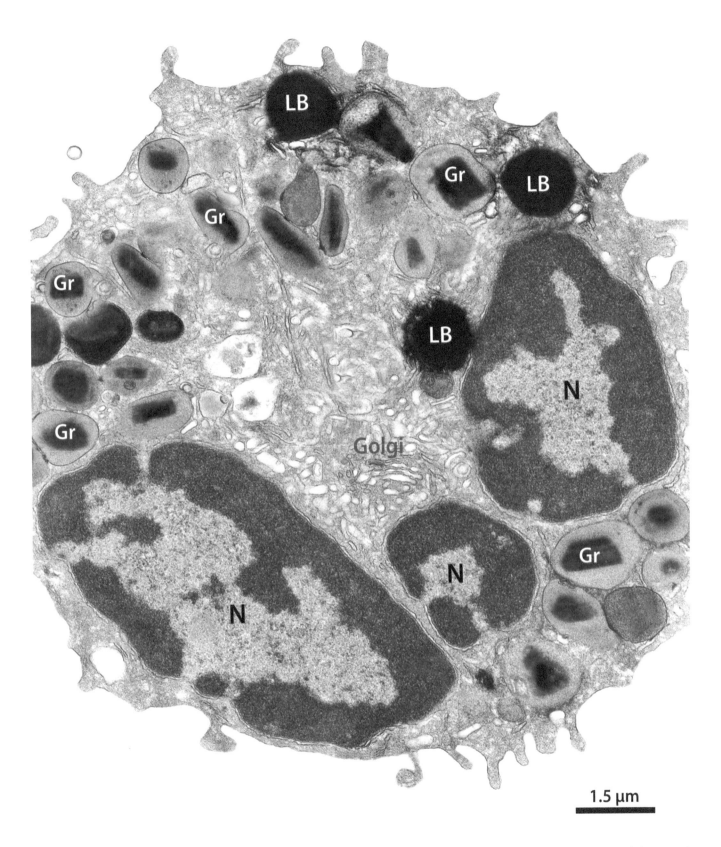

FIG. 2.6 Electron micrograph of a mature human peripheral blood eosinophil showing a three-lobed nucleus (N). The marginal chromatin is characteristically electron-dense. Note osmiophilic lipid bodies (LB), the Golgi complex, and secretory granules (Gr) in the cytoplasm.

A. The cell biology of human eosinophils

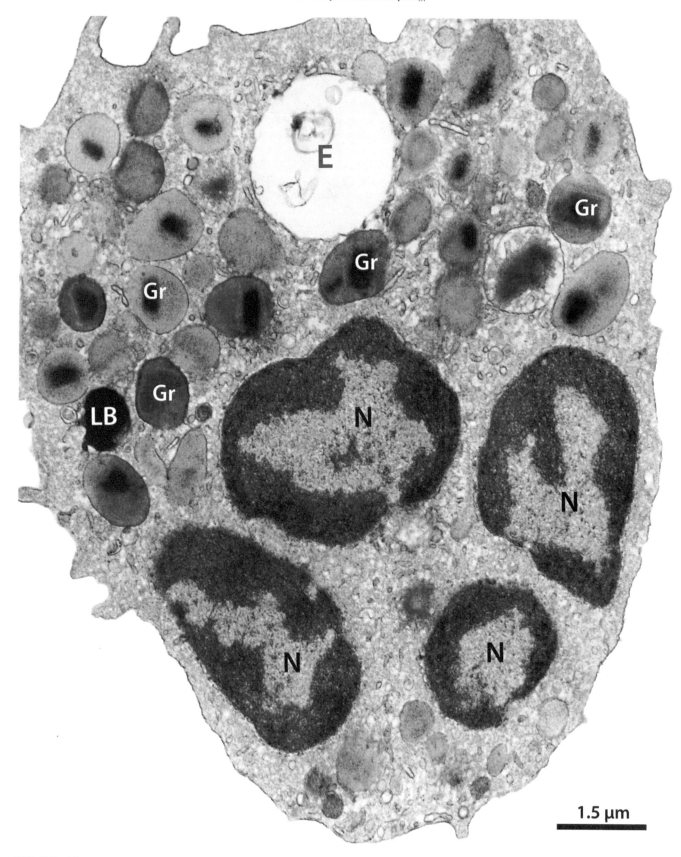

FIG. 2.7   Electron micrograph of a mature human peripheral blood eosinophil showing a four-lobed nucleus (N). An endosome-like large vacuole (E) is observed in the cell periphery. *Gr*, specific granules; *LB*, lipid body.

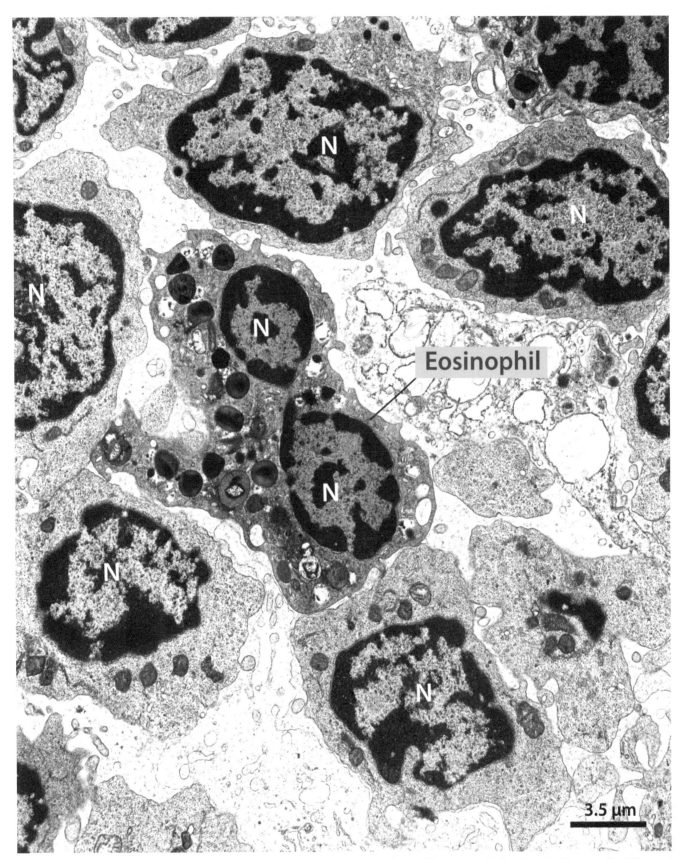

FIG. 2.8   A mature eosinophil, surrounded by several lymphocytes, is seen in a human biopsy from the lymph node. Compare the differential morphology of the nuclei (N) between eosinophil (bilobed) and lymphocytes (mononuclear).

A. The cell biology of human eosinophils

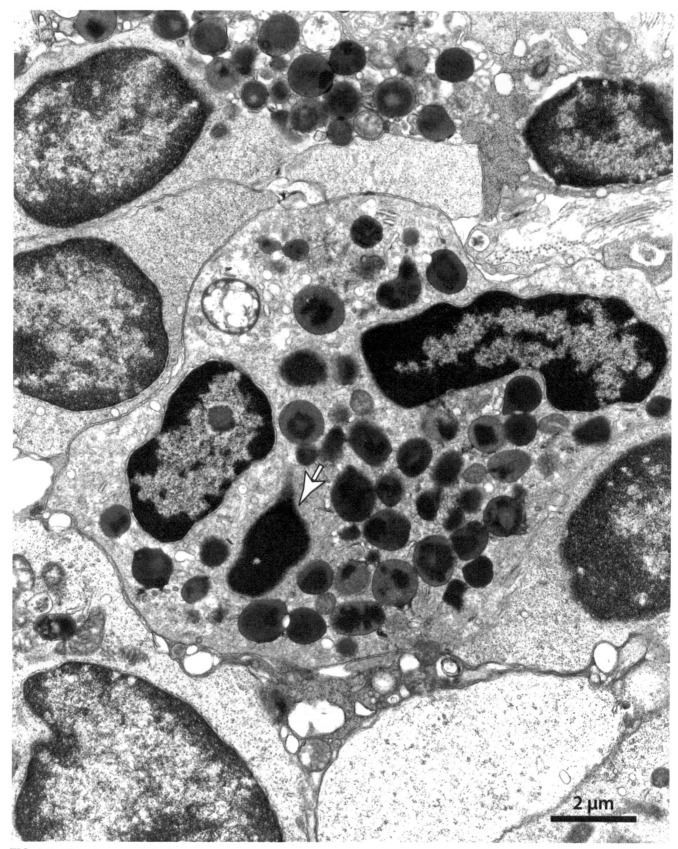

FIG. 2.9   Electron micrograph of a mature tissue eosinophil showing a three-lobed nucleus *(colored in purple)*. Just a fraction of one of the lobes is observed *(arrow)*, while the nucleolus *(highlighted in orange)* can be seen in another lobe. The cytoplasm of surrounding cells (lymphocytes) was colored in *green*.

A. The cell biology of human eosinophils

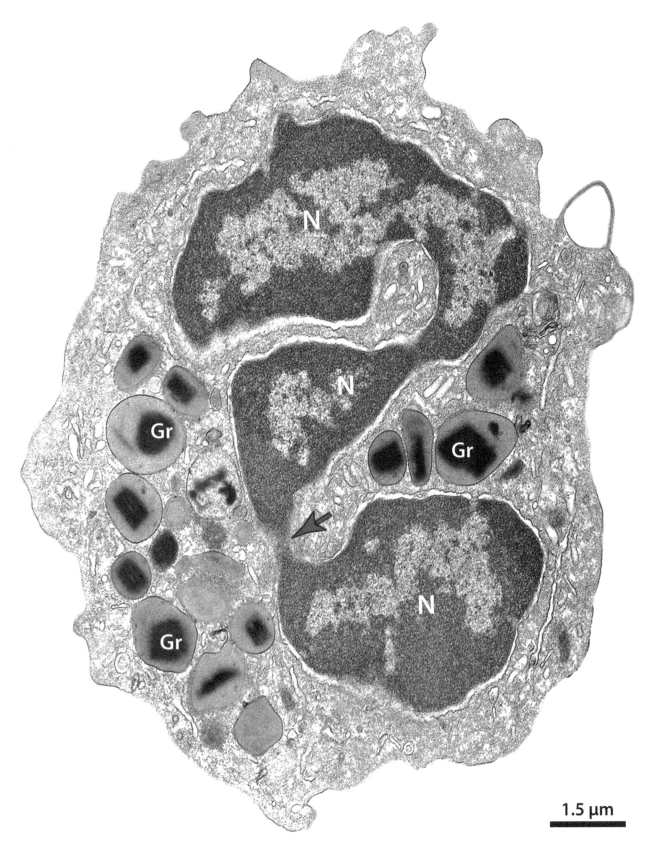

FIG. 2.10   A human peripheral blood eosinophil showing the polylobed nucleus (N) and specific granules (Gr). Note the clear distinction between heterochromatin and euchromatin and the thin bridge of chromatin (arrow) that connects the nuclear lobes in an area of deep constriction.

A. The cell biology of human eosinophils

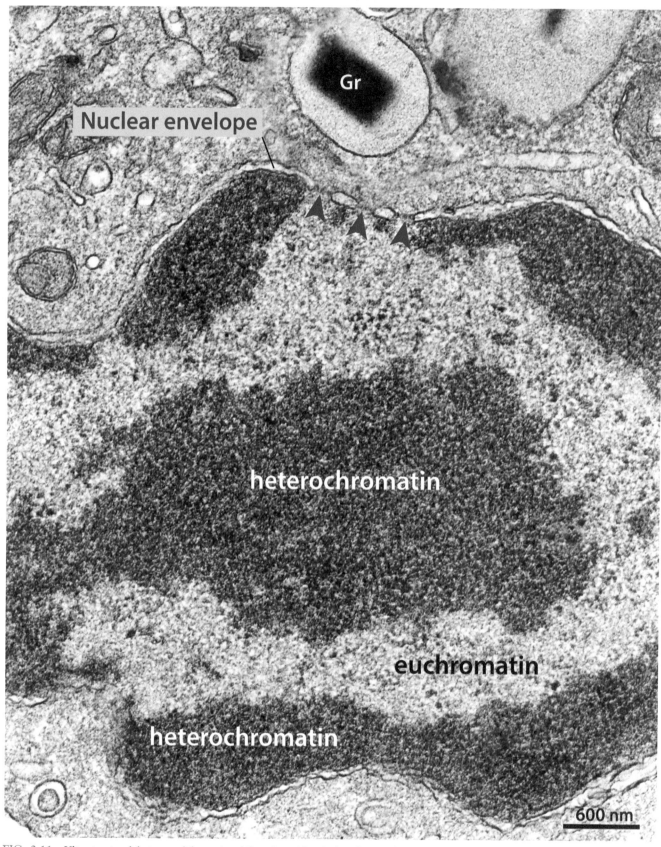

FIG. 2.11 Ultrastructural features of the eosinophil nucleus. Marginal and part of the central chromatin are condensed (heterochromatin), while the remaining chromatin is uncondensed (euchromatin). Note the two concentric membranes and nuclear pores *(arrowheads)* of the nuclear envelope. A specific granule (Gr) with its typical morphology is observed.

The nuclear morphology is an important characteristic that helps to distinguish mature from immature eosinophils, which appear as mononuclear cells (single round nucleus).[40] Consequently, the finding of eosinophils showing a round nucleus in the peripheral blood or tissues may be indicative of an exacerbated recruitment of these cells, thus leading to the release of a cell population from the bone marrow still in the process of maturation/terminal differentiation. In tissues, the presence of eosinophils with a round nucleus may additionally denote a local differentiation of eosinophils or a process of "delobulation" observed during cell death through ETosis, in which the eosinophil nucleus progressively loses its shape in parallel to chromatin decondensation (Chapter 6). Therefore, to characterize a complete mature human eosinophil, the nucleus must show typical segmentation combined with the presence of crystalline granules in the cytoplasm (Figs. 2.1, 2.2, 2.4, and 2.5). However, special attention should be taken when interpreting the nuclear appearance considering that, as noted, the nucleus can be sectioned at different planes. The morphological characteristics of immature eosinophils are presented in Chapter 7.

In routine preparations of eosinophils for TEM, the structural organization of the nuclear envelope can be clearly identified in high magnification and appear typically as two concentric bilayers surrounding the chromatin and interrupted in some points by nuclear pores, observed in all eukaryotic cells (Fig. 2.11). When osmium potassium ferrocyanide or reduced osmium is used during processing for EM, all cellular membranes are extremely well contrasted while other structures/organelles may look less contrasted.[61] Thus, the nuclear envelope can be accentuated with these techniques becoming more discernible while nuclear chromatin is poorly contrasted compared with other routine techniques (Fig. 2.12). When the fixation process is defective, nuclear envelopes as well as other endomembranes are not recognizable or look as distorted structures. Therefore, one way to check if the chemical fixation of eosinophils was adequate after routine EM procedures is to visualize the presence of a morphologically well-preserved nuclear envelope (Figs. 2.11 and 2.12).

As a general cell biological feature, the nuclear envelope is considered an endoplasmic reticulum (ER) domain, and as such, it possesses ER-specific molecules.[62–65] This is the case of the protein disulfide isomerase (PDI), classically involved in the oxidative folding of proteins in the ER[66–68] and detected as a predictable strong pool at the nuclear envelope of human eosinophils by immunonanogold EM[69] (Fig. 2.13).

2.3 Secretory granules

Human eosinophils contain a single population of secretory granules termed specific granules. As a typical eosinophil characteristic, secretory granules from mature eosinophils exhibit a centrally located crystalloid electron-dense core and an outer less dense matrix (Fig. 2.14).[9,18] Because of this particular morphological feature, eosinophil-specific granules are frequently referred to as crystalline granules or cored granules.[18]

Eosinophil secretory granules are delimited by a typical membrane. Since phospholipid membranes universally appear under TEM with a "trilaminar" aspect in which the hydrophilic phosphate "heads" are electron-dense and the hydrophobic fatty acids "tails" are electron-lucent,[70,71] this "trilaminar" structure can be clearly seen around eosinophil secretory granules at high magnification (Fig. 2.15).

In healthy donors, the amount of specific granules in eosinophils isolated from the peripheral blood corresponds to almost 100% of the granules. Eosinophil-specific granules are large, generally ellipsoid (> 0.5 μm in the long axis), and occupy a significant part of the cytoplasm in mature eosinophils (Figs. 2.1 and 2.6).[18] In electron micrographs from thin sections, these granules can also be seen as round and/or occasional coreless structures (Fig. 2.14), which can result from different sectioning planes or represent few immature granules. However, the occurrence of true core-free (immature) granules in mature eosinophils is rare, considering that some oblique sections of specific granules may not exhibit cores. It is also important to emphasize that activated mature eosinophils (Chapter 4) can show emptying coreless granules due to mobilization of intragranular contents and, therefore, EM images from eosinophil granules should be carefully analyzed.

Immature granules (lacking a crystalline core) can be at times observed in eosinophils from peripheral blood and tissues in situations in which there is an overproduction of these cells, thus releasing cells from the bone marrow with immature morphological features.[18] These coreless granules were considered in the past as a separate granule population referred to as "primary" granules,[72] analogously to that found in mature neutrophils in which primary (azurophilic) granules persist in mature cells.[73,74] Thus, two major types of large granules were considered to exist within mature human eosinophils: "primary" (coreless granules) and "secondary" (specific, cored granules).[10,40,56] However, accumulated data provide evidence that "primary" granules are indeed early "secondary" cored granules (reviewed in[18]). Consequently, granules of mature human eosinophils are of a single type derived by the transition from spherical coreless granules.[18] We, therefore, consider that human mature eosinophils contain just a single population of

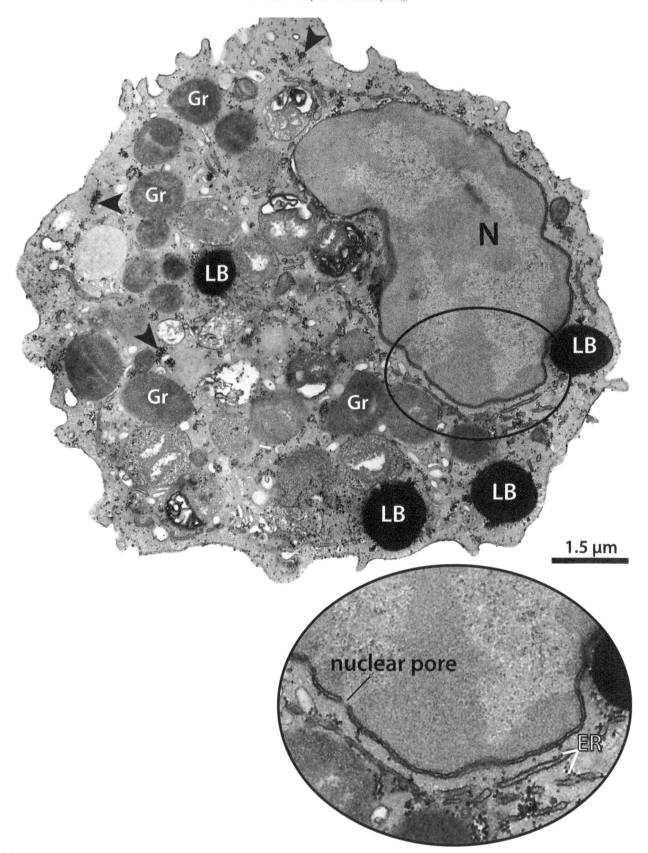

FIG. 2.12   A human blood eosinophil prepared with a reduced osmium technique for TEM. Lipid bodies (LB) and membranous structures are highly contrasted, while the chromatin and other organelles appear less contrasted. A nuclear pore and endoplasmic reticulum (ER) cisternae are seen in higher magnification. This technique also enables the visualization of glycogen particles seen as individual electron-dense dots and small aggregates *(arrowheads)* distributed in the entire cytoplasm. *Gr*, specific granules; *N*, nucleus.

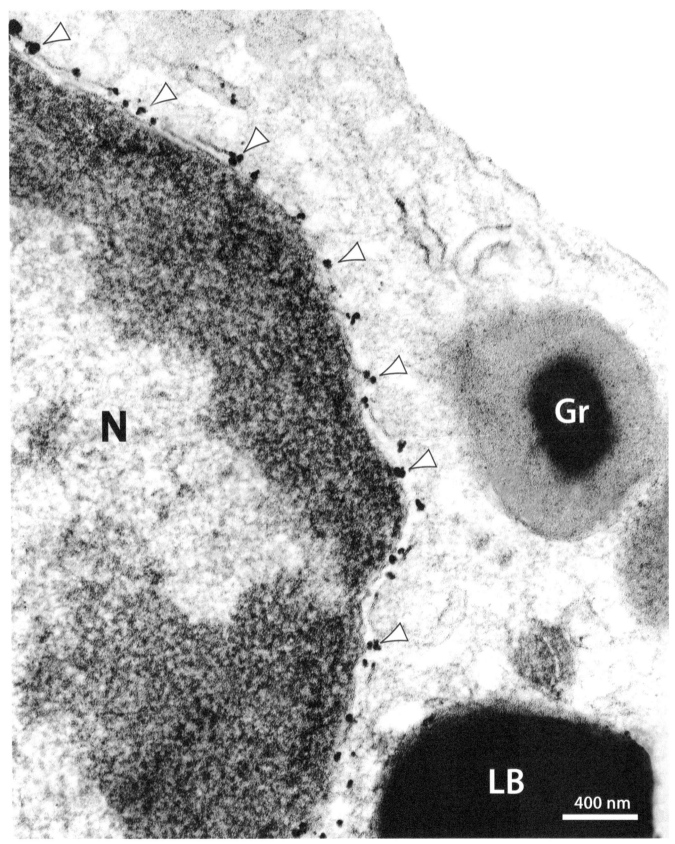

FIG. 2.13 The nuclear envelope of a human eosinophil immunolabeled for protein disulfide isomerase (PDI). Eosinophils isolated from the peripheral blood were prepared for pre-embedding immunonanogold EM.[50] Positivity is indicated by secondary antibodies associated with gold particles *(arrowheads)*. *Gr*, specific granule; *LB*, lipid body; *N*, nucleus.

A. The cell biology of human eosinophils

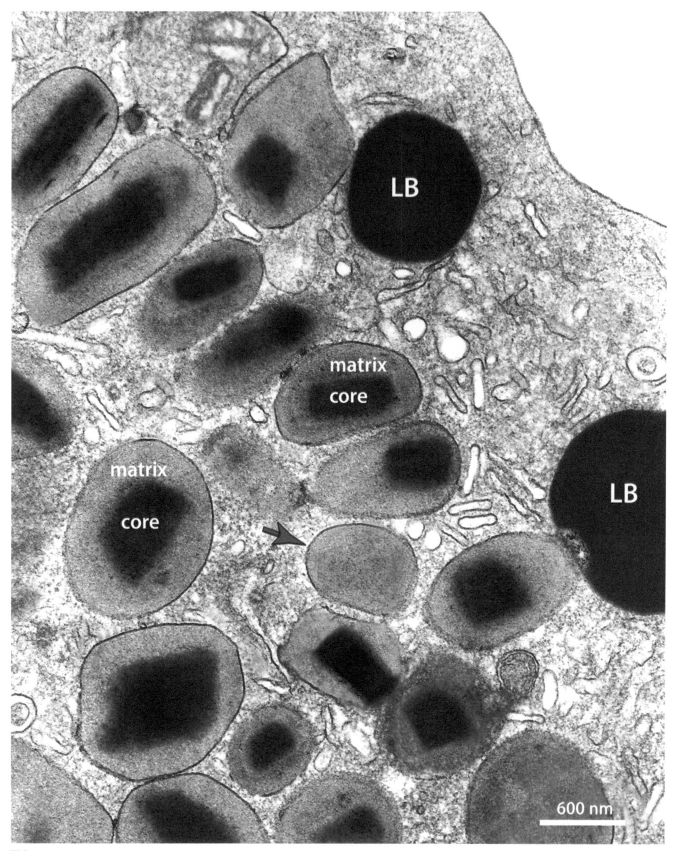

FIG. 2.14   Transmission electron micrograph of a mature human eosinophil. Cytoplasmic specific granules contain electron-dense crystalline cores and outer less dense matrices—morphology unique to eosinophils. In one granule *(arrow)*, the core is not visible in the section plane. *LB*, lipid body.

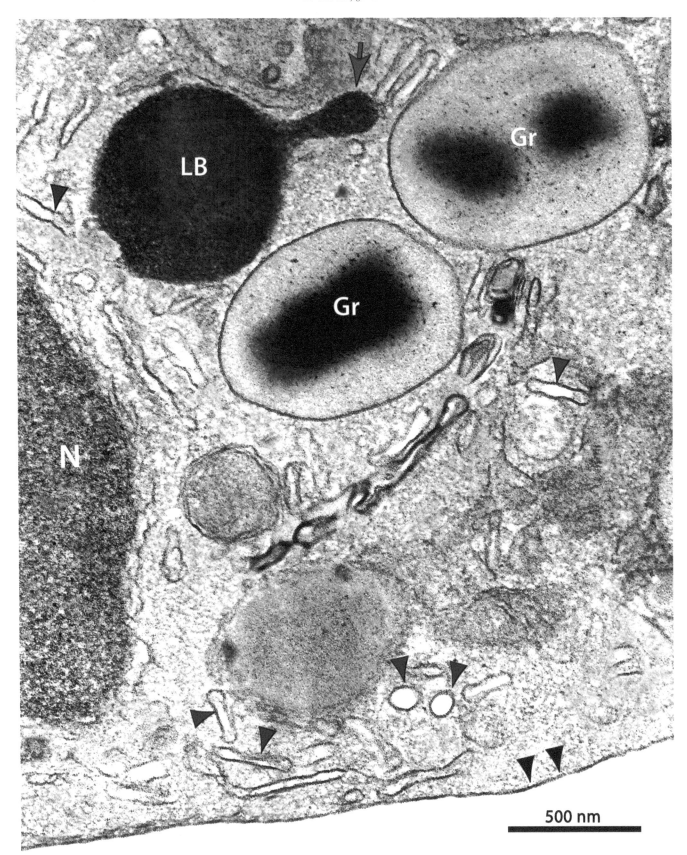

FIG. 2.15 Eosinophil secretory granules (Gr) observed at high magnification by TEM. The trilaminar aspect of the limiting granule membrane is also seen in membrane-bound vesicles *(pink arrowheads)* and plasma membrane *(black arrowheads)*. A lipid body (LB) shows a budding-like structure *(arrow)*. N, nucleus.

A. The cell biology of human eosinophils

granules and that the appropriate nomenclature to refer to them is "specific granules" and not "secondary granules" since what was considered in the past as "primary granules" are in fact immature specific granules. The morphology of immature human eosinophils is shown in Chapter 7.

The specific granules of human eosinophils can occasionally exhibit a reversal of the electron densities in their structural elements, that is, granules in which the crystalline cores appear less dense than the surrounding matrix, a phenomenon also referred to as reversal of the core "staining."[17] We have been observing this particular difference in the electron density of the crystalloid and matrix in both blood (Fig. 2.16) and tissue (Fig. 2.17) eosinophils. This phenomenon is intriguing and has drawn attention for a long time. For example, in 1965, a communication published in *Nature* by Ghadially and Parry[75] reported the occurrence of this reversed electron density in some eosinophils found in both normal and neoplastic conditions. The authors pointed out that the unusual granule appearance was also described in eosinophils from other species and raised a question: "are there in fact two distinct races of eosinophils with different functions or do the two types of granules represent eosinophils before and after fulfilling their function"?[75]

Other authors have evaluated if EM techniques could produce the reversal of the granule electron densities. It is considered that depending on the fixation, embedding, and/or contrasting procedures, the electron density of the core and matrix may vary in relation to each other.[76–79] For example, if the thin sections on the grids are contrasted with phosphotungstic acid, the central cores of the eosinophil granules can decrease their electron density compared with the matrix.[79] In accordance with this observation, an early ultrastructural study of rat blood eosinophils in which thin sections were contrasted with phosphotungstic acid shows an electron micrograph with all cytoplasmic granules depicting reversal of the electron densities.[80] The "stains" used in EM are in fact heavy metals that attach to biological structures, thus promoting contrast and differential electron densities.[70] Currently, uranium (uranyl acetate) and lead (lead citrate) are the most widely used contrasting agents in EM. It was also noted that in some circumstances the cores and matrices of the eosinophil granules can react differently to the contrasting with uranyl acetate[77,78] The reversed electron density reported with this reagent was explained as a possible negative staining effect, that is, failure of the contrasting reagent to penetrate into the crystalline structure or to react with it.[78]

In our EM studies, the reversed density of eosinophil granules has been observed at times. We noticed, for example, that in an experiment in which the same sample of peripheral blood eosinophils was submitted to different fixations for immuno-EM and then treated or not with sodium metaperiodate, a reagent that has been used for antigen unmasking, only the treated groups exhibited reversed density of the granule cores, observed in almost 100% of eosinophil granules from these groups. However, it should be noted that there is still an incomplete understanding why a reversed density of the granules occurs during EM methodological procedures.

While technical issues can undoubtedly induce reversal of the electron densities in the population of eosinophil-specific granules, we and other authors have also noticed that variable electron density of the core and matrix can sometimes occur in adjacent granules within the same cell.[81] We have been observing this situation mostly in eosinophils involved in inflammatory responses or activated in vitro (Figs. 2.18 and 2.19). In this case, the reversed electron density might be associated with biochemical changes indicative of granule activation preceding granule content release. It is well known that not all eosinophil secretory granules are concomitantly responsive to stimuli and different granule morphologies denoting activation and secretion can be observed interspaced with resting granules in a single thin section of an eosinophil.[20] Finally, it is important to highlight that reversal of the granule core densities may be indicative of core content losses associated with PMD and not exactly a reversal of staining within a granule full of contents. In this case, other PMD-related ultrastructural aspects should be considered for accurate interpretation.

Eosinophil-specific granules store a collection of preformed cationic proteins, cytokines, chemokines, and growth factors, as detailed before,[3,8] but these granules are not considered simply storage depots. Secretory granules from human eosinophils are much more complex organelles not only in terms of the molecular composition being storage sites of many receptors[82] but also in terms of structural organization.[21,82] While the classical view of these granules is that of containers with very packed material, the presence of internal membranes was noted sometimes in prior years by conventional TEM application on activated eosinophils during inflammatory responses and pathological situations, as reported in Chapter 4. Compelling evidence for the existence of membranes within secretory granules was provided by electron tomography, a technique that provides 3D views at high resolution, in emptying granules from activated human eosinophils[20] (Chapter 4). The ultrastructural visualization of membranes within secretory granules in resting eosinophils is probably masked by a large amount of protein compacted within them, although analyses of thin sections by TEM occasionally show evidence for the presence of these membranes (Fig. 2.16). In support of this interpretation are findings with specific granules isolated from unstimulated eosinophils by subcellular fractionation, in which an intragranular pool of CD63, a member of the transmembrane-4 glycoprotein superfamily (tetrasp-

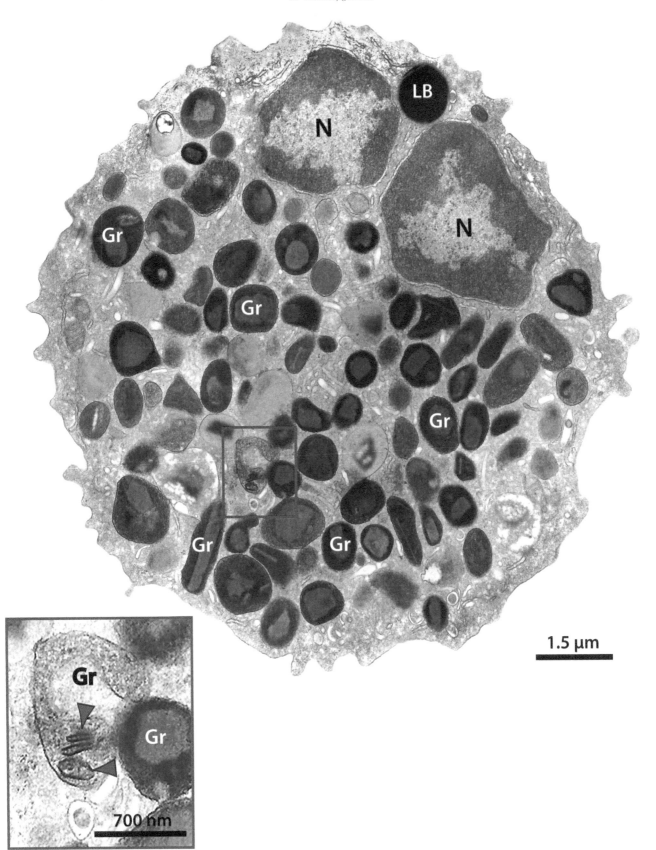

FIG. 2.16 A human peripheral blood eosinophil showing granules (Gr) with reversed electron density. The granule cores appear less dense compared with the granule matrices. Internal membranes *(arrowheads)* are seen in higher magnification in one granule *(boxed area)*. Sample prepared for TEM. *LB*, lipid body; *N*, nucleus.

A. The cell biology of human eosinophils

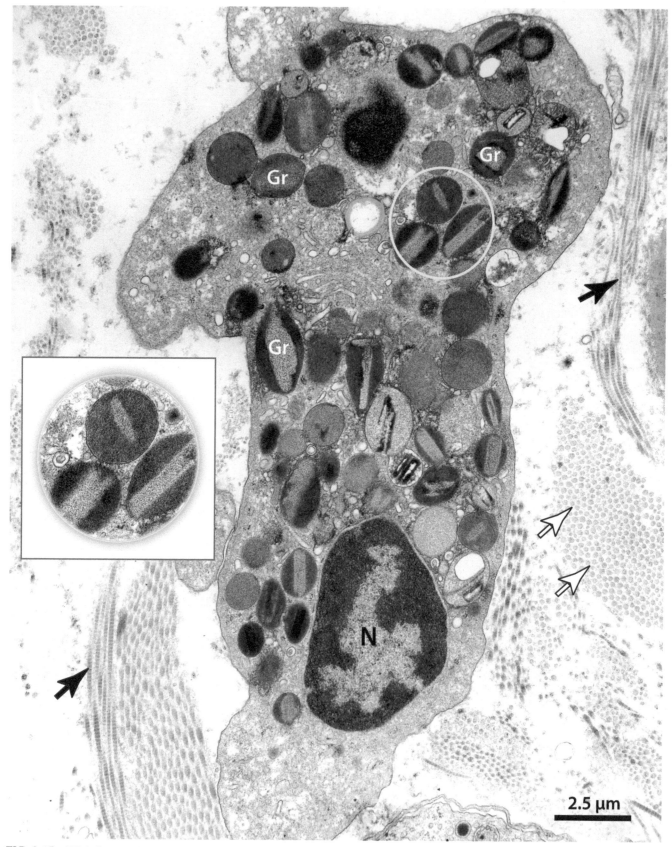

FIG. 2.17 TEM of a tissue human eosinophil showing most specific granules (Gr) with reversed electron density. Note the pale cores and dense matrices (*circle*). Collagen fibrils from the extracellular matrix are seen in longitudinal (*black arrows*) and cross (*white arrows*) sections. *N*, nucleus.

A. The cell biology of human eosinophils

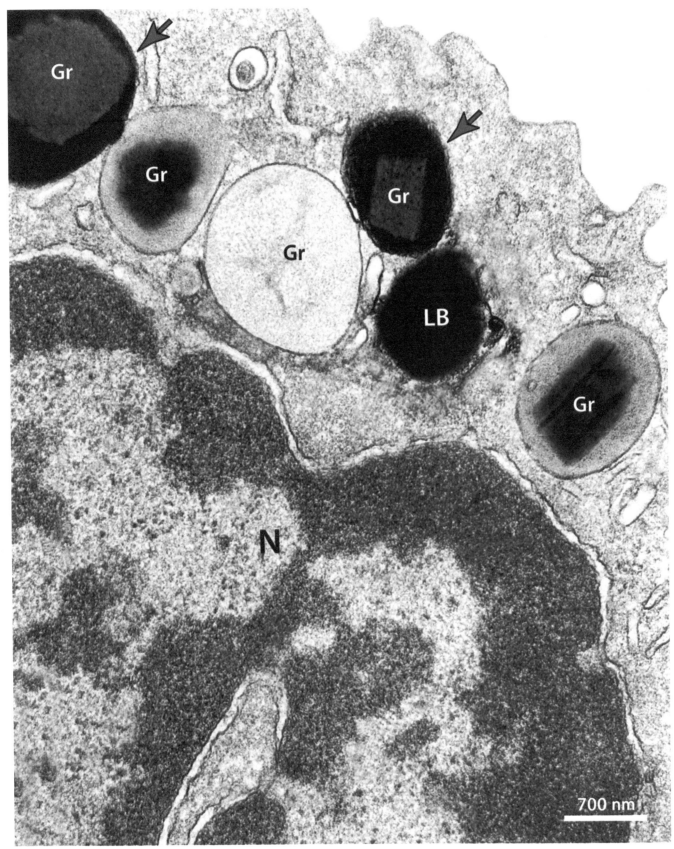

FIG. 2.18 Specific granules (Gr) from an activated human eosinophil show distinct electron densities. *Arrows* indicate granules with electron-dense matrices and less dense cores. A large electron-lucent granule is also observed. Cells were isolated from the peripheral blood, stimulated with CCL5, and processed for TEM.[20] *LB*, lipid body; *N*, nucleus.

A. The cell biology of human eosinophils

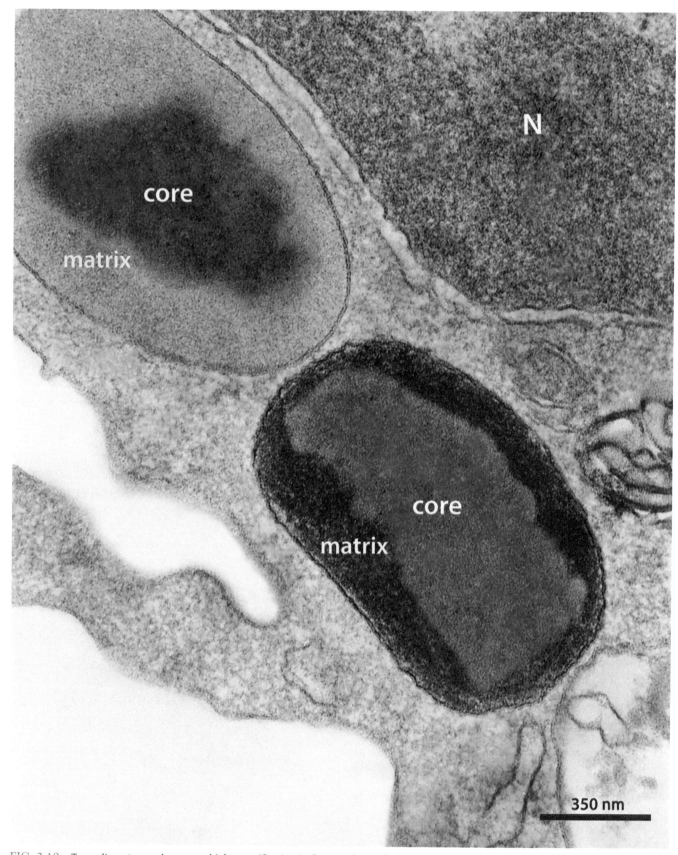

FIG. 2.19   Two adjacent granules seen at high magnification in the cytoplasm of a human eosinophil. One of the granules show reversed electron density with a pale core and a dense matrix. The trilaminar appearance of the granule limiting membrane is clearly observed. Cells were isolated from the peripheral blood, stimulated with CCL5, and processed for TEM[20]. N, nucleus.

anins),[83] was detected by flow cytometry after outer granule membranes undergo saponin permeabilization.[20,84] The immunolocalization of CD63 in both unstimulated and activated human eosinophils is shown in Chapter 5.

2.4 Lipid bodies (LBs)

LBs are lipid-rich organelles, commonly observed in the cytoplasm of human eosinophils (reviewed in[85,86]) (Figs. 2.2, 2.6, 2.12, 2.13, and 2.14). However, the existence of LBs, also known as lipid droplets (LDs), in these cells was unappreciated in the past because of technical issues. Identification of LBs by light microscopy has methodological limitations, as LBs are neither resistant to drying nor to fixation, and commonly used alcohol-based hematological stains solubilize LBs.[86] During the 1970s, applications of osmium to fix and preserve lipids led to the identification of LBs as "regular inclusions of normal eosinophil and neutrophil leukocytes," but LBs were wrongly considered as candidate granules.[87,88] Later, the use of other methodologies, including the use of fluorescent lipophilic dyes, definitively identified LBs as regular organelles within eosinophils and other leukocytes.[85]

Some of the difficulties in the study of LBs in eosinophils as well as in other cells emerge from their unique architecture. In contrast to vesicles and membranous organelles that have an aqueous content surrounded by a phospholipid bilayer membrane, LBs are encircled just by a monolayer of phospholipids (hemimembrane)[85] (Fig. 2.20). LBs, therefore, lack a true delimiting unit membrane structure,[85] which means that, when observed by TEM, a trilaminar pattern, indicative of a phospholipid bilayer membrane, is not seen around them (Fig. 2.21). Moreover, LBs, in general, differ from other membranous organelles in terms of shape; that is, while most cytoplasmic organelles range from tubular, vesicular to cisternal, LBs are typically spherical organelles (Figs. 2.12, 2.14, and 2.20) with most of their volume made of lipid esters.[89]

In conventional TEM using ultrathin sections, eosinophil LBs look roughly round and highly osmiophilic and therefore very electron-dense in electron micrographs (Figs. 2.12 and 2.14).[90] The typical high electron density of LBs, in conjunction with the fact that LBs are not limited by a classical trilaminar structure, enables their unambiguous ultrastructural identification within human eosinophils (Fig. 2.20). LBs are distributed in the entire eosinophil cytoplasm but frequently seen in the peripheral cytoplasm (Figs. 2.6, 2.12, and 2.14) and/or in the proximity of the nucleus (Figs. 2.12 and 2.16), sometimes in direct contact with the nuclear envelope (Fig. 2.22).[86] Images suggestive of LB budding (Fig. 2.15) or fusion of LBs (Fig. 2.23) are occasionally observed in thin sections of human eosinophils.

As a step of sample preparation for TEM, cells are usually treated with osmium tetroxide, and as a resulting reaction, lipids become insoluble and resistant to extraction in the subsequent dehydration process in organic solvents such as ethanol and acetone.[70] Osmium tetroxide reacts primarily with unsaturated bonds of acyl chains and promotes electron density to the specimen.[89,91] Thus, LB lipid composition affects LB electron density.[92] In fact, while LBs in eosinophils are very electron-dense, which indicates a high degree of unsaturated lipids, LBs in other cell types may look electron-lucent or with variable degrees of electron density. Experimental studies using cell lines cultured with fatty acids differing in saturation demonstrated that these lipids give rise to LBs with different electron densities.[89,92] Moreover, there are other factors that can affect LB electron density such as culture conditions or state of cell activation (Chapter 4). Therefore, when analyzing LBs by TEM, it is important to consider not only the number and size of these organelles but also their electron density.

Eosinophil LBs are generally reported as homogeneous organelles (Figs. 2.20, 2.21, and 2.22). However, a comprehensive examination of human eosinophils prepared by conventional TEM has pointed out that LBs in these cells may have a multifaceted organization and are not solely a mass of lipid esters.[93] In fact, careful analysis of thin sections has revealed evidence of membranes within eosinophil LBs (Figs. 2.24 and 2.25). By applying electron tomography followed by 3D-reconstruction and modeling, we demonstrated that LBs from human eosinophils indeed have membranous internal structures organized as a network of tubules (Fig. 2.26) resembling the ER.[94,95] While conventional TEM studies are usually performed on approximately 80-nm-thick sections, the tomographic slices used in our work were only 4-nm thick (digital slices), thus offering a significant advantage over 80-nm serial thin sections for tracking cell structures in 3D.[19]

The membranous system observed within LBs explains how membrane-bound proteins are localized within LB cores and supports a model for LB formation by incorporating cytoplasmic membranes of the ER, instead of the conventional view that LBs emerge from the ER leaflets.[94] Although it is extensively recognized that lipid-rich organelles originate from the ER, it is still a matter of debate on how these organelles are formed and several models have been proposed.[93,96] TEM demonstrated portions of ER at the LB periphery (Fig. 2.24 and 2.25) and even intermingled in the lipid content (Fig. 2.24 and 2.25).[94] Indeed, the association between LBs and typical ER cisternae is frequently seen in electron micrographs (Figs. 2.27 and 2.28) not only within eosinophils but also in different cell types,[89,93] a finding that supports the interrelationship between these two organelles. Moreover, an ER-like membranous network with evidence of lipid accumulation (osmiophilia) is at times found in close apposition to LBs in human eosinophils (Fig. 2.28).

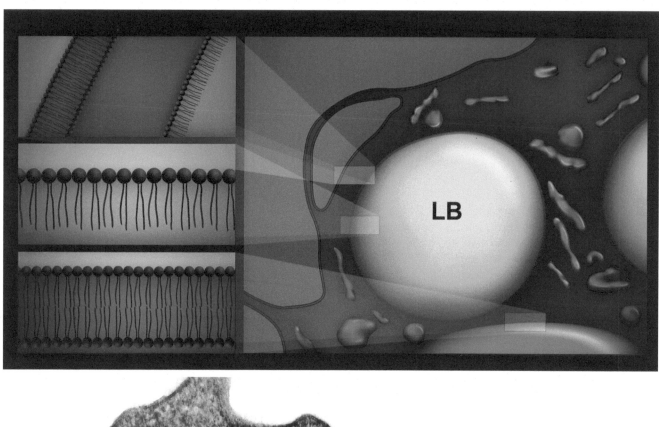

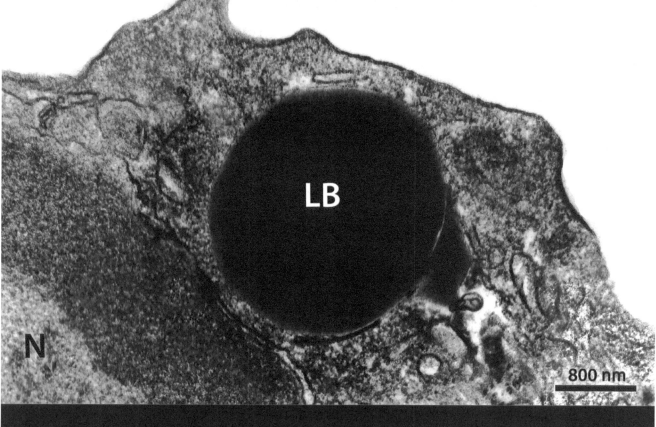

FIG. 2.20   Lipid body (LB) ultrastructure within a human eosinophil. As a general feature, LBs are delimited by a monolayer of phospholipids differing from the classical structural organization (phospholipid bilayer membrane) of all other organelles, cytoplasmic vesicles, and plasma membrane. In electron micrographs of eosinophils obtained by TEM *(bottom panel)*, LBs appear as osmiophilic, very electron-dense organelles. *N*, nucleus.

A. The cell biology of human eosinophils

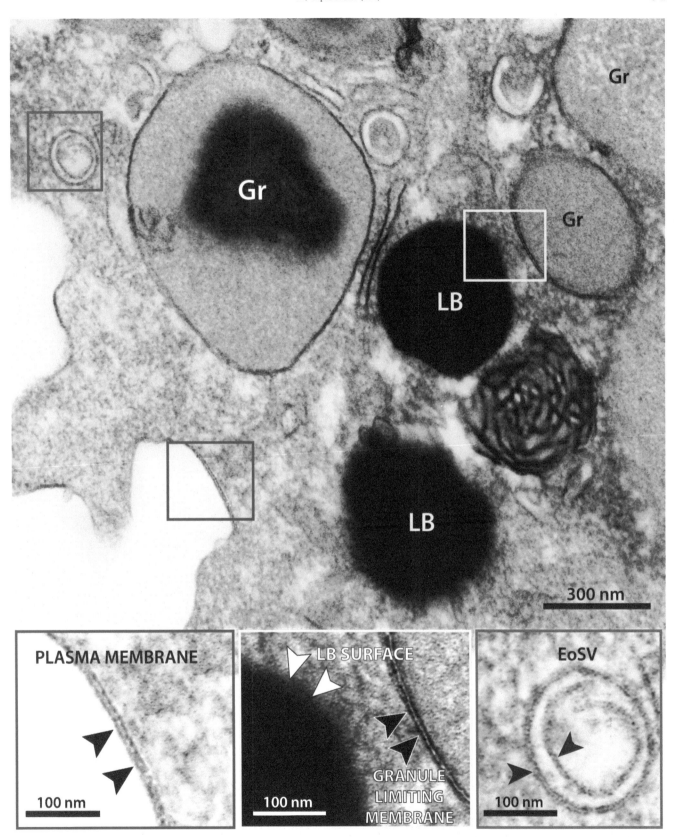

FIG. 2.21 Eosinophil lipid bodies (LB) observed at high magnification by TEM. The LB surface (*white arrowheads*) lacks a trilaminar structure typical of bilayer membranes (*black arrowheads*) as shown by the plasma membrane, surface of specific granules (Gr), and eosinophil sombrero vesicles (EoSV).

A. The cell biology of human eosinophils

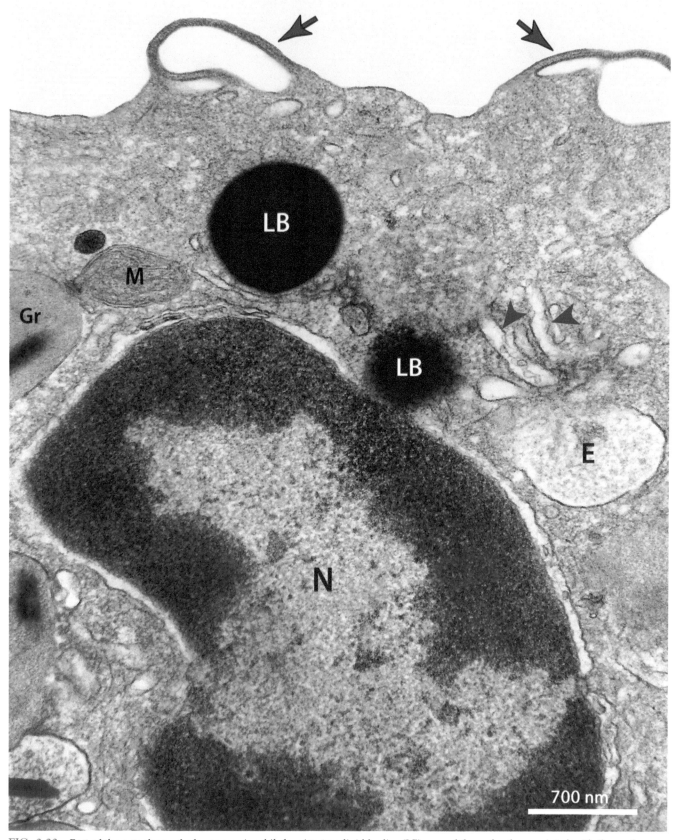

FIG. 2.22   Part of the cytoplasm of a human eosinophil showing two lipid bodies (LB), one of them closely associated with the nuclear envelope. *Arrows* indicate short cell projections at the eosinophil surface. Tubules *(arrowheads)* emanate from an early endosome (E). *Gr,* secretory granule; *M,* mitochondrion; *N,* nucleus.

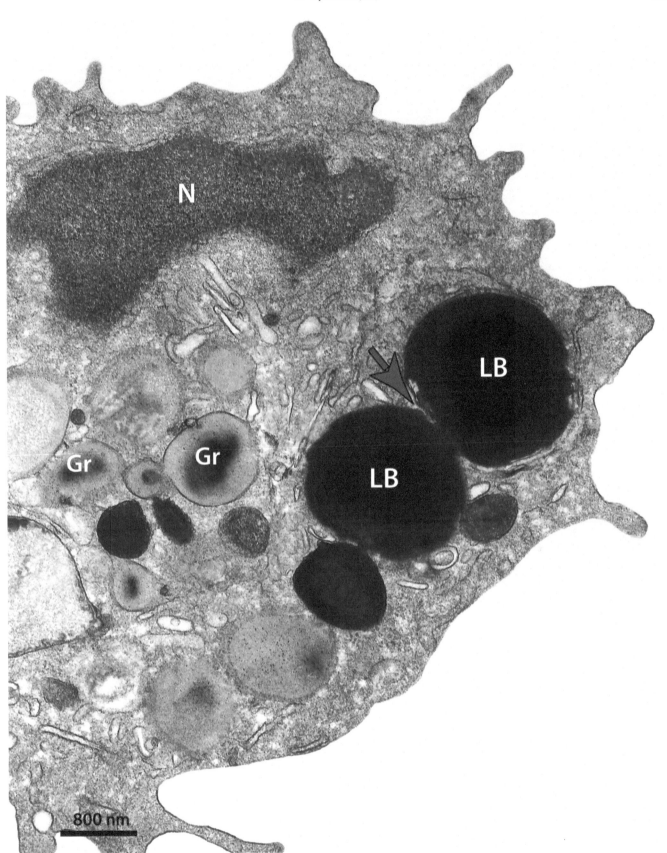

FIG. 2.23   A mature human peripheral blood eosinophil observed by TEM. *Arrow* indicates two fused lipid bodies (LB). *Gr,* specific granules; *N,* nucleus.

A. The cell biology of human eosinophils

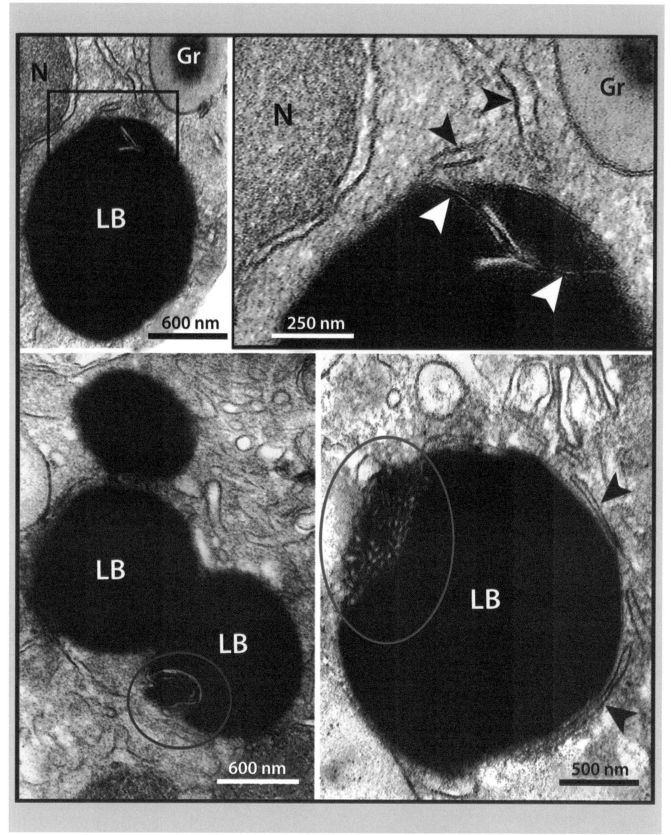

FIG. 2.24 Lipid bodies (LB) of human eosinophils are not structurally homogeneous organelles. *Top panels*: a membrane *(boxed area)* is observed within a LB. Note in higher magnification that the same trilaminar aspect *(white arrowheads)* is also seen in surrounding cisternae of the endoplasmic reticulum (ER) *(black arrowheads)* and around the nucleus (N) and granule (Gr). *Bottom panels* show LBs with heterogeneous areas *(circles)*.

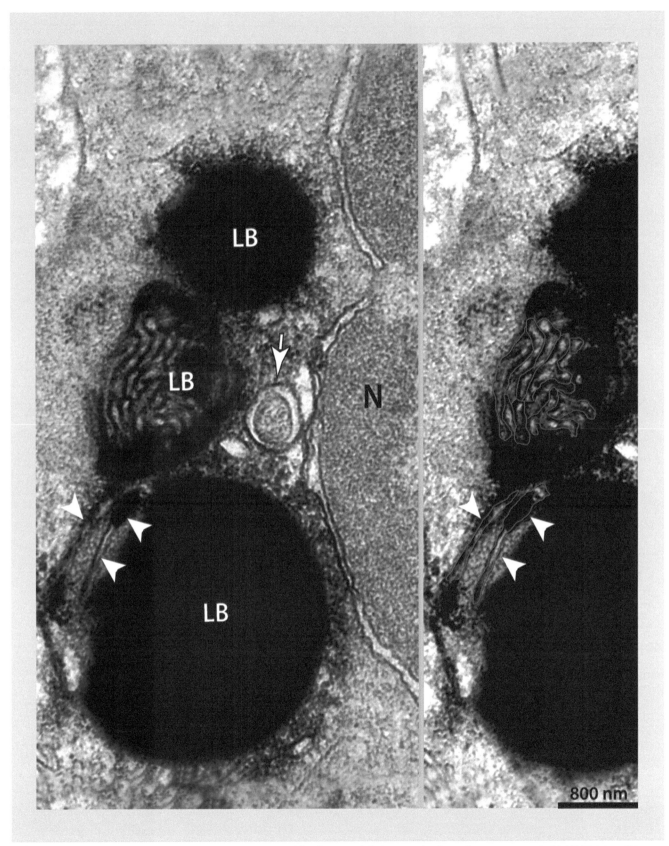

FIG. 2.25 Lipid bodies (LB) in the cytoplasm of a human blood eosinophil. An organized membranous network *(highlighted in red)* is seen within a LB. *Arrowheads* indicate ER cisternae in close contact with a LB and partially intermingled in the lipid content. An EoSV is indicated *(arrow)*. *N*, nucleus.

A. The cell biology of human eosinophils

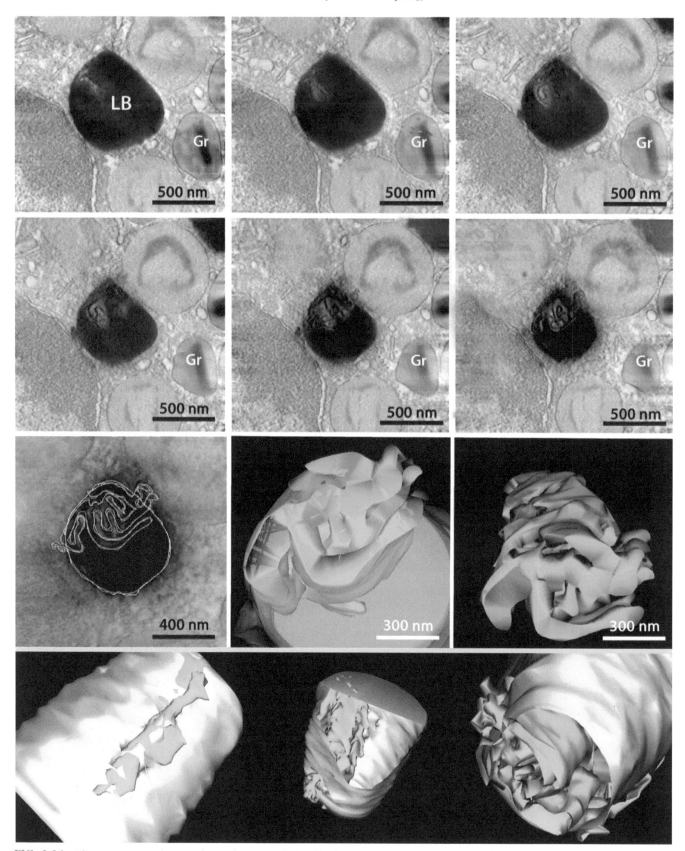

FIG. 2.26   Electron tomography reveals membranes in the lipid body (LB) core. Images are representative 4-nm-digital slices of a tomogram taken from a mature human eosinophil. LB boundaries and internal membranes in consecutive slices were contoured in *yellow* and *blue*, respectively, to produce computer-based 3D models *(bottom panels)*. LB membranes are seen as interconnected tubules resembling the ER. Peripheral blood eosinophils were processed for TEM and analyzed for fully automated electron tomography.[94] *Gr*, secretory granule.

A. The cell biology of human eosinophils

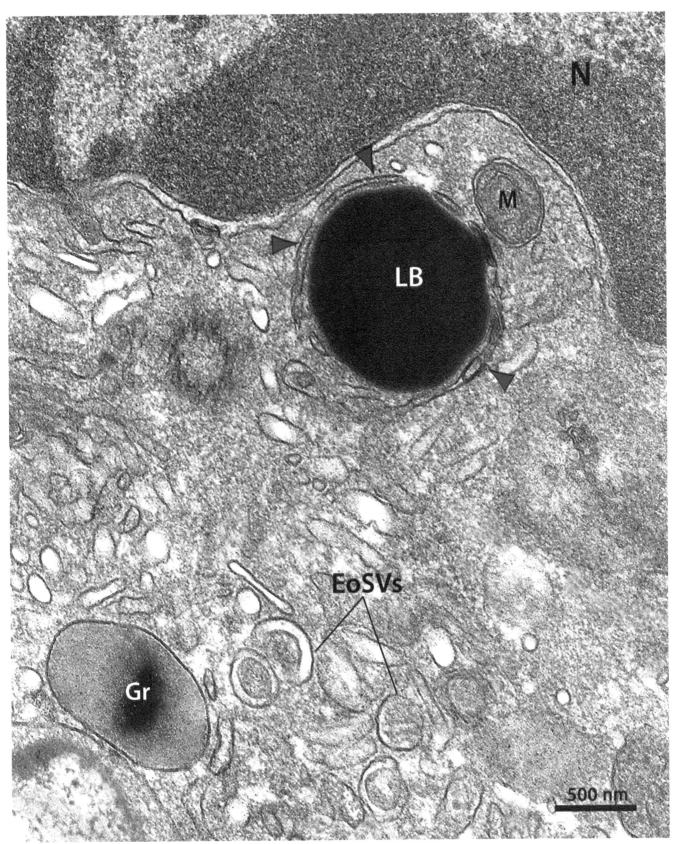

FIG. 2.27   Lipid body (LB) association with the ER in the cytoplasm of a human eosinophil. ER cisternae *(arrowheads)* are observed in close apposition to a LB. EoSVs, large vesiculotubular structures typical of eosinophils, are observed in the proximity of a specific granule (Gr). *M,* mitochondrion; *N,* nucleus.

A. The cell biology of human eosinophils

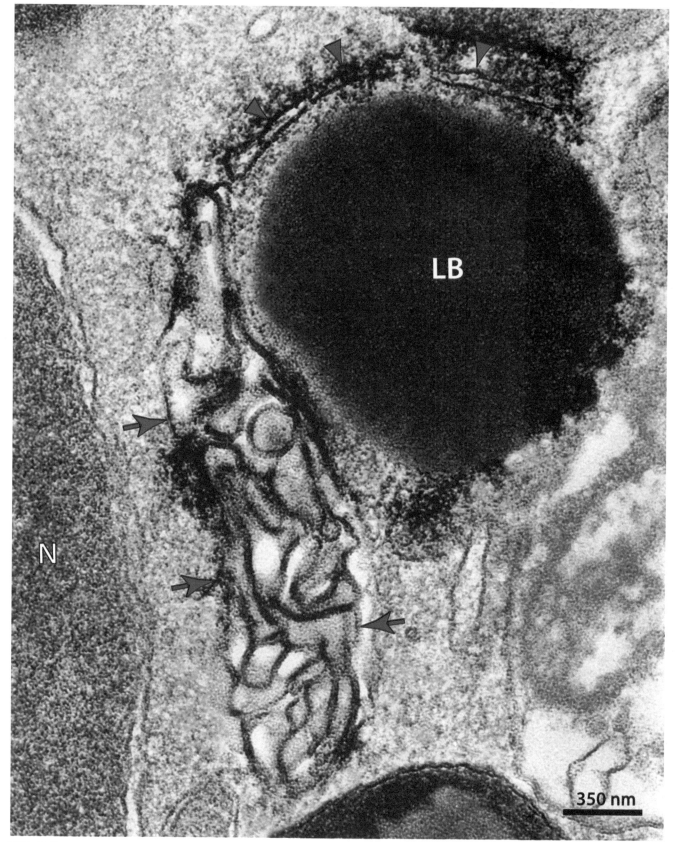

FIG. 2.28　Lipid body (LB) ultrastructural features in a human peripheral blood eosinophil. A membranous network *(arrows)* is observed in close apposition to an apparently homogenous LB. Typical ER cisternae are indicated *(arrowheads). N,* nucleus.

Eosinophil LBs also show considerable variation in number and size. By TEM, under favorable conditions, which may include postfixation with reduced osmium to increase contrast, LBs, as small as 30 nm in diameter, can be identified together with other LBs of greater size, varying up to 1–2 μm in resting eosinophils (Fig. 2.29).[94] Because of this dissimilarity in size, also found in LBs from different cell types, the use of TEM is important to capture LBs on a scale smaller than that provided by light microscopy. The mean number of LBs in resting eosinophils is ~ 1–2 per cell section.[94] On the other hand, eosinophil activation induces a notable increase in the numbers and sizes of cytoplasmic LBs[86] (see Chapter 4).

The dynamic nature of LBs within eosinophils has led to their recognition as highly active organelles with functions, compositions, and structural aspects distinct from classic LDs found in adipocytes. Because of this, we prefer to use the term LBs for lipid-rich organelles from immune cells such as eosinophils, although the terms LBs and LDs have been used to refer to the same organelle. As a feature common to lipid-dominant organelles in diverse cell types,[97,98] eosinophil LBs contain a core rich in neutral lipids, surrounded by a monolayer of phospholipids with associated proteins (Fig. 2.30). LD/LB-specific structural proteins, the PAT family of proteins [renamed to perilipin (PLIN) family proteins]—PLIN/PLIN1,[99] adipose differentiation-related protein (ADRP/adipophilin/PLIN2),[100] and tail-interacting protein of 47 kDa (PLIN3),[101] are constitutively associated with the circumferential rim of these organelles (Figs. 2.30 and 2.31).[102] On the other hand, more specific molecules are present within eosinophil LBs. Remarkably, eosinophil LBs store arachidonic acid (AA), mainly esterified in phospholipids.[103,104] Human eosinophils isolated from the peripheral blood and incubated with tritiated arachidonic acid ([³H]-AA) for 0.5–1 h incorporated AA into the cells as demonstrated by ultrastructural autoradiographic analyses, in which tritium label was found almost exclusively over LBs[103,104] (Figs. 2.31 and 2.32). Interestingly, other compounds, including cytokines and enzymes involved in the enzymatic conversion of AA into eicosanoids, can be expressed in eosinophil LBs in response to cell activation,[86,94] as highlighted in Chapter 4.

2.5 Eosinophil sombrero vesicles (EoSVs)

Vesiculotubular structures have long been recognized in the cytoplasm of mature eosinophils (Fig. 2.33) and attracted attention because of their intriguing morphology and increased numbers in activated cells (reviewed in[21]). This population of large vesicles with their typical ultrastructural morphology, recognizable just by TEM (Figs. 2.25, 2.27, and 2.33), was previously considered as "microgranules" or "small granules" in the earlier eosinophil literature.[21] Only in 2005, with the use of advanced electron microscopic techniques that unraveled the 3D structure (Fig. 2.34) and the functional activities of these vesiculotubular structures, named EoSVs, they received greater consideration due to their remarkable ability to interact with and to bud from specific granules.[19–21] Although EoSVs are distributed throughout the eosinophil cytoplasm, they are frequently seen surrounding or even attached to secretory granules (Fig. 2.33), especially when these granules are in the process of degranulation, as discussed in Chapter 3. This is because EoSVs can originate from these granules and act in the transport of granule-derived products.[18,21] Detection of typical granule products by immunogold EM, such as MBP-1, IL-4, and IFN-γ in EoSVs, is shown in Chapter 5. Our EM data also support a functional role for EoSVs in the translocation of specific proteins, such as CD63, not only from but also to specific granules in response to stimulation to facilitate/regulate secretion (Chapter 3).[23] Thus, EoSVs constitute a highly specialized vesicular system acting in the transport of products within human eosinophils.[21]

EoSVs are generally electron-lucent and easily identifiable by conventional TEM because of their typical "Mexican hat" (sombrero) appearance in cross sections, with a central area of cytoplasm and a brim of a circular membrane-delimited vesicle (Figs. 2.25, 2.27, and 2.33). These vesicles can also show an elongated or curved morphology in thin sections (Figs. 2.27, 2.33, and 2.35).[19,21]

As demonstrated by electron tomography, EoSVs are tubular carriers with substantial membrane surfaces and elevated plasticity (Fig. 2.34).[21] Three-dimensional reconstructions and models generated from serial sections of human eosinophils revealed that individual EoSVs consist of curved tubular structures, with cross-sectional diameters of approximately 150–300 nm, surrounding a cytoplasmic center (Fig. 2.34), while classical round small vesicles are ~ 50 nm in diameter.[19] As visualized by 3D models, along the lengths of EoSVs, there are both continuous fully connected cylindrical and circumferential domains and incompletely connected and only partially circumferential curved domains (Fig. 2.34).[19] These two domains explain both the "C" shaped morphology of these vesicles and the presence of elongated tubular profiles very close to typical EoSVs, as seen in 2D cross-sectional images of entire eosinophils (Figs. 2.27 and 2.33). The morphology of EoSVs provides a higher surface-to-volume ratio system likely suitable for the specific transport of membrane-bound proteins.[21]

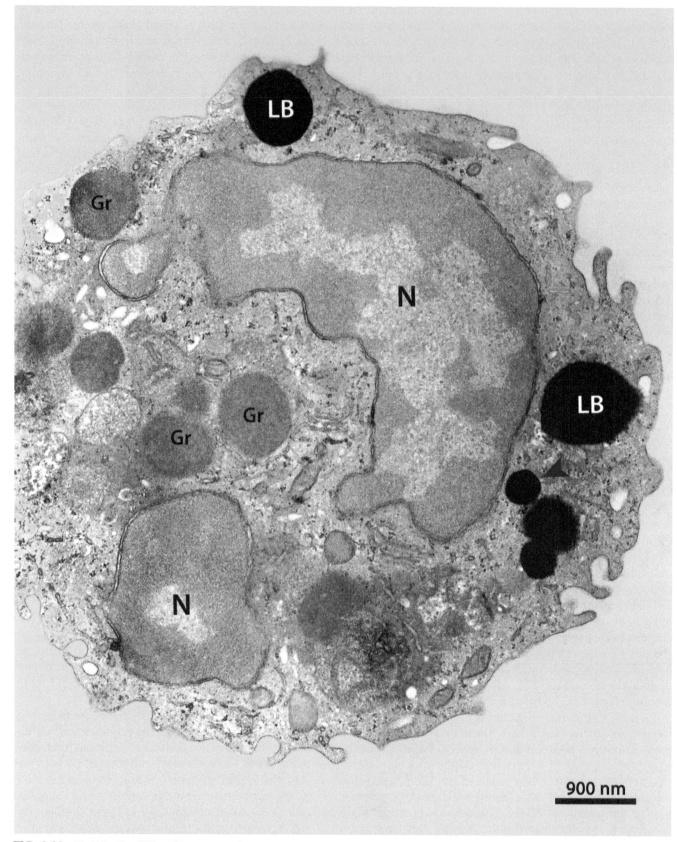

FIG. 2.29 Lipid bodies (LB) within a mature human eosinophil. LBs with different sizes appear as highly contrasted organelles, while the nucleus (N) and secretory granules (Gr) are imaged with less contrast. A small LB is indicated *(arrowhead)*. Cells were prepared for TEM with ferrocyanide-reduced osmium.[94]

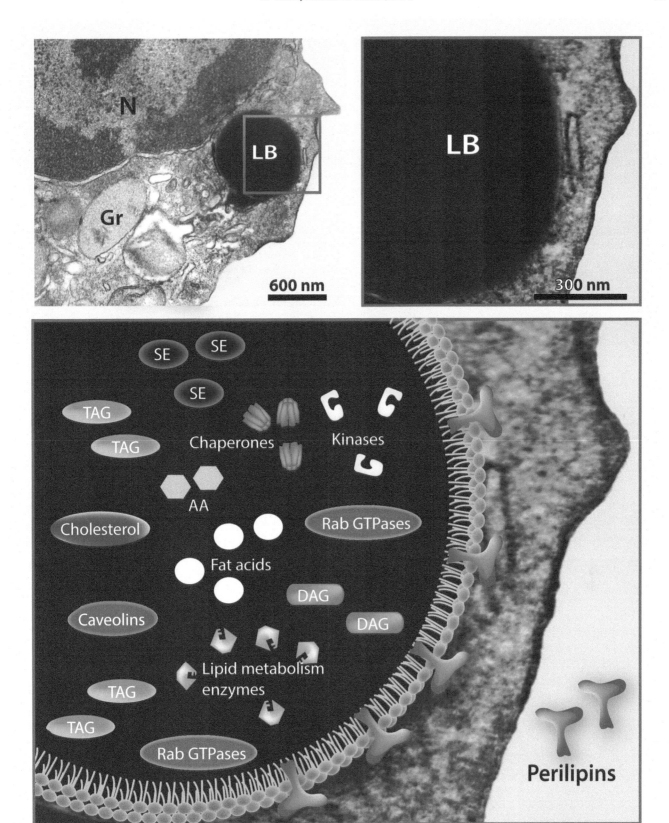

FIG. 2.30 Lipid body (LB) structure and composition in human eosinophils. Structural proteins from the perilipin family are associated with the LB surface, while the LB core contains sterol esters (SE), triacylglycerols (TAG), diacylglycerols (DAG), cholesterol, arachidonic acid (AA), and proteins such as Rab GTPases, lipid metabolism enzymes, kinases, and chaperones. *Gr*, secretory granule; *N*, nucleus.

A. The cell biology of human eosinophils

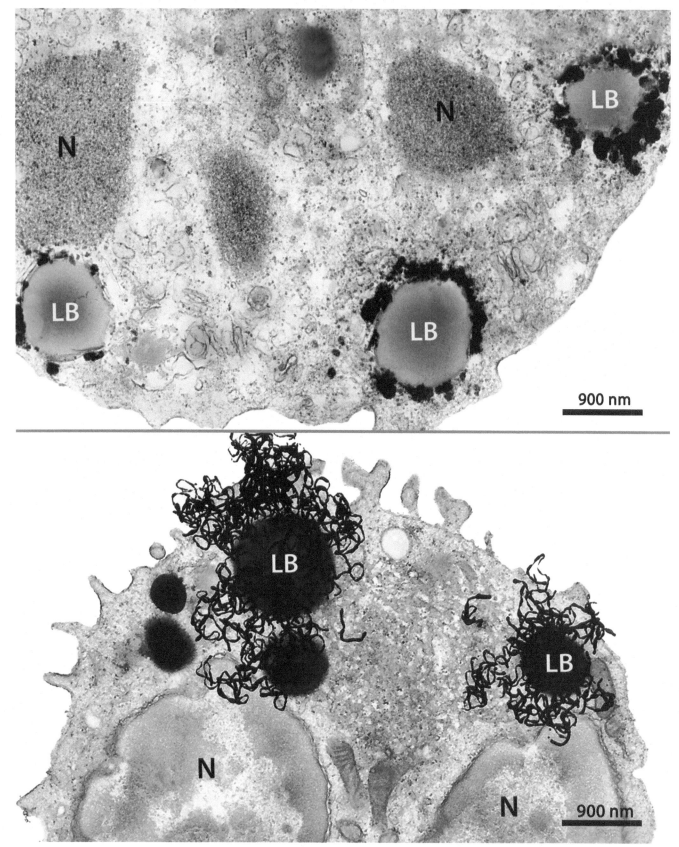

FIG. 2.31   Lipid body (LB) structure and composition in human eosinophils. The LB periphery is heavily labeled for perilipin-2 (PLIN2) with an immunonanogold EM technique[50] *(top panel).* An electron microscopic autoradiograph *(bottom panel)* shows incorporation of tritiated arachidonic acid ([3H]-AA) into LBs.[103] N, nucleus.

A. The cell biology of human eosinophils

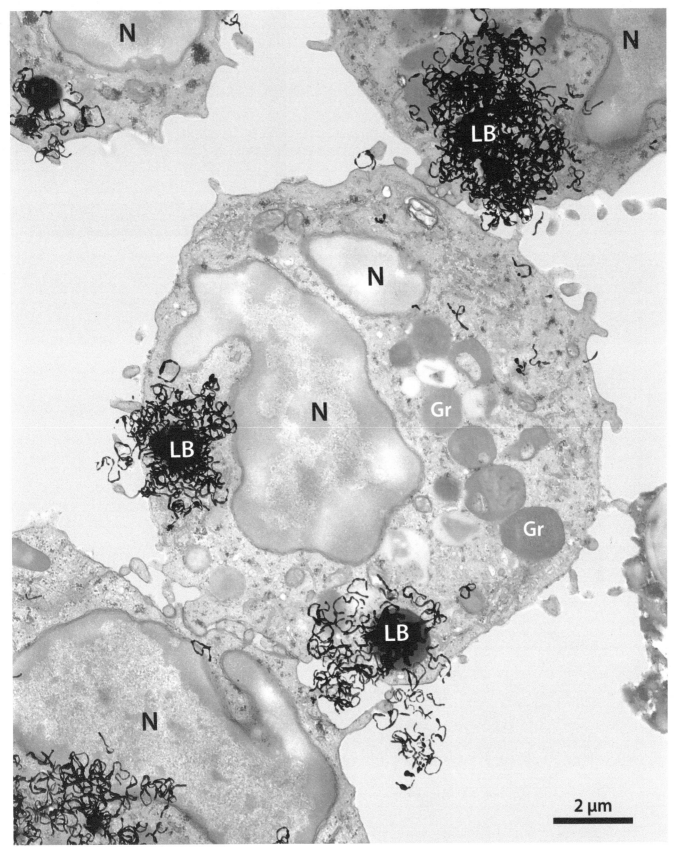

FIG. 2.32   Mature human peripheral blood eosinophils prepared for autoradiography after a pulse of tritiated arachidonic acid ([³H]-AA).[103] Numerous silver grains label lipid bodies (LB). *Gr*, secretory granules; *N*, nucleus.

A. The cell biology of human eosinophils

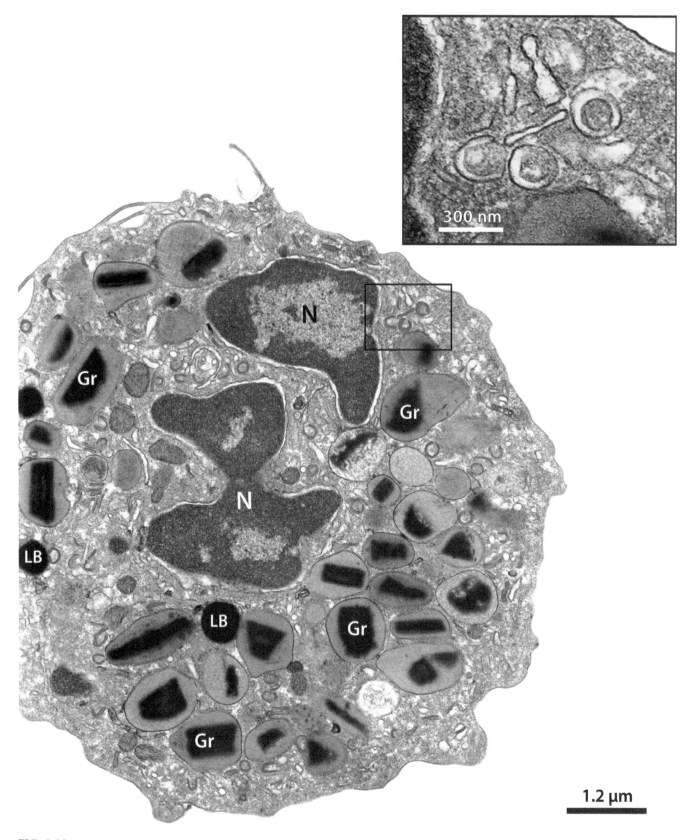

FIG. 2.33 Conventional TEM of a human blood eosinophil. The eosinophil cytoplasm is filled with vesiculotubular structures termed eosinophil sombrero vesicles (EoSVs, *pseudocolored* in pink). Secretory granules (Gr) and lipid bodies (LB) are observed. *N*, nucleus.

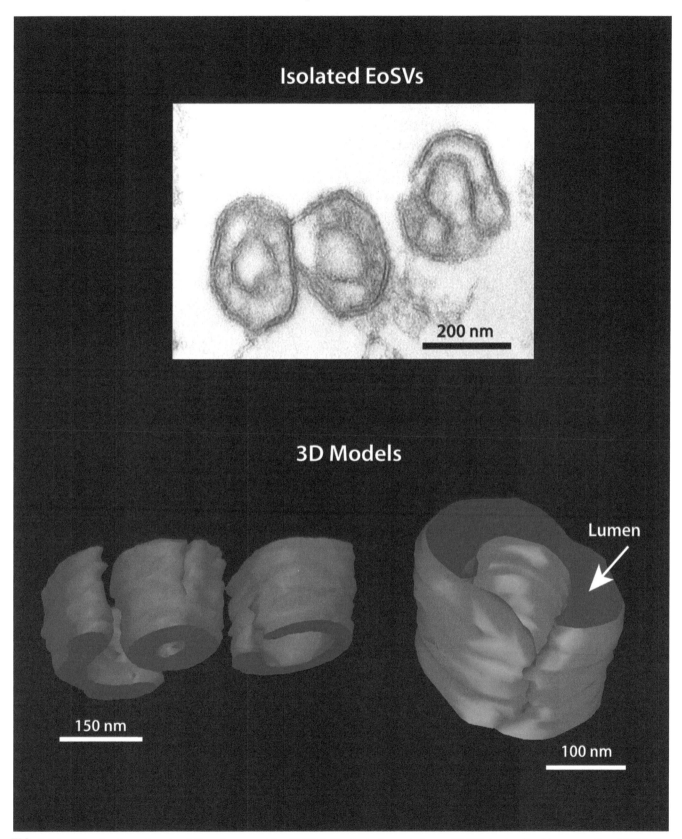

FIG. 2.34 Ultrastructure of eosinophil sombrero vesicles (EoSVs). The typical morphology of EoSVs is visualized by conventional TEM after isolation by subcellular fractionation *(top panel)* and three-dimensional (3D) models *(pink)*. EoSVs are curved tubular and open structures surrounding a cytoplasmic center. The arrow points to the tubular lumen. Note that the trilaminar structure of the vesicle delimiting membrane is clearly observed in the thin section *(top panel)*. 3D models were generated from 4-nm-thick serial slices using fully automated electron tomography at 200 KV[19].

A. The cell biology of human eosinophils

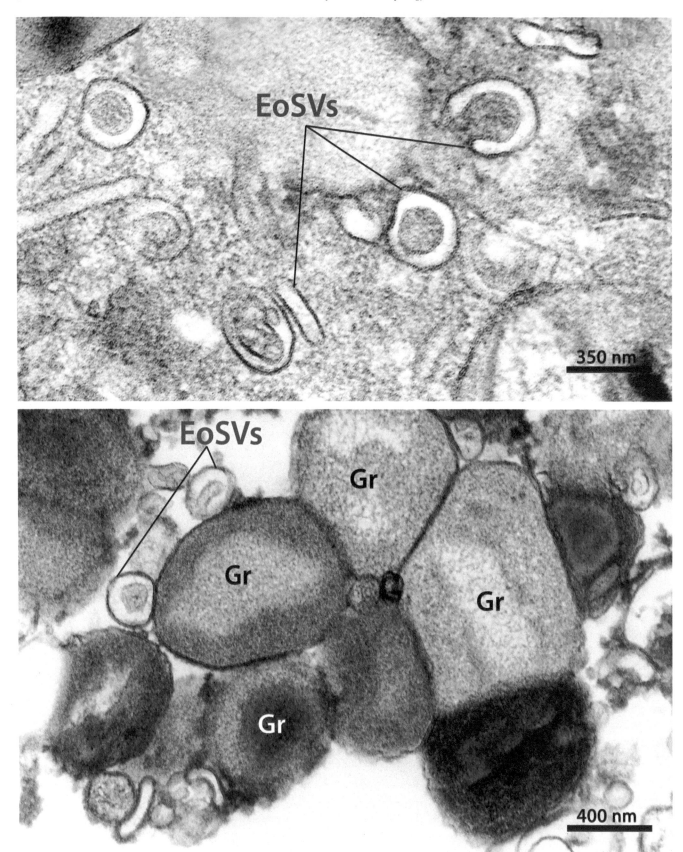

FIG. 2.35 Eosinophil sombrero vesicles (EoSVs) observed by TEM. EoSVs are seen in the cytoplasm *(top panel)* and in a granule-enriched fraction obtained by subcellular fractionation *(bottom panel)* of human eosinophils.[19] EoSVs are seen as circular, curved, and/or elongated membrane-bound structures. *Gr*, specific granules.

Resting human eosinophils have a mean of ~ 15–20 EoSVs/cell section (Fig. 2.33).[21] These vesicles can be adequately isolated by subcellular fractionation being found in a fraction slightly less dense than granule-containing fractions,[19] in contrast to small vesicles, which localize to more buoyant light fractions. Isolated EoSVs, analyzed by conventional TEM after optimal fixation, show the same morphology as seen in intact cells; that is, they appear as circular, elongated, or "C-shaped," which facilitates their identification by form and size (Figs. 2.34 and 2.35). EoSVs and granules are so related to each other that a population of these vesicles can be found even in granule-enriched fractions, especially from agonist-stimulated cells in which EoSVs are frequently seen in contact with the granule limiting membrane (Fig. 2.35).[19,22] This may be explained by the fact that activation leads to granule-EoSVs interaction and EoSVs formation from secretory granules (Chapter 4).

The morphology of EoSVs is so unique in eosinophils that their presence in the cytoplasm of granulocytes, devoid of specific granules, is useful for lineage assignment of granule-poor activated cells (Fig. 2.36).[105] As noted, activated eosinophils participating in many pathological situations or activated in vitro with agonists show increased numbers of EoSVs[21] (Chapter 4). Moreover, the presence of persistent, free EoSVs is a common finding in biopsies after eosinophil lysis[17] and might also be used as a morphological marker for eosinophil activity.

2.6 Other organelles and structures

2.6.1 Endoplasmic reticulum (ER)

The ER comprises an elaborated network of membrane cisternae and tubules that extents throughout the cell and occupies a large fraction of the cytoplasmic volume.[106] Mature human eosinophils are not rich in peripheral ER.[52,56] When these cells are observed by conventional TEM, they show scattered ER cisternae in the cytoplasm. Moreover, ER cisternae with no attached ribosomes on their membranes (smooth ER) are more frequently seen than ribosome-studded cisternae (rough ER).

By applying ultrastructural immunolabeling for a typical RER protein that catalyzes protein folding (PDI),[69] we did not find clear positivity in the peripheral cytoplasm but predominantly at the nuclear envelope, which constitutes, as noted, a specialized ER domain[62–65] (Fig. 2.13). To evaluate the level of PDI labeling, we measured the entire extension of the nuclear envelope of 30 cell sections and counted the number of gold particles. Our quantitative analyses revealed dense labeling for this protein on the nuclear envelope with a mean level of labeling of 33.2 ± 9.5 gold particles (mean $\pm$ SEM) per linear micrometer of nuclear envelope while labeling on peripheral cisternae was almost absent.[69] These immuno-EM findings are in accordance with earlier ultrastructural studies of human eosinophils, describing minimal amounts of RER in mature cells in contrast to immature eosinophils in which there are extensive dilated cisternae of RER (Chapter 7).[56,72] Thus, classical organelles involved in the synthesis and packing of secretory proteins, that is, RER and Golgi complex, as described below, are considered minor cytoplasmic organelles in mature human eosinophils.[52,72] Although these organelles are diminished in mature eosinophils, these cells are specialized in secretion (Chapter 3) and regularly use a nonclassical pathway to export proteins to the cell surface independent of the ER/Golgi.[26]

2.6.2 Golgi complex

The Golgi complex is not considered a prominent organelle in mature human eosinophils.[40,52,56,72] Reduction in Golgi structures is reported during eosinophil maturation (Chapter 7),[40,56,72] and this organelle is not frequently seen in electron micrographs from mature cells. When observed in mature eosinophils by TEM, the Golgi complex appears as small cytoplasmic organelles with typical morphology—a pile of flat membranes, the cisternae, which together form the Golgi stack[107] (Figs. 2.6 and 2.37)—but a certain degree of dilation of Golgi cisternae can be sometimes observed in activated cells (see Fig. 4.14 in Chapter 4 and Fig. 8.26 in Chapter 8).

Cytochemistry and immuno-EM applied to human eosinophils (Chapter 5) to ascertain intracellular localization of some proteins associated with the secretory pathway have not detected substantial Golgi labeling in mature cells. For example, cytochemical methods applied to human eosinophils in process of development showed that EPO is localized in the Golgi and RER structures, including the nuclear envelope, of large immature eosinophils (eosinophilic myelocytes),[108] but not of mature cells.[109] Accordingly, MBP-1 immunolabeling (immunonanogold EM) of mature eosinophils isolated from the peripheral blood was not related to the Golgi complex but with a secretory route transporting MBP from specific granules.[22] We also investigated the distribution of syntaxin 17 (STX17), a SNARE (soluble N-ethylmaleimide-sensitive factor attachment protein receptor) protein linked to constitutive secretion,[110] and

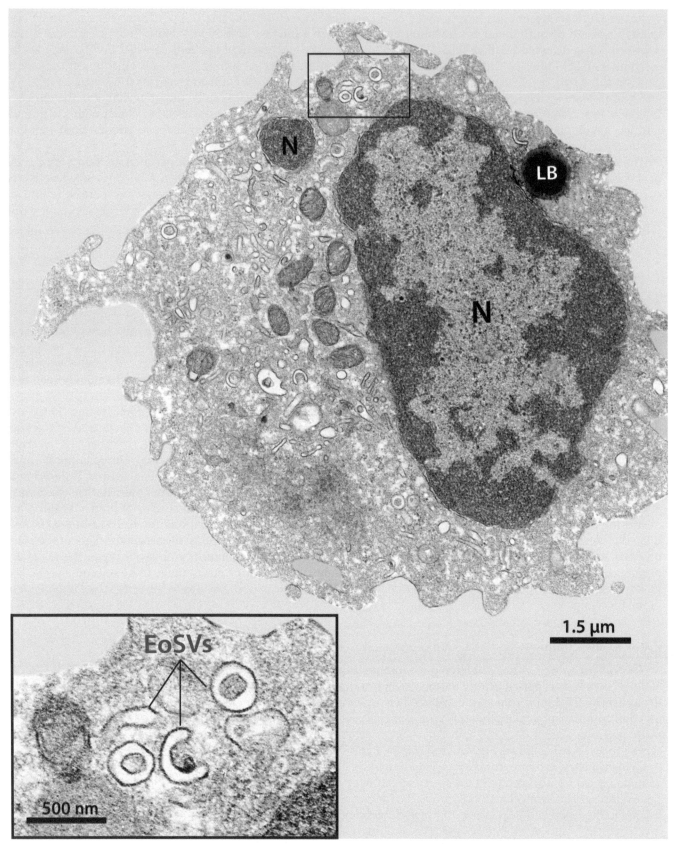

FIG. 2.36   Electron micrograph of a human eosinophil devoid of specific granules. Eosinophil sombrero vesicles (EoSVs), a hallmark of this cell, are distributed in the cytoplasm. An osmiophilic lipid body (LB), typical of eosinophils, is also observed. Mitochondria were colored in *green*. N, nucleus.

A. The cell biology of human eosinophils

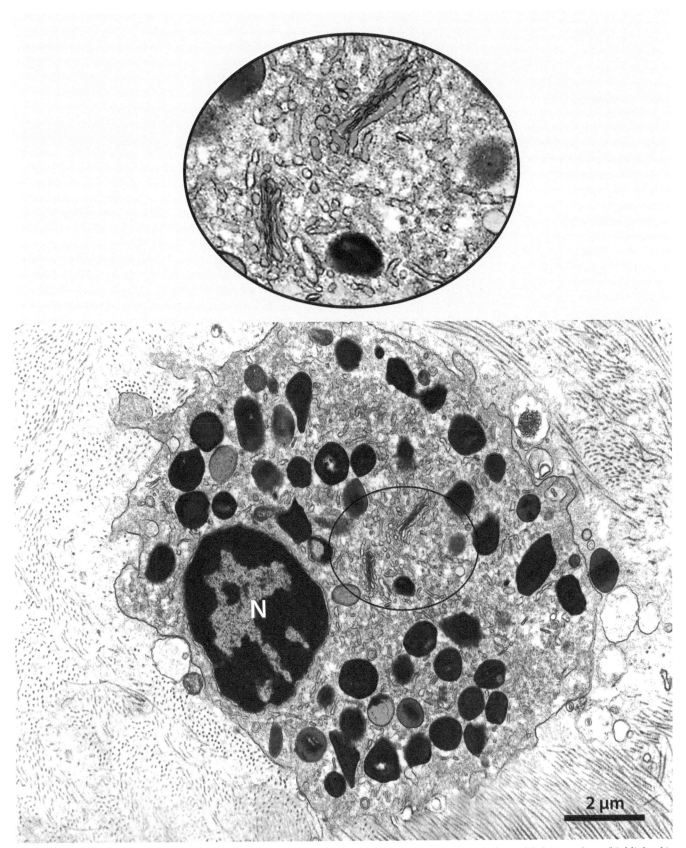

FIG. 2.37 Transmission electron micrograph of a tissue human eosinophil showing the typical morphology of Golgi complexes (highlighted in *purple* in higher magnification). *N*, nucleus.

A. The cell biology of human eosinophils

anterograde transport from the ER to Golgi[111,112] and found just negligible Golgi labeling in both resting and activated human blood eosinophils while secretory granules were consistently labeled (see Fig. 5.15 in Chapter 5).[113] Of note, treatment of human blood eosinophils with brefeldin-A (BFA), a Golgi-disrupting agent,[114] leads to disarrangement of this organelle, with clear ultrastructural morphological changes[20] (Fig. 2.38).

2.6.3 Endosomes

Endosomes are vacuolar/tubular compartments of the endocytic system distributed in the cytoplasm, showing varied size and composition (reviewed in[115–117]). The term endosome relates to the "receptacle-like" nature of these organelles, which receive cargo molecules endocytosed from the plasma membrane.[116]

Despite several internalization routes, the early endosome is the initial destination for the endocytosed materials, such as lipids and proteins including different receptors, which are transported from the plasma membrane to this organelle by endocytic vesicles.[118] The early endosome is also referred to as "sorting endosome" because it serves as a sorting station that determines the subsequent fate of the internalized molecules, which can be redirected to late endosomes and subsequently to lysosomes for degradation or be returned to the plasma membrane.[116,118]

The endosomal system has a complex organization with a multitude of tubular and vesicular membrane-bound structures, including recycling endosomes, which seems to be formed from tubules that emanate from the early endosome.[116] Distinct protein markers have been used to characterize endosome types,[116] with the Rab GTPases, specifically Rab 5 and Rab 7 considered key markers for early and late endosomes, respectively.[119] However, it is not easy to characterize endosomal compartments by light microscopy because of their small size and morphological heterogeneity and also because endosomal markers, such as Rab proteins, are not simply markers of endosomes.[116] Thus, visual observation using EM techniques is necessary to fully understand the complex membrane structure of endosomes.[116,117,120]

Endosomes are identified by TEM as morphologically heterogeneous compartments delimited by a bilayer membrane and with a varied number of membrane-bound vesicles in the lumen.[117] The early endosome, as seen in the cytoplasm of human eosinophils (Fig. 2.39; see also Fig. 4.28 in Chapter 4) and other cells, is a pleiomorphic organelle generally identified as a main electron-lucent vacuole located at the cell periphery and containing few intraluminal vesicles.[117] Tubules can be observed extending from the early endosome surface (Fig. 2.22), which can also show invaginations and fusion profiles of incoming and budding vesicles (Fig. 2.39). Another feature of the early endosome is the presence of a cytoplasmic clathrin coat, which partially covers its delimiting membrane.[117] Clathrin is a protein classically involved in a type of endocytosis (clathrin-mediated endocytosis), in which cargo from the cell surface is packaged into vesicles with the aid of a clathrin coat.[121,122] Clathrin has a typical ultrastructure when observed by conventional TEM: it resembles bristle-like spikes, an appearance generated by sectioning of its lattice-like polygonal array.[122] For this reason, clathrin structures can be characteristically identified by TEM at the borders of invaginations of the plasma membrane and around vesicles in the cytoplasm. Since clathrin-coated vesicles fuse with early endosomes in the endocytic pathway, these vacuoles can exhibit images of clathrin at some areas of their surfaces.[117,123] A clathrin coat on early endosomes has been observed with different frequencies in many cell types, including human eosinophils (Fig. 2.39). However, clathrin-coated areas are not always visualized in the plane of the section. Moreover, although a clathrin coat is predominantly present on early endosomes, this protein can also be found at the surface of a small percentage of late endosomes (~ 6.5%).[123]

Late endosomes are characteristically round to oval compartments containing many intraluminal vesicles. Because of this typical appearance, they are also commonly termed multivesicular bodies (MVBs) or multivesicular endosomes.[117] Here, we use the term MVB as a synonym for the late endosome. Late endosomes mature from early endosomes, and the increasing number of intraluminal vesicles (50–100 nm in diameter) has been considered an ultrastructural feature to position a compartment within the endolysosomal pathway.[117,124] For example, in an ultrastructural study in which HeLa cells were incubated with the endocytic tracer BSA conjugated with gold particles, after 5 min of BSA-gold uptake, a high percentage of positive endosomes (85%) had less than five intraluminal vesicles while endosomes containing more internal vesicles required additional time to be reached by the tracer.[123] Thus, based on a body of work, maturation from early endosomes to late endosomes leads to the accumulation of internal vesicles in the lumen of the endosomal compartments and the number of intraluminal vesicles has been used as one of the morphological criteria to define endosomal organelles.[123,124] Vacuolar organelles showing electron-lucent lumen with up to five intraluminal vesicles were classified as early endosomes while vacuoles with more than five internal vesicles as late endosomes/MVBs.[123] The destiny of intraluminal vesicles of MVBs is either to be sorted for cargo degradation into lysosomes or to be secreted as exosomes into the extracellular medium (Chapter 3).[117]

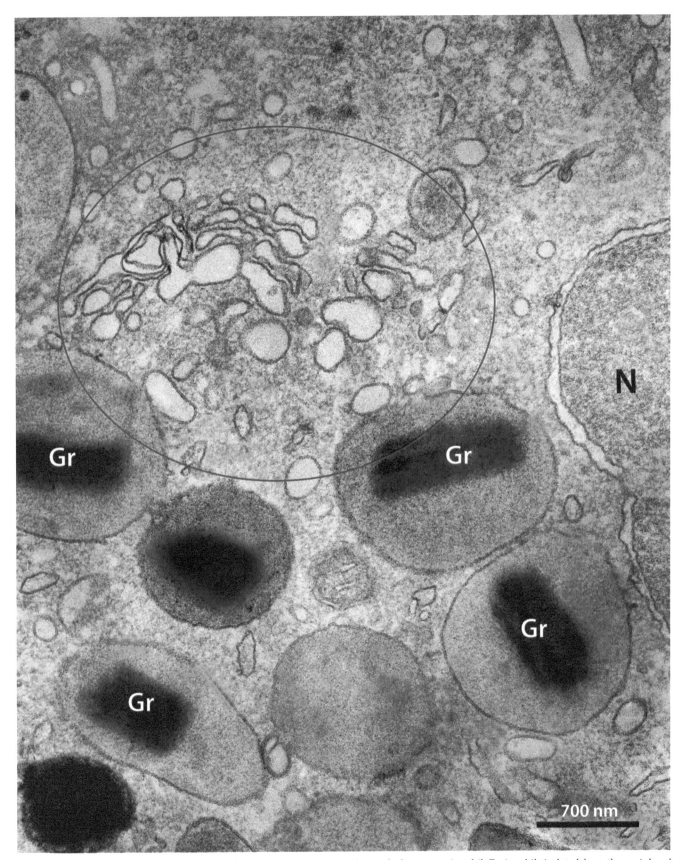

FIG. 2.38 A disrupted and enlarged Golgi complex *(circle)* in the cytoplasm of a human eosinophil. Eosinophils isolated from the peripheral blood were treated with brefeldin-A and processed for TEM.[20] *Gr*, secretory granules; *N*, nucleus.

A. The cell biology of human eosinophils

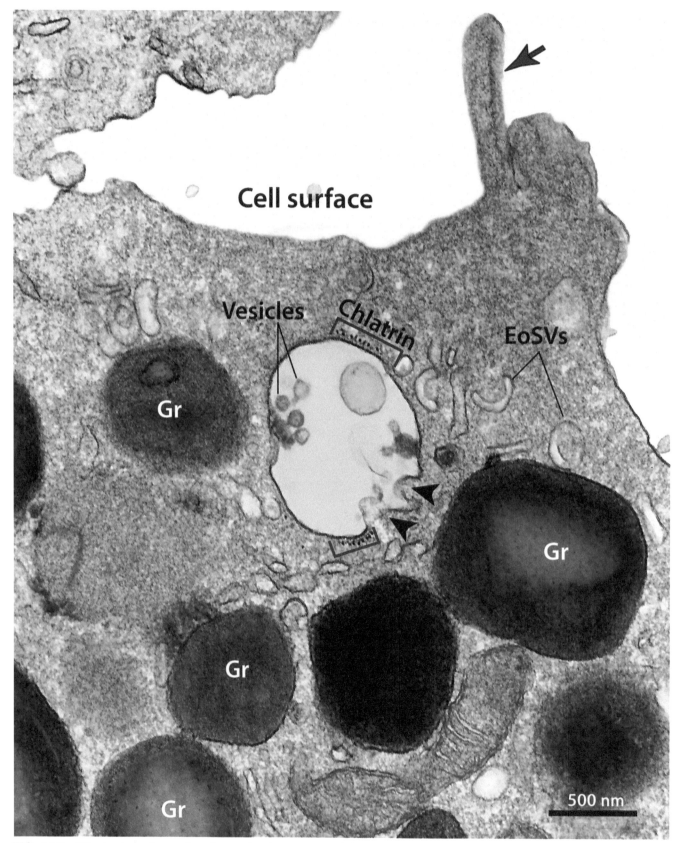

FIG. 2.39   Ultrastructure of an early endosome in a human eosinophil. This membrane-bound vacuole *(colored in orange)* contains a small number of luminal vesicles and clathrin profiles at its surface. *Arrowheads* indicate areas of vesicular fusion events. Note EoSVs, an elongated mitochondrion *(green)*, and a cell surface projection *(arrow)*. *Gr*, secretory granules.

In human eosinophils, the late endosome displays typical morphology with numerous intraluminal vesicles occupying the vacuolar content (Figs. 2.40 and 2.41). Vesicles can also be seen surrounding and/or attached to the limiting membrane of both early and late endosomes (Fig. 2.41).

Late endosomes as well as early endosomes have varied sizes (100 nm to 1 µm)[117] and are not seen in all cross sections of human eosinophils. By analyzing 110 electron micrographs of human blood eosinophils in clear cross-cell sections exhibiting the entire cell profile, intact plasma membranes, and nuclei, we identified typical MVBs (Figs. 2.40 and 2.41) in ~ 25% of these cells (1–3 MVBs per cell section).[125] By our experience, most MVBs identified in the cytoplasm of human eosinophils were seen as vacuoles smaller than secretory granules (Figs. 2.40 and 2.41). However, expanded endosomal vacuoles (> 1 µm), with ultrastructure resembling mainly early endosomes or large endocytic vacuoles, are occasionally observed in the eosinophil cytoplasm (Figs. 2.7 and 2.42).

The protein tetraspanin CD63 is a well-established component of the endosomal system, predominantly distributed in late endosomal and lysosomal compartments and also secreted into exosomes following enrichment in intraluminal vesicles of MVBs.[83,126] In human eosinophils, endosomal compartments are densely labeled for CD63 (see Fig. 5.18 in Chapter 5), which is also detected as a major pool in association with eosinophil specific granules.[23] Additional information on the localization of CD63 in both resting and activated human eosinophils is provided in Chapter 5.

Eosinophils activated in vitro can produce large endocytic vacuoles in response to specific stimuli.[37,127] For example, by stimulating peripheral blood eosinophils with phorbol myristate acetate (PMA) for 15 min and processing them in situ for TEM, that is, directly on the surface of chamber slides after stimulation, we detected increased formation of large electron-lucent vacuoles derived from the plasma membrane in the cytoplasm (Fig. 2.43).[37] The presence of Gal-10 in these endocytic vacuoles is shown in Chapter 5. In vivo, tissue eosinophils participating in inflammatory diseases display at times large electron-lucent endosome-like vacuoles probably formed in response to cell activation (Fig. 2.42). The ultrastructure of human eosinophils involved in diseases is presented in Chapter 8.

2.6.4 Mitochondria

Mature human eosinophils contain few mitochondria. By light microscopy, earlier studies using vital staining of mitochondria with Janus green and diethylsafranin,[128] as well as more recent studies using mitochondria markers,[129] have identified a low population of these organelles in the cytoplasm of mature eosinophils. While the number of mitochondria in mature human eosinophils range from 23 to 36 per cell as found by light microscopic evaluations,[129] cells rich in mitochondria such as hepatocytes show around 1300 mitochondria per cell.[130] However, quantification of mitochondria requires special attention since these organelles, as a general feature, have a remarkable ability to change their morphology as well as to divide (fission) and fuse with each other, events collectively known as mitochondrial dynamics, which is associated with mitochondrial reorganization in response to the continuous cell needs.[131,132]

Mitochondria in different cell types, including eosinophils (Fig. 2.44), are typically seen by TEM as compartmentalized organelles delimited by two phospholipid bilayers and can show an assorted number of cristae and sizes.[131] Because mitochondrial form and function are intimately connected, the ultrastructure of these organelles thus varies considerably between tissues, organisms, and the physiological state of cells.[132]

Our group has been studying the mitochondrial ultrastructure of human eosinophils from the peripheral blood and tissues. In thin sections of mature eosinophils, mitochondria appear as round (Figs. 2.27, 2.36, and 2.44), ellipsoid (Fig. 2.22), and/or elongated (Figs. 2.39 and 2.45) organelles and, in accordance with light microscopy findings,[128,129] the number of mitochondria profiles per cell section is considered low (range of 5–16 profiles per cell section) in resting eosinophils isolated from the peripheral blood. However, because of the mitochondrial capability to change morphology and to fuse, which can generate very elongated mitochondria profiles in thin sections, quantification by number does not seem a sufficient parameter to estimate the mitochondrial population in eosinophils or any other cell. Therefore, in our ultrastructural studies, we also evaluate the area occupied by these organelles in the cytoplasm in addition to other mitochondrial features such as circularity and morphological aspects of the cristae. For example, in mature resting eosinophils isolated from the peripheral blood, we found that their mitochondria show predominantly lamellar cristae (Fig. 2.44) and occupy a small area of the cytoplasm (mean of 0.9 µm²/cell section). This corresponds to ~ 1.8% of the entire cytoplasm per cell section. On the other hand, immature eosinophils show a higher amount of mitochondria, as discussed in Chapter 7.

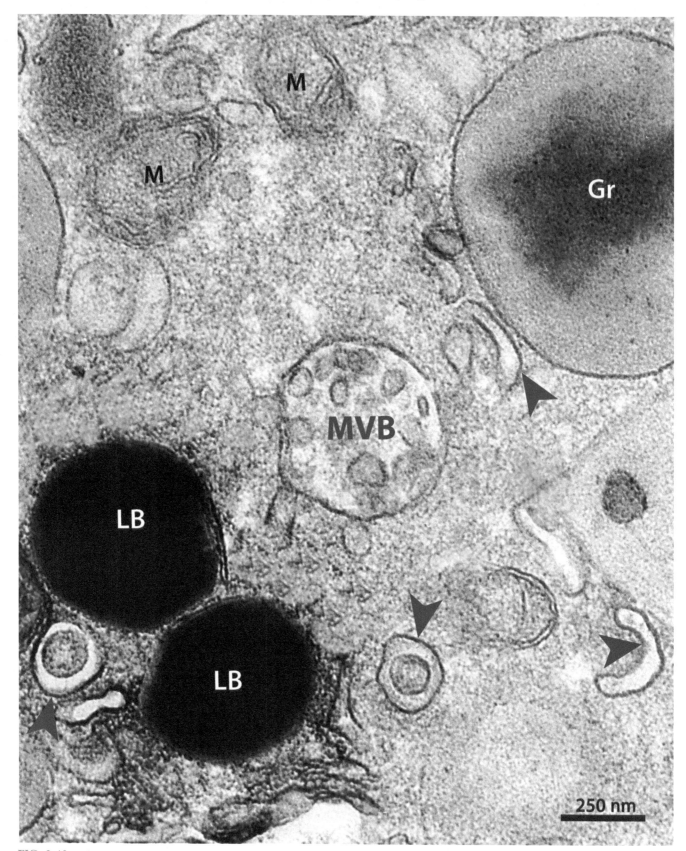

FIG. 2.40   Ultrastructure of a late endosome in a human peripheral blood eosinophil. This vacuole, also termed multivesicular body (MVB), shows several luminal vesicles. EoSVs are indicated by *arrowheads*. *Gr*, secretory granule; *LB*, lipid bodies; *M*, mitochondria.

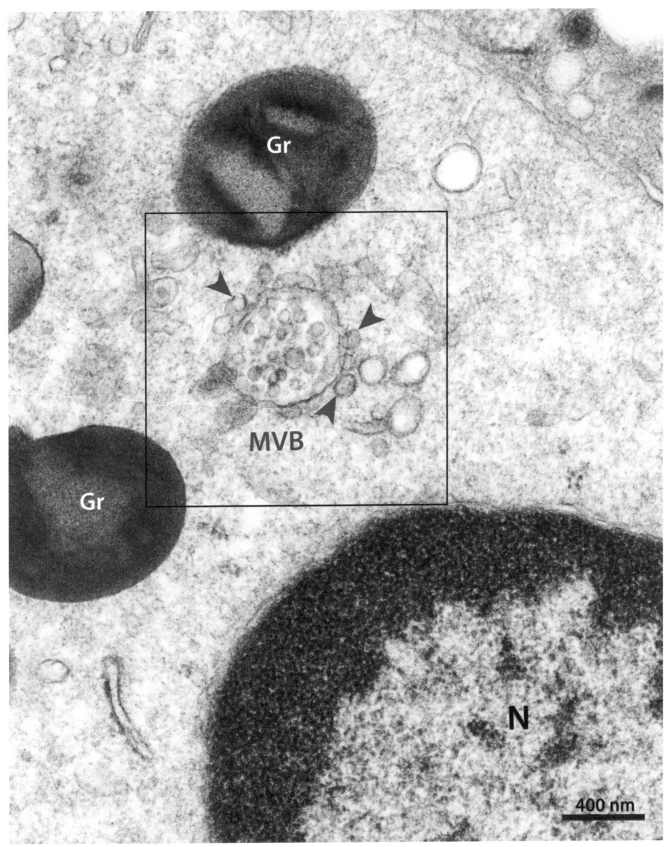

FIG. 2.41   A typical multivesicular body (MVB), characterized by multiple luminal vesicles, is observed in a human tissue eosinophil by TEM. Vesicles are also seen in the proximity and attached to its surface *(arrowheads)*. *Gr*, specific granules; *N*, nucleus.

A. The cell biology of human eosinophils

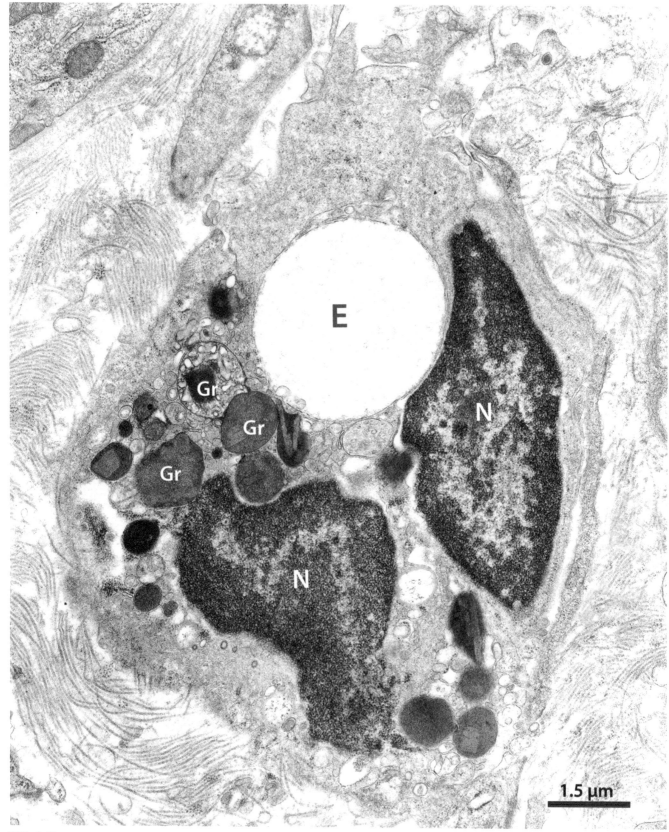

FIG. 2.42   Transmission electron micrograph of a tissue human eosinophil showing an expanded endosomal compartment (E) in the cytoplasm. *Gr*, specific granules; *N*, nucleus.

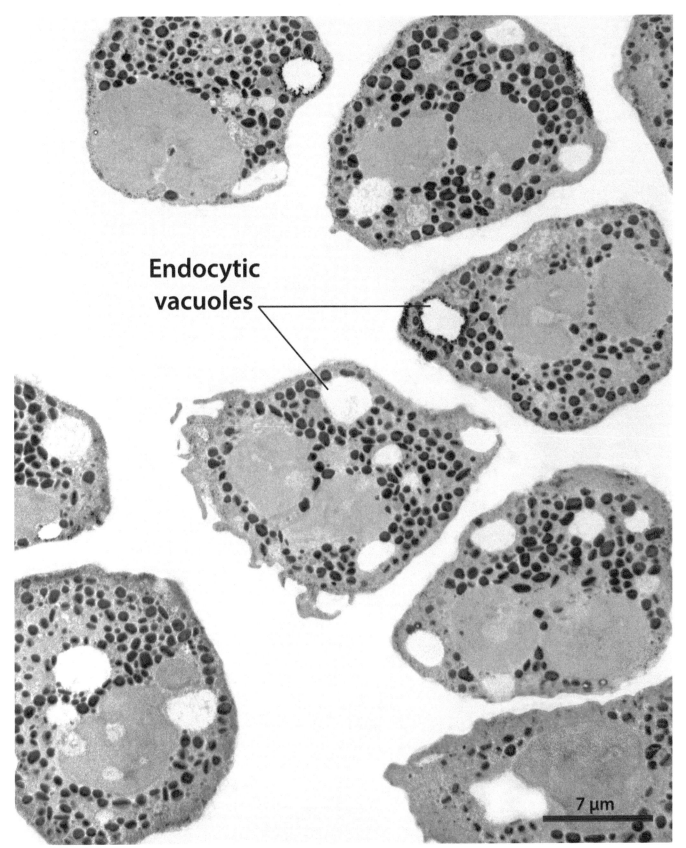

Endocytic vacuoles

7 μm

FIG. 2.43 Endocytic vacuoles observed by TEM at low magnification in the cytoplasm of human eosinophils *(pseudocolored)*. Note multiple vacuoles mainly in the cell periphery. Eosinophils isolated from the peripheral blood were stimulated with phorbol myristate acetate (PMA) for 15 min.[37]

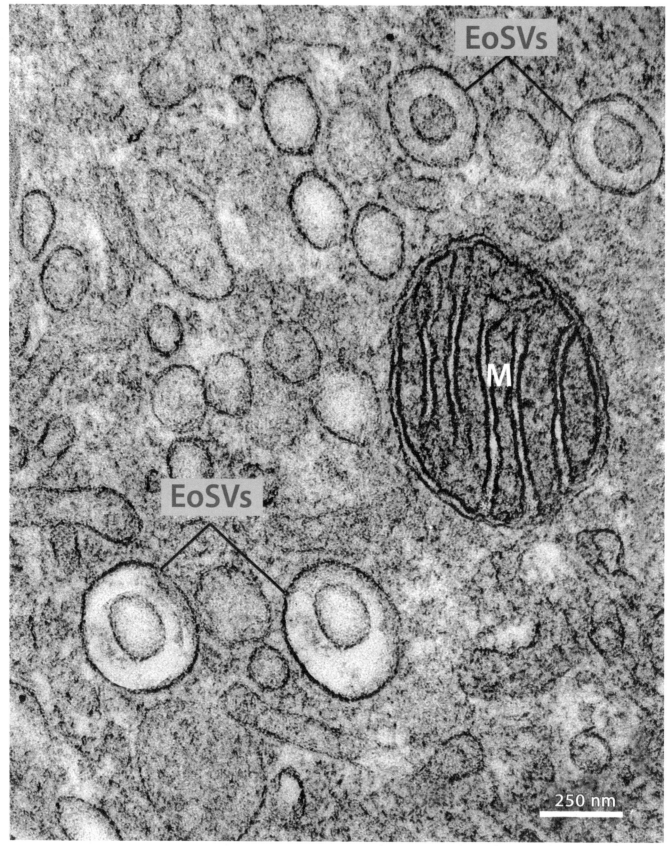

FIG. 2.44   High magnification TEM showing a mitochondrion (M) with lamellar cristae (*green*) in the cytoplasm of a human tissue eosinophil. Typical EoSVs are also observed.

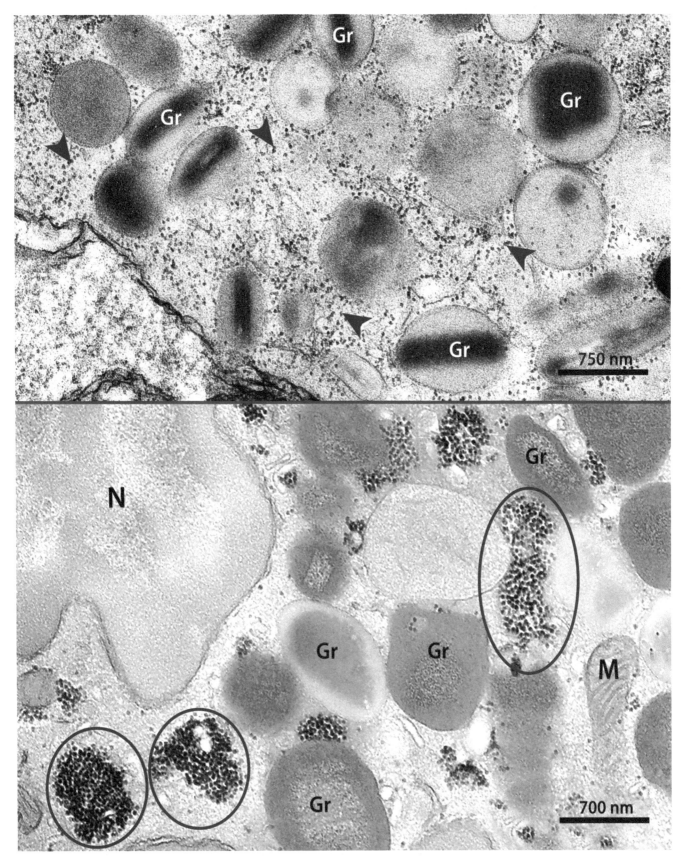

FIG. 2.45 Ultrastructure of glycogen particles within human eosinophils. Glycogen appears as tiny electron-dense dots *(arrowheads)* distributed in the cytoplasm and can form large electron-dense aggregates *(circles)*. *Gr*, secretory granules; *M*, mitochondrion; *N*, nucleus.

A. The cell biology of human eosinophils

2.6.5 Glycogen particles

Human eosinophils, as other leukocytes, contain considerable quantities of glycogen in their cytoplasm.[133] Not all conventional EM methodologies are able to demonstrate glycogen particles. Therefore, the visualization of these particles requires the use of appropriate preparation for EM. We routinely perform postfixation with ferrocyanide-reduced osmium to demonstrate glycogen in eosinophils.[42-44] Thus, glycogen particles are seen as tiny, individual electron-dense "dots" or very small aggregates uniformly distributed in the cytoplasm (Fig. 2.45). In certain situations, accumulations of glycogen can be observed within eosinophils (Fig. 2.45) and other leukocytes and are indicative of pathological alterations associated with glycogen metabolism and/or inflammation.[134] Glycogen agglomerates are visualized with proper EM processing as noted above or indicated by "blank" spaces in the cytoplasm of leukocytes prepared with some routine EM protocols. Therefore, if electron micrographs from eosinophils or other leukocytes show large "empty" areas in the cytoplasm, these findings may be suggestive of glycogen accumulations.

3

Eosinophils as secretory cells

3.1 Overview

In cells from the human immune system such as eosinophils, mechanisms governing secretion underlie different biological activities and responses to allergic and inflammatory diseases.[7,8] Eosinophils are a source of lipid mediators, distinct cationic proteins, and a large number of chemokines and cytokines that have wide-ranging effects on health and disease. Thus, the role of eosinophils markedly relies on their ability to release their immune compounds, a process collectively referred to as degranulation.[8,135]

Different from other immune cells, such as most lymphocytes that must exclusively synthesize proteins prior to secretion, many immune mediators are additionally stored as preformed pools within eosinophil secretory granules.[8,135] Thus, while human eosinophils can also generate immune compounds based on transcriptional activation and de novo protein synthesis, these cells are capable of storing and rapidly releasing an array of mediators from their specific granules in response to external stimuli.[3,8] Therefore, to appreciate eosinophil functions, it is central to understand how and in which situations these cells degranulate.

Since the pioneer works from Palade and collaborators who characterized, for the first time, organelles and interactions between various compartments of the secretory pathway,[136,137] TEM has been used as a robust tool to comprehend cell secretion.[24] These earlier studies elegantly identified the classical ER-Golgi secretory pathway, with vesicular secretory traffic directly from Golgi to the plasma membrane (constitutive secretion), which is well established in most cells. This route includes protein synthesis within the ER, transport of cargo inward toward the Golgi apparatus, and then through the Golgi and trans-Golgi network on the way to the plasma membrane, all carried by transport vesicles.[138] As with other nucleated cells, de novo synthesis of cytokines and other mediators within the ER and secretion through the Golgi apparatus occurs within eosinophils. However, in human eosinophils, immune mediators are mostly transported en route to the plasma membrane in an ER-Golgi-independent manner; that is, messengers are mobilized and released in response to cell activation directly from secretory granules.[3,8]

Eosinophil mediators are released in a tightly orchestrated manner involving not only these granules but also a distinct population of vesiculotubular structures (EoSVs), which actively participate in intracellular secretory trafficking from granules.[21] In activated human eosinophils undergoing secretion, specific granules undergo distinct ultrastructural changes while EoSVs significantly increase in number and act in concert with these granules in an intricate secretory pathway.[3,18,21] Therefore, the ultrastructural morphology of human eosinophils does not support the occurrence of a very active constitutive secretion, which would bypass granule storage sites. Accordingly, as detailed in Chapter 2, ER and Golgi are not prominent organelles in mature eosinophils, although these cells are highly specialized in secretion.

Human eosinophils can show different secretory patterns in vivo in response to a specific eosinophilic disease or condition. Because TEM is crucial to identify differential secretory modes, the application of this technique has greatly contributed to capturing the diversity of structural mechanisms of secretion in human eosinophils.[24–26] Preformed, granule-stored products can be released by three main secretory processes: classical and compound exocytosis, PMD, and cytolysis with granule release, of which PMD and cytolysis are more relevant to eosinophils.[3,8] Secretion through the release of extracellular vesicles (EVs) was also identified in human eosinophils.[125,139]

61

3.2 Classical and compound exocytosis

Exocytosis is uncommon in vivo eosinophil degranulation event while it can be observed during in vitro interaction of eosinophils with different parasitic helminths[140] or environmental fungi.[141] This mode of secretion is identified by conventional TEM by fusion of individual secretory granules with the plasma membrane (termed classical exocytosis) (Figs. 3.1, 3.2, and 3.3) and/or fusion of specific granules with each other, thus leading to the formation of channels in the cytoplasm for granule cargo release (compound exocytosis) (Figs. 3.1, 3.4, 3.5, and 3.6).[3] In this case, connectivity between fused granules and the cell surface is observed through the first granule of the channel, which may appear fused with the plasma membrane in electron micrographs (Fig. 3.1). The observation of compound exocytosis can be facilitated when samples are prepared for TEM with a step of postfixation with reduced osmium, which increases granule membrane contrast, thus highlighting granule-fusion events (Fig. 3.6).[23]

When eosinophils are stimulated with the pro-inflammatory cytokine tumor necrosis factor-alpha (TNF-α) at high concentrations, compound exocytosis is consistently identified, and, for this reason, we have been using this model to better understand this secretory process using TEM (Figs. 3.4–3.7).[23,25,125] In addition to showing granule fusions, compound exocytosis, as a general feature of degranulating eosinophils, is characterized by the presence of specific granules structurally disarranged and more electron-lucent, ultrastructural evidence for the release of granule products (Figs. 3.4–3.7). Moreover, thin sections of human eosinophils exhibiting compound exocytosis show, in optimally fixed samples, one granule projecting their limiting membrane into another granule, thus generating firmly connected structures (Figs. 3.4 and 3.6). As a result of compound exocytosis, enlarged granules and completely empty chambers can be observed in electron micrographs (Fig. 3.5). This is because exocytosis can trigger the wholesale discharge of granule-stored compounds, which means that this kind of secretion corresponds to a more drastic way to release eosinophil products. The finding of multiple crystalloid cores within a single granule or large membrane-bound chamber (Figs. 3.7 and 3.8) is also considered indicative of compound exocytosis.

In vivo, tissue eosinophils showing compound exocytosis are rarely documented by electron microscopic examination, but this secretory process can be found in certain conditions. For example, by analyzing 117 intestinal biopsies from patients with ulcerative colitis, Crohn's disease, rectal polyposis, and rectal carcinoma, 22% of the total biopsies had extensive evidence of compound exocytosis in eosinophils (Figs. 3.8 and 3.9) in association with other secretory processes.[142] These ultrastructural aspects related to the involvement of eosinophils in human diseases will be considered in more detail in Chapter 8. In vitro, we also observed that structural changes indicative of compound exocytosis might coexist with other secretory processes within the same cell in response to activation. This is the case of human eosinophils stimulated with TNF-α, which show predominantly compound exocytosis but also PMD to a lesser degree.[23] This secretory process is discussed below.

3.3 Piecemeal degranulation

PMD is a term coined by our group in the mid-1970s to describe for the first time a distinct morphology of secretion characterized by progressive degranulation of human basophils in contact allergy skin biopsies.[143,144] Evidence for the in vivo occurrence of PMD in eosinophils was initially provided in 1980 from electron microscopic studies of intestinal tissue eosinophils in patients with Crohn's disease.[145] By analyzing biopsies from these patients, it was noted the presence of lucent areas corresponding to content losses in the cores of specific granules, thus representing an in vivo demonstration of the specific release of MBP, the main eosinophil granule core constituent, in human disease.[145] Subsequent to these studies, we have observed similar core losses from specific granules in activated tissue eosinophils in skin samples of hypereosinophilic syndrome (HES) and focal losses in the granule matrices in eosinophils participating in other diseases (Chapter 8).[40] Our earlier works also showed in vitro ultrastructural evidence of partial emptying of granules in both mature and immature eosinophils.[108,146,147] Termed PMD because of a "piece by piece" release of granule contents, this secretion mode is now recognized as a central secretory process of human eosinophils (reviewed in[148]).

In contrast to granule exocytosis, PMD is characterized by progressive losses of granule contents in the absence of substantial granule-granule and/or granule-plasma membrane fusions (Figs. 3.10–3.14).[148] The morphologic hallmark of PMD is thus the presence of nonfused granule containers with varying degrees of content emptying, an event identified by TEM as a continuum of morphologies such as disassembled cores/matrices, lucent areas within their internal structure, reduced electron density, and residual cores (Figs. 3.10–3.14). These morphologic changes reflect the ability of human eosinophils to release individual materials packaged in the specific granule matrix and/or core.[40,148] The crystalline core, which is a tightly compacted structure in resting granules, undergoes, as a result, a disarrangement process leading to different ultrastructural views while the granule limiting membrane is typically kept intact (Figs. 3.15, 3.16, and 3.17). Specific granules undergoing PMD are generally larger than resting granules within the same cell (Figs. 3.16–3.19),

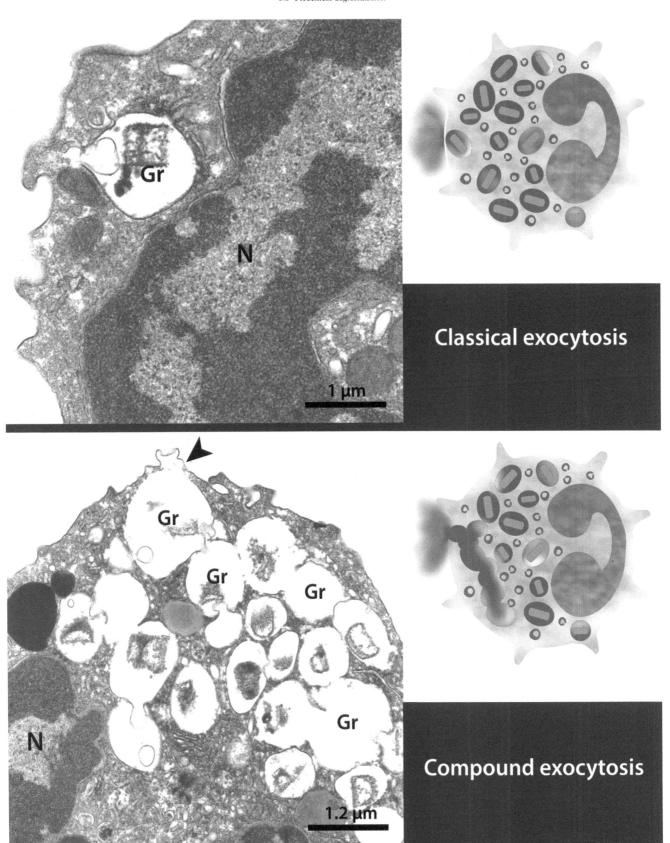

FIG. 3.1 Exocytosis processes observed by TEM in human eosinophils. In classical exocytosis *(top panel)*, a single secretory granule (Gr) fuses with the plasma membrane (PM) while large channels formed by granule-granule fusions characterize compound exocytosis. Note fusion of the first granule from the channel *(arrowhead)* with the PM. *N*, nucleus.

A. The cell biology of human eosinophils

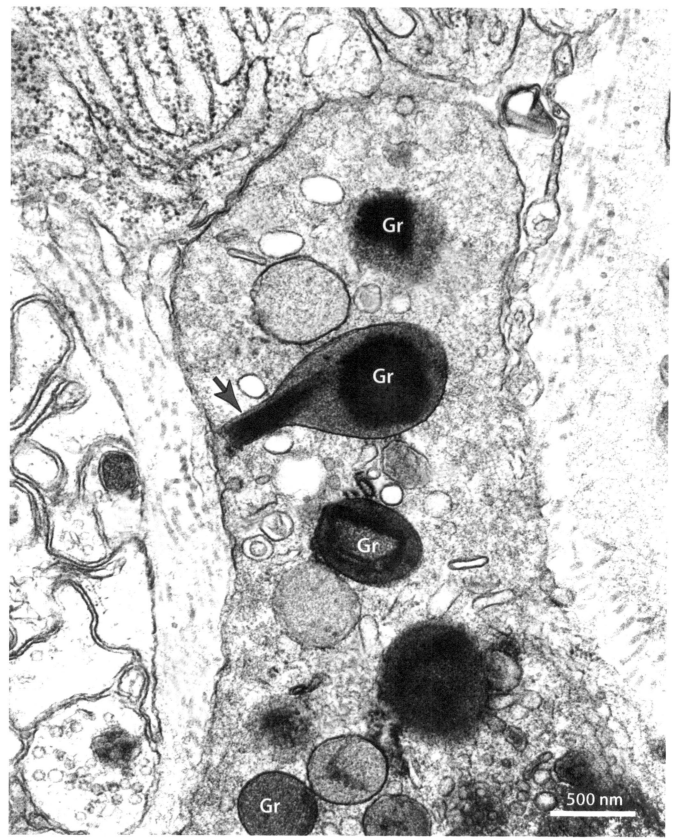

FIG. 3.2   Classical exocytosis in a human tissue eosinophil. Note a single specific granule (Gr) in the process of elongation *(arrow)* and fusion with the plasma membrane. Intestine sample processed for TEM.

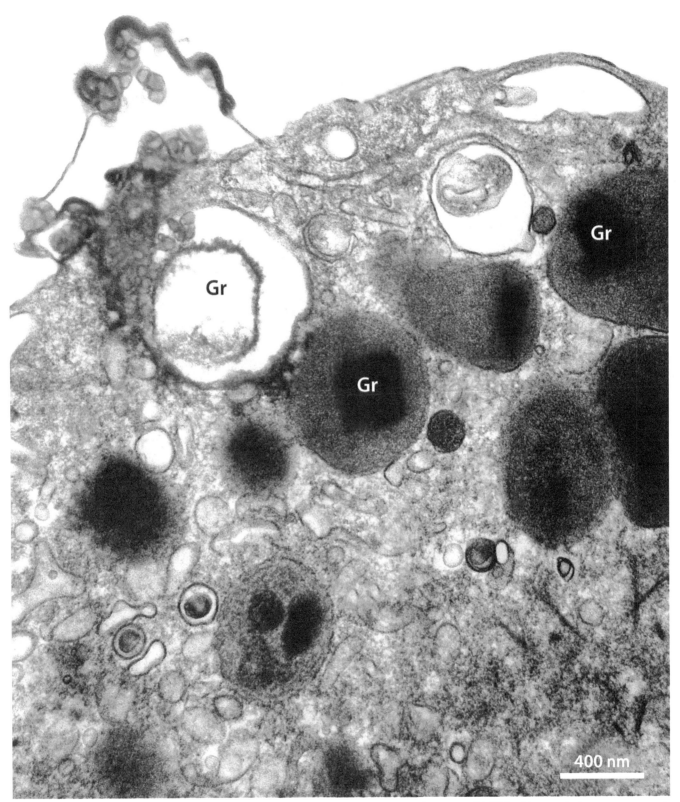

FIG. 3.3 TEM of a human peripheral blood eosinophil showing classical exocytosis. The content *(highlighted in orange)* of a secretory granule (Gr) is being released at the cell surface after fusion with the plasma membrane.

A. The cell biology of human eosinophils

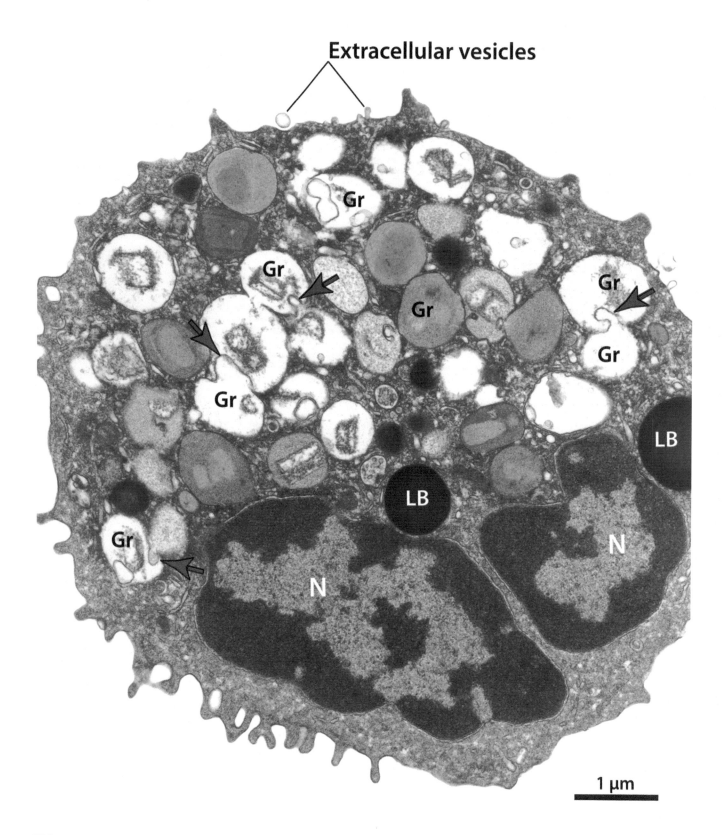

FIG. 3.4   Compound exocytosis in a human peripheral blood eosinophil stimulated with TNF-α. Granule-granule fusions and content losses are observed. Note protrusions *(arrows)* between secretory granules (Gr) and extracellular vesicles at the cell surface. Cells were processed for TEM.[23]

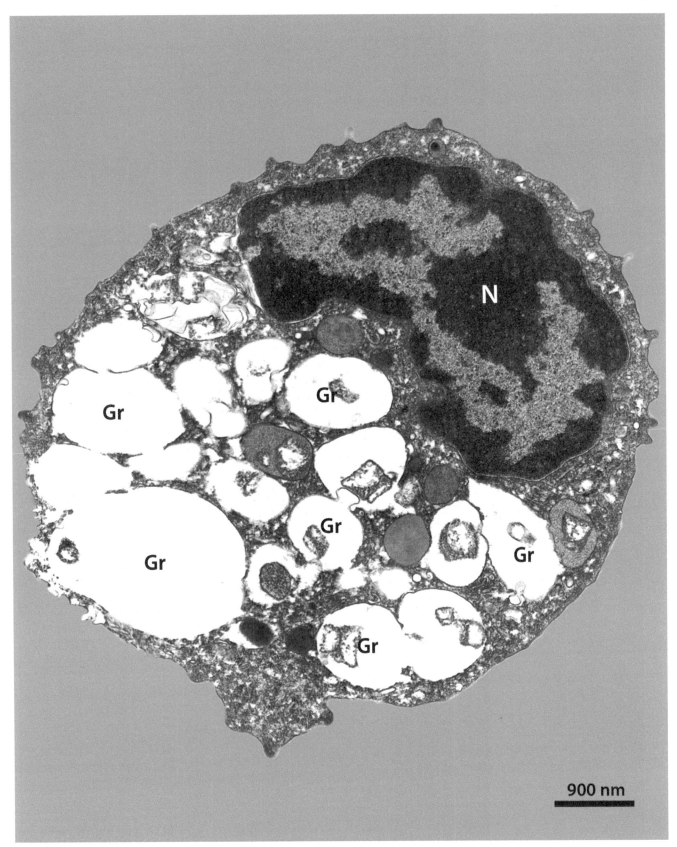

FIG. 3.5   Ultrastructural view of compound exocytosis in an activated human eosinophil. Note large chambers formed by fused granules (Gr) with extensive losses or no content. Eosinophils were isolated from the peripheral blood, stimulated with TNF-α, and processed for TEM.[23] *N*, nucleus.

A. The cell biology of human eosinophils

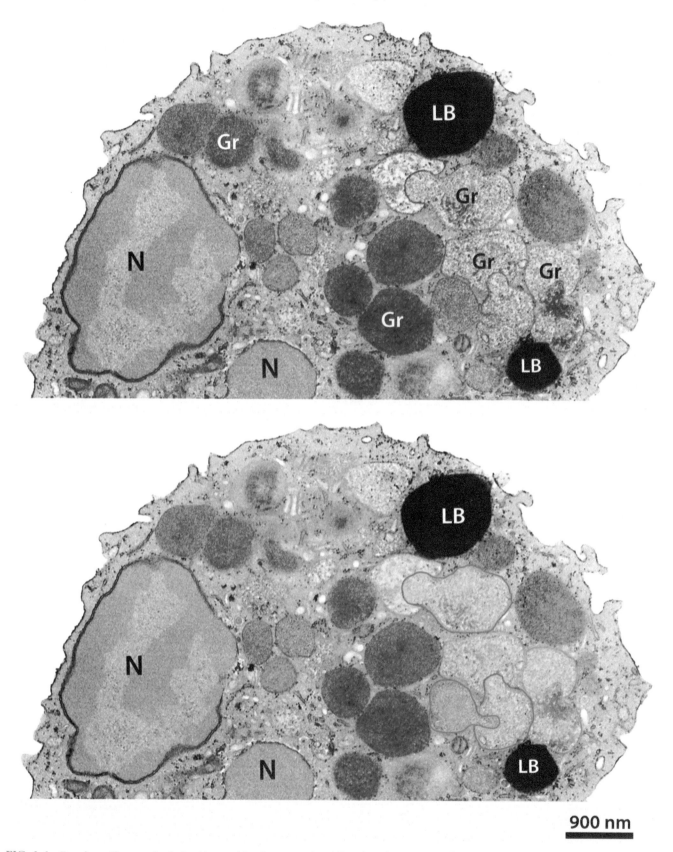

FIG. 3.6 Fused specific granules *(colored in green)* in a human eosinophil undergoing compound exocytosis. TNF-α-stimulated cells were post-fixed for TEM with ferrocyanide-reduced osmium to emphasize membranes.[23] *LB*, lipid body; *N*, nucleus.

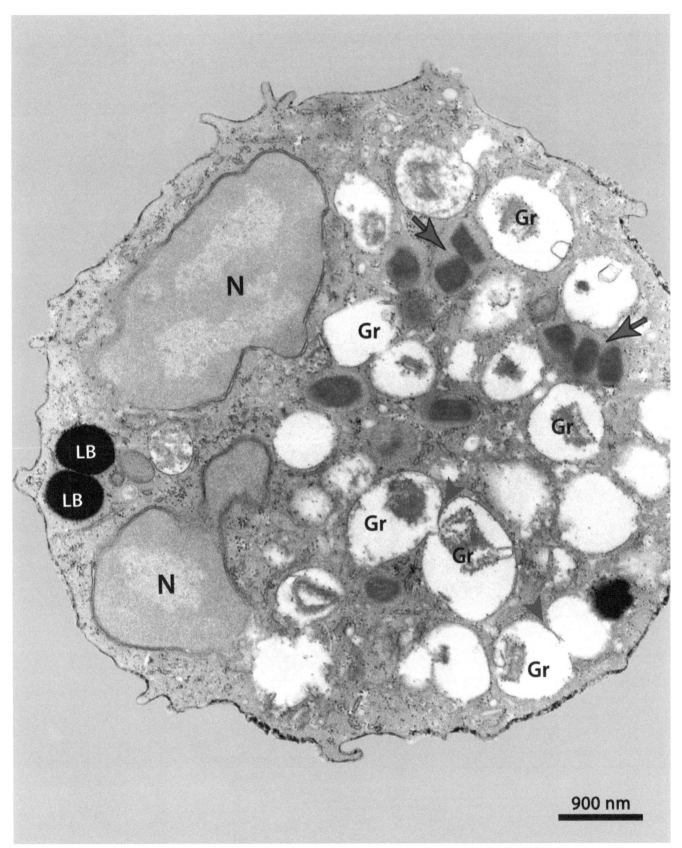

FIG. 3.7 An activated human eosinophil showing ultrastructural evidence of compound exocytosis. Note secretory granules (Gr) with multiple cores *(arrows)* and typical fusions *(arrowheads)*. TNF-α-stimulated eosinophils were prepared for TEM.[23] *LB*, lipid body; *N*, nucleus.

A. The cell biology of human eosinophils

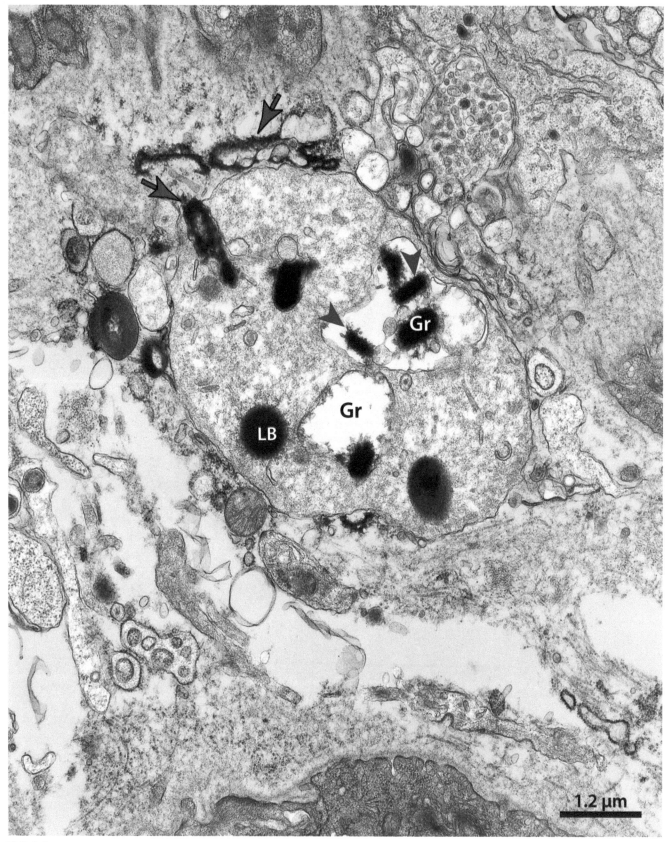

FIG. 3.8 Part of the cytoplasm of a human tissue eosinophil showing compound exocytosis. Note multiple cores *(arrowheads)* in a large chamber formed by fused granules (Gr) and released material at the cell surface *(arrows)*. Extracellular matrix (ECM) colored in *green*. Biopsy (bacterially infected ileum) prepared for TEM. *LB*, lipid body.

A. The cell biology of human eosinophils

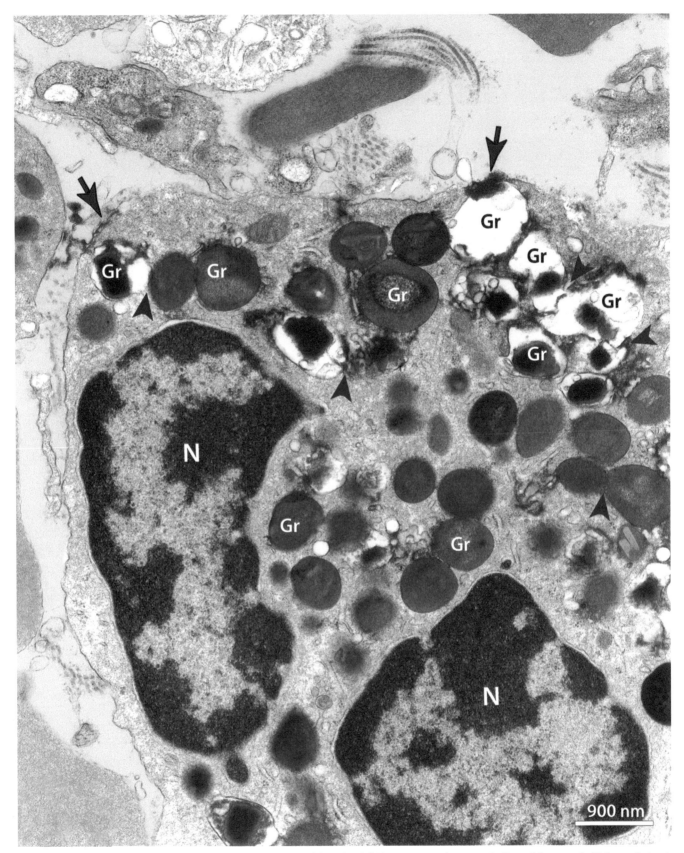

FIG. 3.9   TEM of human tissue eosinophil (bacterially infected ileum) exhibiting compound exocytosis. Part of the granule (Gr) population is fused *(arrowheads)*. Note released granule contents at the cell surface *(arrows)*. ECM colored in *green. N*, nucleus.

A. The cell biology of human eosinophils

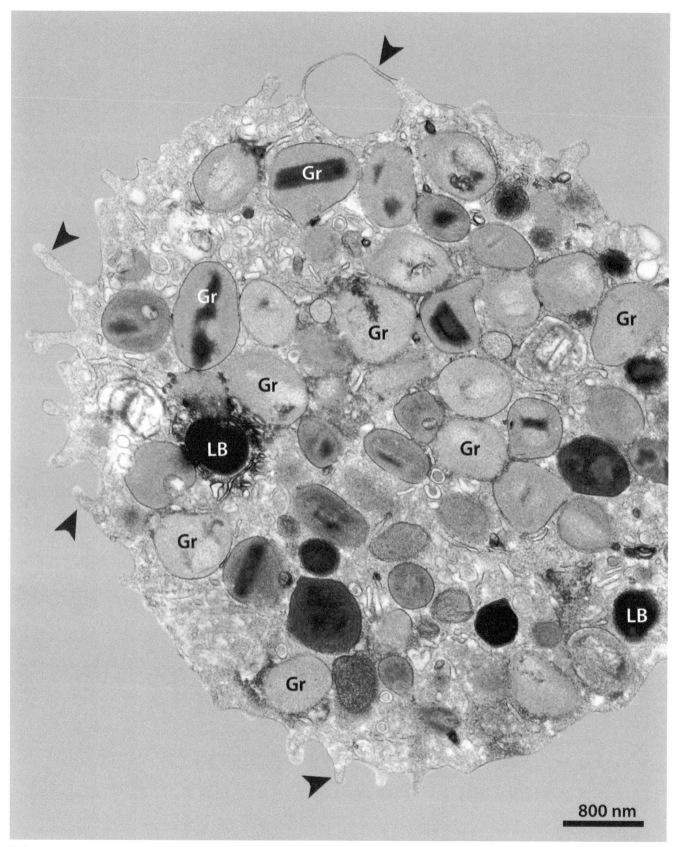

FIG. 3.10 Human peripheral blood eosinophil showing PMD in response to CCL11. Most secretory granules (Gr) are not fused and show reduced electron density. Note protrusive surface structures *(arrowheads)*. The nucleus is not visible in the section plane. Cells were isolated, stimulated, and prepared for TEM.[20] *LB*, lipid body.

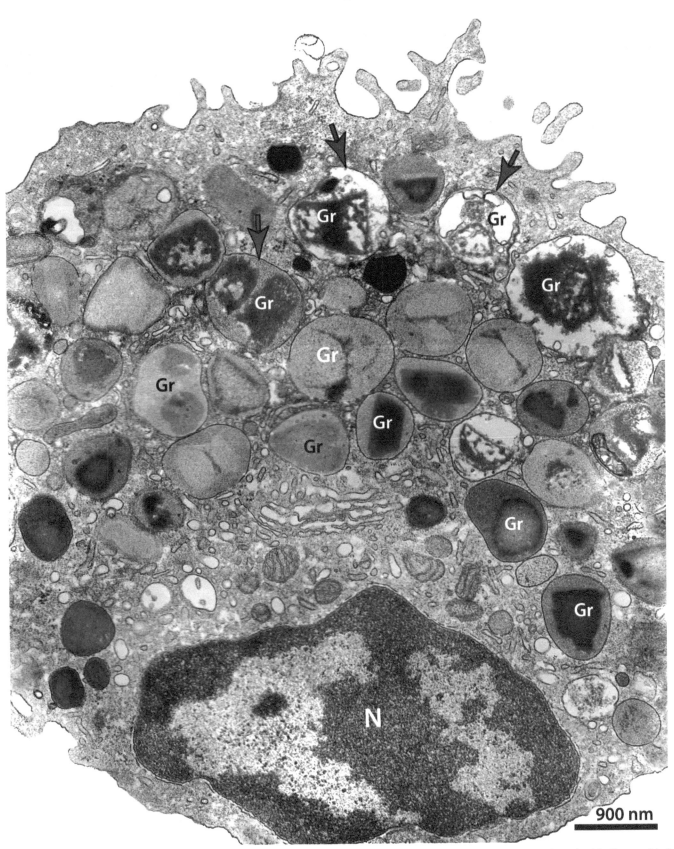

FIG. 3.11 Human peripheral blood eosinophil showing PMD in response to PAF. Secretory granules (Gr) are enlarged with disassembled cores/matrices and lucent areas *(arrows)*. The Golgi area is evident *(highlighted in yellow)*. Cells were isolated, stimulated, and processed for TEM.[20] N, nucleus.

A. The cell biology of human eosinophils

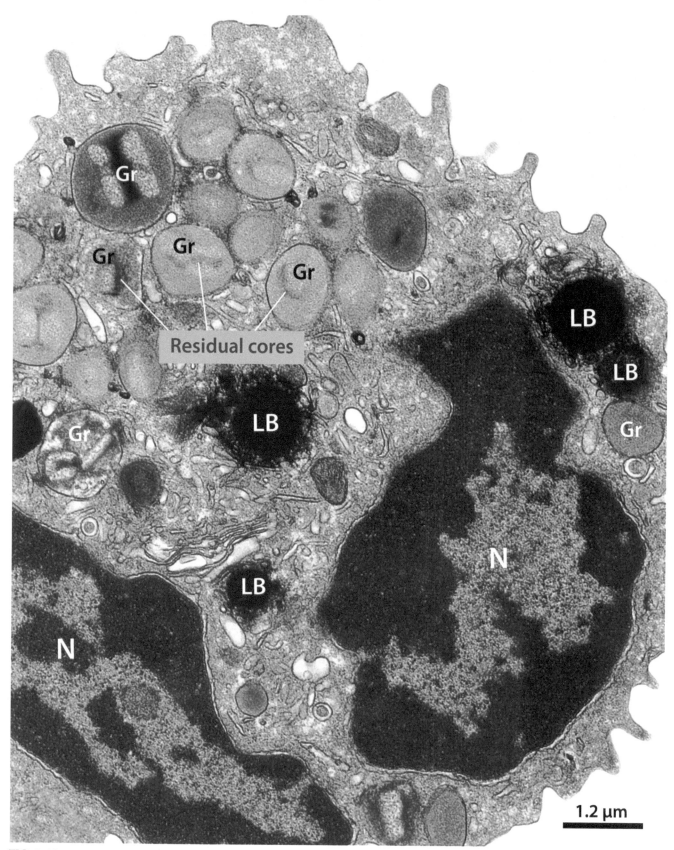

FIG. 3.12   Human peripheral blood eosinophil showing PMD in response to CCL5. Emptying nonfused granules (Gr) with residual cores are observed. Cells were isolated, stimulated, and processed for TEM.[20] *LB*, lipid body; *N*, nucleus.

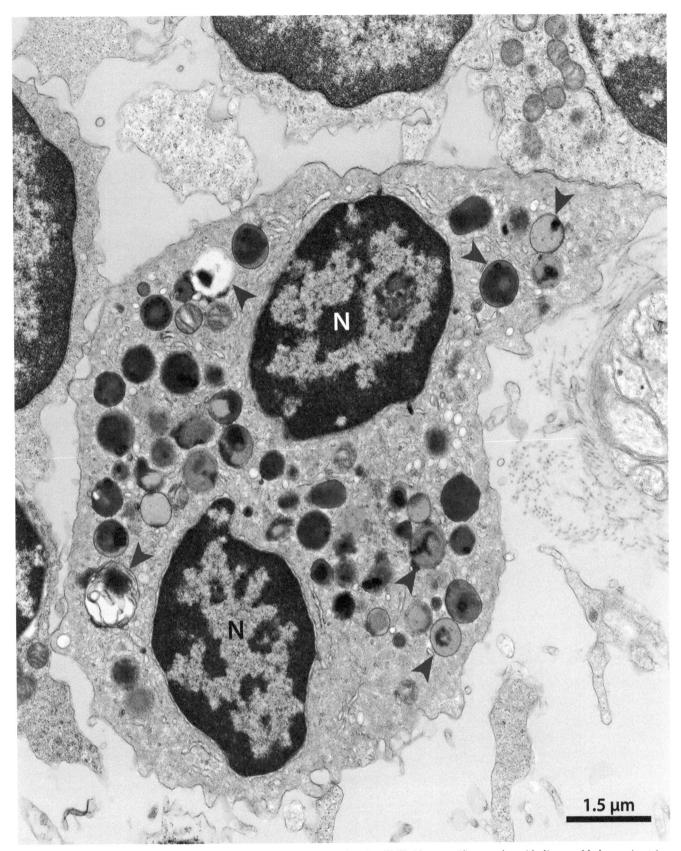

FIG. 3.13   Tissue human eosinophil in the intestinal lamina propria showing PMD. Note specific granules with disassembled cores/matrices and content losses *(arrowheads)*. ECM colored in *green*. Biopsy (familial polyposis) prepared for TEM. *N,* nucleus.

A.   The cell biology of human eosinophils

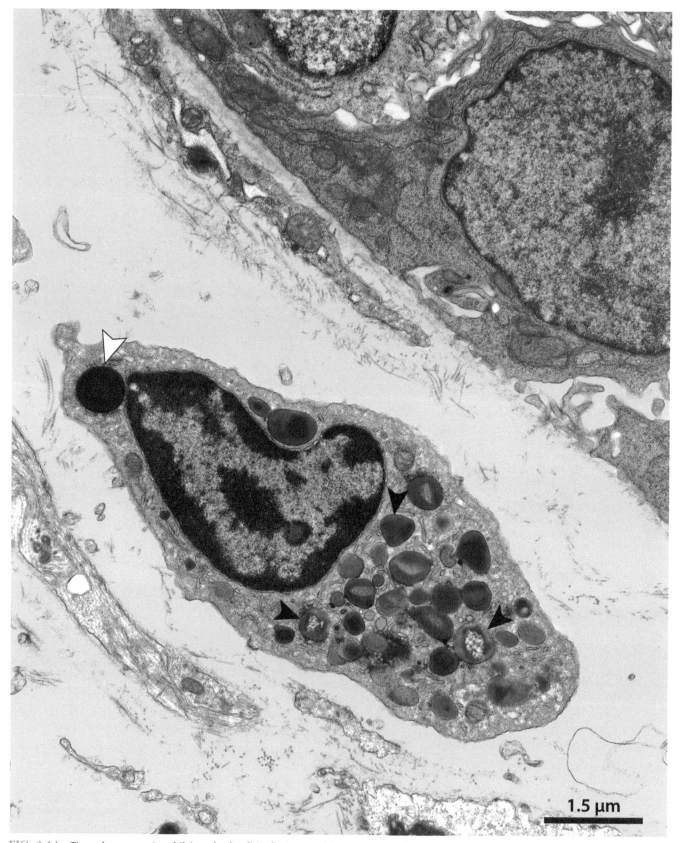

FIG. 3.14   Tissue human eosinophil *(pseudocolored)* in the intestinal lamina propria showing PMD and polarization (nucleus and granules in opposite sides). Specific granules *(yellow)* display disassembled cores/matrices and content losses *(black arrowheads)*. Note a LB *(white arrowhead)*. The intestinal epithelium is partially seen, and the ECM was colored in *green*. Biopsy (familial polyposis) prepared for TEM. Cytoplasm *(pink)*; Nucleus *(purple)*.

A. The cell biology of human eosinophils

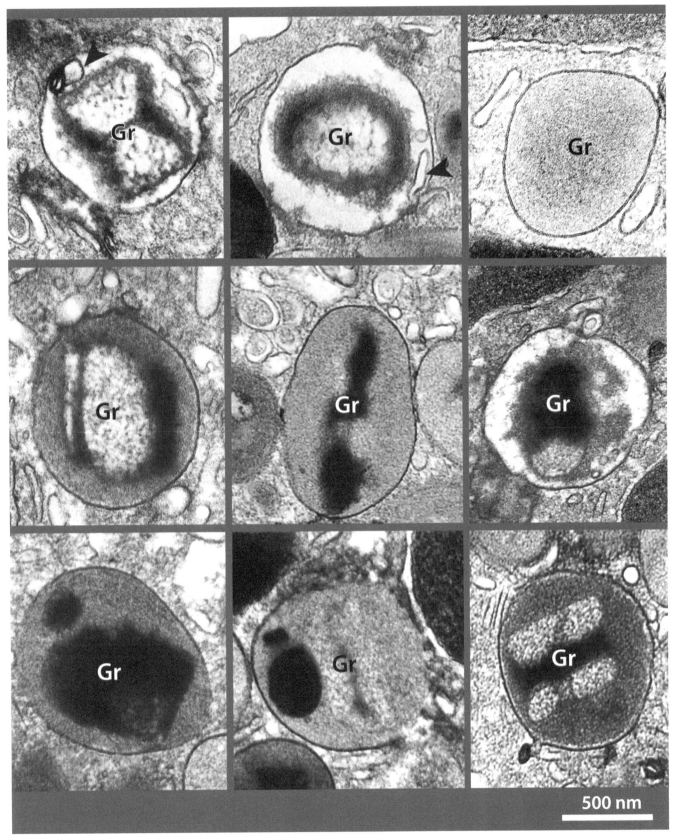

FIG. 3.15   Ultrastructural views of specific granules in human peripheral blood eosinophils undergoing PMD. Granules (Gr) show intact limiting membranes and varied degrees of content losses. *Arrowheads* indicate internal membranes. Cells were stimulated with CCL11, CCL5, or PAF.[20]

A. The cell biology of human eosinophils

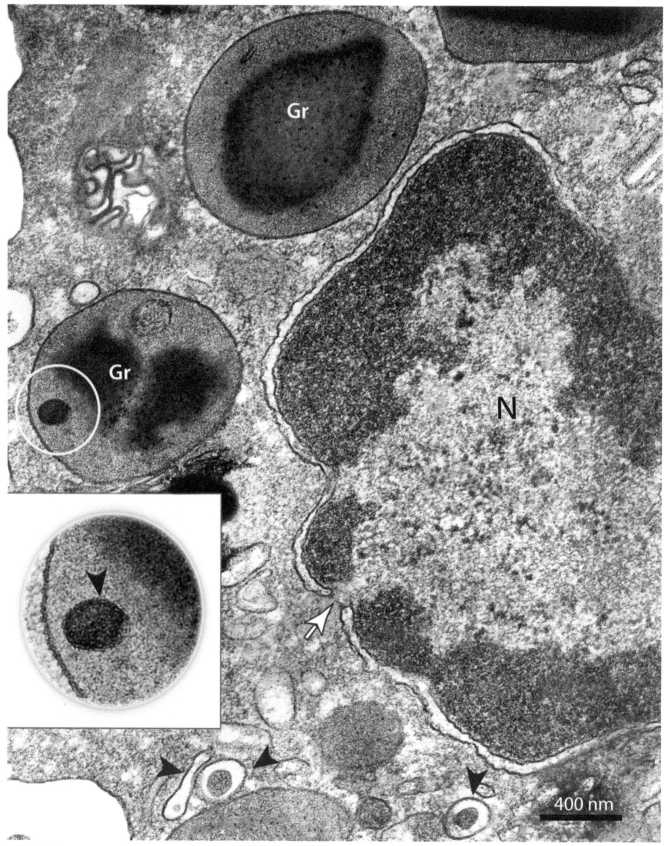

FIG. 3.16 A CCL11-stimulated human eosinophil. Secretory granules (Gr) are enlarged and show the cores in the process of disarrangement. Part of the core content *(circle)* is surrounded by bilayer a membrane *(black arrowhead)*. Note a nuclear pore *(arrow)* and EoSVs *(arrowheads)*. Eosinophils isolated from the peripheral blood were stimulated and prepared for TEM.[20] *N*, nucleus.

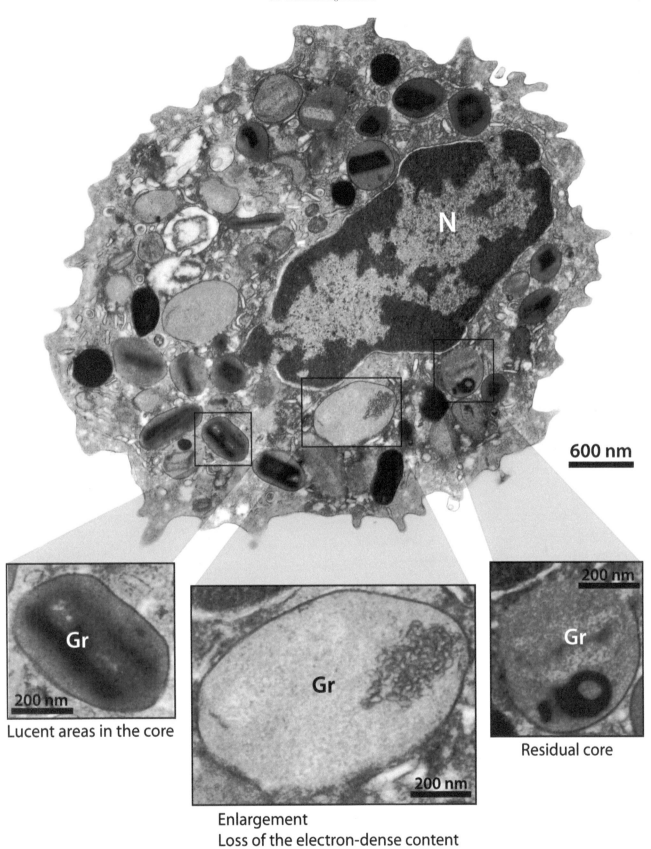

FIG. 3.17 Ultrastructure of a CCL11-activated human eosinophil. After stimulation, specific granules (Gr) exhibit different morphologies indicative of PMD. Eosinophils isolated from the peripheral blood were stimulated and processed for TEM.[20] N, nucleus.

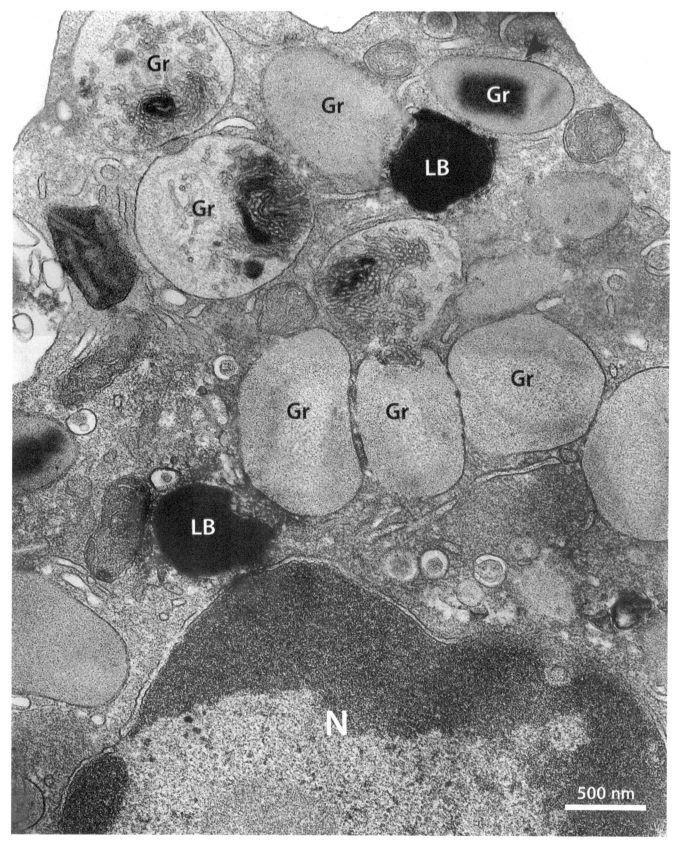

FIG. 3.18   Emptying specific granules in a CCL5-activated human eosinophil. Granules (Gr) appear as large organelles with reduced electron density and disarranged cores. Note a resting granule with electron-dense core *(arrowhead)*. Cells were processed for TEM.[20] *LB*, lipid body; *N*, nucleus.

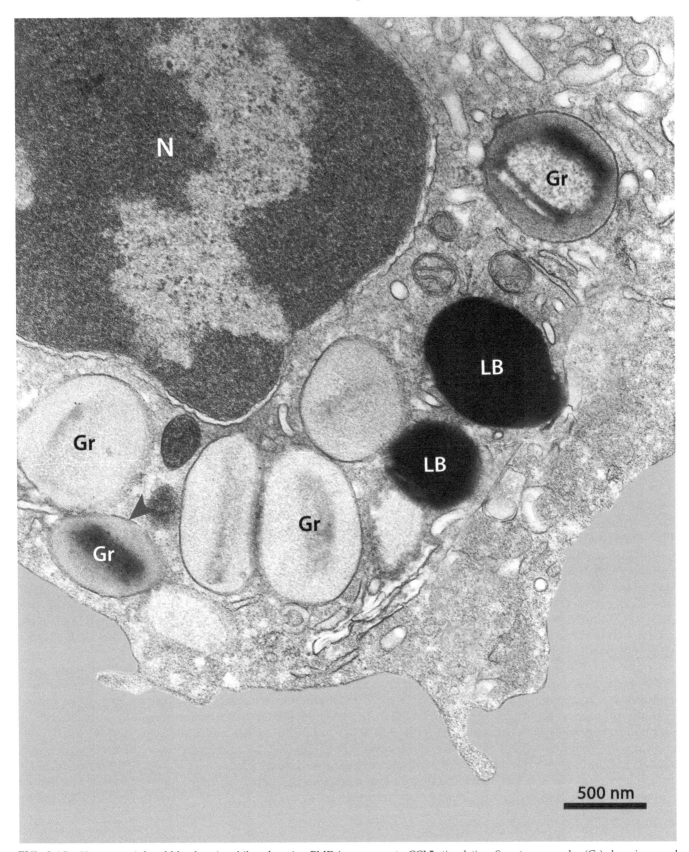

FIG. 3.19   Human peripheral blood eosinophil undergoing PMD in response to CCL5 stimulation. Secretory granules (Gr) show increased size and content losses compared with a resting granule *(arrowhead)*. Cells were stimulated and processed for TEM.[20] *LB*, lipid body; *N*, nucleus.

A.   The cell biology of human eosinophils

a phenomenon likely due to the structural and biochemical changes that occur within granules in response to degranulating stimuli.[20] Moreover, intragranular membranes become more apparent as the granule contents are mobilized and released (Figs. 3.15 and 3.16).

Rarely, fully empty membrane-bound containers persist in tissue eosinophils undergoing PMD. As noted, the most usual representation of PMD as seen by TEM in different conditions is the presence of a diversity of granule morphologies denoting different quantities of product release (Figs. 3.10–3.15). This is because human eosinophils select specific products to be liberated.[7] Moreover, the variable electron density of the granule core and matrix has been associated with PMD.[17,149]

A common view of PMD is the occurrence of emptying nonfused granules interspersed, in the same cell section, with resting, nonmobilized granules (Figs. 3.18, 3.19, and 3.20). Our findings from numerous EM analyses support the fact that not all eosinophil-specific granules are uniformly, coordinately, and simultaneously responsive to stimuli (reviewed in[148]). PMD sustains a pool of intact secretory granules (Fig. 3.20) that may contribute to the special capability of eosinophils to rapidly release their products under different or repetitive stimuli.[20,148]

Although the molecular mechanisms involved in the process of PMD are not fully understood, the participation of cytokine receptors in the release of granule-derived cytokines was demonstrated.[31] Eosinophils contain substantial intracellular amounts of several granule- and vesicle-associated cytokine receptors, including IL-4, IL-6, and IL-13 receptors as well as C–C chemokine receptor type 3 (CCR3).[82] Both IL-4 and IL-4 receptor alpha chain (IL-4Rα) colocalized in eosinophil granules; after eotaxin stimulation, IL-4Rα, bearing bound IL-4, was mobilized into secretory vesicles.[31] Furthermore, membranous compartments, revealed by different EM approaches, including conventional TEM (Figs. 3.15 and 3.16) and electron tomography (see Fig. 4.10 in Chapter 4), are likely involved in the process of product compartmentalization and mobilization upon cell activation.[20]

PMD is a frequent mode of eosinophil degranulation in vivo and is well documented in a diversity of inflammatory and allergic disorders (Chapter 8) including asthma,[150] nasal polyposis,[151,152] allergic rhinitis,[151153] ulcerative colitis,[151] Crohn's disease,[151] familial polyposis (Figs. 3.13 and 3.14), atopic dermatitis,[154] gastric carcinoma,[155] functional dyspepsia,[149,156] shigellosis,[157] and cholera.[158] In vitro, human eosinophils stimulated during one hour with the classical eosinophil agonists chemokine C-C motif ligand 11 (CCL11/eotaxin-1) (Figs. 3.10, 3.15, and 3.16), platelet-activating factor (PAF) (Fig. 3.11), or chemokine C-C motif ligand 5 (CCL5) (Figs. 3.12, 3.18, and 3.19) consistently showed PMD as the main secretory process as demonstrated by quantitative TEM studies while fused granules were just occasionally observed with these stimuli.[20]

Early aspects of stimulus-induced eosinophil PMD, when most granules have not yet shown signs indicative of content losses, can be observed after 30 min of stimulation with CCL11.[20] At this time, granules develop into irregular structures with progressive protrusions from their surfaces, preferentially present on intact granules that had an ill-defined core and matrix (Fig. 3.21).[20]

PMD is accomplished by vesicular transport of small packets of materials from the secretory granules to the cell surface.[148] EoSVs shuttle granule-derived proteins, such as MBP,[22] IL-4,[19] and IFN-γ[25] from the granules to the plasma membrane for secretion. As a result, these vesiculotubular structures are frequently seen around, attached, and/or budding from emptying granules by conventional TEM in activated human eosinophils (Figs. 3.21, 3.22, and 3.23).[21,148] Ultrastructural events reflecting vesicle formation include the occurrence of clouds of amorphous dense material around the granule limiting membranes together with granule-associated EoSVs (Fig. 3.21). All of these structural events are fully observed in vivo, as discussed in Chapter 8.

When high-resolution 3D analysis by electron tomography was applied to CCL11-stimulated eosinophils as a model to understand PMD and to track vesicle formation from specific granules, a clear budding of EoSVs from granules was demonstrated (Fig. 3.24).[19] Notably, electron tomography revealed that EoSVs emerge from mobilized granules through a "tubulation" process (Fig. 3.24). This may explain why EoSVs are frequently seen attached to emptying granules by conventional TEM (Figs. 3.21, 3.22, and 3.23). Images indicating processes of "tubulation" from specific granules can also be seen by conventional TEM (Figs. 3.25 and 3.26). Moreover, the findings of EoSVs around or budding from specific granules and carrying granule-derived products, as shown by ultrastructural cytochemistry (Fig. 3.27 and 3.28) and immunogold EM (addressed in Chapter 5), demonstrate that these vesicles are being formed from specific granules as part of the secretory pathway.[19,21,22,25]

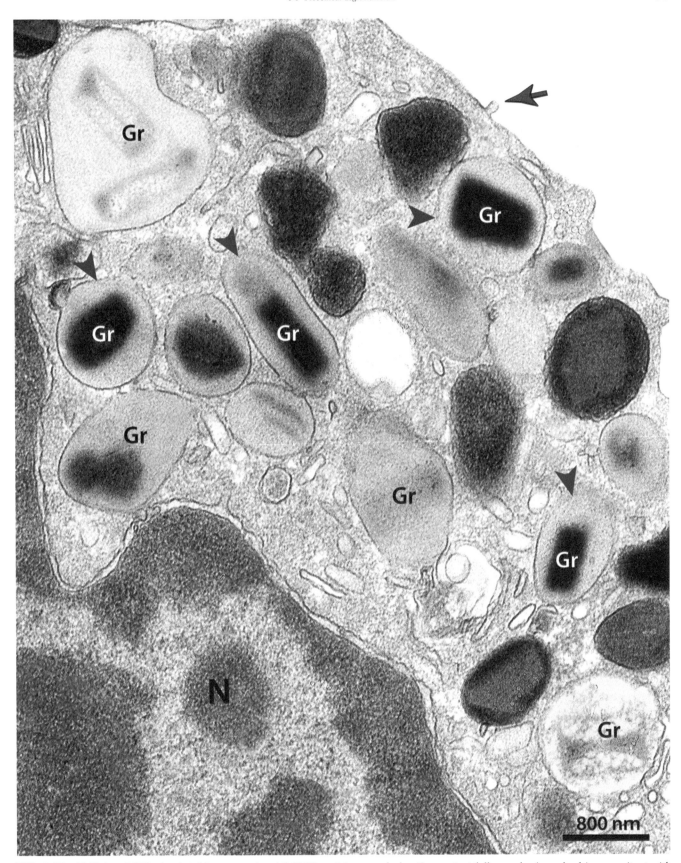

FIG. 3.20 PMD in a CCL5-stimulated human eosinophil. PMD sustains a pool of resting content-full granules *(arrowheads)* concomitant with emptying enlarged granules *(colored in yellow)*. Note a tiny extracellular vesicle being formed at the plasma membrane *(arrow)*. Cells were processed for TEM.[20] *Gr*, secretory granules; *N*, nucleus.

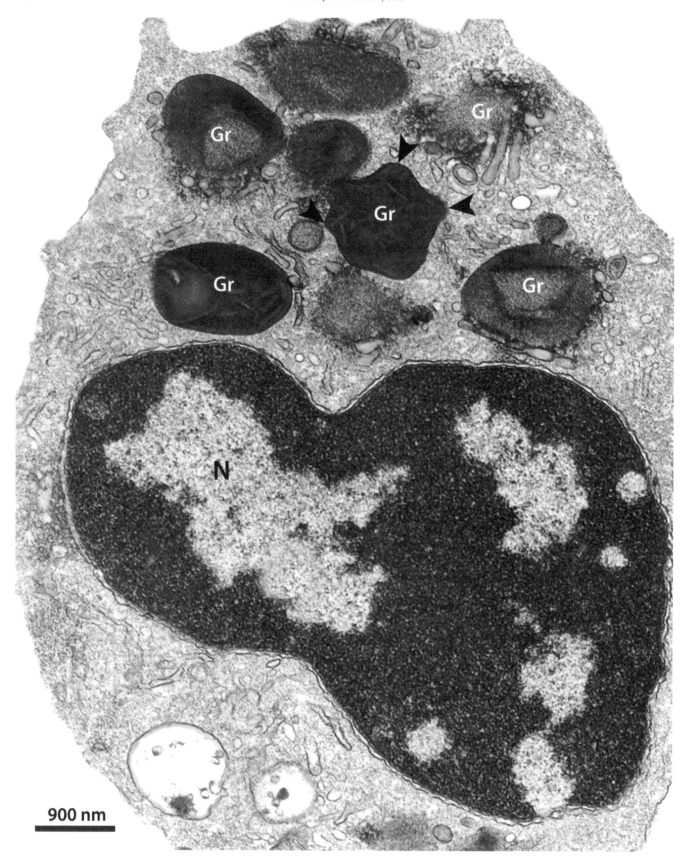

FIG. 3.21   TEM showing granule-associated EoSVs *(colored in pink)* within a CCL11-stimulated human eosinophil. Clouds of dense material accompany EoSV formation. Note granule protrusions *(arrowheads)* and two endosomes *(colored in orange)*. Cells prepared as described.[20] *N*, nucleus.

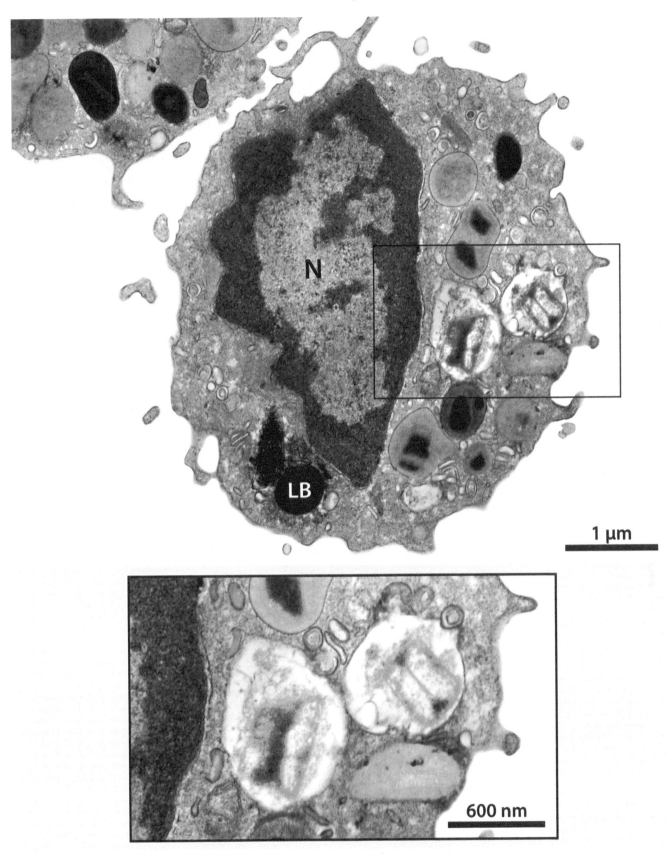

FIG. 3.22 EoSVs associate with secretory granules. TEM of a CCL11-activated human eosinophil reveals EoSVs *(colored in pink)* in contact, surrounding and/or budding from enlarged specific granules *(colored in yellow)* undergoing PMD. Cells prepared as described.[20] N, nucleus.

A. The cell biology of human eosinophils

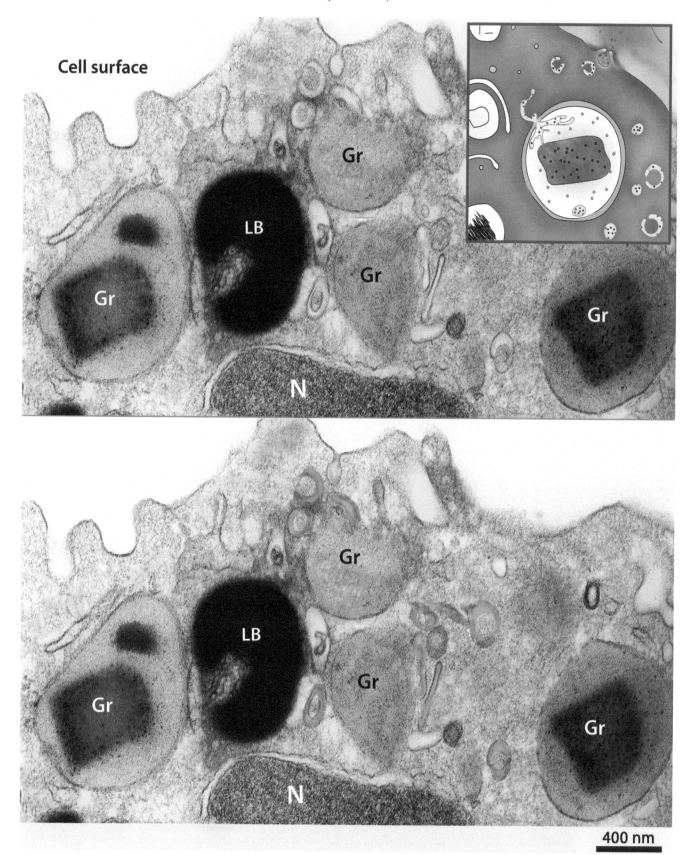

FIG. 3.23 Ultrastructural view of cytoplasmic EoSVs (*highlighted in pink*) in a human CCL5-stimulated eosinophil. Activation leads to EoSV formation from specific granules (Gr), as shown by the diagram. Note a lipid body (LB) with heterogeneous content. Cells were prepared for TEM as before.[20] N, nucleus.

A. The cell biology of human eosinophils

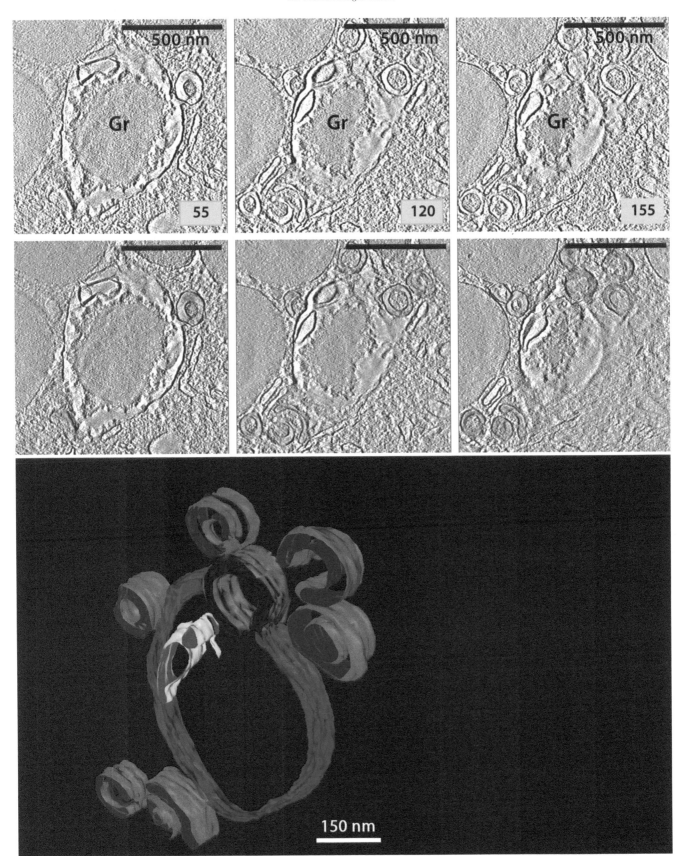

FIG. 3.24   Representative serial tomographic slices (4 nm thick), obtained from automated electron tomography of an emptying granule (Gr) within a CCL11-stimulated human eosinophil, show substantial structural changes associated with EoSV formation.[19] EoSVs *(highlighted in pink in identical panels)* are imaged as open, tubular-shaped structures. Numbers on the lower right corners of the *top panels* indicate slice numbers through the tomographic volume. The *bottom panel* shows a 3D model based on the serial slices. EoSVs *(pink)*; Granule limiting membrane *(purple)*; Intragranular membranes *(green)*.

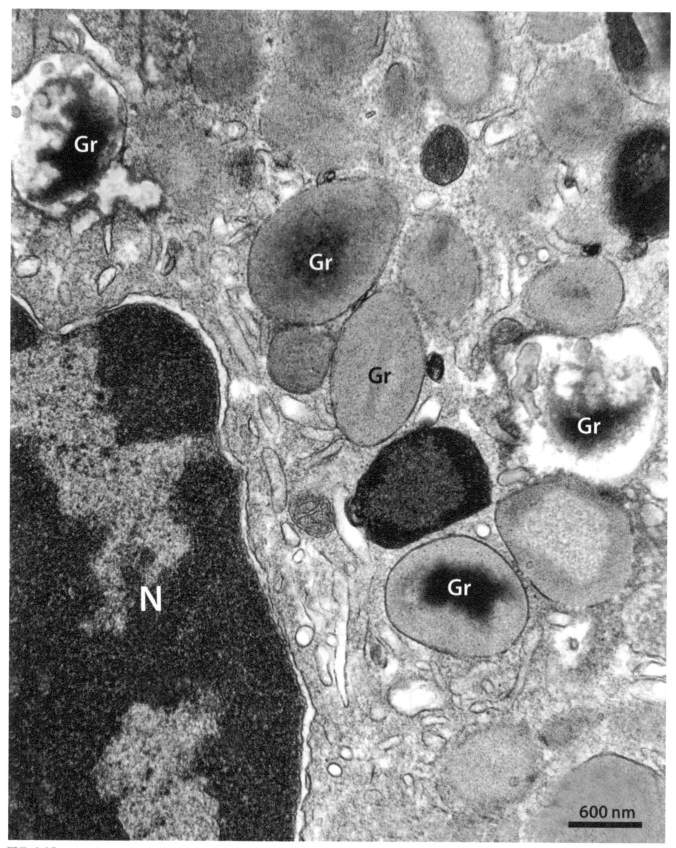

FIG. 3.25 Membranous vesiculotubular structures *(colored in pink)* within and projecting from the surface of emptying specific granules (Gr). Eosinophils isolated from the peripheral blood were stimulated with CCL11 and processed for TEM.[20] N, nucleus.

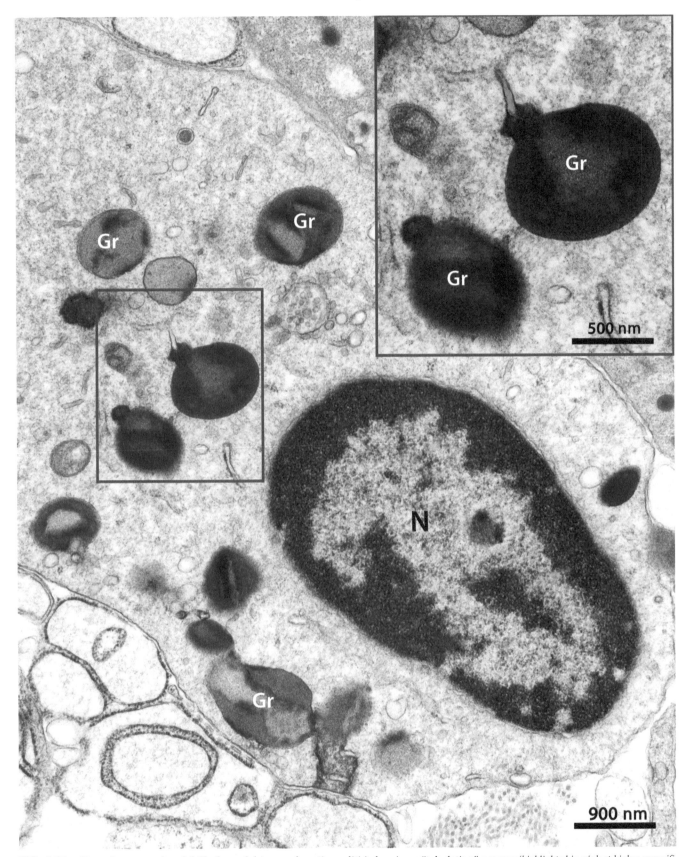

FIG. 3.26   Tissue human eosinophil (ileal pouch biopsy, ulcerative colitis) showing a "tubulation" process *(highlighted in pink at higher magnification)* from a secretory granule (Gr). A late endosome was colored in *orange*. *N*, nucleus.

A. The cell biology of human eosinophils

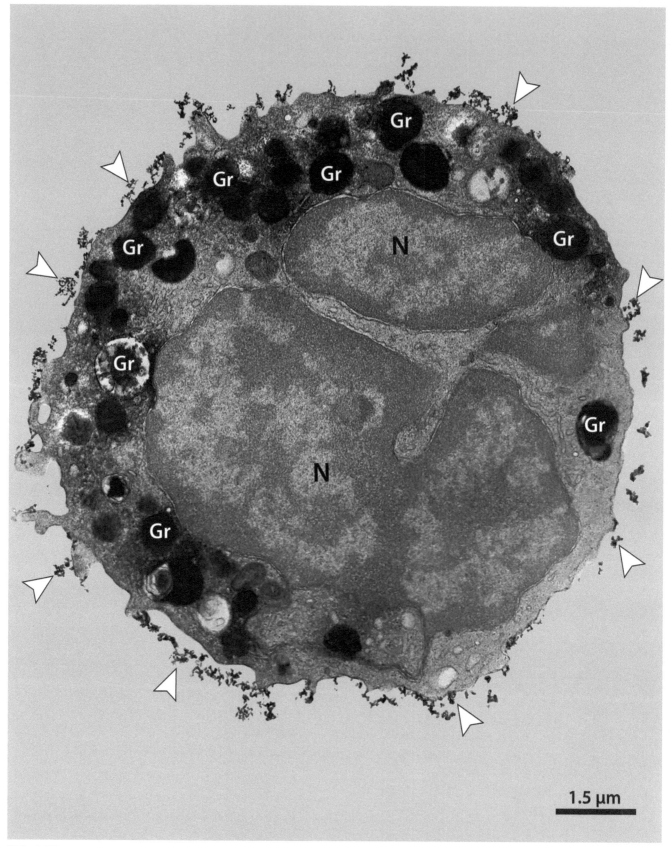

FIG. 3.27   Electron micrograph of a human activated eosinophil showing peroxidase-positive specific granules (Gr). Eosinophil peroxidase (EPO) is a granule-derived cationic protein imaged by EM as a dense reaction product following a standard cytochemistry procedure.[108,147] Note focal deposits of released peroxidase *(arrowheads)* attached to the cell surface. *N,* nucleus.

A. The cell biology of human eosinophils

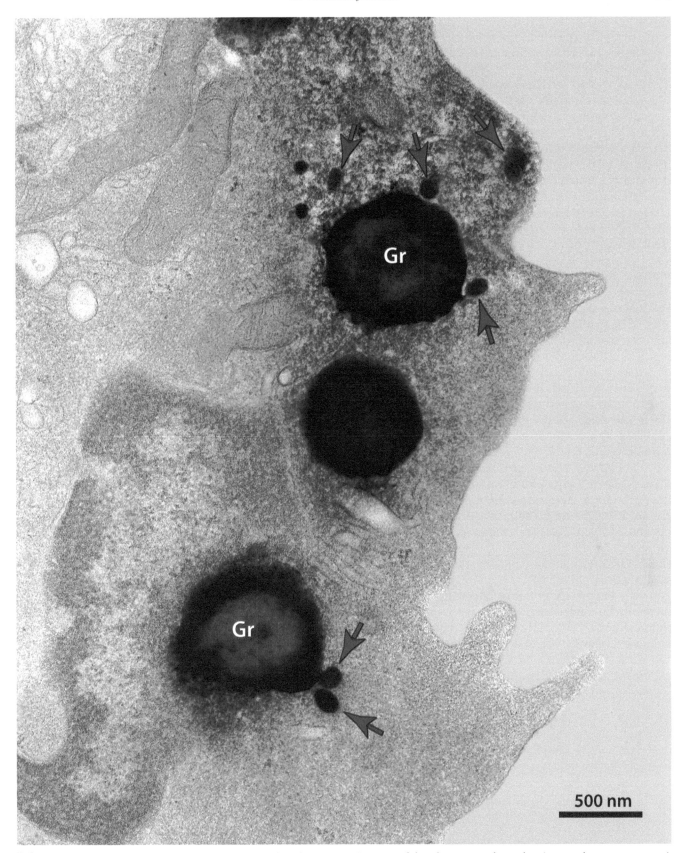

FIG. 3.28   High magnification of a human eosinophil undergoing PMD prepared for ultrastructural cytochemistry to demonstrate peroxi-dase.[108,147] Specific granules (Gr) with dense peroxidase activity are surrounded by peroxidase-loaded EoSVs *(arrows)*.

A. The cell biology of human eosinophils

3.4 Cytolysis

Cytolysis, like PMD, is a common secretory process found in vivo in tissue eosinophils participating in inflammatory responses (reviewed in[3,159]). Eosinophil cytolysis is characterized by the deposition of membrane-bound granules in tissues and secretions after cell death.[3,159]

Cytolytic eosinophils can be observed by conventional TEM as cells with a range of morphological signs indicative of cell lysis such as initial loss of the plasma membrane integrity (Fig. 3.29); complete rupture of the plasma membrane (Fig. 3.30) associated with chromatin decondensation and dissolution (chromatolysis) (Fig. 3.31); or cells that underwent a high degree of dissolution with the disappearance of most organelles (Fig. 3.32). In all of these situations, intact specific granules still bound by their granule delimiting membranes persist in tissue sites and are observed individually and as clusters of FEGs (Figs. 3.29–3.32).[3] Therefore, for eosinophils, a cytolytic cell death (detailed in Chapter 6) frequently represents a degranulation (secretory) process.[3,159–161]

Intact EoSVs are also frequently seen in vivo by TEM in tissues close to eosinophil FEGs (Fig. 3.33). [17] We[8] and others[17] have been observing EoSVs with typical morphology deposited within the tissue alongside FEGs in different diseases, for example, eosinophilic esophagitis,[17] inflammatory bowel disease,[8] and eosinophilic chronic rhinosinusitis (ECRS) (Figs. 3.32 and 3.33). Whether these vesicles are functionally active after being expelled from cytolytic eosinophils or generated in situ from activated FEGs remains to be established. Moreover, in a variety of eosinophilic diseases, lytic eosinophils and FEGs are often observed adjacent to CLCs in sites of human tissues (Fig. 3.34), as shown in Chapter 8.

By SEM, cytolytic eosinophils appear in 3D as disrupted cells with a clear observation of the granule population seen mostly as intact organelles while the nucleus and other cytoplasmic components appear disarranged (Fig. 3.35). Eosinophil cytolysis may also be associated with the release of filamentous chromatin structures termed eosinophil extracellular traps (EETs)[46,49,162] clearly seen by SEM (Fig. 3.36), which characterizes the process of cell death by ETosis (Chapter 6). Recent evidence shows that eosinophil cytolytic degranulation, that is, the release of intact, membrane-delimited granules that arises from the cytolysis of eosinophils, occurs mainly through ETosis, meaning death with a cytolytic profile and extrusion of extracellular traps, which has increasingly been identified in several eosinophilic diseases.[46,49,163] Eosinophil ETosis (EETosis) has also been associated with the formation of CLCs.[41,164,165]

Eosinophil FEGs in tissues likely act as independent, ligand-responsive, secretion-competent structures.[16,166,167] Isolated human eosinophil granules express membrane receptors, including for a chemokine (CCR3), a cytokine (IFNγ receptor α chain, also known as IFNGR1), and cysteinyl leukotrienes (CysLTR1, CysLTR2, and the purinergic receptor P2Y12). Additionally, ligand-stimulated cell-free granules can differentially secrete their content of cationic proteins (ECP and EPX) and cytokines.[16] Thus, eosinophil FEGs are considered functionally active organelles, while cytolytic cell death of eosinophils is recognized as a mechanism for the regulated secretion of eosinophil granule proteins into tissues that may contribute to eosinophil-mediated inflammation and immunomodulation in the absence of intact eosinophils.[16,166,167]

In vivo, the presence of extracellular, membrane-bound granules with the signature crystalline core of eosinophils has been documented by TEM in biopsies obtained from many diseases of the respiratory tract, such as asthma,[160,161,168,169] allergic rhinitis,[170–172] nasal polyps,[173,174] eosinophilic pneumonia,[175–177] and other eosinophil-associated disorders, such as atopic dermatitis[154] and eosinophilic esophagitis,.[17,178] We also observed extensive clusters of FEGs in intestinal biopsies from a patient with schistosomiasis mansoni (Chapter 8). The ultrastructural aspects of cytolytic eosinophils and the occurrence of degranulated eosinophils/FEGs in the context of diseases will be discussed in more detail in Chapters 6 and 8, respectively.

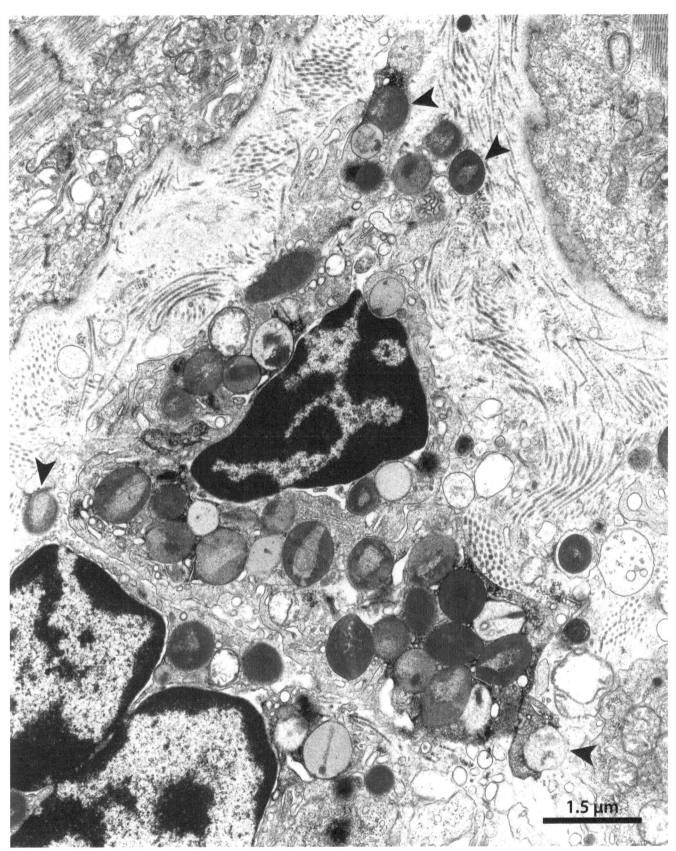

FIG. 3.29 Electron micrograph showing tissue human eosinophils *(pseudocolored)* in the early stage of cytolysis with a partially disrupted plasma membrane. Free extracellular granules (FEGs) are indicated by arrowheads. Lip biopsy (oral carcinoma) prepared for TEM. Cytoplasm *(pink)*; Nucleus *(purple)*; Secretory granules *(yellow)*.

A. The cell biology of human eosinophils

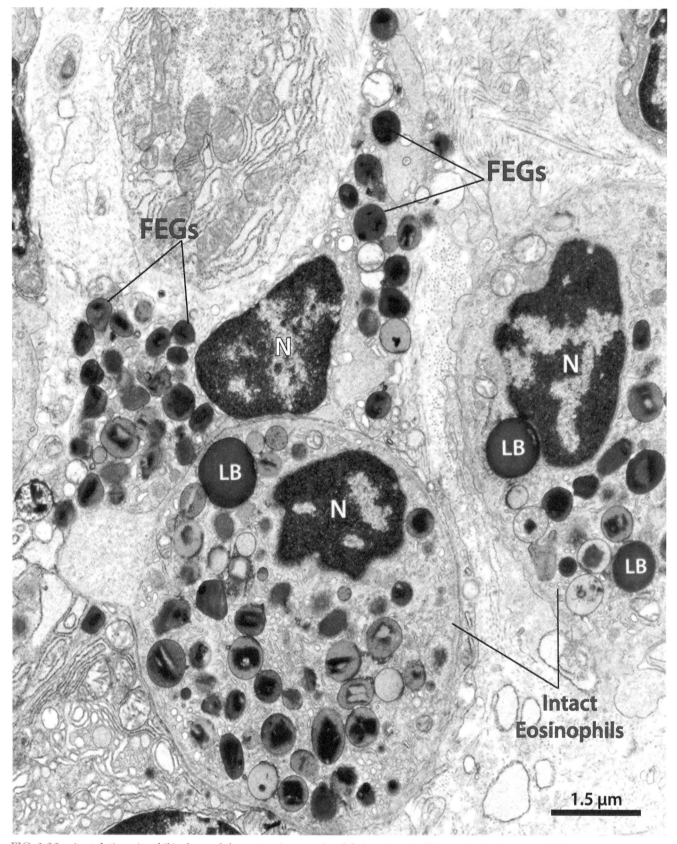

FIG. 3.30   A cytolytic eosinophil is observed close to two intact eosinophils in an intestinal biopsy from a patient with ulcerative colitis. The cell lost its plasma membrane and FEGs are deposited in the ECM. *LB*, lipid body; *N*, nucleus.

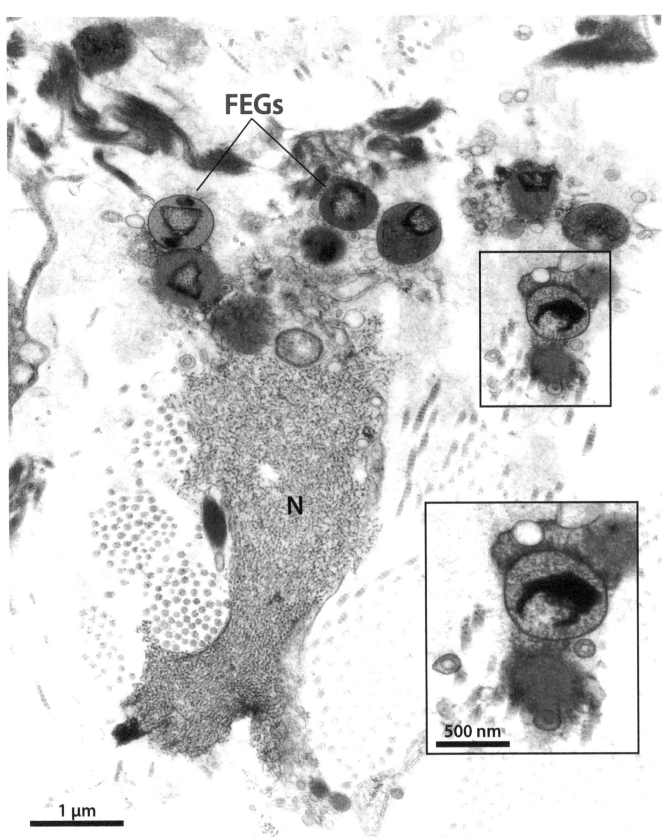

FEGs

N

1 µm

500 nm

FIG. 3.31   A human tissue eosinophil undergoing cytolysis in an inflammatory site (skin). Note the disrupted and highly decondensed nucleus (N), FEGs, and intact free EoSVs *(colored in pink in higher magnification)*. Sample prepared for TEM.

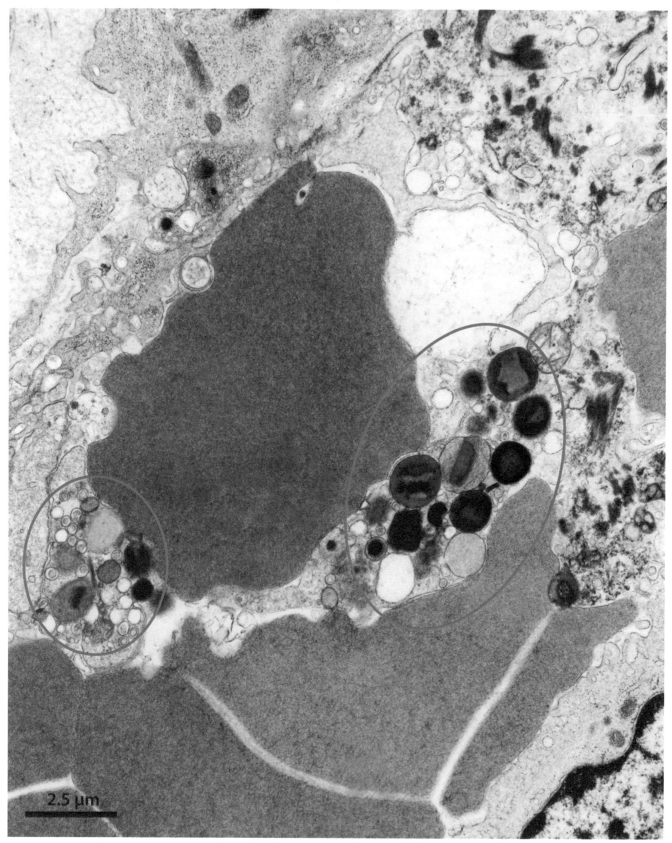

FIG. 3.32   FEGs and intact EoSVs *(circles)* are observed within a small blood vessel in a biopsy (frontal sinus) from a patient with eosinophilic chronic rhinosinusitis (ECRS). Note red blood cells *(colored in red)* and part of the endothelium *(yellow)*. Sample prepared for TEM.

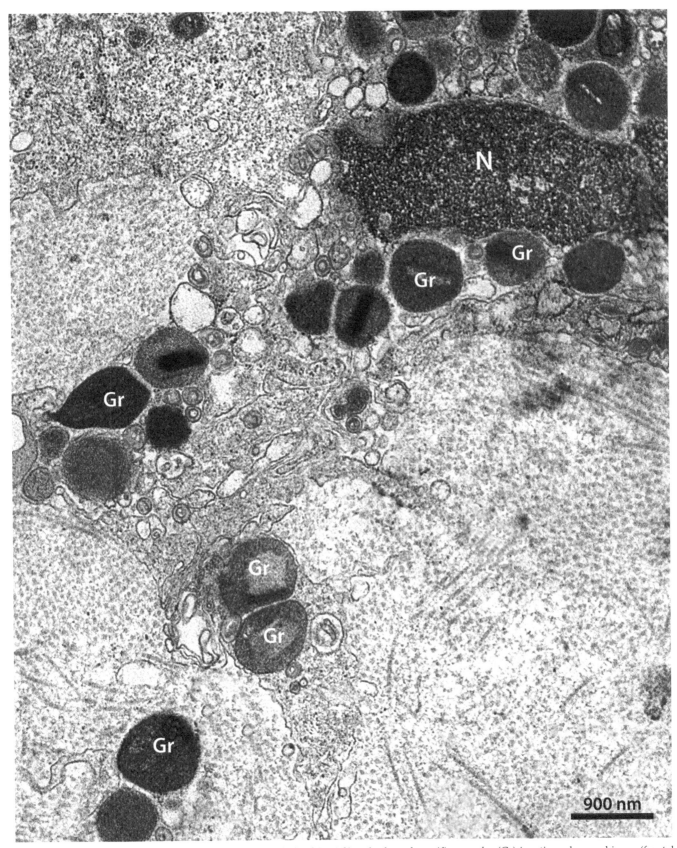

FIG. 3.33 Electron micrograph showing intact EoSVs *(colored in pink)* and released specific granules (Gr) in a tissue human biopsy (frontal sinus, ECRS). Note the chromatolytic eosinophil nucleus (N). Sample prepared for TEM.

A. The cell biology of human eosinophils

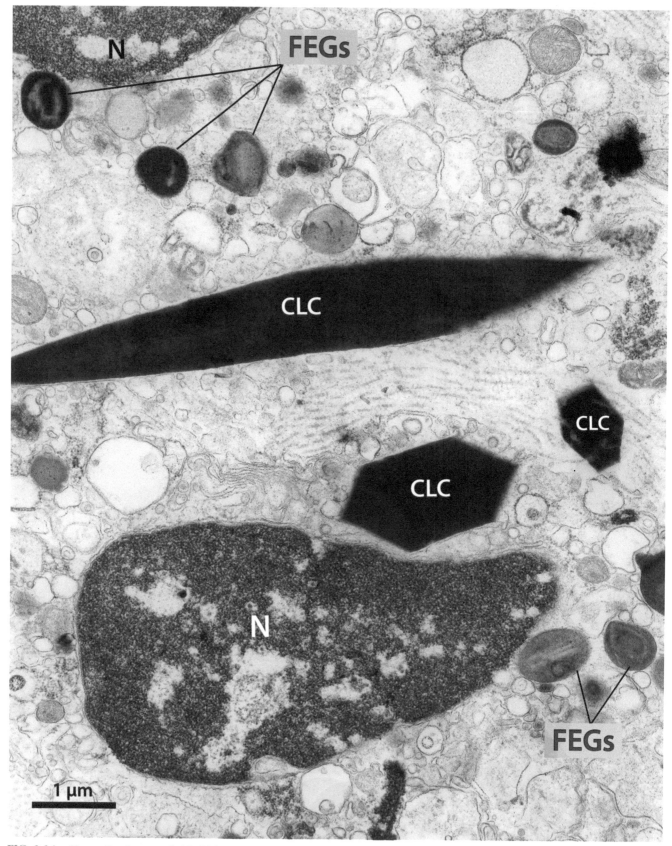

FIG. 3.34   Charcot-Leyden crystals (CLCs) from different sizes are observed in the proximity of cytolytic human eosinophils in inflamed tissue (nasal sinus). Note FEGs and eosinophil nuclei (N). ECRS biopsy prepared for TEM.

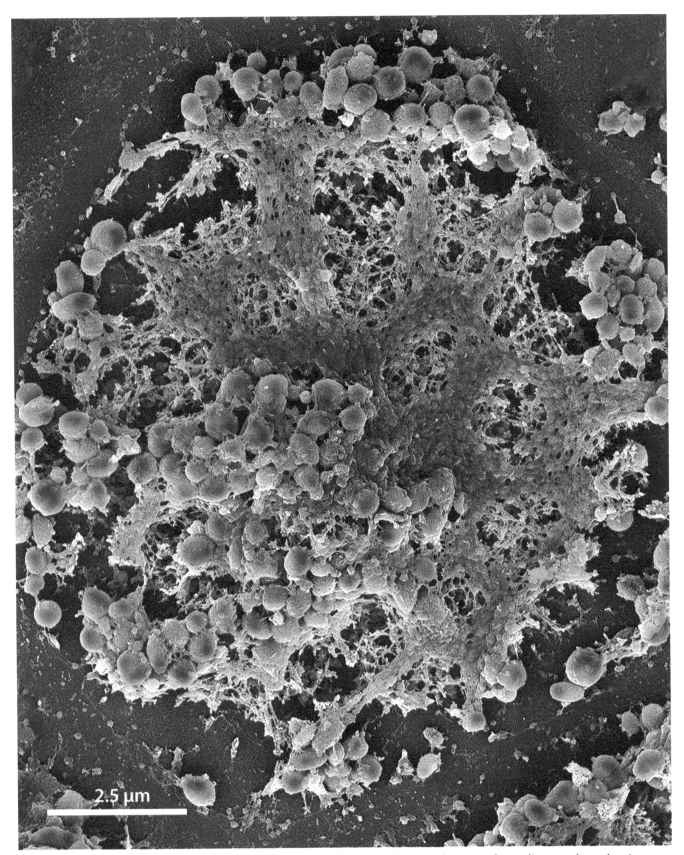

FIG. 3.35 SEM showing a cytolytic human eosinophil *(pseudocolored)*. Note the disrupted plasma membrane, disarranged cytoplasmic components *(blue)*, and spilled intact secretory granules *(orange)*.

A. The cell biology of human eosinophils

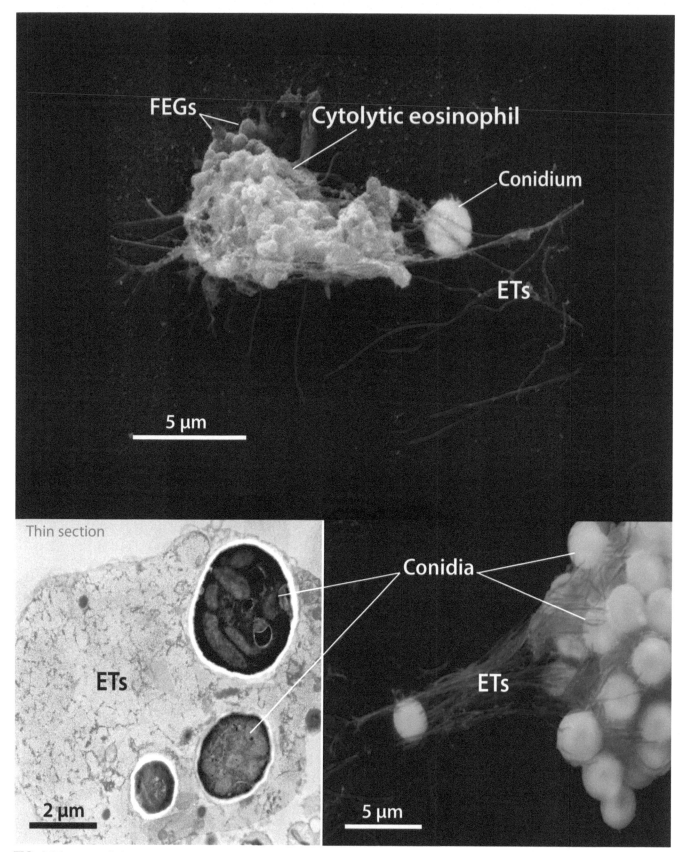

FIG. 3.36   Extracellular DNA traps (ETs, *colored in purple*) released from a cytolytic eosinophil entrapping conidia from *Aspergillus fumigatus*. Human eosinophils were stimulated with *A. fumigatus* conidia and processed for SEM and TEM.[47] Note in the thin section that the fungal wall is completely surrounded by eosinophil ETs. *FEGs*, free extracellular granules.

3.5 Extracellular vesicle production

The secretion of membrane-bound vesicles, collectively termed extracellular vesicles (EVs), is an important biological process of cells from the immune system, including human eosinophils.[25,125,139] Although the functions of EVs secreted by immune cells are still poorly understood, it is believed that these vesicles can act as carriers of cell-cell communication mediators such as cytokines and lipid mediators and potentially contribute to inflammation and/or immunomodulation (reviewed in[179,180]).

Various names, including exosomes and microvesicles (MVs), have been given to secreted EVs. While the term exosomes is used for referring to a population of EVs, which are released from cells when MVBs fuse with the plasma membrane, the term MVs has been generally used for EVs formed by budding and shedding from the plasma membrane (reviewed in[181–183]). The ability to secrete both exosomes[139,184] and MVs[25,125] has been demonstrated in human eosinophils during inflammatory responses.

TEM is considered a gold standard technique to characterize individual EVs and to distinguish them from non-membranous particles of similar size, as endorsed by the International Society for Extracellular Vesicles (ISEV) in an effort to provide minimal requirements for EV definition.[185] This is because EVs are nanostructures delimited by a phospholipid membrane, which can be unambiguously imaged by TEM as a "trilaminar" structure in which the hydrophilic phosphate "heads" are electron-dense and the hydrophobic fatty acids "tails" are electron-lucent.[70,71] For this reason, the use of TEM provides the most direct evidence for EV production. Moreover, the ISEV recommends that, for better characterization of the vesiculation event, TEM images should show a "wild field" encompassing multiple vesicles in addition to close-up images of single vesicles at the cell surface.[185]

By studying the ultrastructure of human eosinophils isolated from the peripheral blood, we noticed, for the first time, the presence of MVs budding from the cell surface when the cells were kept alive in medium (Fig. 3.37).[125] Stimulation with CCL11 or TNF-α, which are known to induce eosinophil activation and secretion,[23] produces the increased formation of MVs by human eosinophils, an event captured by a quantitative EM approach (detailed in Chapter 4), which enables clear characterization and quantification of nascent EVs at the cell surface (Figs. 3.4, 3.20, and 3.38).[125] The diameters of MVs from human eosinophils vary from 20 to 1000 nm (Figs. 3.38 and 3.39), with most MVs showing diameters between 20 and 200 nm.[125] The ability to release MVs is also observed in tissue eosinophils (Fig. 3.40) after careful analysis at high magnification.

Thus, in addition to the secretory processes largely described for human eosinophils (PMD, exocytosis, and cytolysis), these cells have the competence to secrete EVs and these vesicles likely underlie eosinophil immune responses. MVs released by human eosinophils may potentially carry different cargos and mediate different effects on other cells, depending on the stimulus/pathological condition. By applying immunonanogold EM, we identified the presence of IFN-γ in these vesicles,[25] as well as CD9 and CD63, which are considered markers for EVs,[186] as shown in Chapter 5. Moreover, extensive production of Gal-10-containing EVs was identified in human eosinophils in the process of cytolysis (Chapter 5). However, identification of other biomolecules within eosinophil-released EVs as well as the mechanisms by which these molecular cargos are transferred to recipient cells awaits further investigation to get insights into the functional roles of eosinophil EVs.

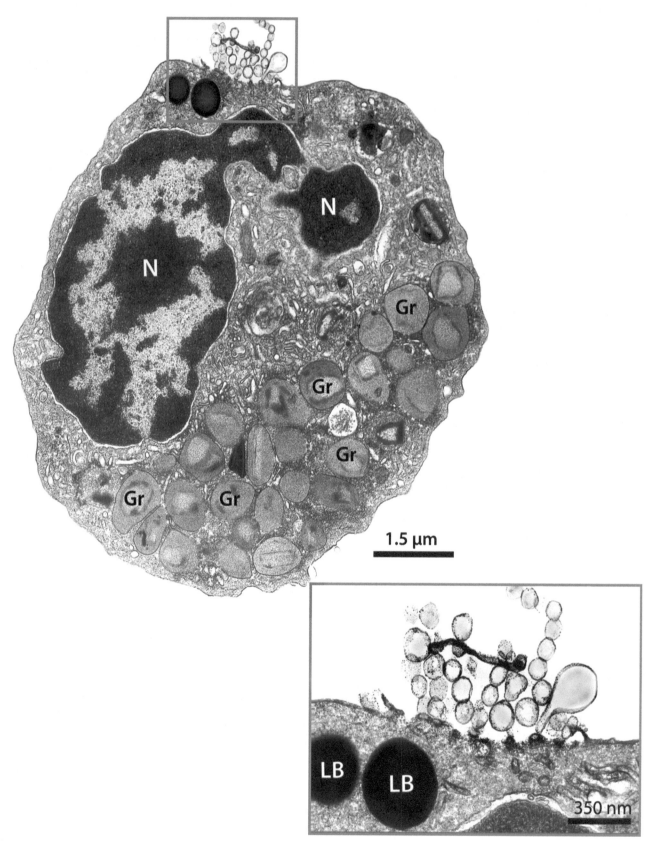

FIG. 3.37   TEM showing production of extracellular vesicles (microvesicles) from the plasma membrane of a human eosinophil. Microvesicles are highlighted in blue in higher magnification *(boxed area)*. Two electron-dense lipid bodies (LB) are observed in the cell periphery. *Gr*, secretory granules; *N*, nucleus.

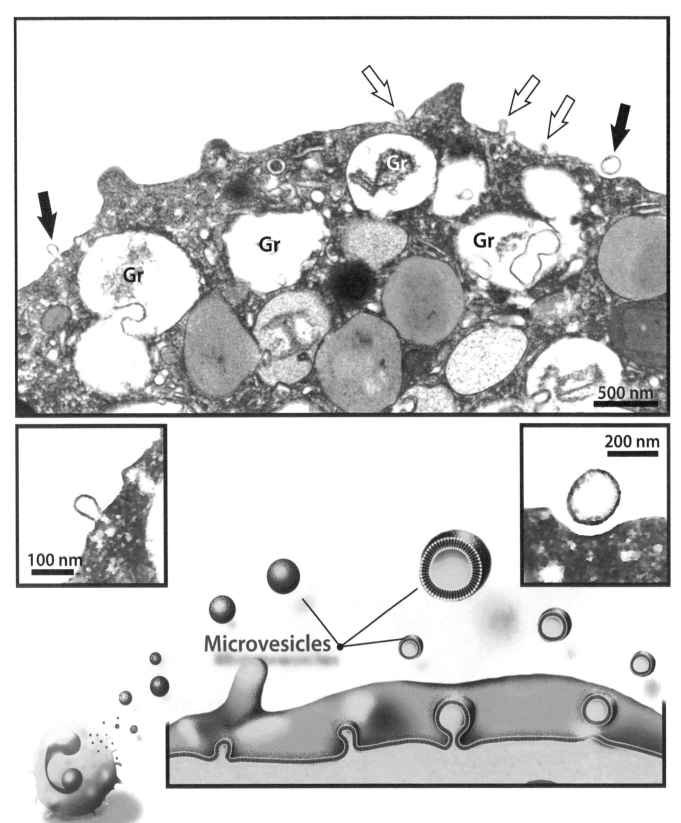

FIG. 3.38 Release of microvesicles (MVs) by an activated eosinophil. A TNF-α-stimulated human eosinophil shows MVs *(highlighted in blue, arrows)* in different steps of budding at the cell surface and secretory granules (Gr) exhibiting losses in the cytoplasm. MVs indicated by black arrows are seen in higher magnification. The illustration depicts the process of MV formation from an eosinophil. Eosinophils isolated from the peripheral blood were stimulated and processed for TEM as described.[125]

A. The cell biology of human eosinophils

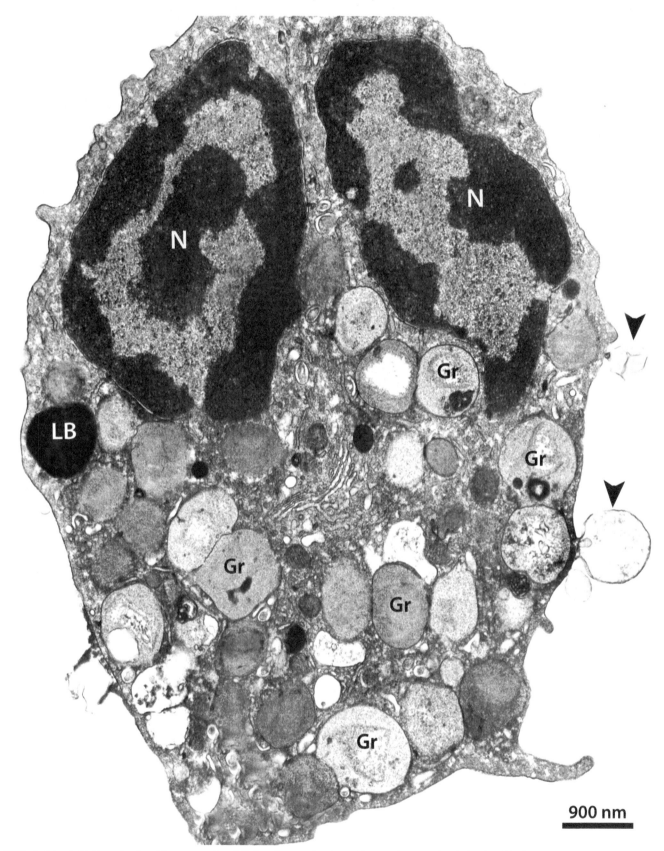

FIG. 3.39   A TNF-α-stimulated human eosinophil showing production of large microvesicles *(colored in blue/arrowheads)* at the cell surface. Secretory granules (Gr) exhibit content losses. *LB*, lipid body; *N*, nucleus. Cells were stimulated and processed for TEM.[23]

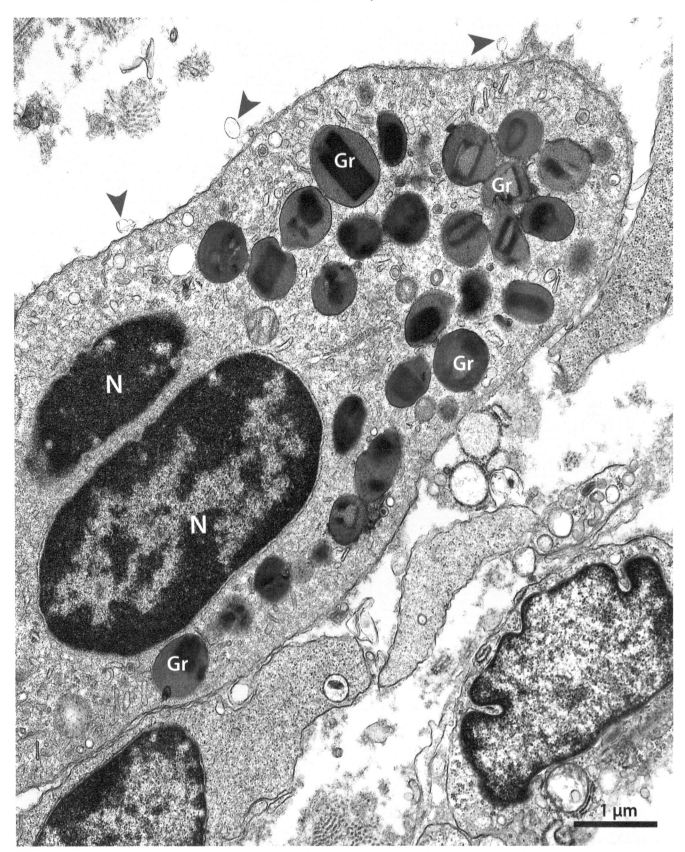

FIG. 3.40   Tissue eosinophil (ileal pouch biopsy, ulcerative colitis) showing formation of microvesicles *(arrowheads)* at the cell surface. *Gr*, secretory granules; *N*, nucleus. Sample prepared for TEM.

A. The cell biology of human eosinophils

4

Eosinophil activation

4.1 Overview

As discussed in Chapter 3, eosinophils are remarkable secretory cells that release a large and varied collection of immune mediators upon cell activation (reviewed in[3,7,135]). The activation state of human eosinophils has been studied under different viewpoints.[40,187,188] For example, several proteins on the cell surface have been suggested to mark eosinophil activation and are associated with allergic and other inflammatory diseases.[188] However, the molecular machinery related to eosinophil activation during different conditions/diseases is still mostly unknown.

Ultrastructural evidence of eosinophil activation encompasses a multitude of events such as substantial morphological alterations of specific (secretory) granules, including a significant increase in the numbers of granules undergoing content disarrangement and losses,[20,23] accentuated formation of EoSVs,[19,22,23,25] interaction between these vesicular compartments and secretory granules,[19,23] increased production of cytoplasmic LBs,[94] cell shape changes,[105,189] and amplified release of extracellular vesicles.[125]

In vivo, morphologic evidence of activation generally becomes more apparent in tissue eosinophils after their migration from the vascular compartment.[190] In certain circumstances, such as in HES, activation morphologies are apparent in eosinophils still circulating in the blood.[40]

Eosinophils, like all circulating leukocytes, must go through a series of well-coordinated steps involving leukocyte and endothelial adhesion molecules in order to exit the circulation and enter any extravascular compartment. Eosinophil-rich inflammatory reactions in the skin, lung, gastrointestinal tract, and lymph nodes as well as in the inflammatory sites associated with tumors provide ample opportunity to assess activation morphologies in these situations.[40] In situ activation can also lead to eosinophil-associated formation of CLCs and extrusion of DNA extracellular traps events generally connected with the release of competent FEGs after eosinophil cytolysis, as discussed in Chapters 3 and 6.[41,46,48,49] In fact, cytolysis is considered an "ultimate activation" mechanism leading to the secretion of membrane-bound FEGs,[160,168,171] which persist active in the tissue microenvironment.[16,167]

We have been studying ultrastructural changes in subcellular organelles that characterize activated human eosinophils in tissues. Moreover, we have also been using blood eosinophils activated with varied stimuli to understand the morphological aspects of eosinophil activation. Our EM analyses are combined with quantitative assessments, which we term quantitative TEM that has proved useful to demonstrate the cell activation status with important insights into eosinophil functional capabilities. In this chapter, we address the ultrastructural aspects of cell activation in intact eosinophils (with no disrupted plasma membrane) with an emphasis in the TEM quantitative analyses of structures and organelles particularly changed and/or formed in response to cell activation.

4.2 Ultrastructural changes of secretory granules

The specific granules of human eosinophils undergo remarkable morphologic changes in response to both in vitro activation and diseases. As previously highlighted, these granule alterations can be identified in detail only at high resolution by TEM and reveal the intricate secretory activity of eosinophils (Chapter 3). Therefore, "degranulation" is the first aspect to be analyzed when eosinophil activation is investigated under TEM. Because specific granules have a unique ultrastructural morphology, considerable size (~ 700–1000 nm), and constitute a large population in the eosinophil cytoplasm,[18] as described in Chapter 2, EM can feasibly capture the structural alterations of these granules indicative of cell activation.

Eosinophil Ultrastructure. https://doi.org/10.1016/B978-0-12-813755-0.00004-6

107

It is important to emphasize that the handling of human eosinophils can easily and inappropriately activate these cells. Therefore, eosinophil isolation from the peripheral blood and tissues requires steps that need to be performed with great attention to prevent cell activation. By our experience, even unstimulated cells kept in a medium can present a low morphological degree of activation when imaged by TEM. To minimize potential cell activation, we fix eosinophils for TEM immediately after isolation or experimental procedures when the cells are still in suspension and pellet the cells just after fixation. In experiments in which eosinophils are kept on glass slides, including chamber slides, we perform EM procedures directly on the slide surface to avoid any interference with the cell morphology.

In TEM studies aiming to identify eosinophil degranulation/activation, the following morphological parameters have been evaluated in specific granules within intact cells: (i) occurrence of signs indicative of PMD mostly represented by partly empty granules with preserved delimiting membranes and significant absence of granule extrusion,[20,150,153] (ii) occurrence of compound/classical exocytosis represented by clear granule-granule and granule-plasma membrane fusions,[20,142,153] and (iii) quantitative evaluation of these morphological aspects.[17,20,153]

We perform quantitative TEM of eosinophil degranulation/activation by counting the total number of specific granules, the number of intact granules, and the number of granules showing signs of PMD and/or exocytosis, thus establishing the mean number and percentage of "degranulating" granules per cell section/group. Quantitative analyses are done in electron micrographs showing clear cross sections of eosinophils in which the entire cell profile and nucleus are apparent.[20,23] For example, after one hour of stimulation of peripheral blood eosinophils with CCL11, CCL5, or PAF, specific granules show clear changes in ultrastructure compared with those in unstimulated cells (Fig. 4.1).[20] In resting eosinophils, granules are mostly seen as round or elliptical structures with their classical morphology and full of contents (Fig. 4.1). Upon stimulation, granules exhibit structural disarrangement, enlargement, and reduced electron density in the absence of granule fusion events, all features indicative of activation and content release classically associated with PMD (Figs. 4.1 and 4.2). Both chemokines and PAF induced the same morphological changes indicative of PMD within granules (Fig. 4.1). In counting a total of 3945 specific granule numbers, whereas only 8% of granules in inactivated eosinophils had granules undergoing PMD, 43% (CCL11), 25% (CCL5), and 34% (PAF) showed emptying granules with signs of PMD.[20] Moreover, the responses of eosinophils were not uniformly distributed. For example, in recording the numbers of granules that exhibited loss of granule contents indicative of PMD, in unstimulated cells, 70% of eosinophils had < 10% of granules with losses, whereas CCL11 elicited a marked heterogeneity of granule-emptying responses within eosinophils such that > 15% of eosinophils had > 90% of their granules exhibiting content losses. The fusion of granules with each other or with plasma membranes was only occasionally seen and did not increase significantly with cell activation using these agonists.[20]

We also investigated the ultrastructure of human blood eosinophils stimulated for one hour with other cytokines known to induce eosinophil activation, such as stem cell factor (SCF/cKit ligand)[191] (Fig. 4.3) and TNF-α (Fig. 4.4).[192,193] Quantitative evaluation showed that SCF/cKit ligand led to a significant increase of specific granules numbers with evidence of PMD (Fig. 4.3). In another study, EM analyses of a total of 3259 specific granules from samples obtained from the same donor/experiment and activated with TNF-α or CCL11 revealed a differential response of human eosinophils in terms of their morphological secretory profiles.[23] Whereas PMD was the predominant event found in CCL11-activated cells, TNF-α induced mostly compound exocytosis, with PMD being found in a lesser degree (Fig. 4.4).[23] In both CCL11 and TNF-α-activated eosinophils, the numbers of CD63-positive granules significantly increased compared with unstimulated cells,[23] as addressed in Chapter 5.

Degranulation patterns based on qualitative and quantitative evaluation of secretory granules using TEM have also been studied in infiltrating tissue eosinophils found in different diseases such as asthma,[151] atopic dermatitis,[154] allergic rhinitis,[151] nasal polyposis,[151] Crohn's disease (Fig. 4.5),[151] ulcerative colitis,[151] eosinophilic esophagitis,[17] and gastric dyspepsia.[156] In all of these pathological conditions, secretory granules showed ultrastructural evidence of content release associated with eosinophil activation.[17,151,153] Quantitative analyses included enumeration of the granules with intact morphology (intact core and matrix) and with signs of degranulation (showing variable content losses).[17,151,153] Moreover, the pixel intensity profile per granule has also been used as a parameter to detect structural signs of degranulation in eosinophils, for example, in duodenal biopsies from patients with functional dyspepsia.[156]

Quantitative TEM studies of tissue eosinophils also comprised identification of eosinophil cytolysis[17,151,154] and reveal that eosinophils showing different degranulation modes can coexist in the same situation or disease.[17,142] For example, by evaluating skin biopsies obtained at sites of subcutaneous administration of recombinant human SCF/cKit ligand, within approximately one to two hours after SCF injection to treat breast carcinomas,[194] we found infiltrating granulocytes in the inflammatory sites (Fig. 4.6) with eosinophils undergoing PMD (53%) and cytolysis (47%) (Fig. 4.7).

Ultrastructural changes in specific granules thus reflect the activation state of these organelles, which can show different morphological aspects (from disarrangement of the matrix and core to more advanced stages up to almost

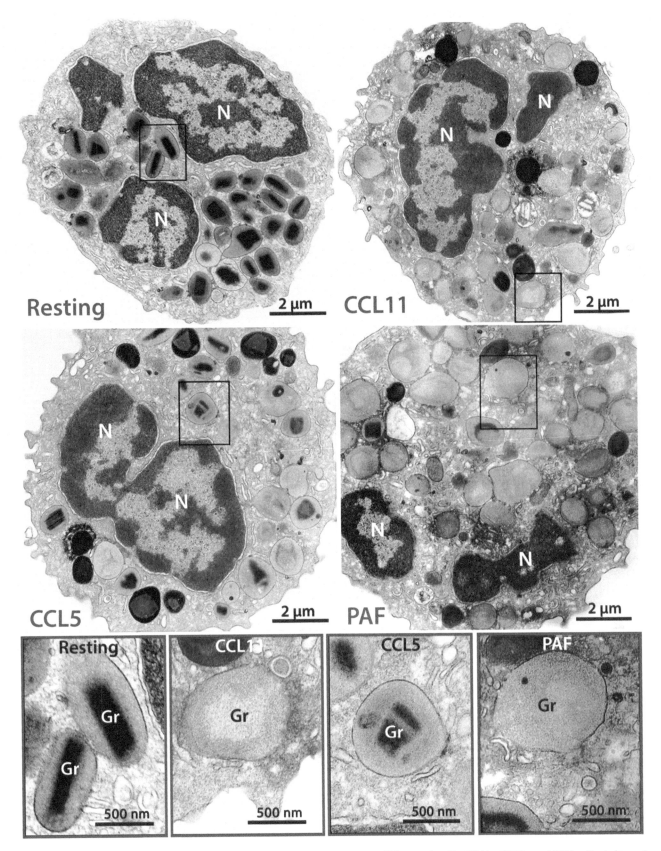

FIG. 4.1 Human peripheral blood eosinophils undergoing PMD in response to different stimuli. CCL11-, CCL5-, and PAF-activated secretory granules (Gr) are enlarged, nonfused, and show reduced electron density, features indicative of content mobilization and release, compared with resting granules. Cells were stimulated for one hour as described.[20] N, nucleus.

A. The cell biology of human eosinophils

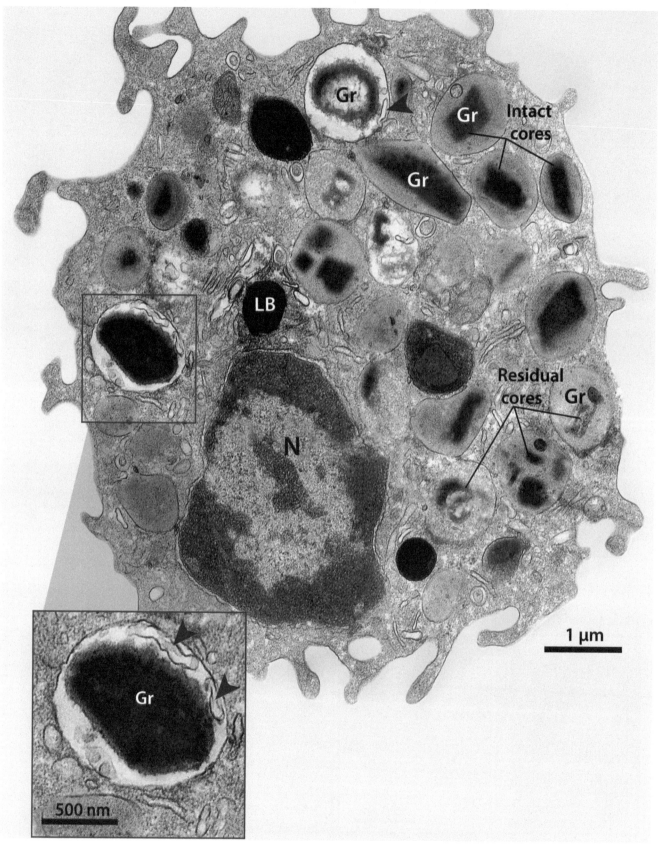

FIG. 4.2   CCL5-activated human eosinophil exhibiting ultrastructural changes of PMD. While a small population of specific granules (Gr) sustains their classical morphology with intact cores, most granules are activated showing variable degrees of content losses. Note residual cores. *Arrowheads* indicate intragranular membranes. Cells were stimulated for one hour as described.[20] *LB*, lipid body; *N*, nucleus.

A. The cell biology of human eosinophils

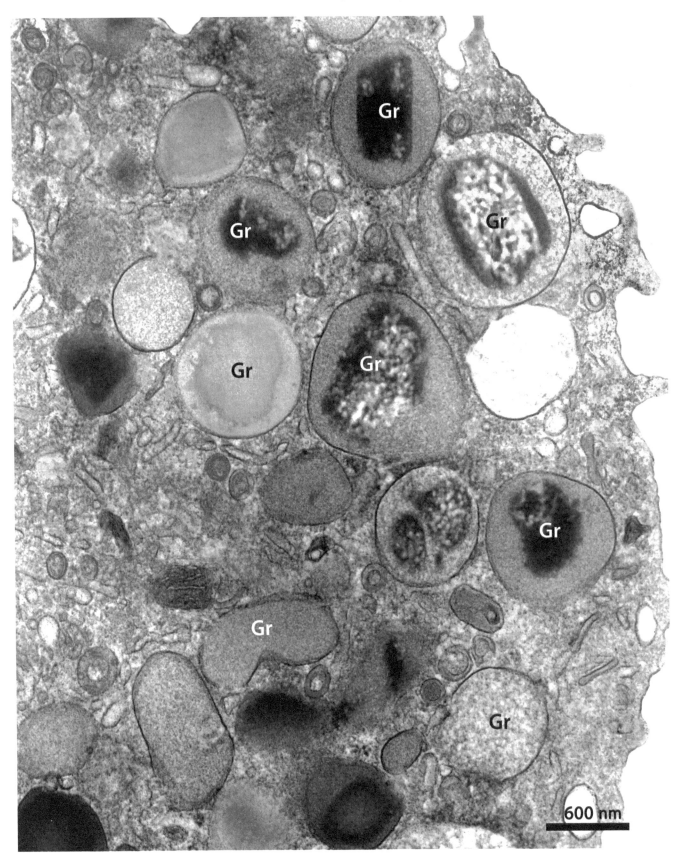

FIG. 4.3 High magnification electron micrograph showing an eosinophil stimulated with stem cell factor (SCF/cKit ligand). Note emptying nonfused specific granules (Gr) and EoSVs *(colored in pink)*. Cells were isolated from the peripheral blood, stimulated with SCF for one hour, and prepared for TEM.

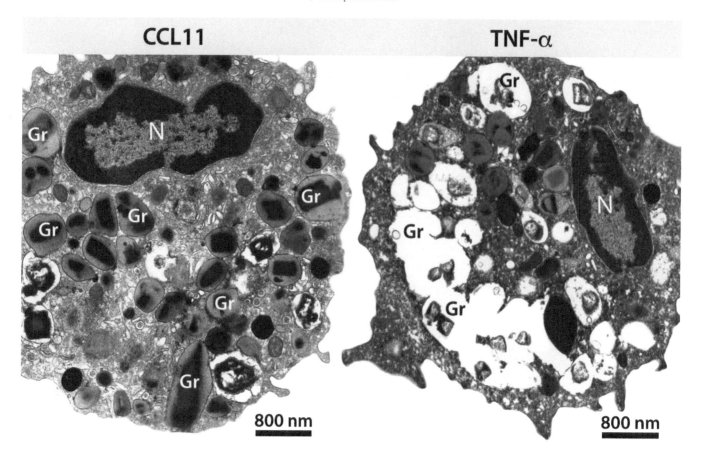

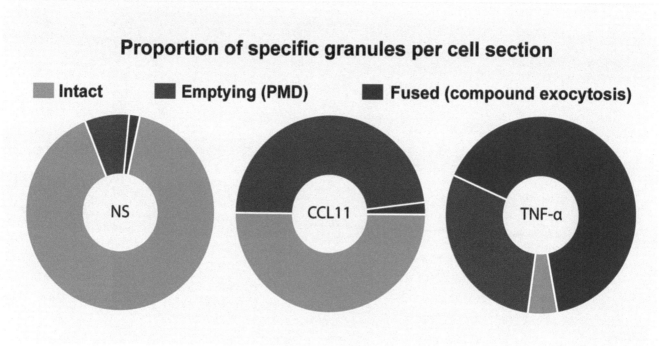

FIG. 4.4   Ultrastructure of human eosinophils activated with CCL11 or TNF-α. Piecemeal degranulation (PMD) is observed in response to CCL11, while compound exocytosis, characterized by large channels formed by granule-granule fusions, is the predominant mode of secretion induced by TNF-α. Significant increases in numbers of emptying nonfused (PMD) or fused (compound exocytosis) granules occurred after activation with CCL11 or TNF-α, respectively, compared with nonstimulated (NS) cells. Eosinophils were isolated from the peripheral blood, stimulated, and processed for TEM as described.[23] *Gr*, secretory granules; *N*, nucleus.

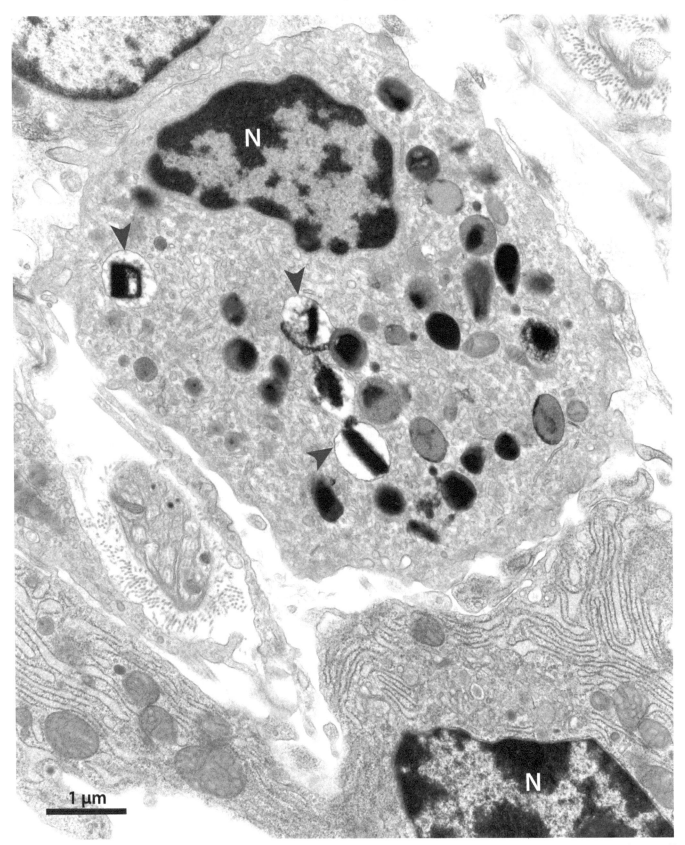

FIG. 4.5 Tissue eosinophil from a patient with Crohn's disease showing secretory granules *(arrowheads)* undergoing PMD. A plasma cell *(pseudocolored)*, which is characterized by abundant RER *(light blue)* is close to the eosinophil. Mitochondria were colored in *green* and cytoplasm in *pink*. N, nucleus.

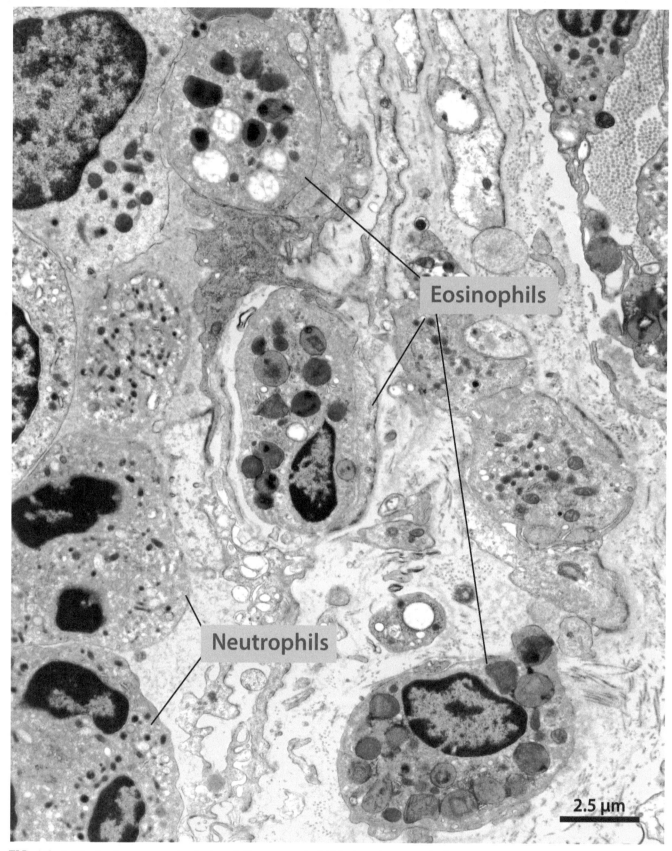

FIG. 4.6 Infiltrating human granulocytes in an inflammatory site (skin) of a patient treated subcutaneously with SCF/cKit ligand. Eosinophils show PMD features with enlarged secretory granules. Extracellular matrix (ECM) colored in *green*. Samples were prepared for TEM as described.[194]

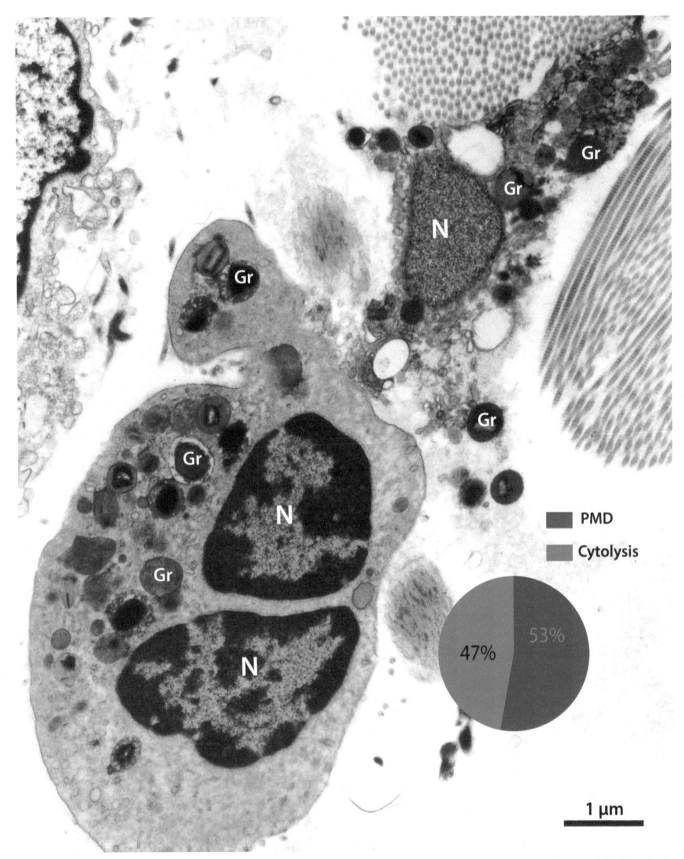

FIG. 4.7   Skin biopsy from a patient treated subcutaneously with SCF/cKit ligand showing representative activated eosinophils. Quantitative TEM analyses revealed a high proportion of eosinophils undergoing cytolysis (*colored in purple*) and PMD (*green*). *Gr*, secretory granules; *N*, nucleus.

A. The cell biology of human eosinophils

complete emptying). As highlighted, for human eosinophils, cytolysis is not considered just a degenerating process, but a critical degranulation mode, culminating with the release of intact FEGs,[160,168,171] which are considered to remain active in the tissue microenvironment through the local release of preformed cytokines and other proteins (Chapter 3).[16,167] In addition to exhibiting content losses, in general, activated specific granules are rounder and larger than resting granules within the cytoplasm of human eosinophils undergoing degranulation through PMD and/ or exocytosis. For example, in an ultrastructural study of human eosinophils isolated from the blood, a significant increase of the granule profile areas was demonstrated 10 min after PAF stimulation.[189] Our quantitative EM analyses also detected a significant increase of the granule areas after one hour of activation with CCL11 and TNF-α in comparison with unstimulated human eosinophils (kept just in medium) (see Fig. 5.25 in Chapter 5).[23] Enlarged activated granules undergoing emptying of their contents are seen in both in vitro activated-blood eosinophils (Figs. 4.1–4.4) and tissue eosinophils participating in diseases (Figs. 4.5 and 4.6).

One interesting structural aspect of human eosinophils is the presence of internal membranes within specific granules.[20] As pointed out in Chapter 2, these membranous compartments are likely an integral structural component of these granules but are mainly seen after eosinophil activation when packed material inside the granule is disarranged and released (Figs. 4.2 and 4.8). Imaging of these membranes is highly dependent on both optimal sample preparation and meticulous EM analysis.

Evidence of membranes within specific granules, seen as vesiculotubular structures in thin sections of human eosinophils, was noted in vivo in tissues from patients affected with nasal allergy,[195] Hodgkin lymphoma,[196] Crohn's disease,[197] eosinophilic gastroenteritis,[198] and carcinoma.[199] Moreover, vesiculotubular structures were also reported within specific granules in peripheral blood eosinophils from normal patients after stimulation with calcium ionophore[200] and patients with asthma[189] and helminthic infections[201] after stimulation with PAF or aerosol, respectively.

By applying conventional TEM, immunonanogold EM to label membrane-associated proteins, and 3D electron tomography, we demonstrated that specific granules from human eosinophils indeed contain internal membranes mainly in the granule matrix. Conventional TEM, after careful analysis of eosinophils stimulated with CCL11, CCL5, or PAF,[20] revealed membranous subcompartments within specific granules with clear trilaminar structure, which is typical of phospholipid membranes (Figs. 4.2 and 4.8). Quantitative TEM analyses showed that, in parallel with a significant increase of emptying granule numbers, there were significant increases in numbers of granules showing internal membrane domains in response to these agonists (Fig. 4.8).[20] Moreover, the collapse of these intragranular membranes was observed after treatment with BFA, which affects vesicular trafficking.[20] Ultrastructural analyses showed that BFA acts within granules to elicit the formation of distinct electron-dense lipid deposits (Figs. 4.8 and 4.9) in addition to affecting the Golgi complex,[20] its classical organellar target (see Fig. 2.38 in Chapter 2).[202] Quantitative evaluation showed that the numbers of specific granules exhibiting internal lipid deposits after BFA treatment increased significantly compared to the stimulation with CCL11 alone.[20]

Application of electron tomography brought a new view of specific granules as organelles with membranous subcompartments organized as an aggregate of flattened tubular networks and tubules with interconnections in some planes as well as connections with the granule limiting membrane (Fig. 4.10). This vesiculotubular network is likely involved in the formation of EoSVs from specific granules.

4.3 EoSV formation

Amplified production of EoSVs is a feature of activated eosinophils.[18,21] The presence of an increased number of these vesicular compartments within tissue eosinophils triggered by human diseases has been attracting attention for a long time. In the early eosinophil literature, EoSVs, referred in the past to as "microgranules," "small granules," or simply "tubulovesicular structures," were described with a marked increase in the cytoplasm of tissue eosinophils involved in Hodgkin lymphoma[196] and Crohn's disease[197] and also in blood eosinophils, stimulated with PAF[203] or cultured with human immunodeficiency virus-1 (HIV-1).[204]

Qualitative and quantitative TEM studies conducted by our group clearly demonstrated that different stimuli such as CCL11 (Figs. 4.11 and 4.12),[19,23] TNF-α (Fig. 4.11),[23,25] PAF (Fig. 4.13), SCF/cKit ligand (Figs. 4.3, 4.14, and 4.15), and CCL5 (Fig. 4.16) lead to a significant formation of EoSVs within the cytoplasm of human eosinophils compared with control cells. Thus, it is now clear that EoSVs are useful morphological markers for human eosinophil activation being found in increased numbers even within naturally activated eosinophils from patients with HES when compared with normal donors.[22] By counting 958 EoSVs in 35 electron micrographs of human eosinophils showing the entire cell profile and nucleus, we found that the total number of EoSVs in HES eosinophils was significantly higher than that in resting eosinophils [35.4 ± 11.4 EoSVs per cell section in HES versus 21.3 ± 4.9 EoSVs in normal donors (mean $\pm$ SEM, $P < 0.05$)].[22]

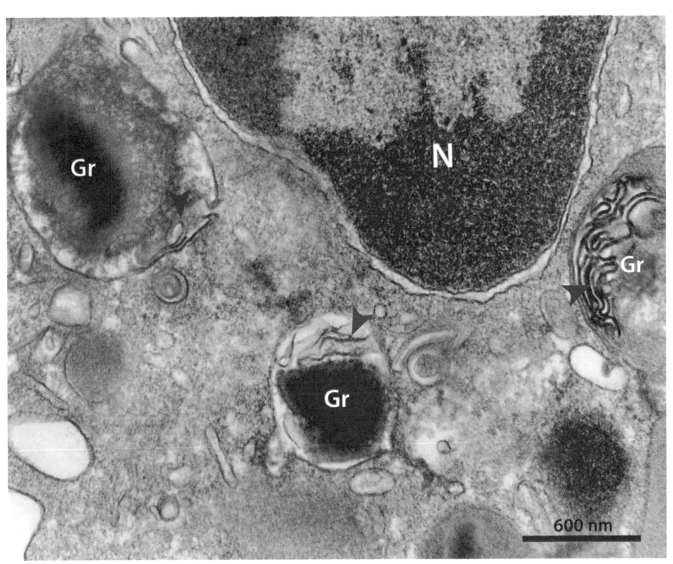

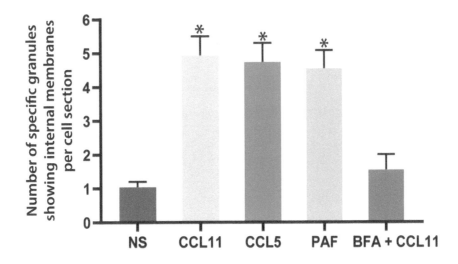

FIG. 4.8 Specific granules (Gr) exhibit internal membranous domains *(arrowheads)* in agonist-stimulated human eosinophils. The numbers of granules with internal membranes significantly increased after stimulation with CCL11, CCL5, or PAF compared with nonstimulated (NS) eosinophils (*$P < .05$), except in brefeldin A (BFA)-pretreated cells. Eosinophils were isolated from the peripheral blood, stimulated, and processed for TEM as described.[20] *N*, nucleus.

A. The cell biology of human eosinophils

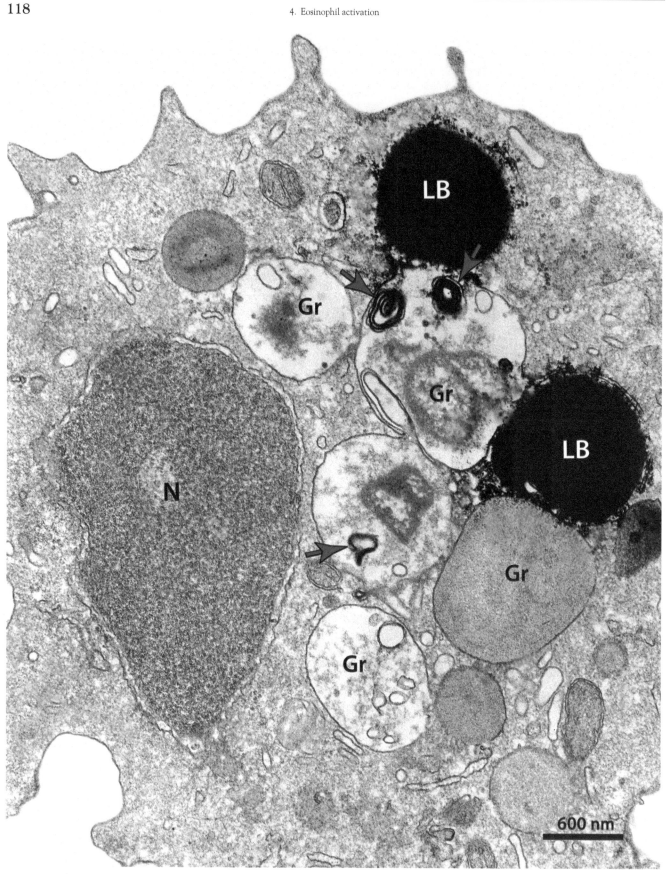

FIG. 4.9 Specific granules (Gr) of a human eosinophil show collapse of internal membranes after treatment with brefeldin-A (BFA). *Arrows* indicate intragranular electron-dense deposits. Cells were treated as described and processed for TEM.[19] *LB*, lipid body; *N*, nucleus.

A. The cell biology of human eosinophils

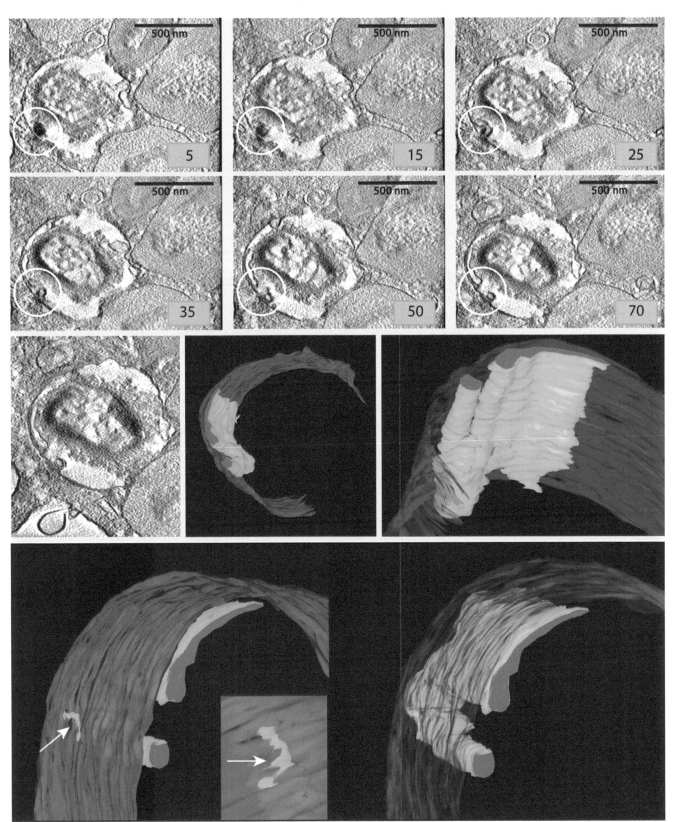

FIG. 4.10 Tomographic slices and three-dimensional (3D) models from an activated secretory granule seen in a CCL11-stimulated human eosinophil. Circles indicate the same subcompartment surrounding part of the electron-dense content that is relocated to the granule outer membrane. Numbers on the lower right corners indicate slice numbers through the tomographic volume. Intragranular membrane domains *(blue)* are organized as a tubular network. Arrow indicates a continuity area between the intragranular compartments and the limiting granule membrane *(red)*. Cells were stimulated and prepared for electron tomography.[20]

A. The cell biology of human eosinophils

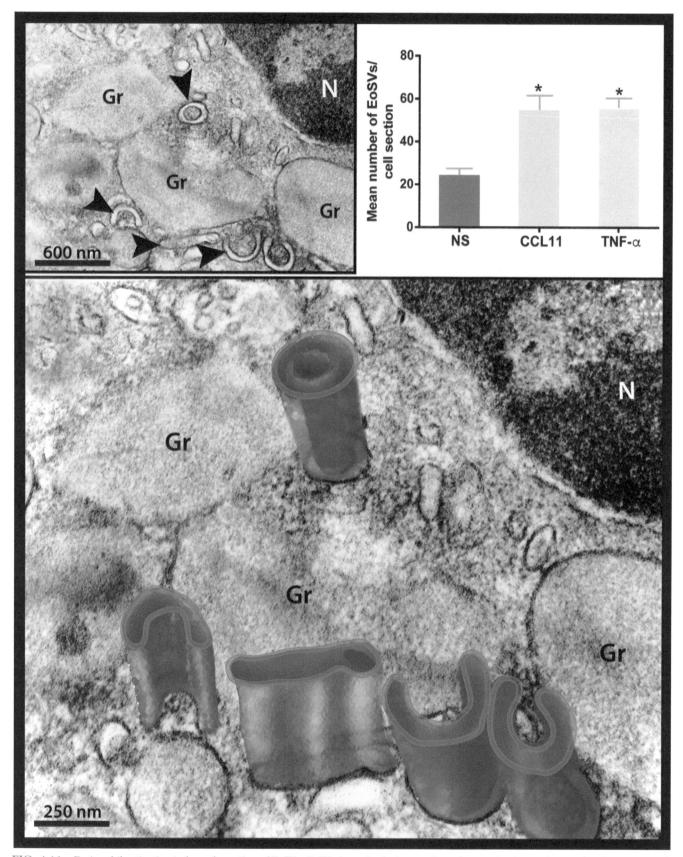

FIG. 4.11   Eosinophil activation induces formation of EoSVs. EoSVs *(arrowheads, top panel)* are frequently observed around or in contact with emptying granules (Gr). The 3D morphology of EoSVs is shown in higher magnification *(bottom panel, pink)*. Different stimuli trigger a significant increase in EoSV numbers compared with nonstimulated (NS) eosinophils (*$P < .001$). Cells were stimulated and prepared for TEM.[25] *N*, nucleus.

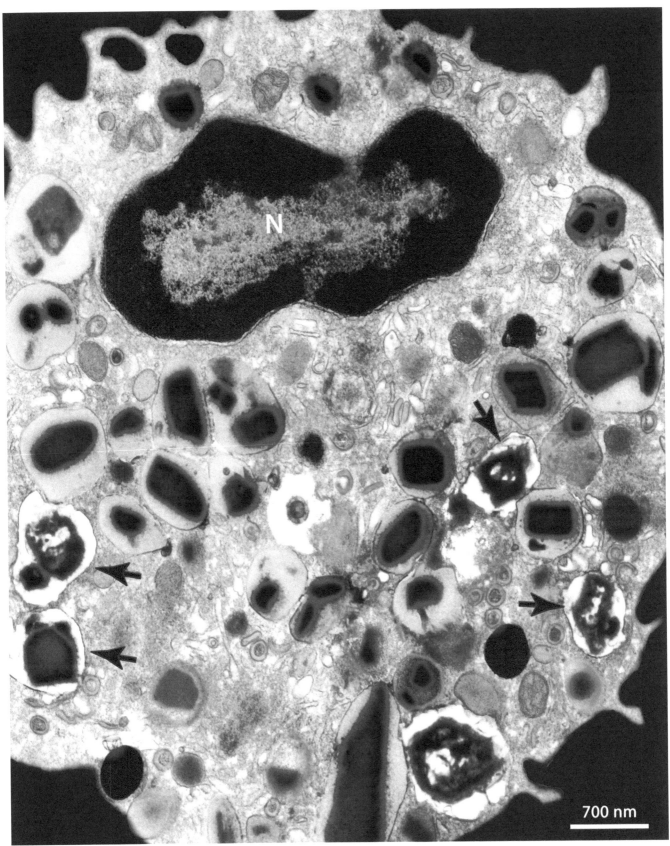

FIG. 4.12 CCL11-activated human eosinophil. Granules were pseudocolored to indicate PMD changes in the cores *(purple)* and matrix *(yellow)*. Note the association between EoSVs *(pink)* and emptying granules *(arrows)*. Eosinophils isolated from the peripheral blood were stimulated as before.[20] N, nucleus.

A. The cell biology of human eosinophils

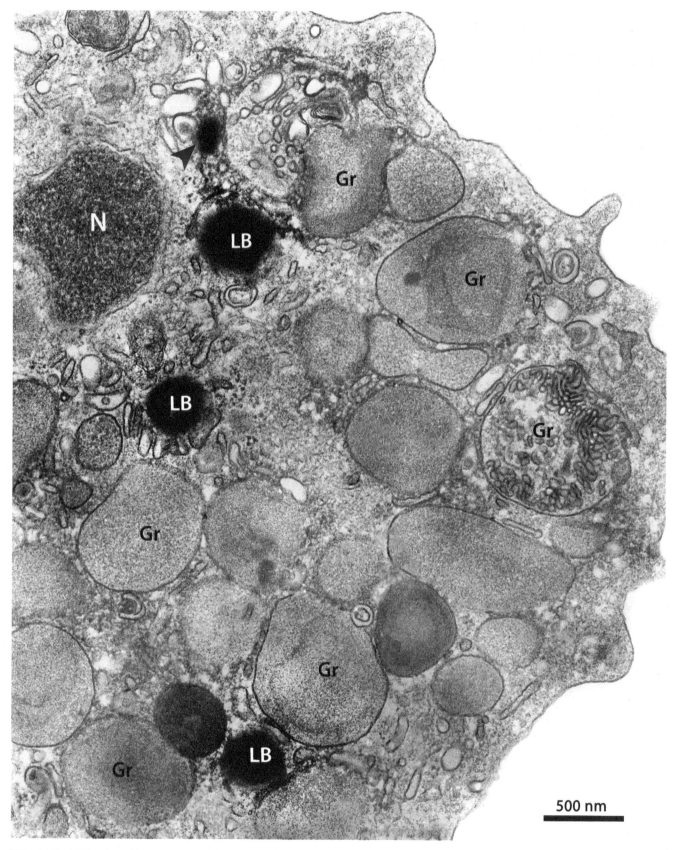

FIG. 4.13 PAF-activated human eosinophil. EoSVs *(highlighted in pink)* surround or contact enlarged granules (Gr) undergoing PMD. LBs, including a very small one *(arrowhead)*, are formed in response to cell stimulation. Eosinophils were isolated from the peripheral blood, stimulated with PAF for one hour, and processed for TEM.[20] *N*, nucleus.

A. The cell biology of human eosinophils

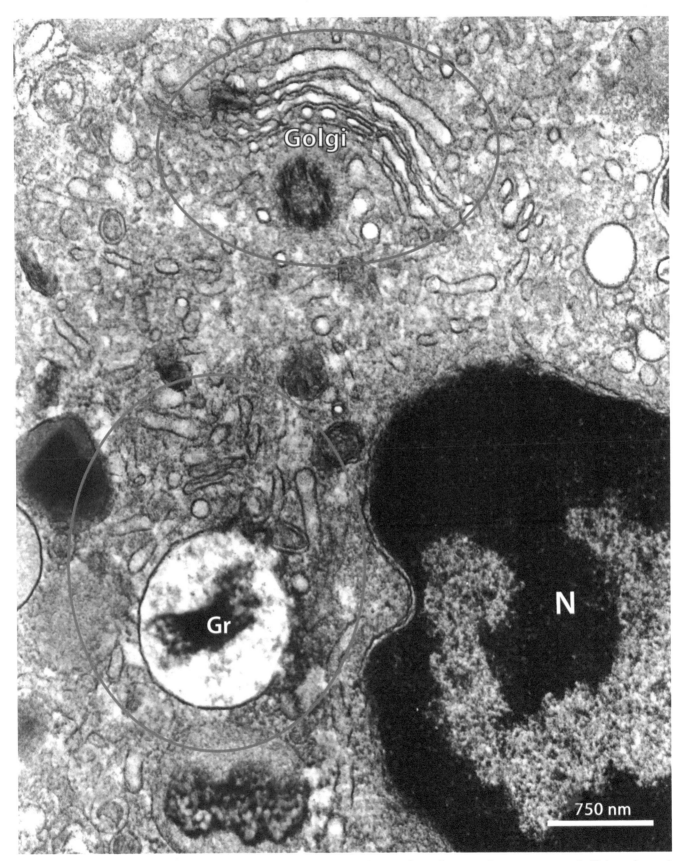

FIG. 4.14 SCF/cKit ligand-activated human eosinophil. An enlarged Golgi complex and an emptying secretory granule (Gr) are observed. Note EoSVs in close interaction with the granule *(circle)*. Eosinophils isolated from the peripheral blood were stimulated with SCF for one hour and prepared for TEM. *N*, nucleus.

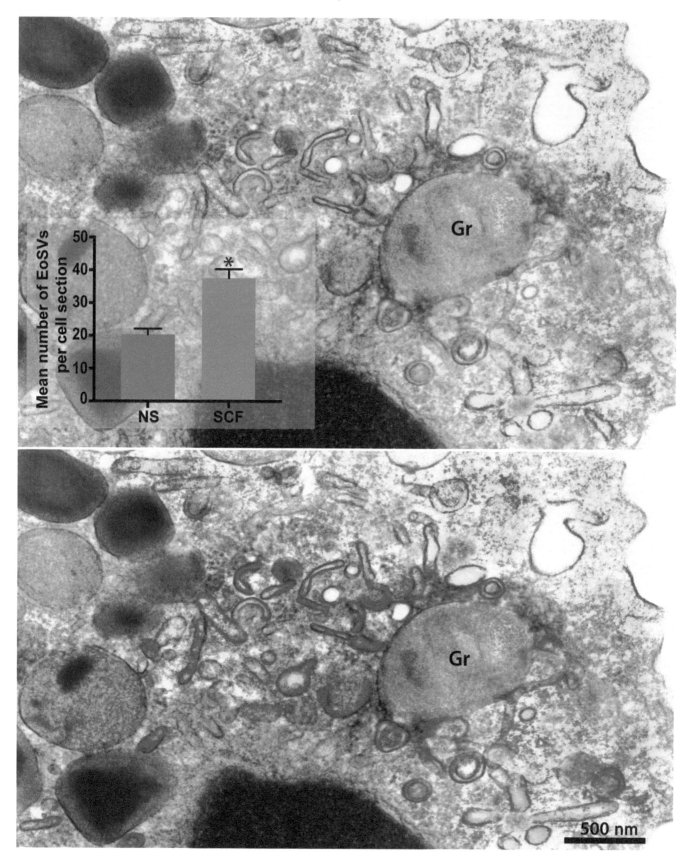

FIG. 4.15   A profusion of EoSVs *(highlighted in pink)* associates with an activated secretory granule (Gr) in a SCF/cKit ligand-stimulated human eosinophil. Note granule enlargement and loss of the electron-dense content. EoSV numbers increase after stimulation compared to nonstimulated (NS) cells (*P < .001). Eosinophils isolated from the peripheral blood were stimulated with SCF for one hour and prepared for TEM.

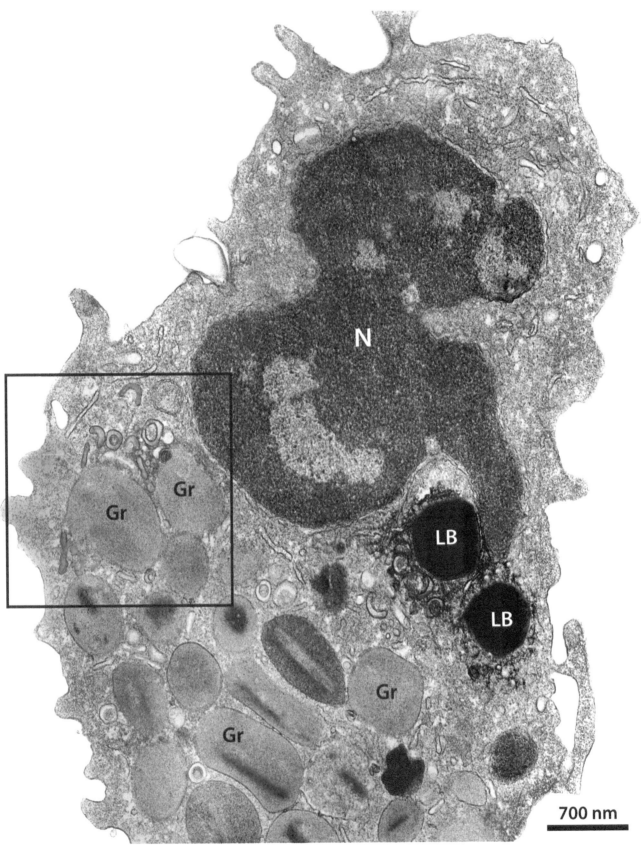

FIG. 4.16 CCL5-activated human eosinophil. EoSVs *(highlighted in pink)* are seen in contact with specific granules (Gr) *(boxed area)* and around LBs. Eosinophils isolated from the peripheral blood were stimulated and prepared for TEM.[20] *N*, nucleus.

Quantification of EoSVs by TEM also demonstrated that not only the total number of EoSVs per cell section expands in activated eosinophils but also the number of EoSVs in contact with specific granules, mainly with those undergoing release of their contents, increases (Figs. 4.11–4.15).[19,23]

EoSVs constitute a large population of membrane-bound tubular vesicles involved in the intracellular transport of granule-stored products in human eosinophils.[21,24] This vesicular system is consistently immunolabeled for granule-derived proteins, as shown in Chapter 5. For example, single-cell analysis at high resolution revealed that the number of EoSVs positive for IFN-γ, a cytokine that is a major product of human eosinophils,[193] is increased in response to cell activation with CCL11 or TNF-α.[25] Moreover, quantitative evaluations after ultrastructural immunolabeling for CD63, a tetraspanin involved in eosinophil secretion, demonstrated that the total number of CD63-positive EoSVs increases in response to activation with CCL11 or TNF-α (Chapter 5), thus associating EoSVs with secretory events.[23] In activated eosinophils, typical EoSVs can also be seen around LBs (Fig. 4.16), but the meaning of this possible interaction remains to be investigated.

Altogether, in response to activation of human eosinophils, EoSVs (i) increase in number in the cytoplasm, (ii) interact with and bud from specific granules undergoing secretion, and (iii) transport cytokines and other granule-stored products. All of these events were consistently demonstrated by the application of different TEM approaches.[19,21,23,25]

4.4 Lipid body formation

A major advance in LB biology is the recognition of LBs as inducible, newly formable organelles, elicitable in response to inflammatory stimuli.[85,205,206] Different from cells associated with lipid storage and lipid-laden cell lineages that have a large number of cytoplasmic LBs, resting leukocytes have few LBs, but these cells, when activated, form new LBs.[85,86,205–207]

Therefore, LBs in leukocytes are remarkably linked to inflammatory responses and considered structural markers of inflammation/activation. The association of LBs with inflammation/activation in leukocytes, such as eosinophils, was demonstrated by the following: (1) accentuated and rapid formation of these organelles in response to inflammatory stimuli or allergic/inflammatory human diseases, including infections with parasites and bacteria; (2) presence of arachidonyl phospholipids, which serve as precursors for the synthesis of inflammatory mediators (eicosanoids), within LBs; (3) correlative associations of LB formation (or inhibition) with levels of eicosanoids secretion; (4) localizations of eicosanoid-generating enzymes within LBs; and (5) in situ synthesis of eicosanoids (prostaglandins and leukotrienes) in activated cells.[86,206,208]

In EM studies of human tissue removed for diagnostic purposes, we have observed that eosinophils frequently contain numerous LBs (Fig. 4.17). Activated eosinophils with increased LB numbers in the cytoplasm were also noted in bronchoalveolar lavage fluid from patients with eosinophilic pneumonia[209] and lung adenocarcinoma (Chapter 8) and in peripheral blood eosinophils from HES patients.[210] It is now well documented that, in response to multiple inflammatory diseases and stimuli, the number and size of LBs increase within eosinophils, and these events can be detected and quantitated by EM.[86] In TEM studies, enumeration of these organelles can be done in electron micrographs displaying the entire cell profile and the total number of LBs is expressed in terms of cell section. Eosinophils seen in normal peripheral blood buffy coats and bone marrow or after careful isolation from the peripheral blood using negative selection exhibit LBs infrequently (0 to 1 LB per thin section). This number is consistently increased in eosinophils infiltrating in inflammatory lesions or after stimulation in vitro.[85,86] LBs are rapidly formed within eosinophils after stimulation, and TEM enables easy detection of LBs because, as noted in Chapter 2, they appear as nonmembrane-bound, highly contrasted (very electron-dense) organelles. For example, higher LB formation was documented ultrastructurally within human blood eosinophils after stimulation with PAF (Fig. 4.13),[203] CCL5 (Fig. 4.16),[94] CCL11 (Fig. 4.18),[94,211] and with the proinflammatory cytokines TNF-α (Figs. 4.19 and 4.20)[94] and IFN-γ[94] as demonstrated by quantitative analyses (Fig. 4.20).[94]

Increased numbers of LBs are detected within eosinophils from HES patients in both peripheral blood (Fig. 4.21) and tissue (Figs. 4.22 and 4.23). HES eosinophils are naturally activated, showing several functional and biochemical measures indicative of activation, including enhanced production of leukotriene C_4 (LTC4).[209,210,212–214] Morphologically, evidence of activation in HES eosinophils includes not only increases in the LB numbers (Figs. 4.20–4.23)[103] but also amplified EoSV formation (Fig. 4.24) and losses of the granule contents (Figs. 4.22, 4.23, and 4.24).[22] LBs also increase in size within eosinophils activated in vivo in HES (Fig. 4.25) or in vitro after stimulation with cytokines (Fig. 4.26), another morphological sign of cell activation.[86,215] The ultrastructural features of eosinophils in the context of HESs will be addressed in Chapter 8.

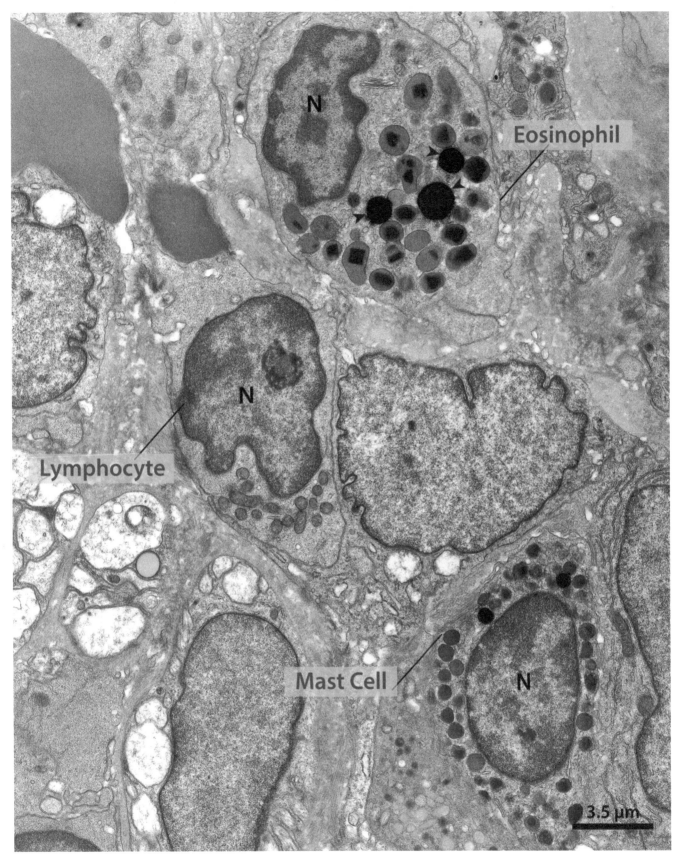

FIG. 4.17   An eosinophil and other inflammatory cells in the human intestinal lamina propria (ileal biopsy, Crohn's disease). Note osmiophilic LBs *(arrowheads)* interspaced with specific granules in the cytoplasm. Samples were prepared for TEM.[197] *N*, nucleus.

A. The cell biology of human eosinophils

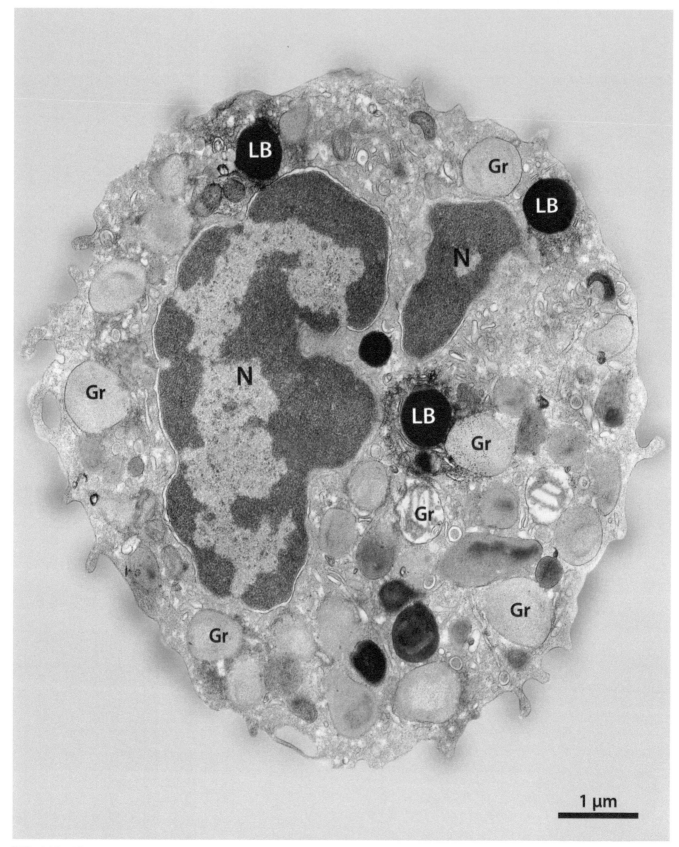

FIG. 4.18   Electron micrograph of a human peripheral blood eosinophil stimulated with CCL11. LBs increase in number in parallel to the emptying of secretory granules (Gr). Eosinophils were stimulated and prepared for TEM as before.[94] N, nucleus.

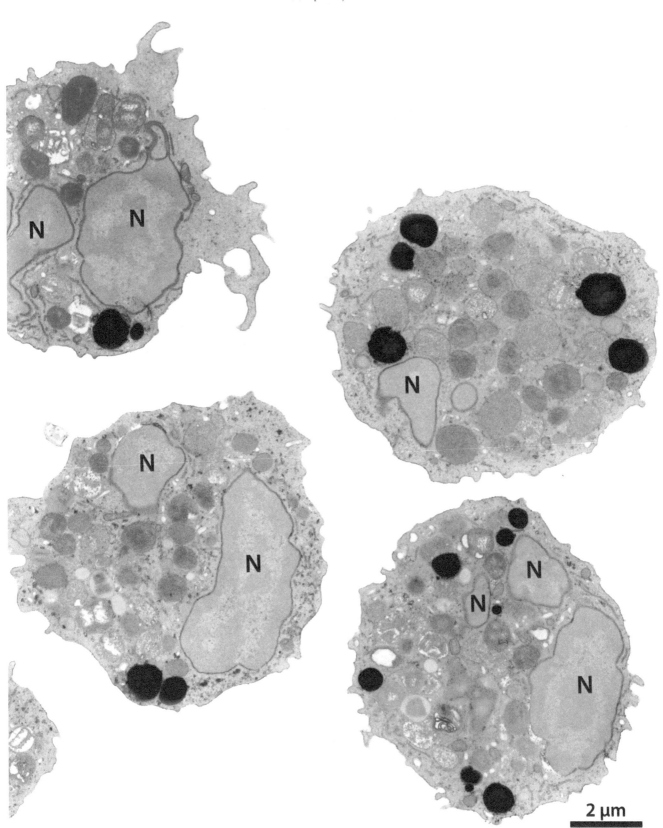

FIG. 4.19   Low-magnification electron micrograph showing TNF-α-activated human eosinophils. Note the high number of LBs seen as highly contrasted organelles in the cytoplasm while nuclei (N) and other organelles appear with less contrast. Cells were stimulated and prepared for TEM as described.[94]

A. The cell biology of human eosinophils

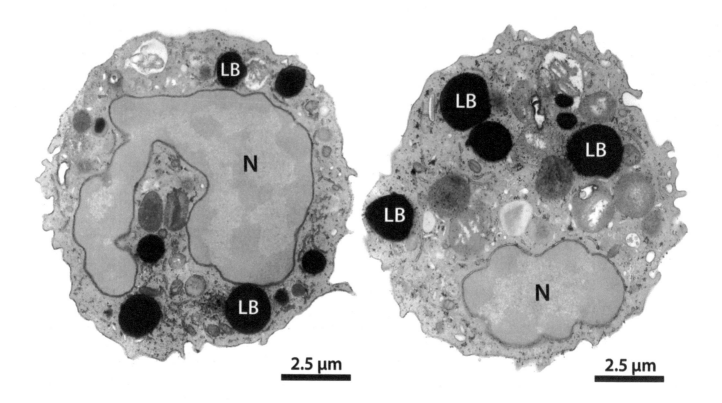

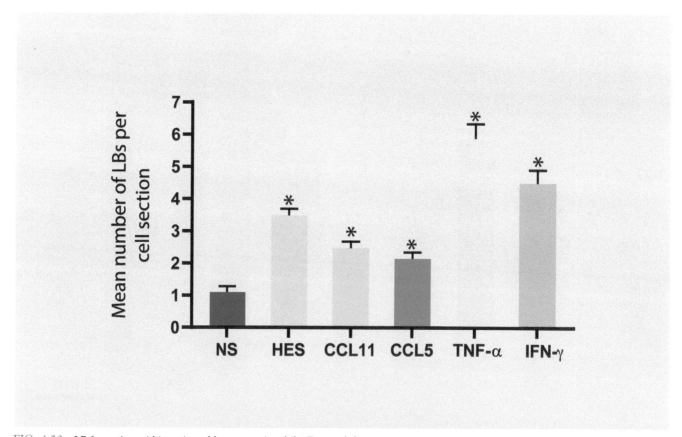

FIG. 4.20 LB formation within activated human eosinophils. *Top panel* shows representative TNF-α-stimulated eosinophils with several cytoplasmic LBs. Stimulation with inflammatory chemokines and cytokines or hypereosinophilic syndrome (HES) induces significant LB formation (*$P \leq .05$) compared to nonstimulated (NS) cells. Peripheral blood eosinophils, stimulated or not were processed for TEM with ferrocyanide-reduced osmium postfixation.[94] The numbers of LBs were quantitated in electron micrographs showing the entire cell profile and nucleus. *N*, nucleus.

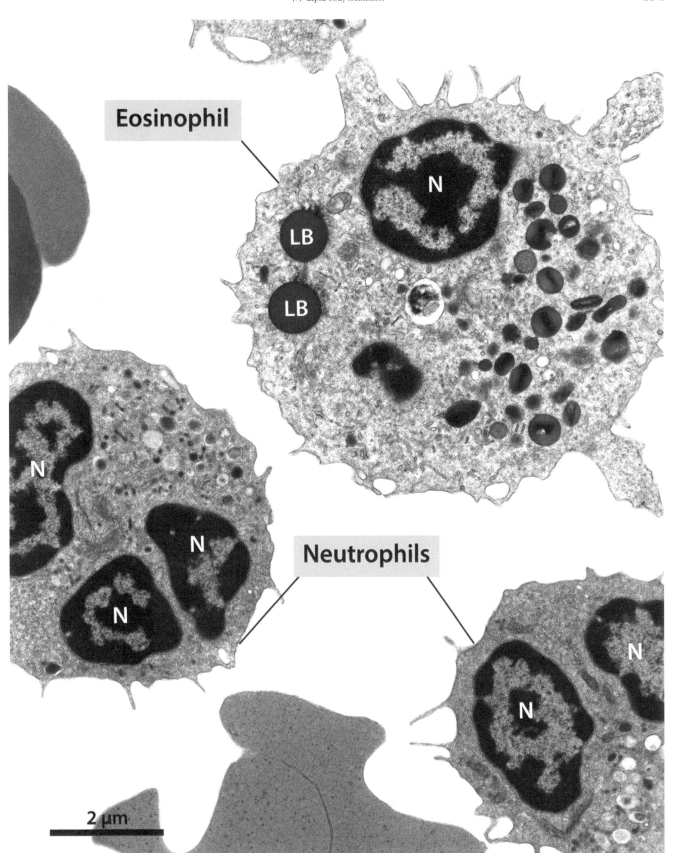

Eosinophil

Neutrophils

N

LB

LB

N

N

N

N

N

N

2 μm

FIG. 4.21 Granulocytes in the peripheral blood buffy coat from a patient with HES. LBs are seen in the eosinophil cytoplasm. Note neutrophils and red blood cells *(colored in red)*. Sample prepared for TEM. *N*, nucleus.

A. The cell biology of human eosinophils

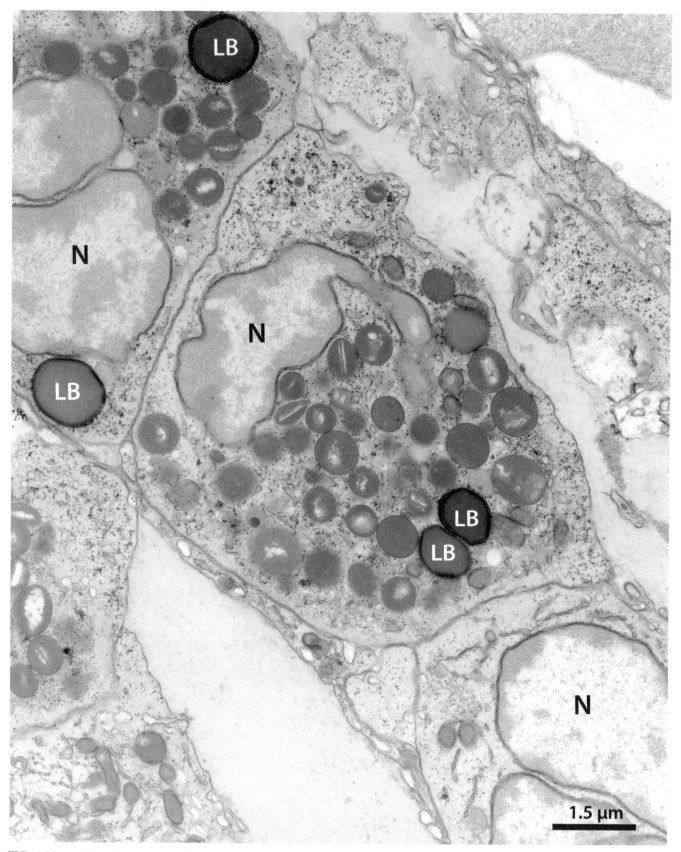

FIG. 4.22   Activated tissue eosinophils in a skin biopsy of a patient with HES. LBs are frequently observed in the cytoplasm. Cells were post-fixed for TEM with ferrocyanide-reduced osmium. ECM colored in *green*. N, nucleus.

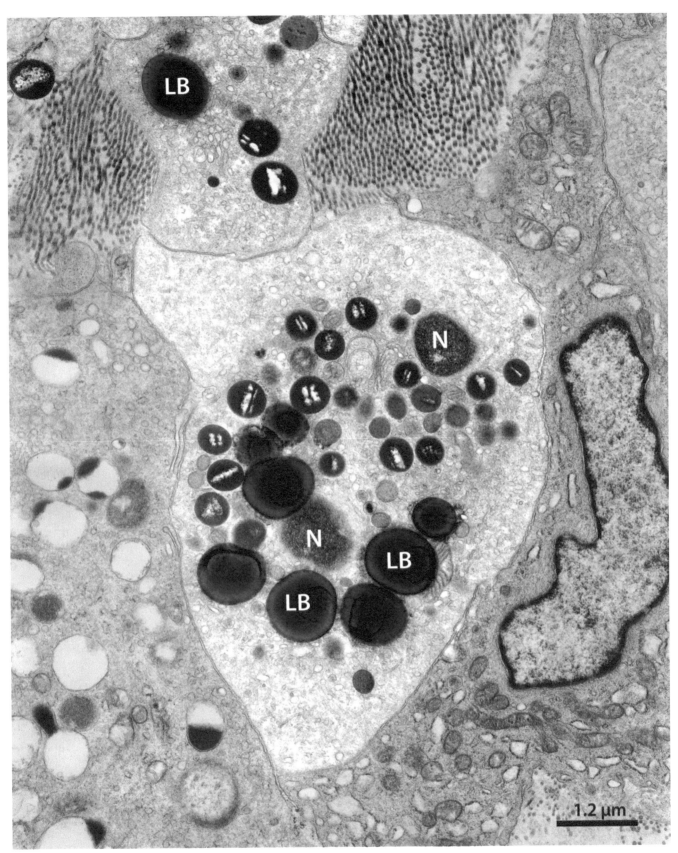

FIG. 4.23   TEM of tissue eosinophils (skin biopsy, HES). Note a high number of LBs and specific granules *(colored in orange)* with core losses. The surrounding ECM and cells were colored in *green. N,* nucleus.

A. The cell biology of human eosinophils

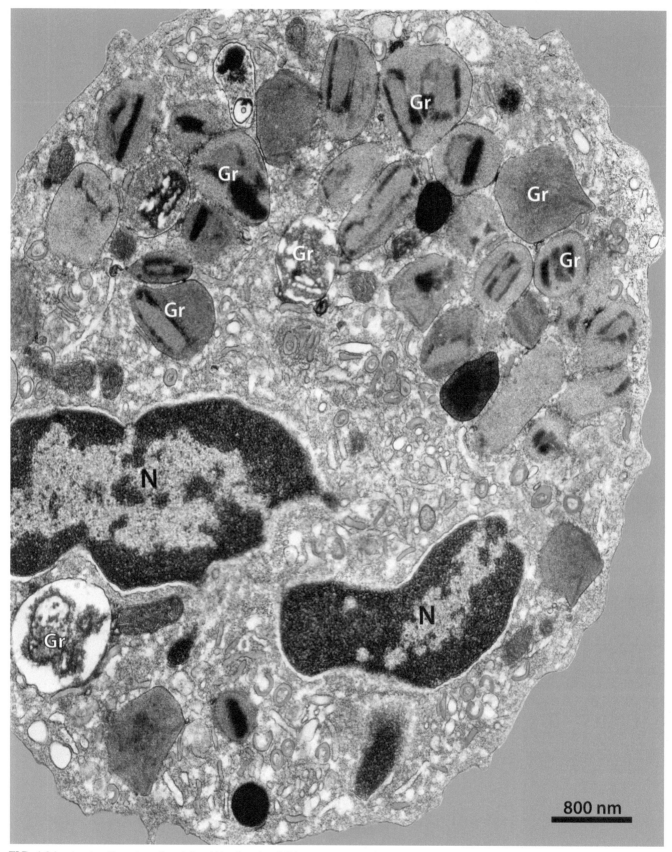

FIG. 4.24   A naturally activated peripheral blood eosinophil from a patient with HES. Note in the cytoplasm, the high amount of EoSVs *(colored in pink)* and specific granules (Gr) showing core losses. *N*, nucleus.

A. The cell biology of human eosinophils

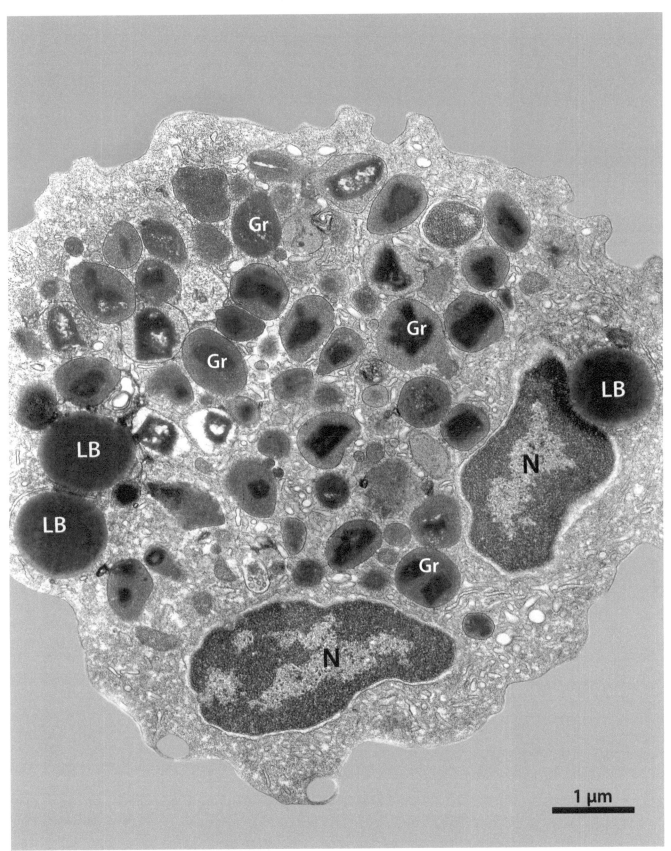

FIG. 4.25 Electron micrograph showing large LBs in the cytoplasm of a peripheral blood eosinophil isolated from a patient with HES. Note one LB in close contact with the nucleus (N). *Gr*, secretory granules.

A. The cell biology of human eosinophils

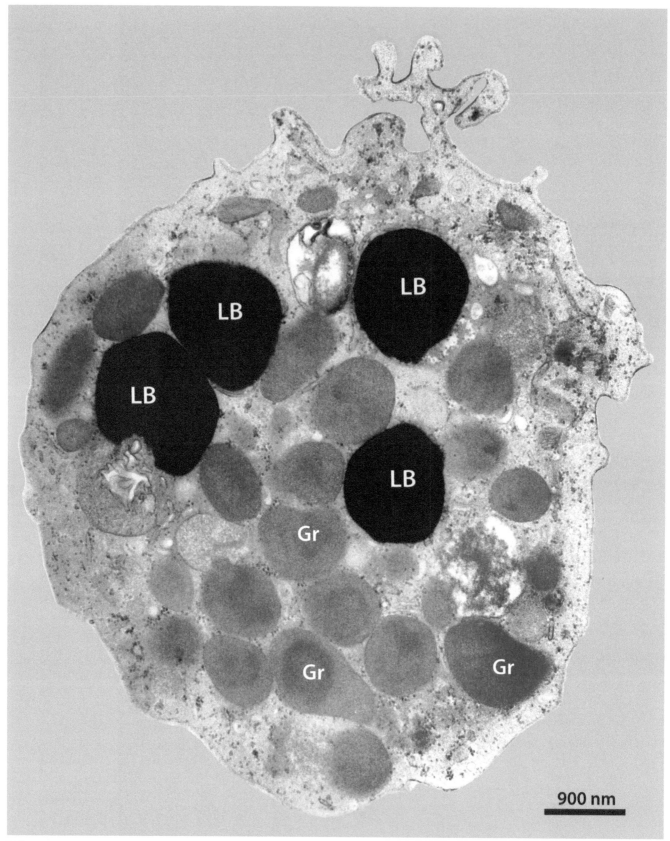

FIG. 4.26   IFN-γ-activated human eosinophil showing large cytoplasmic LBs. Specific granules (Gr) appear with less contrast. The nucleus is not visible in the section plane. Eosinophils isolated from the peripheral blood were stimulated and prepared for TEM using postfixation with ferrocyanide-reduced osmium.[94]

A. The cell biology of human eosinophils

One intriguing ultrastructural aspect of LBs is their electron density that is dependent on the cell type and can change in response to cell activation.[85] For example, great alterations in LB electron density are documented in macrophages triggered by infectious diseases.[207,215,216] In human eosinophils, such ultrastructural alteration is rarely observed. However, TEM of mature, blood-derived eosinophils cultured with HIV-1 revealed not only increased formation of LBs[204] but also a remarkable change of their electron density (Figs. 4.27 and 4.28). As noted, eosinophil LBs are osmiophilic organelles and thus typically appear as very electron-dense organelles in thin sections, but in cultures of HIV-infected eosinophils, LBs were also seen as electron-lucent organelles with a peripheral electron-dense rim (Figs. 4.27 and 4.28). This change in the ultrastructural aspect was also documented in LBs formed within macrophages during in vivo infection with mycobacteria in parallel to the increased formation of these organelles.[216] Within HIV-activated eosinophils in cultures, LBs also showed a remarkable increase in size with the formation of very large LBs (Figs. 4.29 and 4.30), some of them presenting evidence of fusion (Fig. 4.29) or budding (Fig. 4.30).

Cell activation-mediated LB morphological changes may reflect differences in lipid composition, stages of formation of new LBs, and/or the neutral lipids/phospholipids ratio within LBs.[215] Based on the fact that osmium tetroxide binds preferentially to the unsaturated bonds of fatty acids,[217,218] it was demonstrated that the electron density of existing LBs of fibroblasts increases gradually by adding polyunsaturated fatty acids to the cell culture medium.[92] Electron density was highest in the cells treated with docosahexaenoic acid, followed by those treated with linoleic acid and oleic acid, in accordance with the number of unsaturated bonds of these fatty acids.[92] Therefore, lipid composition affects LB electron density, and TEM can be used to distinguish triglycerides synthesized from fatty acids of different saturation.[92]

Within activated eosinophils and other cells from the immune system, alterations in electron density may indicate the cascade of events involved in the synthesis of inflammatory mediators within LBs formed in response to cell activation. Under stimulation, AA, a fatty acid stored in LBs in association mainly with phospholipids, is released from its esterified pool and acts as a substrate for enzymatic conversion into lipid mediators.[86,205,208] Changes in electron density induced by inflammatory signals and other stimuli, in conjunction with alterations in LB numbers and size, highlight the fact that LBs are dynamic and active organelles, able to modify their structure in concert with cell activation.

In addition to TEM, we used wet SEM technology to study LB formation within human activated eosinophils (Fig. 4.31).[211] In contrast to the routine SEM, which is based on secondary electron detection and limits visualization to the cell surface, the wet SEM technique uses backscattered electron detection, enabling the analysis of defined sample depth in a fully hydrated system,[211] which is particularly important for LB studies, because these organelles are highly sensitive to dehydration processes.[90] Wet SEM combines the advantages of light microscopy (i.e., rapid sample preparation; the complete processing requires ~ 1.5h) and can be a good alternative (compared with methods using fluorescence microscopy) for large-scale LB imaging and scoring within leukocytes and other cells under different conditions. Our quantitative analysis using this approach and computerized image processing easily scored LB numbers, which appears as electron-lucent organelles after wet SEM, and revealed a significant increase of them in CCL11-stimulated eosinophils (Fig. 4.31) compared with unstimulated cells [8.5 ± 2.9 LBs/cell in unstimulated cells compared with 20.4 ± 4.7 LBs/cell in CCL11-stimulated eosinophils (mean $\pm$ SD; $n = 50$ cells/group)]. Although this technique enables visualization of LBs at the EM level, it is important to keep in mind that the resolution is much lower compared with TEM.[211] This means that detailed observation of LBs including interaction with other organelles is not possible using this methodology.

Since eosinophil LBs compartmentalize several proteins and lipids involved in the control and biosynthesis of inflammatory mediators, these organelles, formed in response to cell activation, act as competent organelles for modulation of immune responses in leukocytes.[86,205] Enzymes linked to the synthesis of inflammatory mediators are expressed in activated human eosinophils.[206] Activated LBs are also potential sites for cytokines.[219,220] The demonstration of these proteins using immunogold EM will be addressed in Chapter 5.

4.5 Shape changes

Leukocytes can change their morphology when they undergo activation, migration, and cell-to-cell interactions. The shape change is a prominent feature of human eosinophils responding to chemotactic/activation agents.[187,221–225] Several agonists are known to induce morphological changes in human eosinophils, including complement factor 5a,[187] PAF,[187,189,203] activators of protein kinase C (PKC),[222] IL-5 family cytokines,[223–225] and CCL11.[105,224,226]

Diverse methodologies have been used to assess eosinophil shape changes, such as flow cytometry, laser turbidimetry, light microscopy, and EM. We and other authors have been documenting eosinophil shape changes with both TEM and SEM.[40,53,105,189,203] Unstimulated/resting eosinophils frequently have a spherical shape, and their surface exhibits short membrane projections (ruffles) (Fig. 4.32). Morphological events initially detectable in response to cell activation are cell flattening and elongation, which can be seen within five minutes[189] up to one hour after cell

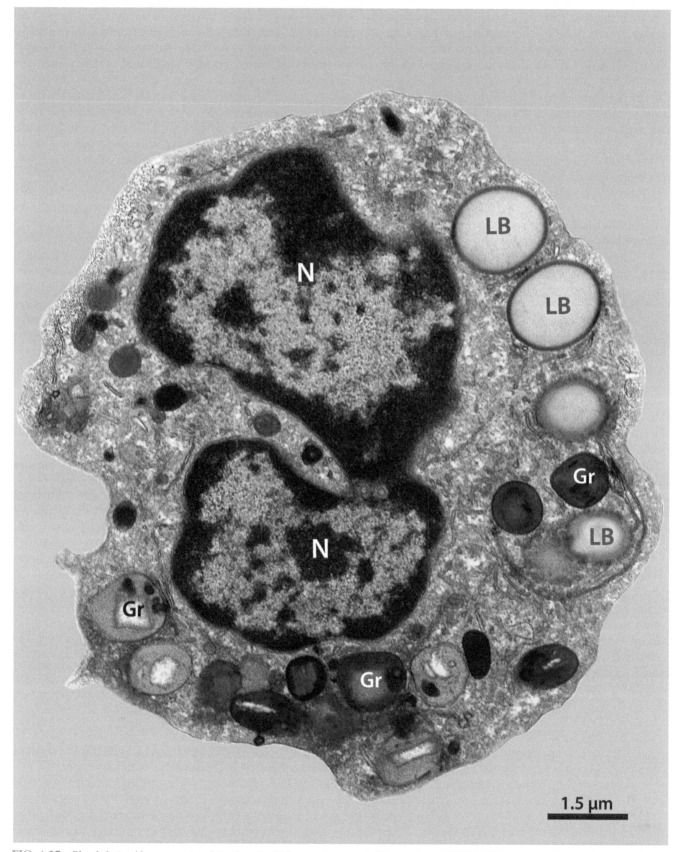

FIG. 4.27   Blood-derived human eosinophil cultured with human immunodeficiency virus-1 (HIV-1). LBs appear as electron-lucent organelles with an electron-dense peripheral rim. Cultures prepared as described.[204] *Gr*, secretory granules; *N*, nucleus.

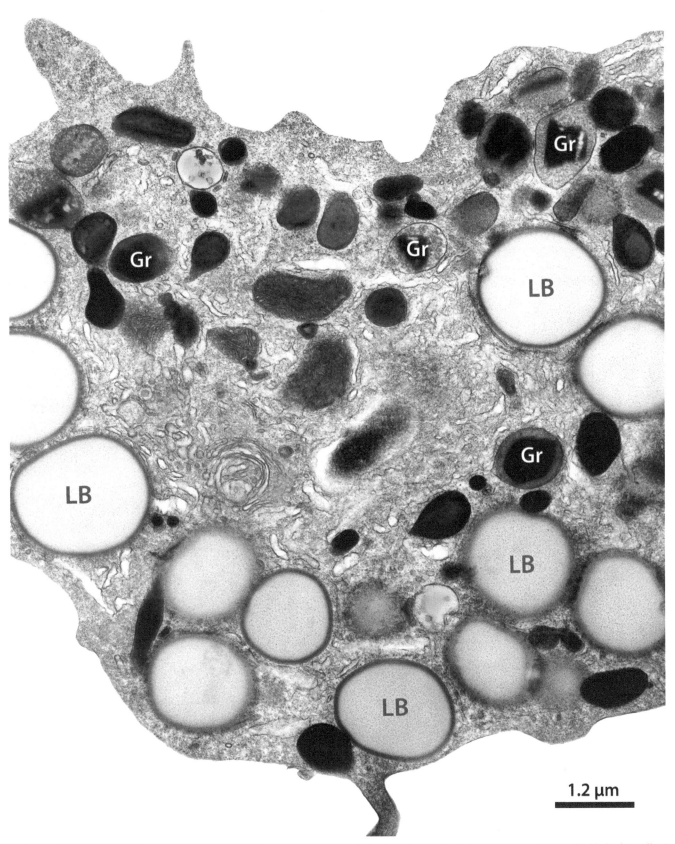

FIG. 4.28 Increased number of electron-lucent LBs in an human eosinophil cultured with HIV-1. Note the Golgi region *(highlighted in yellow)* and early endosomes *(colored in orange)*. Cultures prepared as described.[204] *Gr*, secretory granules.

A. The cell biology of human eosinophils

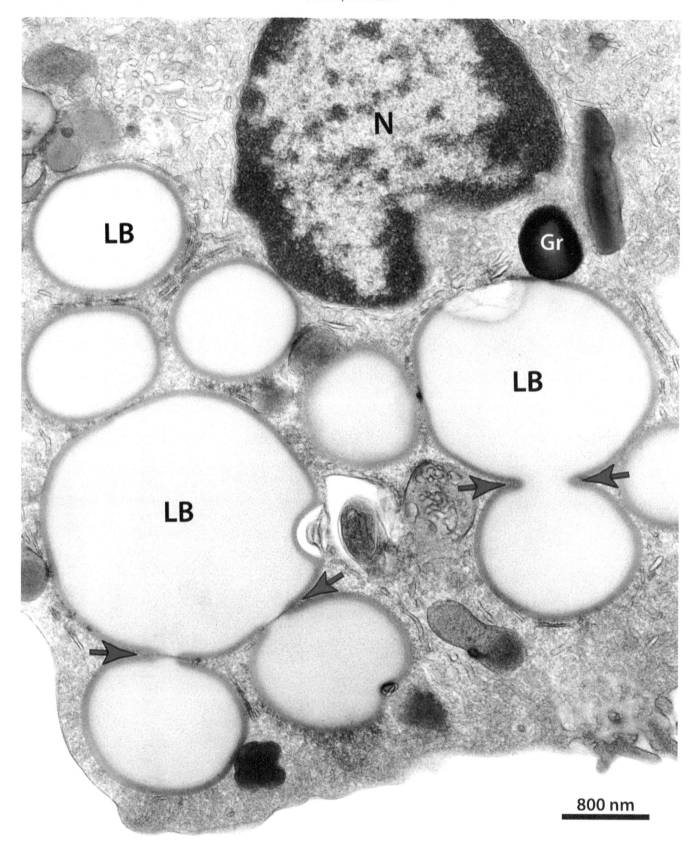

FIG. 4.29 TEM showing a blood-derived human eosinophil cultured with HIV-1. Note large, electron-lucent, and fused LBs *(arrows)* in the cytoplasm. Cultures prepared as described.[204] *Gr*, secretory granule; *N*, nucleus.

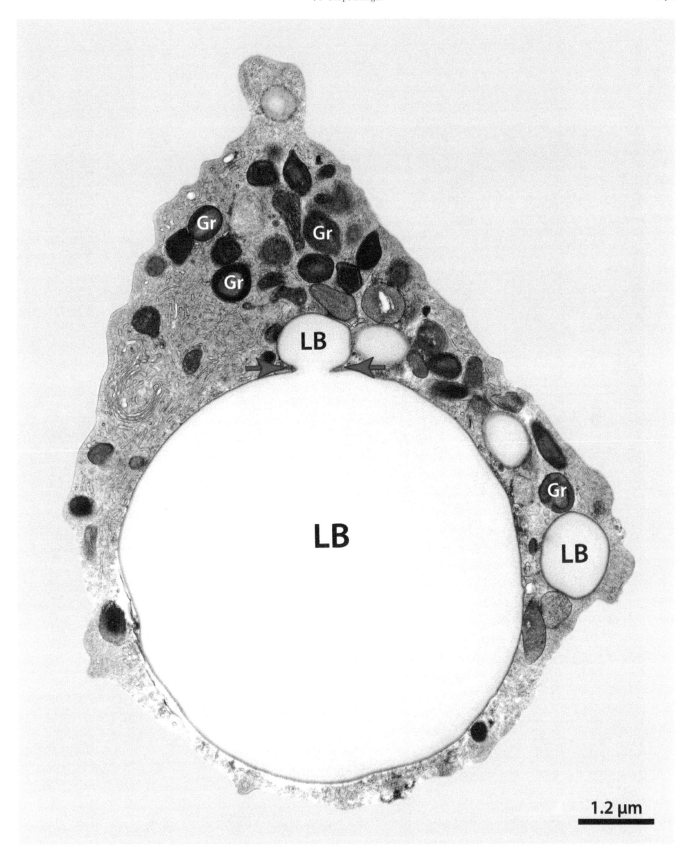

FIG. 4.30   A giant LB and small LBs occupy most of the cytoplasm in a human eosinophil cultured with HIV-1. Note a point of LB budding *(arrows)*. Cultures prepared as described.[204] *Gr,* secretory granules.

A. The cell biology of human eosinophils

CCL11-stimulated

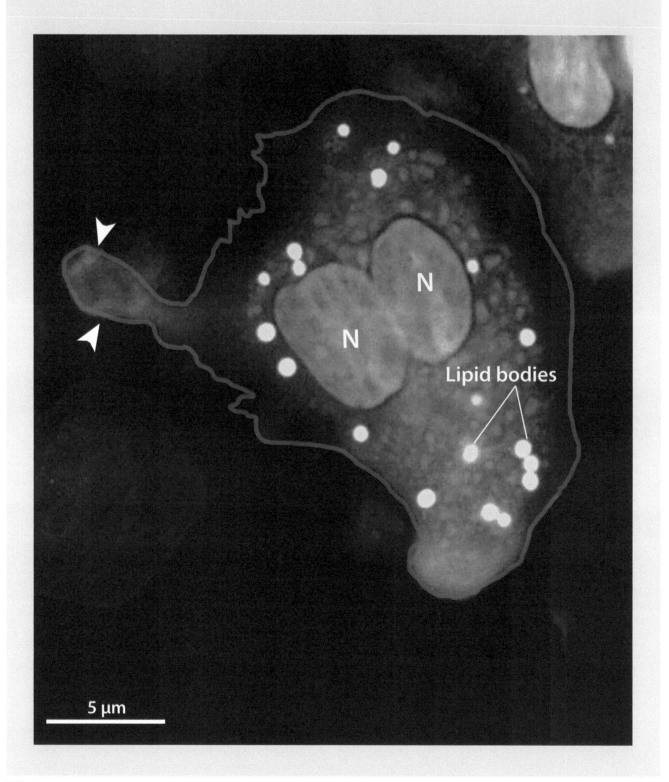

FIG. 4.31 CCL11-activated human eosinophil (outlined in *pink*) viewed with wet SEM. LBs appear as electron-lucent organelles in a fully hydrated system.[211] Note the bilobed nucleus (N) and a surface protrusion (uropod arrowheads) typical of activated cells. Eosinophils isolated from the peripheral blood were stimulated with CCL11 for one hour.

Unstimulated

IL-5-stimulated

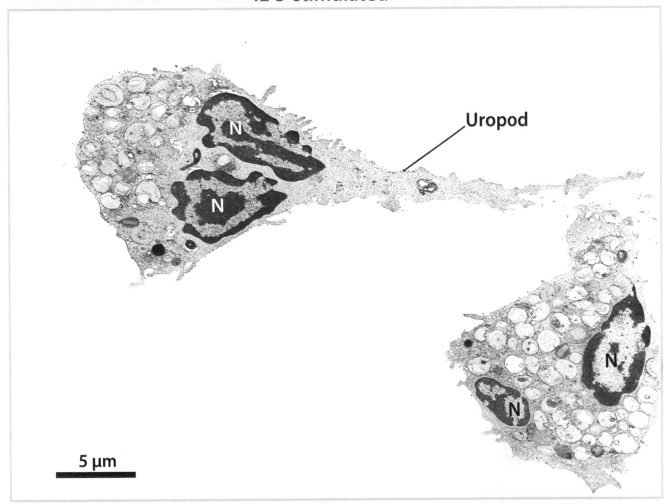

Uropod

FIG. 4.32 Shape changes in IL-5-activated human eosinophils. Unstimulated/resting eosinophils (*top panel*) are round and exhibit short surface projections. Stimulation leads to cell expansion and elongation. Note one polarized eosinophil (nucleus and granules in opposite sides) with an agranular protrusion (uropod). Cells were stimulated for one hour with IL-5 in chamber slides and adhered cells processed in situ for TEM. *N*, nucleus.

stimulation[105,203] (Figs. 4.31–4.34). These morphological changes can be observed in eosinophils stimulated either in suspension (Figs. 4.31 and 4.33) or after cell adhesion (Figs. 4.32 and 4.34). When eosinophils are stimulated in vitro with agonists and processed for TEM directly on a slide surface, there is clear visualization at a high resolution of the prominent eosinophil shape changes, and a polarized morphology with secretory granules and nucleus located in opposite directions can also be observed (Figs. 4.32 and 4.34).

In vivo, shape changes are frequently documented in human tissue eosinophils in the process of activation and/or migration (Figs. 4.35–4.38). For example, we have been observing eosinophils undergoing shape changes in the intestine (Figs. 4.35 and 4.36) , heart (Fig. 4.37), and skin (Fig. 4.38). In addition to elongation and flattening exhibited by eosinophils mainly in inflammatory sites from human biopsies, these cells can be seen as polarized leukocytes with nuclei and granules displayed in opposed sites (Figs. 4.36, 4.37, and 4.38) as also observed in vitro by both light microscopy[223] and TEM after stimulation, for example, with IL-5 (Figs. 4.32 and 4.34). While this polarization morphology indicates cell activation, the meaning of this granule reorganization in the cytoplasm in the context of cell secretion is not understood.

Eosinophil shape changes involve the formation of protrusive structures at the cell surface termed lamellipodia and filopodia (Fig. 4.39). A lamellipodium is a thin (100–200 nm thickness), sheet-like protrusion that is filled with a branched network of actin, while filopodia are thin, finger-like structures that are filled with tight parallel bundles of filamentous (F)-actin.[227,228] Filopodia project from the cell surface as tiny (~ 100–300-nm-wide) micrometer-long structures (Figs. 4.39, 4.40, and 4.41) and are frequently found embedded in or protruding from the lamellipodial structures (Fig. 4.39).[228] The elongation of actin filaments in filopodia/lamellipodia pushes the leading cell edge forward, which promotes cell migration or extension. These protrusive structures are involved in diverse cellular processes, including adhesion to the extracellular matrix and guidance toward chemoattractants.[228] At the rear of the cell, a knob-like protuberance ("tail"), generally free of granules, termed uropod, is also frequently observed by both light microscopy[223,226] and TEM[53,105,154] (Figs. 4.32, 4.42, 4.43, and 4.44) in human eosinophils and has been associated with events of activation and cell migration.[229,230] A typical uropod is clearly distinguishable from the other protruding structures present at the cell surface of eosinophils and other leukocytes since the uropod is a huge cell protrusion characteristically longer and thicker as compared with filopodia and lamellipodia.[229] Eosinophil shape change events are also described in Chapter 8 in the context of eosinophilic diseases.

4.6 Release of extracellular vesicles (EVs)

As a secretory ability (discussed in Chapter 3), human eosinophils release EVs in normal conditions and when activated during inflammatory responses.[25,125,139,184] We have been using a TEM approach to study the phenomenon of EV secretion by human eosinophils and other cells and to quantitate nascent EVs at the cell surface.[125] EV formation is evaluated in clear cross-cell sections exhibiting the entire cell profile, intact plasma membranes, and nuclei, and the number of EVs per cell section is determined. For example, by studying a total of 110 electron micrographs (39 from unstimulated, 37 from CCL11-stimulated, and 34 from TNF-α-stimulated cells), we counted 516 EVs and found that while unstimulated cells had 1.4 ± 0.4 EVs/cell section, CCL11- and TNF-α-stimulated eosinophils showed 5.0 ± 0.8 and 8.0 ± 1.0 EVs/cell section (mean $\pm$ S.E.M), respectively, corresponding to an increase of 360% (CCL11) and 570% (TNF-α) (Fig. 4.45).[125] Thus, amplified production of EVs by eosinophils was demonstrated by TEM and it is considered an important event associated with eosinophil activation during inflammatory responses.

In our qualitative and quantitative ultrastructural studies using in vitro activation with CCL11 and TNF-α, EVs appeared mostly as microvesicles (MVs) in both unstimulated and stimulated eosinophils, that is, vesicles shedding directly from the plasma membrane and not vesicles released by fusion of MVBs with the plasma membrane (exosomes).[25,125] Moreover, by scoring the number of MVs, while just 50% of unstimulated cells produced MVs, 90% and 100% of eosinophils formed MVs after stimulation with CCL11 and TNF-α, respectively (Fig. 4.45). Interestingly, in unstimulated eosinophils, most MVs-producing cells released 1–3 MVs per cell section, whereas the majority of cells produced 1–9 MVs and 4–21 MVs per cell section in response to CCL11 and TNF-α stimulation, respectively (Fig. 4.45).[125]

The formation of MVs is a dynamic process, and TEM captures these vesicles in different stages of budding from the plasma membrane or free at the cell surface (Figs. 4.40 and 4.45). In our quantitative analyses, the numbers of MVs in the process of budding (still in contact with the plasma membrane) and free at the cell surface were also evaluated.[125] Interestingly, not only the total number of MVs increased but also the numbers of MVs in different degrees of budding were significantly higher in stimulated than in unstimulated cells, an event that might be related to a faster production of these vesicles in activated cells.[125] Moreover, quantitative TEM also detected that stimulation with TNF-α led to higher production of MVs than CCL11 (Fig.4.45), indicating that the stimulus impacts the amount of EV secretion. The presence of cytokines and tetraspanins within MVs is shown in Chapter 5.

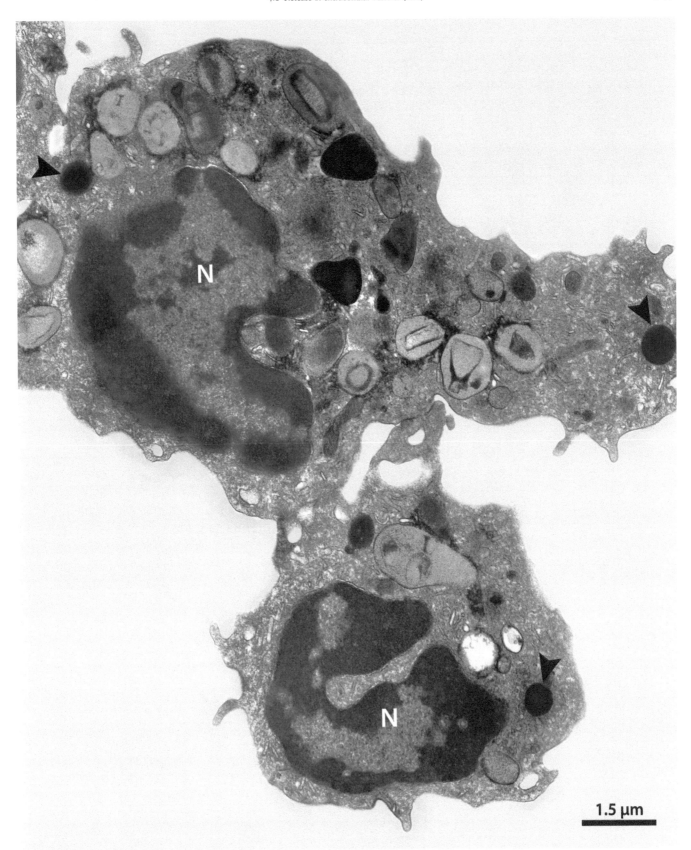

FIG. 4.33   Human eosinophils *(pseudocolored)* activated with CCL11. An elongated eosinophil is observed close to a round one. Note LBs *(arrowheads)* in the cytoplasm. Cells were isolated from the peripheral blood, stimulated in suspension for one hour with CCL11, and prepared for TEM.[20] N, nucleus; Cytoplasm *(pink)*; Secretory granules *(yellow)*.

A. The cell biology of human eosinophils

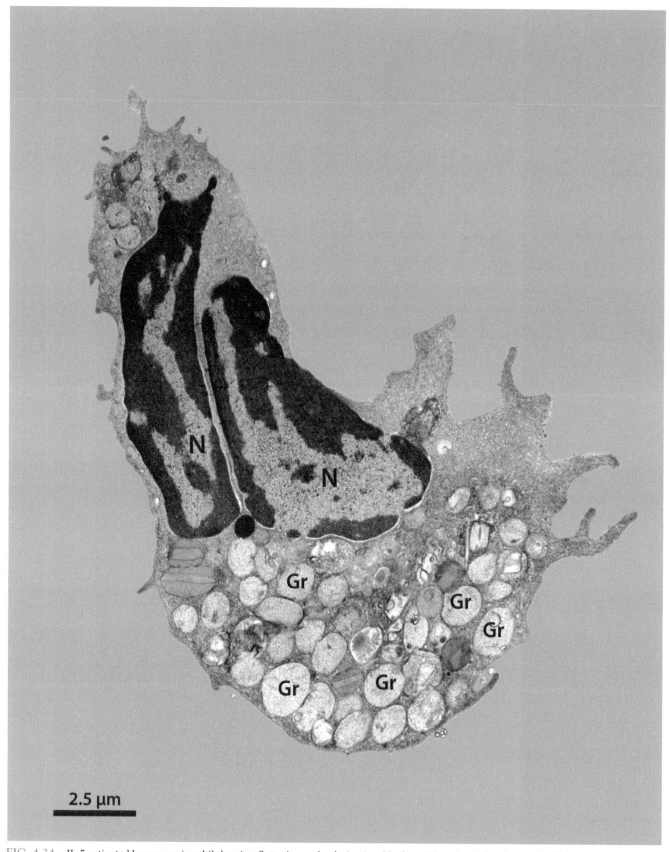

FIG. 4.34   IL-5-activated human eosinophil showing flattening and polarization. Nucleus (N) and mobilized granules (Gr) are seen in opposite sides. Cells were stimulated for one hour with IL-5 in chamber slides and adhered cells processed in situ for TEM.

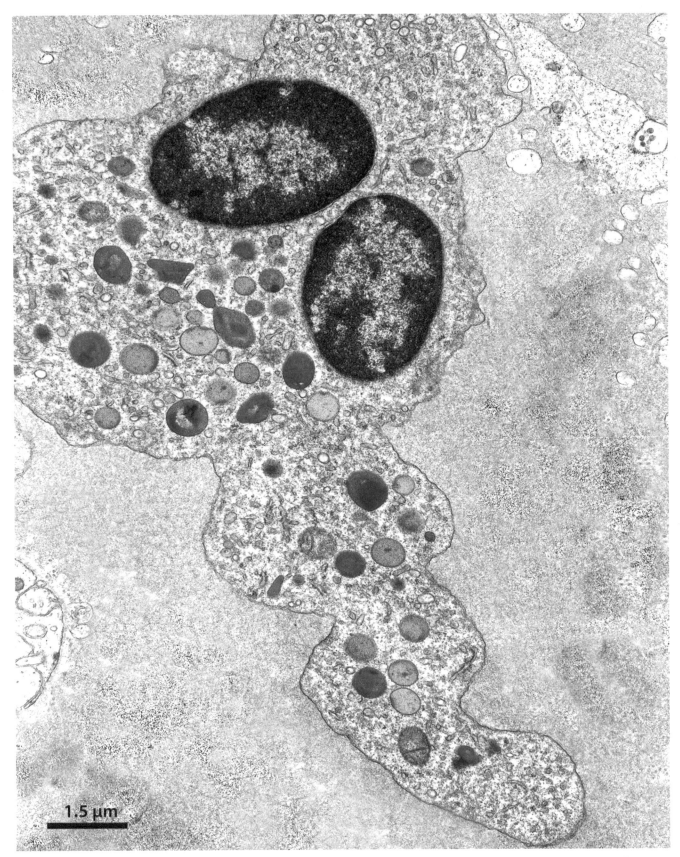

1.5 μm

FIG. 4.35   Transmission electron micrograph showing a migrating, elongated tissue eosinophil (pseudocolored) in the human small intestine (ileum). Note the typical bilobed nucleus *(purple)*. Cytoplasm *(pink)*; Specific granules *(yellow)*.

A. The cell biology of human eosinophils

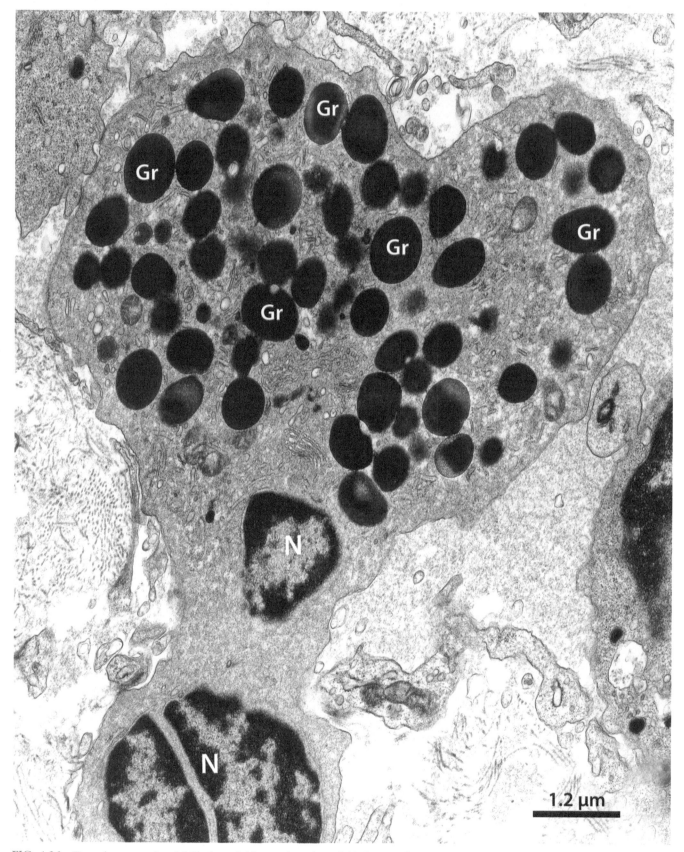

FIG. 4.36   Tissue human eosinophil (ileal pouch biopsy, ulcerative colitis) showing flattening and a clear polarization. Nucleus (N) and secretory granules (Gr) are displayed in opposite directions. Sample prepared for TEM.

A. The cell biology of human eosinophils

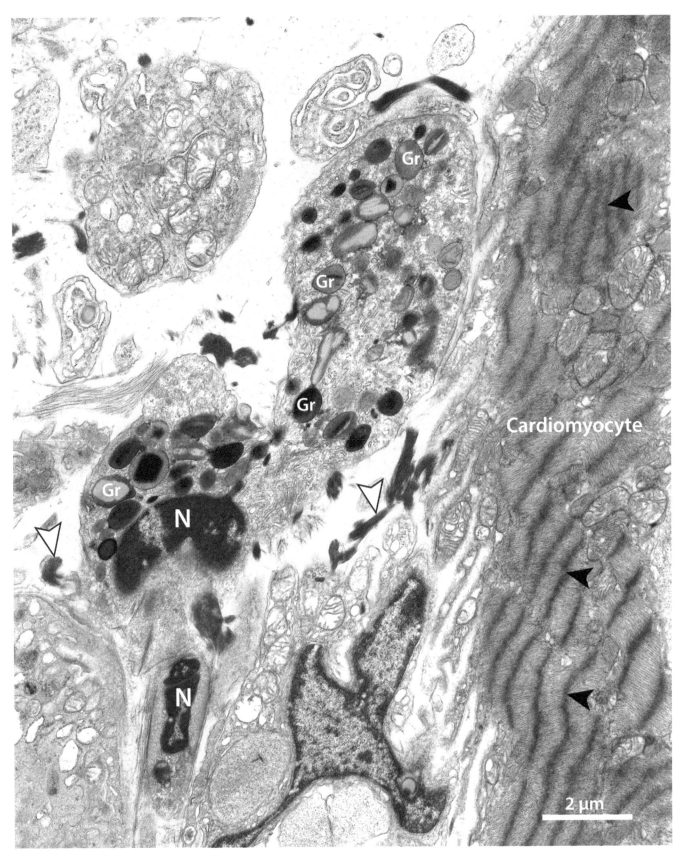

FIG. 4.37 An activated eosinophil *(pseudocolored)* in the heart of a patient with myocarditis. Cell is elongated, polarized (nucleus and granules in opposite sides), and with mobilized granules (Gr). Note an adjacent damaged cardiomyocyte with aggregated sarcomeric Z-lines *(black arrowheads)* and fibrin *(white arrowheads)* spread in the ECM *(green)*. Sample prepared for TEM. Cytoplasm *(pink)*; N *,nucleus*; Secretory granules *(yellow)*.

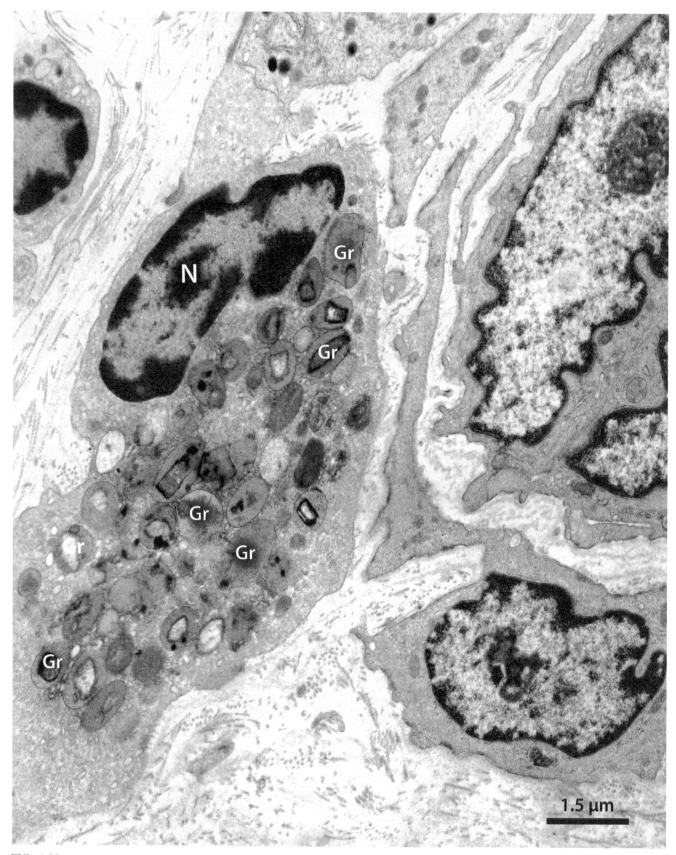

FIG. 4.38   An activated eosinophil in an inflammatory site in the skin of a patient treated subcutaneously with SCF/cKit ligand. Eosinophil exhibits polarization and PMD. Sample prepared for TEM. ECM colored in *green*. *Gr*, secretory granules; *N*, nucleus.

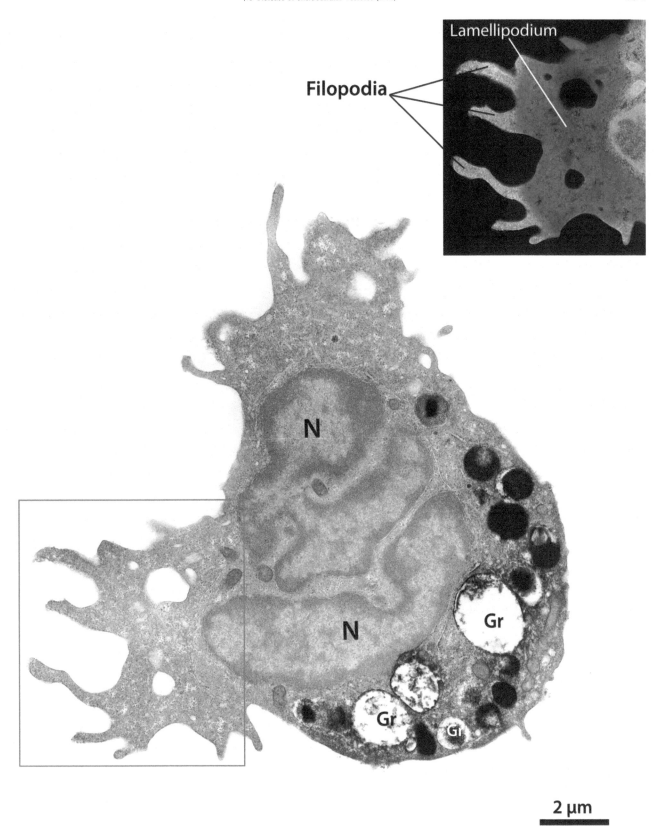

FIG. 4.39   A mature eosinophil developed in a human-cord blood mononuclear cell culture supplemented with IL-5. Shape changes, characterized by formation of protrusive structures at the cell surface (lamellipodia and filopodia, boxed area), cell polarization (nucleus and granules in opposite sides), and enlarged secretory granules (Gr) with content losses denote cell activation. *N*, nucleus.

A. The cell biology of human eosinophils

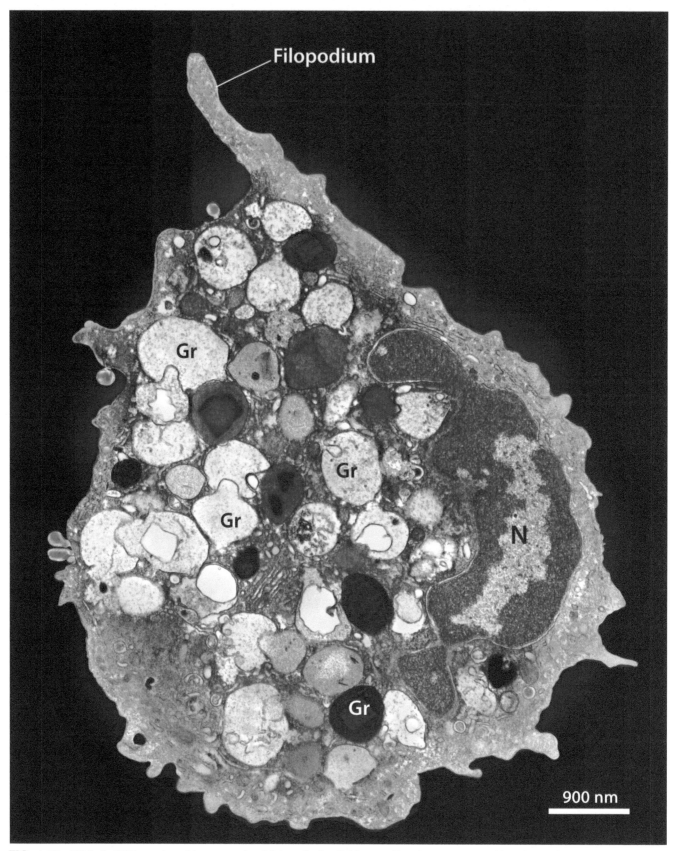

FIG. 4.40   TEM of a TNF-α-activated human eosinophil *(pseudocolored)*. Note a cell surface protrusion (filopodium), fused secretory granules (Gr) with content losses, indicative of compound exocytosis, and release of extracellular vesicles *(colored in blue)*. Cells were isolated from the peripheral blood and stimulated as described.[125] *N*, nucleus.

A. The cell biology of human eosinophils

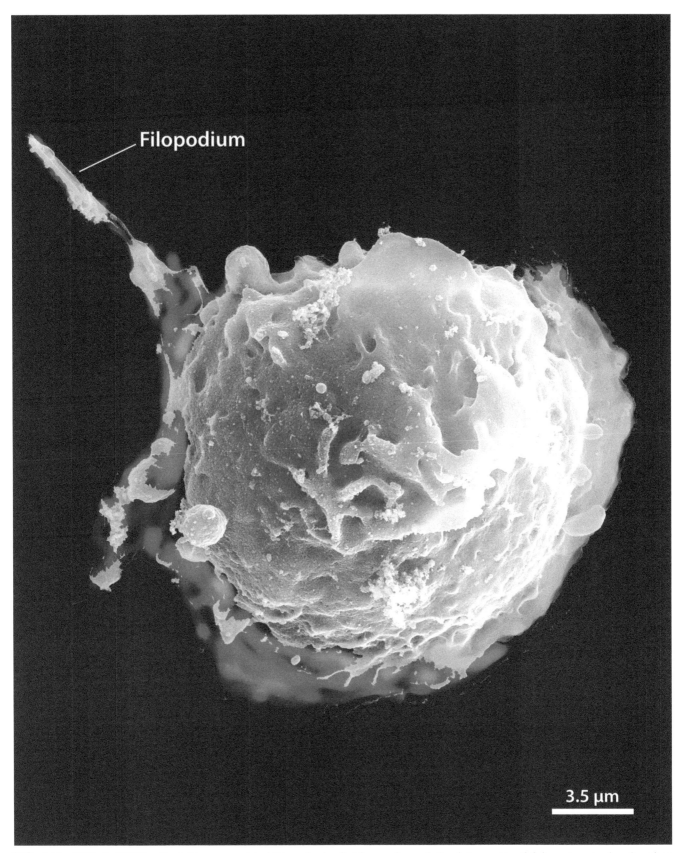

FIG. 4.41 SEM of a CCL11-activated human eosinophil *(pseudocolored)*. After one hour of stimulation, eosinophil shows increased surface short projections and a filopodium. Cells were isolated from the peripheral blood and stimulated as described.[20]

A. The cell biology of human eosinophils

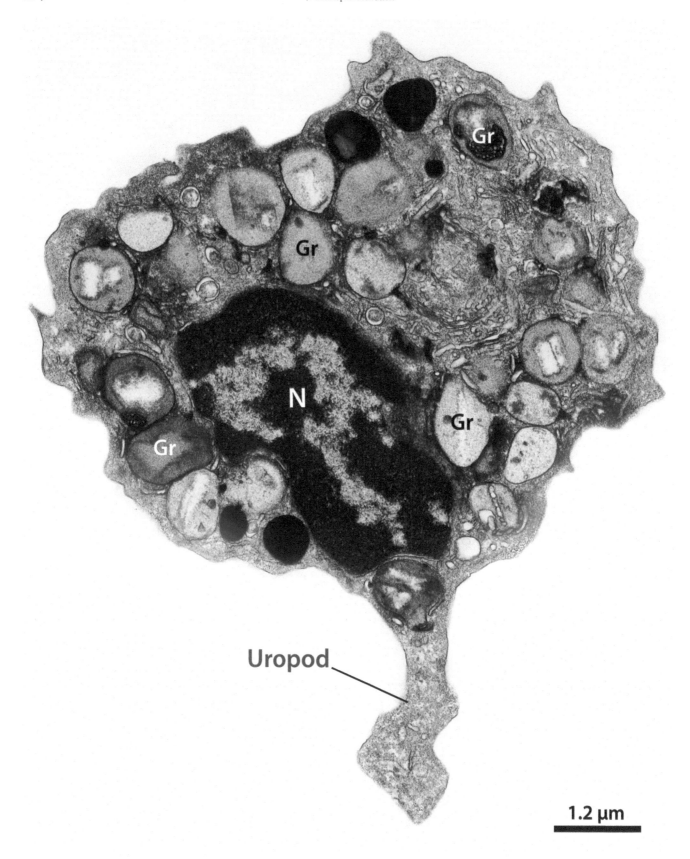

FIG. 4.42   Blood-derived human eosinophil cultured with HIV-1. Cell activation is depicted by the presence of secretory granules (Gr) with content losses and uropod formation. Culture prepared as described.[204] N, nucleus.

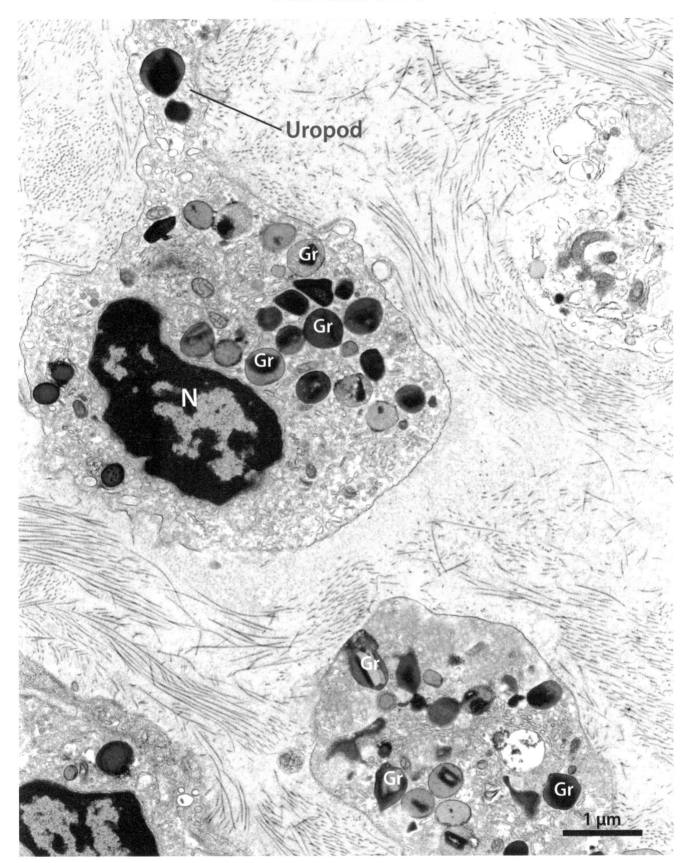

FIG. 4.43   Infiltrating eosinophils in the nasal sinus from a patient with ECRS. A uropod, typical of activated cells, is observed. ECM colored in *green*. Sample prepared for TEM. *Gr*, secretory granules; *N*, nucleus.

A. The cell biology of human eosinophils

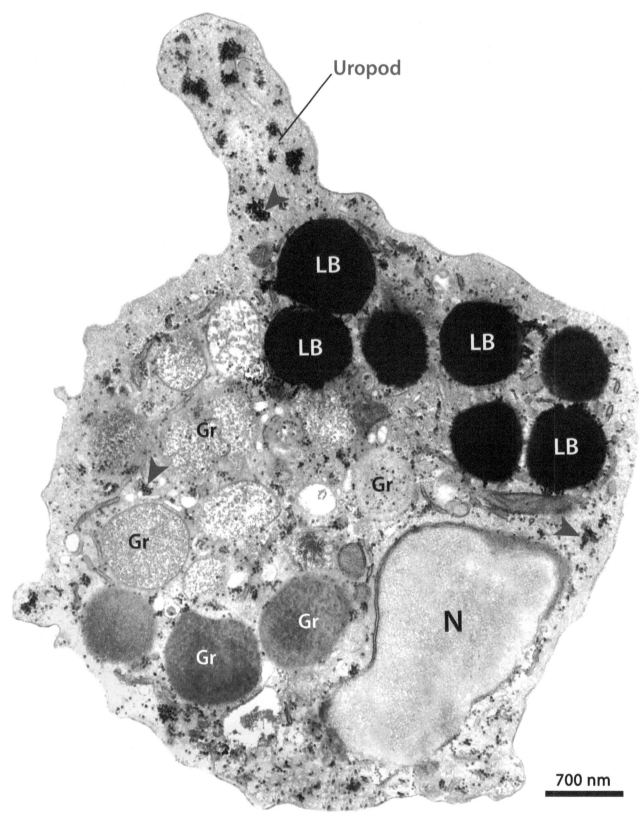

FIG. 4.44   IFN-γ-activated human eosinophil showing a high number of LBs in the cytoplasm and uropod formation. Cells were prepared for TEM using postfixation with ferrocyanide-reduced osmium.[94] Small aggregates of glycogen particles *(arrowheads)*, visualized with this technique, are distributed in the cytoplasm. *Gr*, secretory granules; *N*, nucleus.

A. The cell biology of human eosinophils

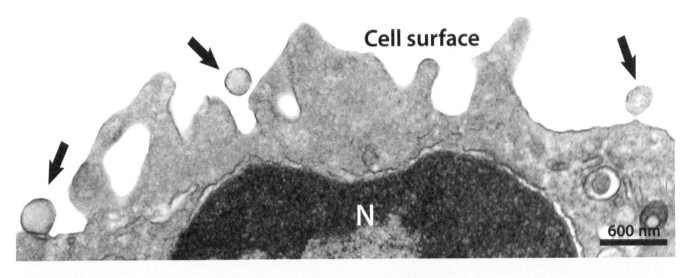

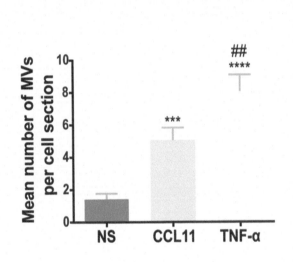

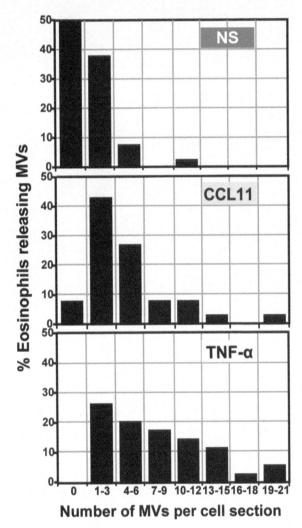

FIG. 4.45 CCL11 and TNF-α induce formation of microvesicles (MVs) by human eosinophils. *Top panel*: High magnification electron micrograph shows MVs *(arrows)* at the surface of a representative eosinophil activated with CCL11. Significant increases of MVs occur after stimulation compared with nonstimulated (NS) cells. Data represent mean ± S.E.M. ***$P < .002$ (CCL11 vs. NS); ****$P < .0001$ (TNF-α vs. NS); ##$P < .02$ (TNF-α vs. CCL11). Counts were derived from three experiments with a total of 516 MVs counted in 110 electron micrographs randomly taken and showing the entire cell profile and nucleus. Cells were stimulated and prepared for TEM as described.[125]

Subcellular localization of immune mediators and other proteins

5.1 Overview

While conventional TEM has been used to unveil ultrastructural features of eosinophils, such as distinct secretory patterns, TEM combined with molecular detection methods has been applied for precise localization of proteins to intracellular compartments.[24,50,231] Visualization of specific proteins at the nanoscale level is achieved with immunogold EM (also referred to as immunonanogold EM), a technique that investigates the protein of interest by performing immunolabeling with a primary antibody against the target molecule followed by a secondary antibody (against primary antibody) conjugated with gold nanoparticles, which have sufficient contrast to be observed by TEM (Fig. 5.1).[50] By performing this technique, it is crucial to consider that the primary antibody is reliable, with recognized specificity, and will provide a reproducible and compartment-selective labeling pattern.[232] Moreover, for each antigen to be immunolabeled, it is necessary to have a control for the primary antibody, which should be replaced by an irrelevant antibody. In this case, negative or negligible labeling after incubation with the secondary antibody validates the immunolabeling process (Fig. 5.2). The omission of the primary and/or secondary antibodies can be additionally used as control strategies.[50]

Immunogold EM has provided substantial insights into the cellular content of biomolecules and eosinophil functional activities, but it involves several technical challenges. The main challenge of immunogold EM is to provide sensitive antigen detection alongside detailed information on the cell structure; however, these are conflicting technical situations. Antigen preservation may be hampered by glutaraldehyde fixation, which is optimal for morphology, and by dehydration and resin embedding, conventional procedures for EM, resulting in weak or negative labeling. Consequently, antigens may not be revealed by post-embedding immunogold EM, that is, by labeling on the surface of a thin section after EM procedures.[233] In addition to impacting antigenicity, labeling is restricted to the thin section surface because antibodies cannot penetrate into the resin.[233,234] Post-embedding immunogold EM has been used since the 1970s and is most convenient for detecting abundant antigens.[235]

An alternative method is pre-embedding immunogold EM, which means antigen labeling before all regular procedures for TEM, thus enabling improved antigen preservation.[50] Pre-embedding immunogold EM has been used with the effective application to different biological systems.[233,236–239] Our group has been working with both post- and pre-embedding immunogold EM approaches for many years. We have localized enzymes involved in inflammatory pathways within leukocytes using post-embedding methods[214,240] but were not able to detect intracellular sites of interleukins in the same cells using this approach.

For optimal molecular imaging of proteins in subcellular compartments and small membranous domains of human eosinophils, we have mostly been applying pre-embedding immunonanogold EM.[23–25,32,50,69,113] Our protocol combines several strategies for ultrastructure and antigen preservation in conjunction with robust blocking of nonspecific binding sites and improved antibody penetration.[50] Visualization is accomplished with electron-dense markers (very small gold particles with 1.4 nm diameter) covalently conjugated with Fab′ fragments, which are only one-third the size of a whole IgG molecule, thus facilitating antibody penetration (Fig. 5.1).[50]

Pre-embedding immunonanogold EM has been used to identify the intracellular localization of cytokines, cationic proteins, immune cell signaling molecules, and tetraspanins in human eosinophils.[24] This approach enabled the first ultrastructural identification of a vesicle-based transport of IL-4 and MBP-1.[19,22] The existence of vesicle-mediated secretion from eosinophil secretory granules has been previously underestimated, likely because of technical issues: inadequate preservation of vesicles and/or inability of antibodies to access them. This technique has also been used

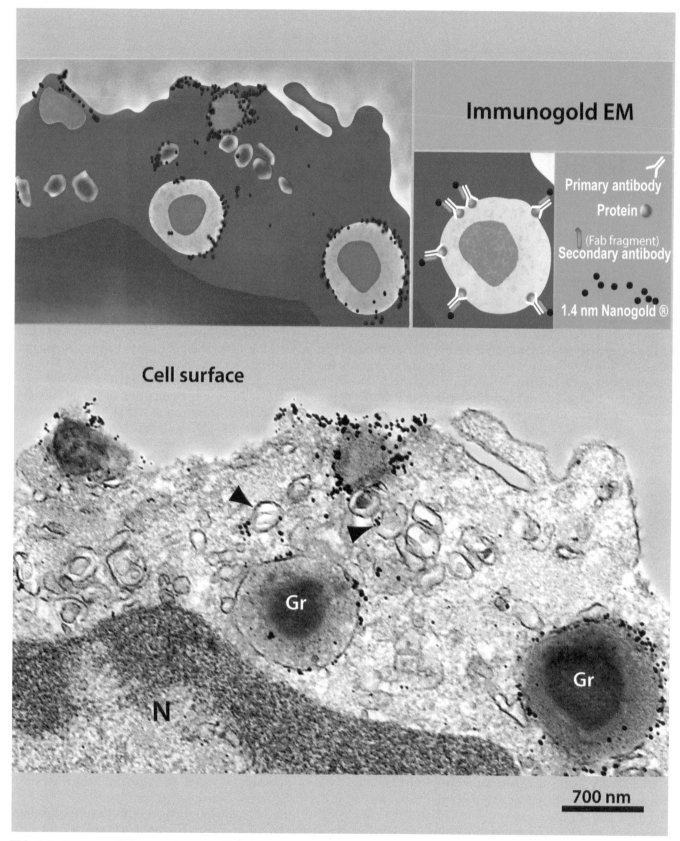

FIG. 5.1   Immunogold electron microscopy (EM) technique. The protein of interest is investigated by immunolabeling with a primary antibody (Ab) against the target molecule followed by a secondary Ab (against primary Ab) conjugated with gold particles, which are electron-dense markers and therefore can be visualized in electron micrographs. In our approach, we use Fab fragments conjugated with 1.4-nm nanogold particles as secondary Abs.[50] The electron micrograph and drawing show an example of immuno-EM application to human resting eosinophils. Cell surface microdomains and cytoplasmic sites—secretory granules (Gr) and EoSVs *(arrowheads)*—are labeled for a protein (CD63). *N*, nucleus.

A. The cell biology of human eosinophils

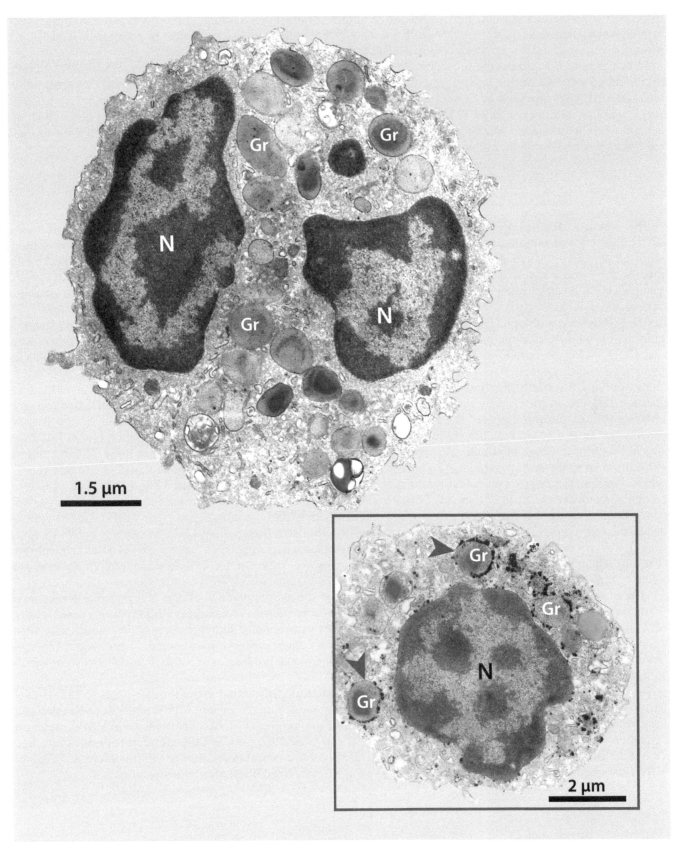

FIG. 5.2   Immunogold electron microscopy technique. A representative human eosinophil *(top panel)* in which the primary antibody (Ab) was replaced by an irrelevant Ab shows negative labeling. In the *bottom panel*, an eosinophil from the same sample incubated with anti-CD63 monoclonal Ab shows dense labeling around secretory granules (Gr, *arrowheads*). Primary Ab was monoclonal mouse anti-human CD63, and the irrelevant Ab was an isotype IgG1 control. Secondary Ab was goat anti-mouse Fab fragment conjugated to 1.4-nm nanogold particles. Eosinophils were isolated from the peripheral blood and immunolabeled as described.[23,50] N, nucleus.

A. The cell biology of human eosinophils

to study the intracellular localization of cytokine receptors, for example, IL-4Rα,[31] and other types of receptors such as Notch signaling molecules[31,32] and SNARE proteins.[113] The intracellular localization of galectin-10 (Gal-10), the protein that forms CLCs, was ascertained in human eosinophils with the same approach.[37]

In this chapter, we address the subcellular localization of different types of immune mediators and other proteins in human eosinophils during resting conditions and immune responses. We also show representative quantitative analyses that have provided additional information on the pattern, frequency, and level of the immunolabeling. Application of immunonanogold EM has helped to appreciate the complex collection of biomolecules, which are constitutively expressed in human eosinophils as well as the secretory pathways involved in the release of these mediators during eosinophil inflammatory responses.

5.2 Major basic protein (MBP)

The release of MBP by human eosinophils is considered a key event in the pathogenesis of allergic airway diseases, such as rhinitis and asthma, and other eosinophil-mediated diseases.[241,242] During the 1970s, Gleich's group demonstrated that MBP is the major constituent of the crystalloid cores of eosinophil-specific granules in both experimental models[28,243,244] and humans.[245] Additional studies from our and other groups have confirmed MBP within specific granules with the application of immunogold EM.[29,154,246] In parallel, early EM investigations of lesional eosinophils in Crohn's disease, eosinophilic gastroenteritis, and HES noted that eosinophils lost their electron-dense cores[145,198,247] and suggested that eosinophil granule core-derived MBP might be released by a mechanism not involving wholesale granule fusion at the plasma membrane. A candidate mechanism for MBP release suggested by these early reports was PMD; that is, eosinophil granule proteins might be mobilized from within intracellular granules into vesicles that traffic to and release extracellularly at the cell surface.[148]

By applying pre-embedding immunonanogold EM during the 2000s, we found, for the first time, consistent vesicular trafficking of MBP-1 within human eosinophils.[20] As noted in Chapter 3, in response to cell activation with multiple stimuli, the granule cores, the only sites in which MBP-1 is localized, undergo remarkable structural disarrangement. Thus, after stimulating eosinophils with agonists such as CCL11 and immunolabeling for MBP-1, granules in progressive stages of mobilization are observed, and MBP-1-immunoreactivity within these altered granules is localized over the entire granule as a consequence of the core disassembling (Fig. 5.3). MBP-1 is also detected within vesicles attached to or surrounding the surface of these emptying granules (Figs. 5.3 and 5.4). Since CCL11 is a recognized inducer of PMD,[20,148] these ultrastructural findings combined with molecular detection of MBP-1 provide compelling evidence that MBP-1 is being transported in vesicles from mobilized granules.[22] MBP-1-loaded vesicles are connected to specific granules and distributed in the cytoplasm (Figs. 5.3 and 5.4). Moreover, EoSVs carrying MBP-1 are frequently seen in the peripheral eosinophil cytoplasm and sometimes fused with the plasma membrane (Fig. 5.5). The presence of MBP-1 within EoSV lumina can also be demonstrated by immunonanogold EM after isolation of these vesicles by subcellular fractionation (Fig. 5.3).

While the association of MBP with the granule cores was consistently demonstrated, the vesicular transport of granule-derived MBP has not been previously detected, possibly because of technical limitations. As noted, earlier MBP localizations by immuno-EM have been evaluated only by post-embedding techniques,[29,154,246] which may affect epitope preservation in membrane microdomains. In addition, these earlier applications of immunogold EM used larger-sized gold particles (around 15 nm) conjugated to polyclonal antibodies[29,154,246] that might also compromise subcellular localization of MBP in vesicular compartments.

Our EM findings demonstrate that secretory vesicles constitute substantial extragranular pools of MBP-1 even within unstimulated human eosinophils.[22] The presence of small storage/transient MBP-1 sites (vesicular pools) in the cytoplasm of resting eosinophils may be relevant for the rapid release of small concentrations of MBP-1 under cell activation without immediate disarrangement of the intricate crystalline cores within eosinophil-specific granules.[22] This is important because it may underlie eosinophil functions as an immunoregulatory cell (reviewed in[248–250]), and MBP-1 may have other functional roles, acting, for example, in the regulation of cytokine responses.[251]

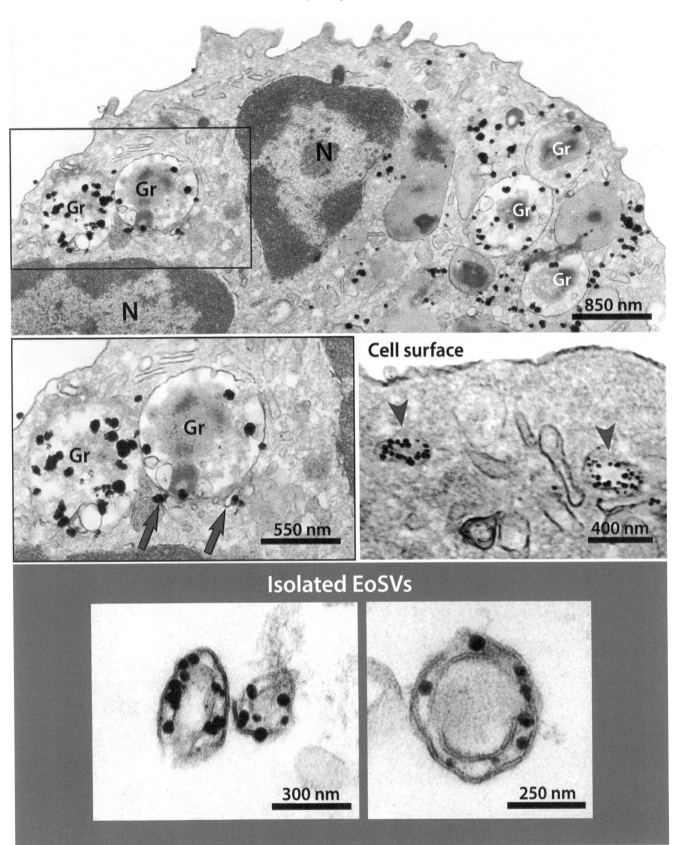

FIG. 5.3 Immunonanogold EM showing MBP-1 in specific granules and vesicles from human eosinophils. In the cytoplasm of a CCL11-activated eosinophil, note structural disarrangement of the granules (Gr) and vesicular trafficking of MBP-1. Vesicles arise from a granule *(arrows, boxed area)* and large carriers (EoSVs, *arrowheads*) are seen underneath the plasma membrane. The *bottom panels* show MBP-1 in the lumen of EoSVs isolated by subcellular fractionation from unstimulated eosinophils. Samples were prepared as described.[19,22] *N*, nucleus.

A. The cell biology of human eosinophils

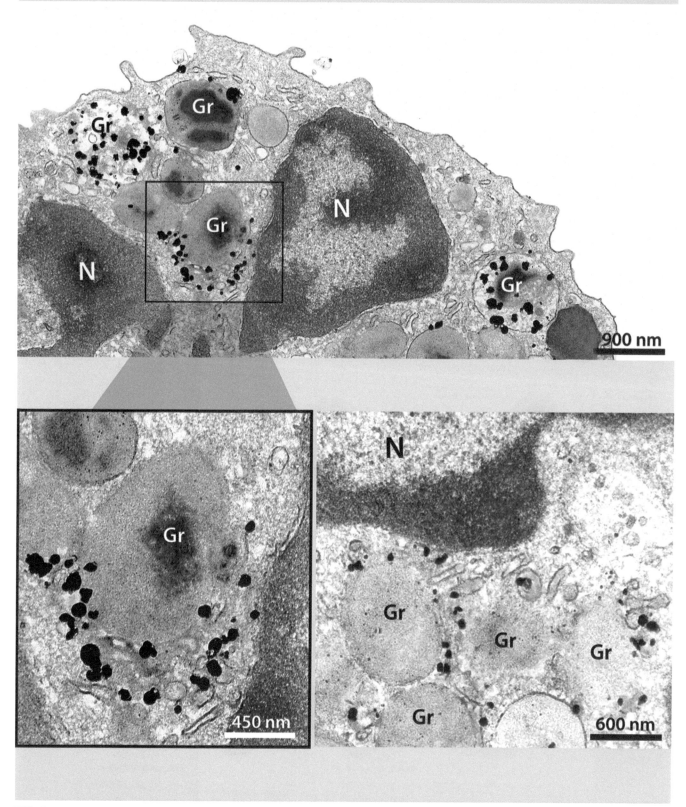

FIG. 5.4 Pools of MBP-1-positive vesicles *(highlighted in pink)* are seen surrounding mobilized secretory granules (Gr) within the cytoplasm of human eosinophils. Cells were isolated from the peripheral blood, stimulated with CCL11 and immunolabeled as described.[22] *N*, nucleus.

A. The cell biology of human eosinophils

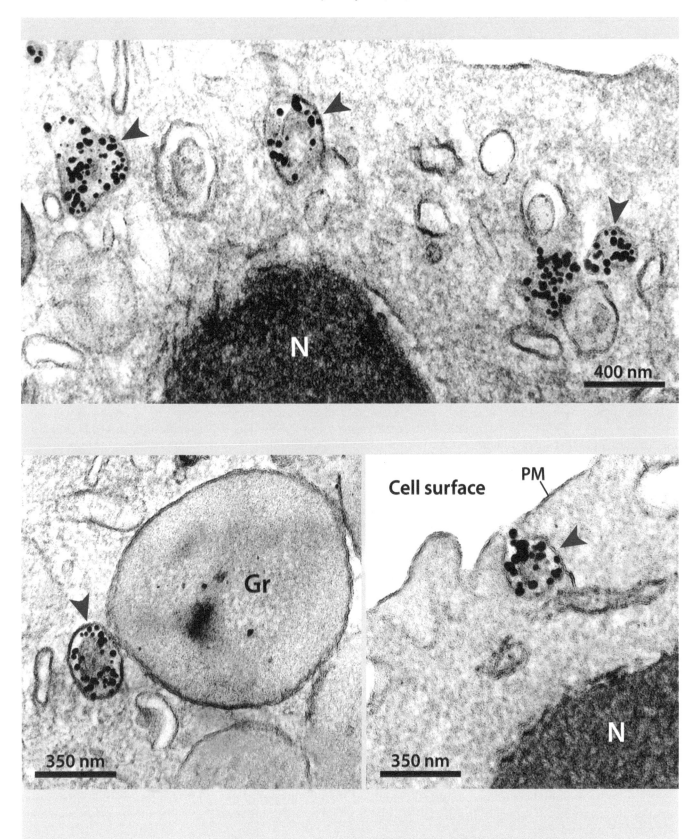

FIG. 5.5 Large tubular carriers (EoSVs) actively transport MBP-1 within human eosinophils. EoSVs carrying MBP-1 *(arrowheads)* are seen in the peripheral cytoplasm *(top panel)*, attached to an enlarged specific granule (Gr) undergoing PMD, and fused with the plasma membrane *(PM)*. Note the trilaminar structure of the delimiting vesicle membrane. Cells were isolated from the peripheral blood, stimulated with CCL11, and processed for immunonanogold-EM.[22] *N*, nucleus.

A. The cell biology of human eosinophils

5.3 Cytokines

Cytokines are a family of glycoproteins with diverse biological activities involved in cell growth, inflammation, immunity, differentiation, and repair. Works conducted from the 1990s have been demonstrating that, in addition to cationic proteins, human eosinophils are equipped with a cache of cytokines with T helper 1 (Th1), T helper 2 (Th2), and regulatory capacities. A detailed list of these molecules and their biological properties in human eosinophils has been summarized in different reviews.[3,7,135,252]

Although expression of cytokines is well documented in both resting and activated human eosinophils, the exact localization of these mediators requires imaging techniques at high resolution, and, as highlighted above, detection of molecular complexes combined with ultrastructural approaches faces several challenges.[50] We provided the first ultrastructural localization of any cytokine in human eosinophils.[219] This initial study localized TNF-α to the matrix of the specific granules from eosinophils obtained from the peripheral blood of patients with HES.[219] Subsequent work from our group detected TNF-α within activated eosinophils from colonic biopsies of patients with Crohn's disease, confirming that specific granules are sources of this cytokine and providing initial evidence that a granule-derived cytokine might be loaded and transported by cytoplasmic vesicles.[220] In activated granules undergoing PMD in these tissue eosinophils, TNF-α was localized within both intragranular and perinuclear vesicles.[220] Other ultrastructural studies from the same decade reported the presence of transforming growth factor-alpha (TGF-α) in secretory granules and vesicular compartments of human eosinophils.[253] However, most ultrastructural studies conducted during the 1990s and beginning of the 2000s have demonstrated cytokines only within specific granules.[219,254-257]

Conclusive evidence that subcellular localization of cytokines within human eosinophils encompasses intricate pathways involving granules and vesicular trafficking came from ultrastructural studies in which better morphological preservation and antigen detection were accomplished.[19,25,31] These investigations have confirmed the presence of cytokines within specific granules and cytoplasmic vesicles in both stimulated and unstimulated eosinophils.[19,25]

As representative cytokines, we studied IL-4 and IFN-γ in blood human eosinophils during different conditions. IL-4 labeling was demonstrated within the cores, matrices, and limiting membranes of specific granules as well as in association with vesicular compartments, both small vesicles and EoSVs (Fig. 5.6). Quantitative analysis showed a higher number of IL-4-loaded EoSVs in CCL11-stimulated than in unstimulated eosinophils,[19] and immunoreactivity for IL-4 was found preferentially at vesicle membranes and not within their internal content as observed for MBP-1 (Fig. 5.7).

A single probe (secondary antibody) conjugated with both a fluorochrome and a gold particle is useful to perform comparative studies using light microscopy and TEM. When this kind of probe was applied to detect pools of IL-4 being mobilized and released from CCL11-activated eosinophils, IL-4 was seen at cell surfaces as intense speckled fluorescence while TEM revealed the presence of this cytokine at vesicles distributed in the cytoplasm and underneath the plasma membrane (Fig. 5.8).[19] Overall, these findings uncover a route from granules to the cell surface for IL-4 mobilization and release via vesicular compartments, as previously indicated by studies in which a system to capture and detect extracellular cytokines was applied at light microscopy level.[258]

Human eosinophils contain high levels of Th1 cytokines including IFN-γ,[193] and secretory granules are remarkably labeled for this cytokine at the ultrastructural level (Fig. 5.9).[25] By enumerating a total of 4095 secretory granules after immunolabeling for IFN-γ (1260 from unstimulated, 1499 from CCL11-stimulated, and 1336 from TNF-α-stimulated eosinophils), we detected that most of them (more than 70%) show positivity for IFN-γ in both resting and activated cells with labeling in the matrices and at the granule limiting membranes (Fig. 5.9).[25] These data provide direct evidence that IFN-γ is constitutively stored in human eosinophils and that a substantial pool of this cytokine is compartmentalized within secretory granules.[25]

In addition to secretory granules, immunogold EM also showed that IFN-γ localizes in vesicles distributed in the cytoplasm and markedly around or in contact with secretory granules (Fig. 5.10). As observed for IL-4, immunolabeling for IFN-γ is associated with the vesicle membranes (Fig. 5.10). Quantitative analyses demonstrated that, in response to cell activation with CCL11 or TNF-α, the numbers of IFN-γ-labeled EoSVs increase more than 200% in the cytoplasm compared with unstimulated eosinophils (Fig. 5.10).[25] Moreover, EoSVs carrying IFN-γ are found in significantly higher numbers in the peripheral cytoplasm than in the rest of the cell.[25] This differential distribution denotes the occurrence of robust traffic of this cytokine from secretory granules to the cell periphery for extracellular release in stimulated cells.[25]

IFN-γ was also localized, for the first time, as cargo in EVs budding from the cell surface of activated eosinophils (Fig. 5.11). As noted (Chapter 4), increased production of EVs is a feature of activated human eosinophils, and released cytokine-containing EVs may be acting as mediators of immune responses, thus influencing target cells.[125]

Conventional TEM

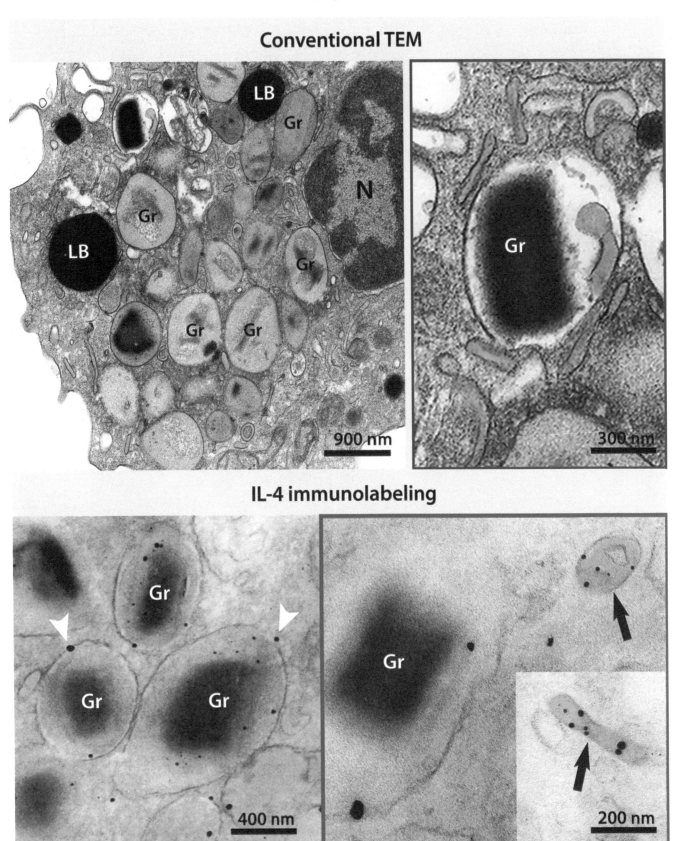

IL-4 immunolabeling

FIG. 5.6 Ultrastructure of an activated human eosinophil and immunolocalization of IL-4. Conventional TEM shows enlarged secretory granules (Gr) with content losses. EoSVs and intragranular membranes were colored in *pink*. IL-4 is localized to granule *(arrowheads)* and vesicle *(arrows)* membranes. Cells were isolated, stimulated with CCL11, and immunolabeled.[19] *LB*, lipid body; *N*, nucleus.

A. The cell biology of human eosinophils

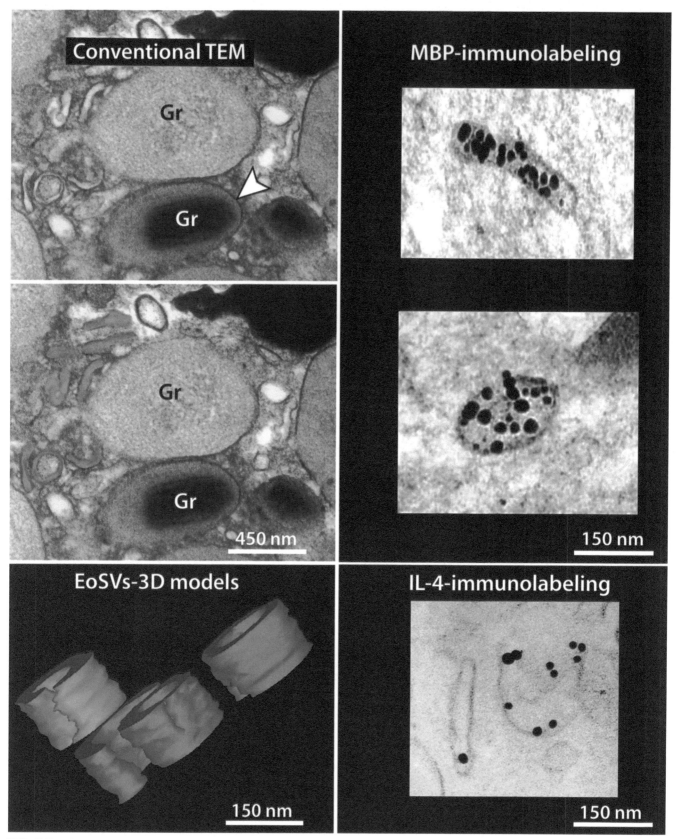

FIG. 5.7   Ultrastructural views of EoSVs by conventional, 3D, and immunonanogold TEM. EoSVs *(colored in pink)* accumulate around an emptying, enlarged secretory granule (Gr). An intact resting granule with a typical electron-dense core is also seen *(arrowhead)*. The morphology of EoSVs with a higher surface-to-volume ratio is shown by 3D models obtained from electron tomographic analyses. Note that while MBP-1 is mostly localized within the vesicle lumen, IL-4 is associated with vesicle membranes.

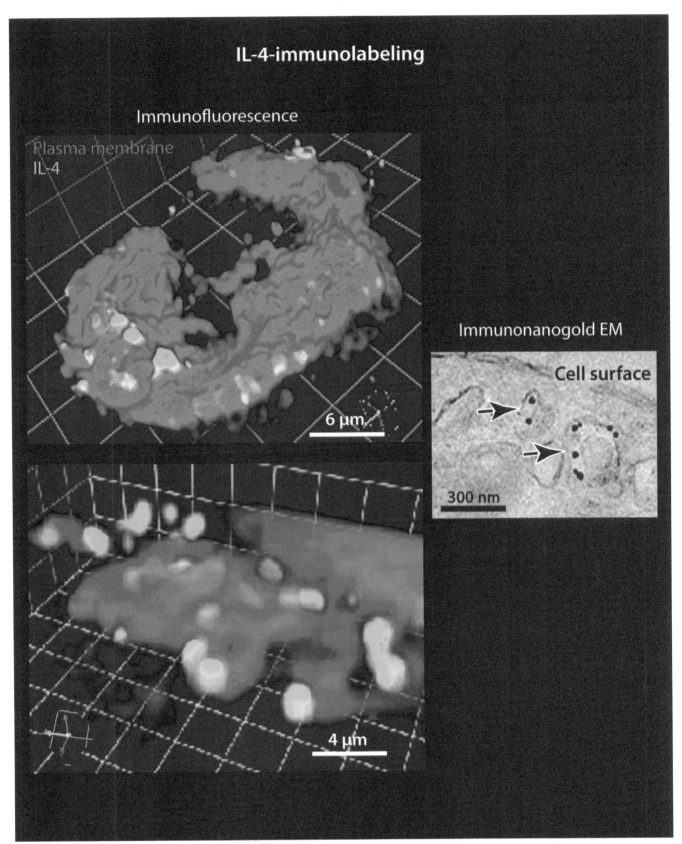

IL-4-immunolabeling

Immunofluorescence

Plasma membrane
IL-4

6 µm

4 µm

Immunonanogold EM

Cell surface

300 nm

FIG. 5.8   Vesicular trafficking of IL-4 in human eosinophils seen by light microscopy and TEM *(immunonanogold EM)*. The same fluoronanogold probe demonstrates IL-4 as focal fluorescent green spots at the plasma membrane *(stained in red)* and in association with EoSVs *(arrows)* underneath the plasma membrane. Cells were stimulated and immunolabeled as described.[19]

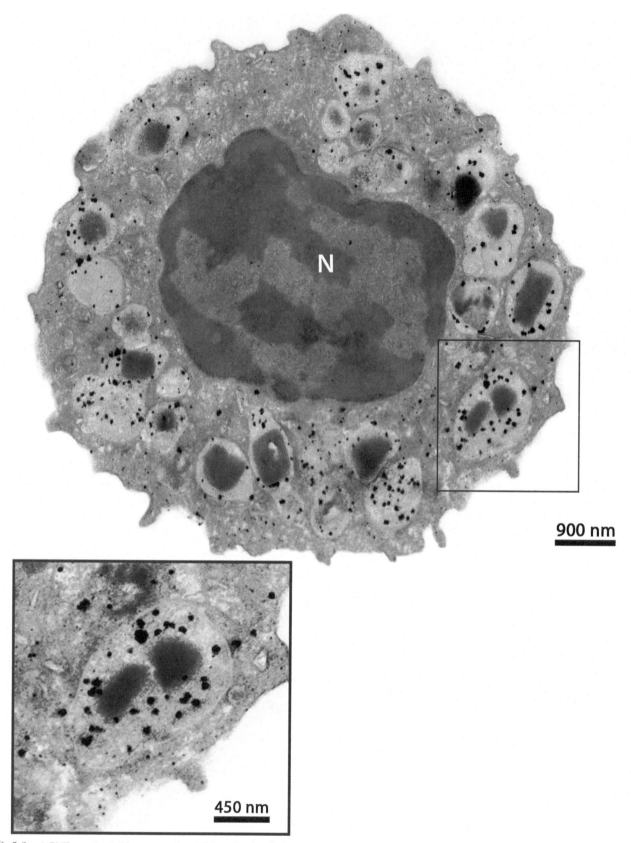

FIG. 5.9   A TNF-α-activated human eosinophil *(pseudocolored)* showing robust IFN-γ immunolabeling within secretory granules *(yellow)*. Note in higher magnification the dense labeling in the granule matrix. Eosinophils were isolated from the peripheral blood, stimulated with TNF-α, and prepared for pre-embedding immunonanogold EM.[25] *N*, nucleus.

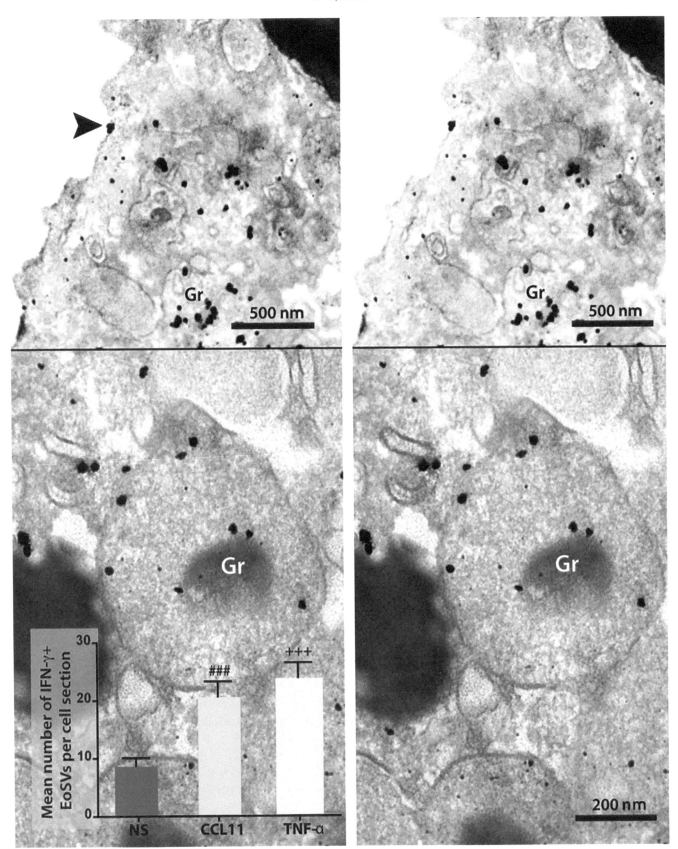

FIG. 5.10 Vesicular trafficking of IFN-γ in the cytoplasm of activated human eosinophils. IFN-γ-positive EoSVs *(highlighted in pink)* are seen in TNF-α- *(top panels)* or CCL11- *(bottom panels)* activated cells in contact with emptying secretory granules (Gr). Labeling is associated with vesicle membranes. *Arrowhead* indicates extracellular release of IFN-γ. The number of IFN-γ-positive EoSVs significantly increases in response to cell activation compared with nonstimulated (NS) cells. ###$P = .0004$; +++ $P = .0003$ vs. NS. Data represent mean ± S.E.M. A total of 1357 EoSVs were counted. Cells were prepared for immunogold EM.[25]

A. The cell biology of human eosinophils

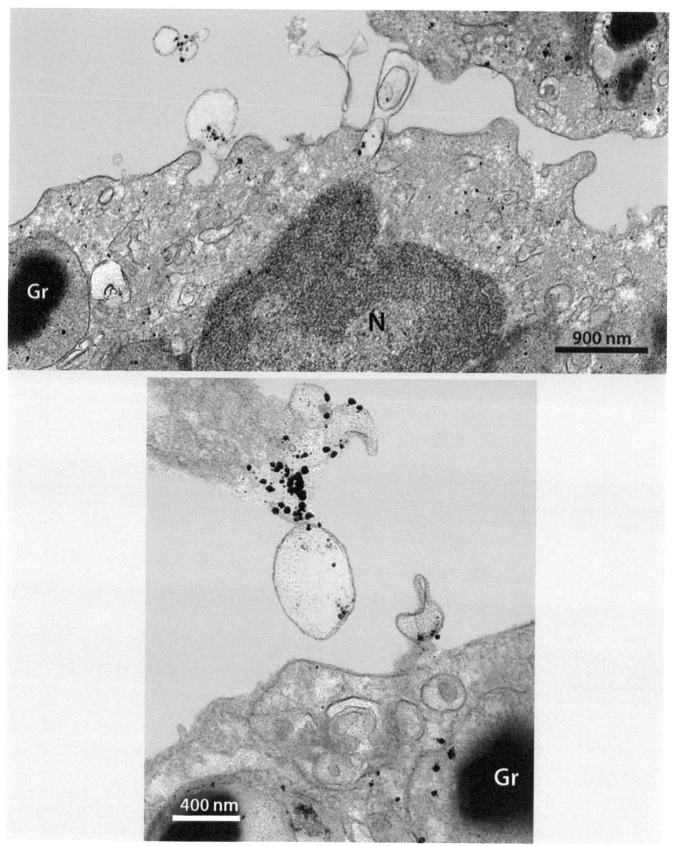

FIG. 5.11 Immunodetection of IFN-γ in extracellular vesicles (EVs) of human eosinophils. IFN-γ-positive EVs are seen as microvesicles *(colored in blue)* budding from the plasma membrane and released at the cell surface. Cells were stimulated with CCL11 and prepared for pre-embedding immunonanogold EM.[25] *Gr*, specific granules; *N*, nucleus.

LBs represent additional intracellular sites for cytokines in leukocytes.[205,259] As detailed in Chapter 2, in addition to their distinct molecular machinery, leukocyte LBs are structurally complex organelles in which protein compartmentalization and management occur during cellular mechanisms of diseases. The first cytokine found within LBs was TNF-α, which was detected in vivo by applying ultrastructural immunolabeling to colonic Crohn's disease biopsies. This cytokine was localized in infiltrating neutrophils, macrophages, and eosinophils.[220] In peripheral blood human eosinophils activated in vitro with TNF-α but not in unstimulated cells, we localized IFN-γ in cytoplasmic LBs (Fig. 5.12), thus providing additional evidence that LBs might act as transient sites for cytokines during inflammatory responses. However, it is not clear yet if cytokines are directly synthesized at LBs or transported from other intracellular sites to LBs.

5.4 Receptors

Eosinophils are cells with multifaceted roles associated with the secretion of cytokines and other immune mediators stored preformed within their specific granules.[3,82] We have documented robust expression in eosinophil granules of the receptors for several preformed cytokines, with IL-4 receptor subunit α (IL-4Rα) being particularly studied.[31]

As shown above, we localized IL-4 within secretory granules as well as to vesicles budding from granules and trafficking through the cytoplasm (Fig. 5.6). In contrast to MBP-1 staining within the lumen of eosinophil secretory vesicles, IL-4 labeling on vesicles was preferentially associated with vesicle limiting membranes (Fig. 5.7). Because IL-4 lacks a membrane-spanning or insertion region, IL-4 association with vesicle and granule membranes suggested the presence of an IL-4-docking molecule. Using several approaches, including immunonanogold EM (Fig. 5.13), pools of IL-4Rα, that is, the binding component of functional IL-4R complexes, were detected intracellularly. When eosinophils were stimulated with CCL11 to secrete IL-4, the amount of intracellular vesicle-associated IL-4Rα increased rapidly, and IL-4 was translocated from granule stores into cytoplasmic secretory vesicles.[3,31] Clusters of IL-4Rα were localized within granules, at granule limiting membranes, and frequently at sites where vesicles appeared to be budding or interacting (Fig. 5.13). Moreover, clusters of IL-4Rα immunoreactivity were prominent beneath the plasma membrane (Fig. 5.13).

The recognition of pools of ligand-binding cytokine receptor chains in eosinophil secretory granules uncovers mechanisms for selective chaperoned release of cytokines.[7] Large tubular vesicles such as EoSVs act as suitable intracellular carriers to accommodate membrane-bound proteins because of their curved and elongated morphology with higher surface-to-volume ratios (Figs. 5.7 and 5.13). In addition to IL-4Rα, secretory granules isolated from human eosinophils also express domains of IFN-γ receptors α chains on their membranes,[16,82] and it is probable that IFN-γ-loaded EoSVs arising from granules (Fig. 5.10) are carrying these receptors. However, the ultrastructural demonstration of IFN-γ receptors on membranes of EoSVs remains to be addressed in future studies.

Other receptors expressed in human eosinophils include Notch receptors. Notch signaling is an evolutionarily conserved pathway that regulates cellular processes, including migration.[260] Immuno-EM revealed a robust signal of the Notch1 receptor localized to secretory granules of blood eosinophils.[32] This study also demonstrated, for the first time, that mature blood eosinophils express Notch ligands and are capable of Notch-mediated autocrine signaling.[32]

SNARE receptors mediate membrane fusion during intracellular trafficking underlying innate and adaptive immune responses by different cells. SNAREs are generally small (14–40 kDa), coiled-coil-forming proteins that are anchored to the membrane via a C-terminal anchor. They were originally classified as v- (vesicle-associated) or t- (target-membrane) SNAREs, on the basis of their locations and functional roles in a typical trafficking step. However, this orientation is not always maintained, and an alternative structure-based terminology has been used, wherein the family is divided into R-SNAREs and Q-SNAREs, on the basis of whether the central functional residue in their SNARE motif is arginine (R) or glutamine (Q). Q-SNAREs are then further classified into Qa, Qb, Qc, and Qb,c subtypes based on where their SNARE domain(s) would sit in an assembled trans-SNARE complex (reviewed in[261,262]).

Human eosinophils express both R- and Q-SNAREs,[263–267] but only the R-SNARE vesicle-associated membrane protein-2 (VAMP2)[265] and the Qa SNARE syntaxin17 (STX17)[113] were previously studied by immuno-EM in these cells. While VAMP2 localizes predominantly on vesicles,[265] STX17 is localized both in secretory granules (limiting membranes and matrices) and EoSVs (Fig. 5.14), indicating that this SNARE may be functionally implicated in membrane trafficking from secretory granules to the plasma membrane.[113] Our quantitative analyses of 1088 secretory granules in a total of 53 electron micrographs from resting and CCL11 and TNF-α-activated human eosinophils showed that ~ 80% of the granules were positive for STX17 in all groups.[113] In other cells, STX17 has been deemed necessary for constitutive secretion[110] and to function as a receptor at the ER membrane that mediates trafficking between the ER and post-ER compartments,[111,112] but its absence or negligible occurrence in eosinophil ER and Golgi (Fig. 5.15) while consistently found in secretory granules indicates a more complex role for STX17 in these granulocytes.[113]

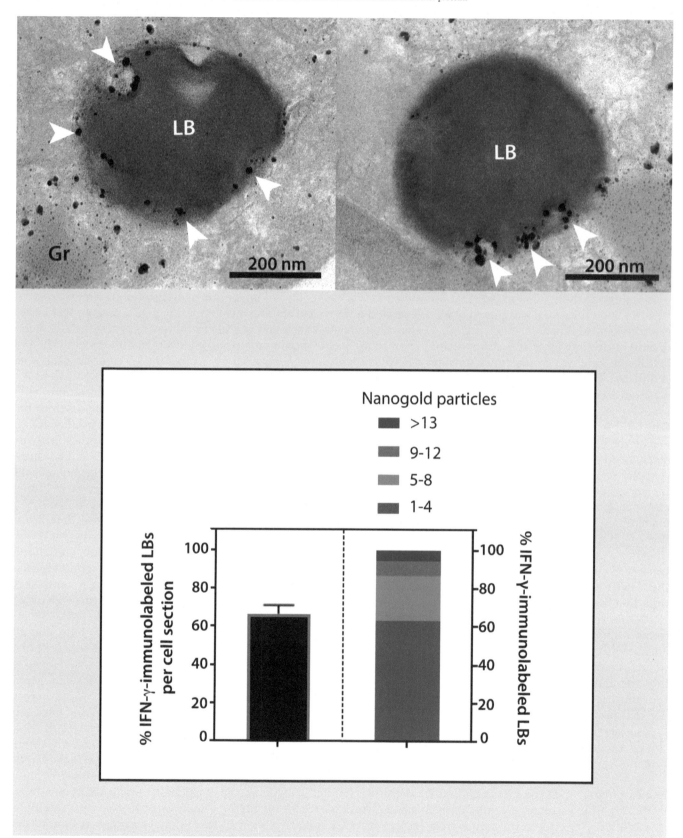

FIG. 5.12   Lipid bodies (LBs) are sites for IFN-γ in activated eosinophils. *Arrowheads* indicate immunolabeling *(immunonanogold EM)* at newly formed LBs within eosinophils stimulated with TNF-α. Quantitative imaging analysis showed that more than 60% of LBs per cell section were immunolabeled. Cells were stimulated as described.[94] *Gr*, secretory granule. *Source: Carmo LAS, Weller PF, Melo RCN. TNF-alpha-induced lipid droplets interact with secretory granules and are sites for IFN-gamma in human eosinophils. Front Cell Dev Biol. 2021 (in press). Licensed under the Creative Commons Attribution 4.0 International (CC BY 4.0). Link to the license: https://creativecommons.org/licenses/by/4.0/.*

A. The cell biology of human eosinophils

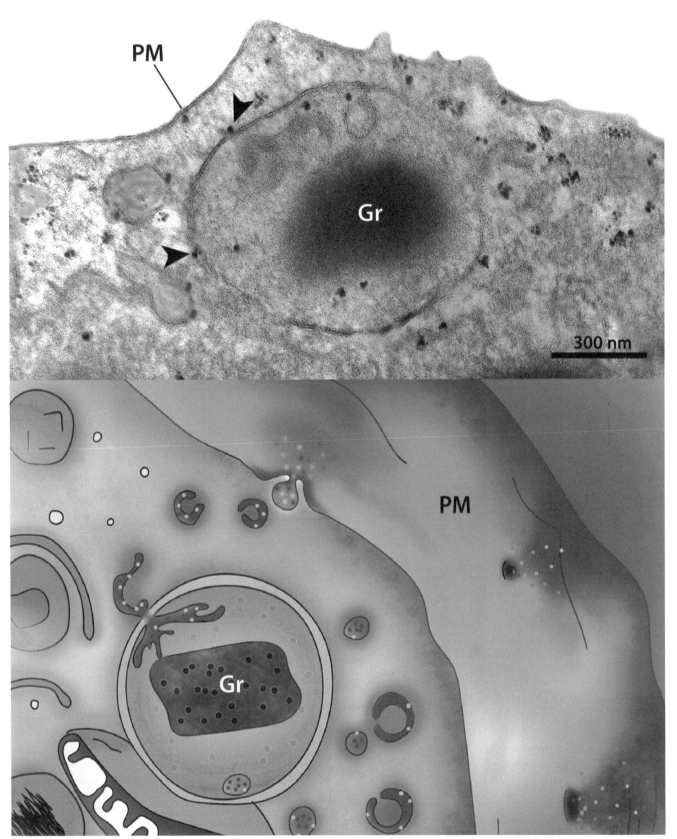

FIG. 5.13 EoSVs *(colored in pink)* being formed from a mobilized specific granule (Gr) in a CCL11-stimulated human eosinophil. *Top panel* is a transmission electron micrograph showing EoSVs immunolabeled for IL-4Rα, also detected within and on granule outer membrane *(arrowheads)* and as clusters beneath the plasma membrane (PM). A drawing *(bottom panel)* illustrates the process of EoSV formation. As a model,[31] EoSV-associated IL-4R sequesters and chaperones its cognate cytokine from intragranular stores to the PM for extracellular release.

A. The cell biology of human eosinophils

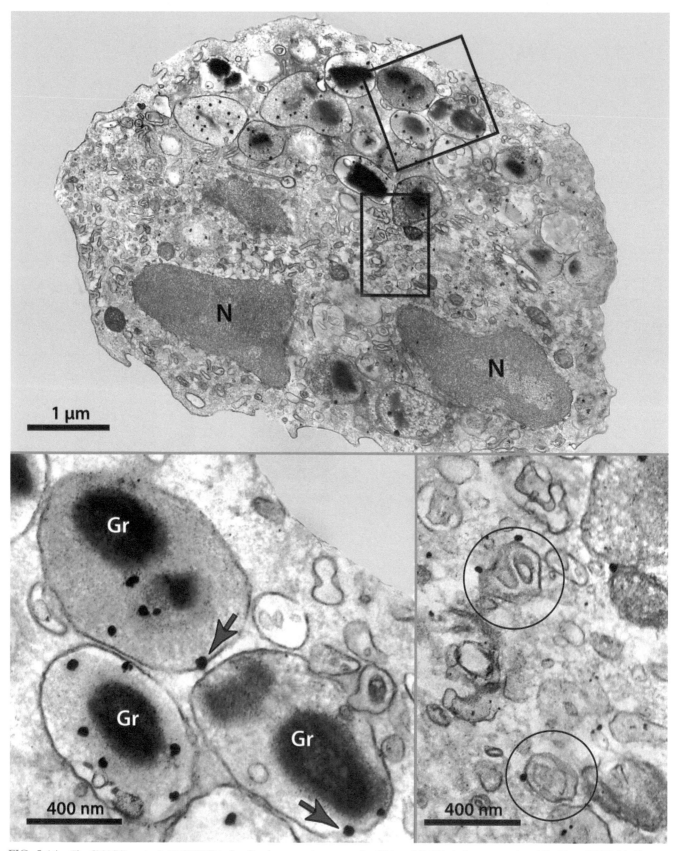

FIG. 5.14   The SNARE syntaxin17 (STX17) is localized on secretory granules (Gr) and EoSVs in human eosinophils. Note STX17 labeling on granule outer membranes *(arrows)* and matrices and EoSVs *(colored in pink, circles)*. Unstimulated peripheral blood eosinophils were processed for immunonanogold EM.[113] N, nucleus.

A. The cell biology of human eosinophils

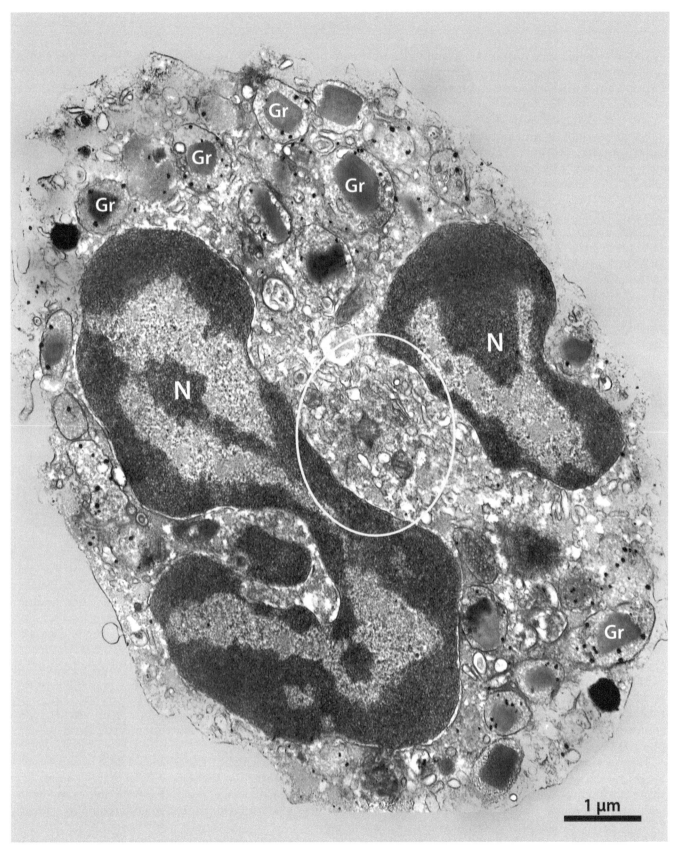

FIG. 5.15 STX17 is not associated with Golgi compartments of human eosinophils. While secretory granules (Gr) are labeled for STX17, the Golgi region *(circle)* shows the absence of labeling for this SNARE. Unstimulated peripheral blood eosinophils were processed for immunonano-gold EM.[113] *N*, nucleus.

A. The cell biology of human eosinophils

5.5 Tetraspanins

The tetraspanins are a large superfamily of membrane proteins widely distributed in eukaryotic cells and involved in the regulation of varied cell processes such as cell adhesion, migration, signaling, and proliferation.[268,269] Tetraspanins have four transmembrane spans linked extracellularly by one large and one small extracellular loop (termed LEL and SEL, respectively).[268,269]

The tetraspanins CD9 and CD63 are extensively expressed in human eosinophils, yet the functions of these molecules in eosinophils remain unclear. We demonstrated that CD9 is abundant on the surface of human eosinophil leukocytes (Figs. 5.16 and 5.17), presenting the first EM data of the ultrastructural immunolocalization of CD9 in human eosinophils, which represents a novel insight into the organization of the antigen presentation complex of these cells.[30] CD9 was notably detected at the cell surface of both resting (Fig. 5.16) and activated (Fig. 5.17) eosinophils, with clusters being observed at the plasma membrane. Moreover, a CD9 intracellular pool was detected in association with EoSVs and secretory granules (Figs. 5.16 and 5.17).[30]

CD63 is associated with secretory granules and used as a surface marker for degranulation in several types of leukocytes, including eosinophils.[83,84,141,270,271] CD63 also constitutes a well-established component of endosomal and lysosomal membranes.[83]

The intracellular distribution and trafficking of CD63 within human eosinophils were investigated in detail by immuno-EM in resting and activated human eosinophils.[23] We confirmed at high resolution that CD63 is prominently linked to resting secretory granules and primarily localizes on the limiting membranes of these organelles, but labeling can also be detected in the granule matrices (Fig. 5.18). EoSVs (Fig. 5.1) and endosomes (Fig. 5.18) are also positive for CD63 in the cytoplasm of these cells.

In stimulated cells, CD63 is strongly labeled on eosinophil granules actively participating in degranulation events, both PMD and compound exocytosis. Thus, CCL11, which is known to induce PMD,[20] leads to CD63 accumulation within granules with different degrees of emptying and uniformly distributed in the cytoplasm (Figs. 5.19 and 5.20).[23] Labeling on the cell surface can be observed (Fig. 5.21), but robust pools of CD63 remain intracellularly, even after eosinophil activation.[23]

Stimulation of eosinophils with TNF-α induces an immunolabeling pattern for CD63 markedly different from that for CCL11 (Fig. 5.22). As discussed in Chapter 4, activation with TNF-α triggers compound exocytosis with secretory granules appearing fused mostly at the cell periphery neighboring the plasma membrane.[23] These fused granules are heavily labeled for CD63 in TNF-α-activated eosinophils, whereas granules localized in the inner cytoplasm are weakly positive or negative for CD63 (Figs. 5.22 and 5.23).[23] Large clusters of CD63 immunoreactivity are associated with channels or enlarged chambers formed by granule-granule fusions at the cell periphery (Figs. 5.22 and 5.24).[23]

We performed a comprehensive immunogold-EM quantitative study to understand the intracellular trafficking of CD63 within resting and CCL11- and TNF-α-activated human eosinophils (Fig. 5.25).[23] These analyses revealed that (i) most secretory granules in both resting and activated cells are positive for CD63. At least 60% of the granules per cell section were labeled for this tetraspanin in all groups (Fig. 5.25); (ii) the number of CD63-positive granules significantly increased in response to cell activation (Fig. 5.25); (iii) CD63 concentrates within granules undergoing active processes of secretion—both PMD and compound exocytosis (Fig. 5.22). By measuring the CD63-immunolabeled area in each granule, we found that this area remarkably increased in granules undergoing losses of their contents after activation compared with unstimulated cells in which the labeling of intact granules was mostly observed at the granule limiting membrane (Fig. 5.25). In stimulated cells, individual granules exhibited an area up to 300 nm² labeled for CD63, which corresponds to more than 50% of the total granule area (Fig. 5.25); and (iv) a shift in the distribution pattern of CD63-positive secretory granules occurs when eosinophils are undergoing compound exocytosis. In this case, 75% of the granules with pools of CD63 localized in the peripheral cytoplasm (in a band of 1-μm wide from the plasma membrane) (Figs. 5.22, 5.23, and 5.24) while most secretory granules in the inner cytoplasm showed an absence or very weak labeling for CD63 (Figs. 5.22 and 5.23).

The association of CD63 with vesicles was also investigated at high resolution. CD63-labeled EoSVs are present in the cytoplasm of both unstimulated and stimulated eosinophils. Quantitative analyses showed that the numbers of CD63-labeled EoSVs significantly increased in eosinophils stimulated with CCL11 or TNF-α compared with unstimulated cells.[23] Many of these CD63-positive tubular carriers were seen contacting granules undergoing secretion by PMD as well as by compound exocytosis (Fig. 5.26).

Altogether, immuno-EM findings demonstrated that CD63 is remarkably linked to different secretory processes in human eosinophils. Moreover, our data indicate that EoSVs are acting in the translocation of CD63 from and to intracellular compartments, particularly secretory granules, in response to cell activation. Thus, this tetraspanin traffics in the eosinophil cytoplasm, likely as a secretion facilitator/regulator molecule chaperoning both PMD and compound exocytosis.[23]

Because the tetraspanins CD63 and CD9, in general, have been proposed as "universal" markers for EVs,[186] we also investigated their presence on EVs secreted by human eosinophils (Fig. 5.27).[125] Our immunonanogold EM

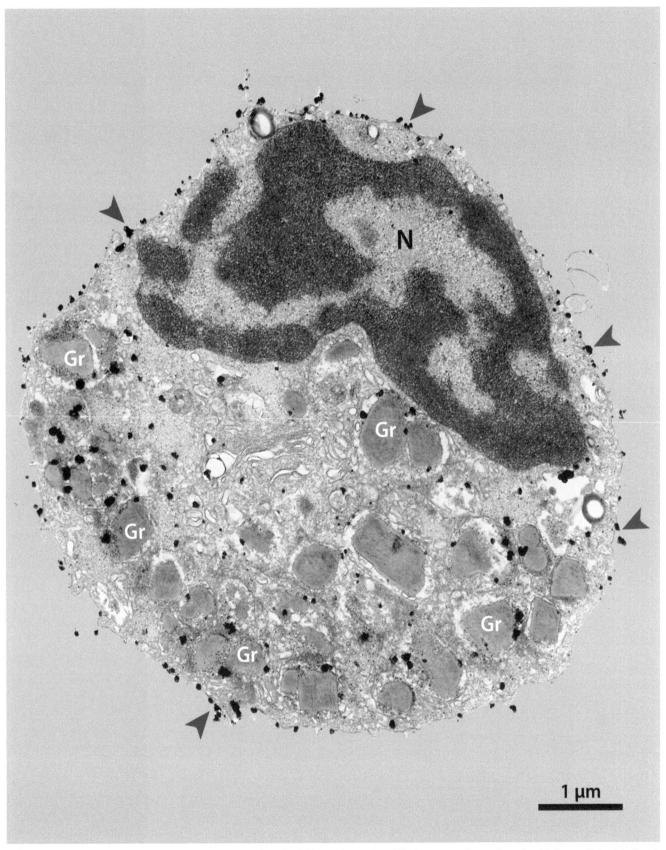

FIG. 5.16   Immunonanogold EM shows a substantial pool of CD9 at the eosinophil plasma membrane *(arrowheads)*. Intracellular labeling is also detected in association with secretory granules (Gr). Unstimulated human eosinophils isolated from the peripheral blood were processed as described.[30] *N*, nucleus.

A. The cell biology of human eosinophils

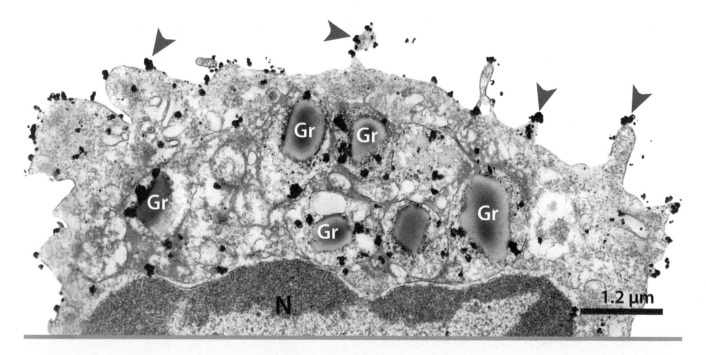

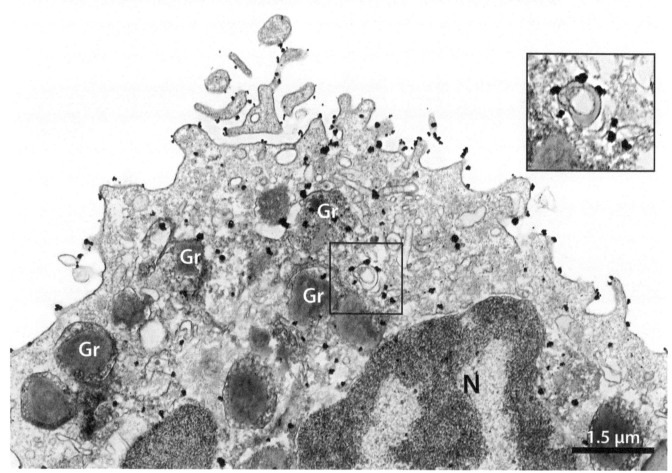

FIG. 5.17   CD9 forms clusters *(arrowheads)* at the surface of human eosinophils. Secretory granules (Gr) and EoSV *(colored in pink, boxed area)* are also labeled for CD9. CCL11-stimulated peripheral blood eosinophils were processed for immunonanogold EM.[30] Pseudocolored eosinophil *(top panel). N*, nucleus.

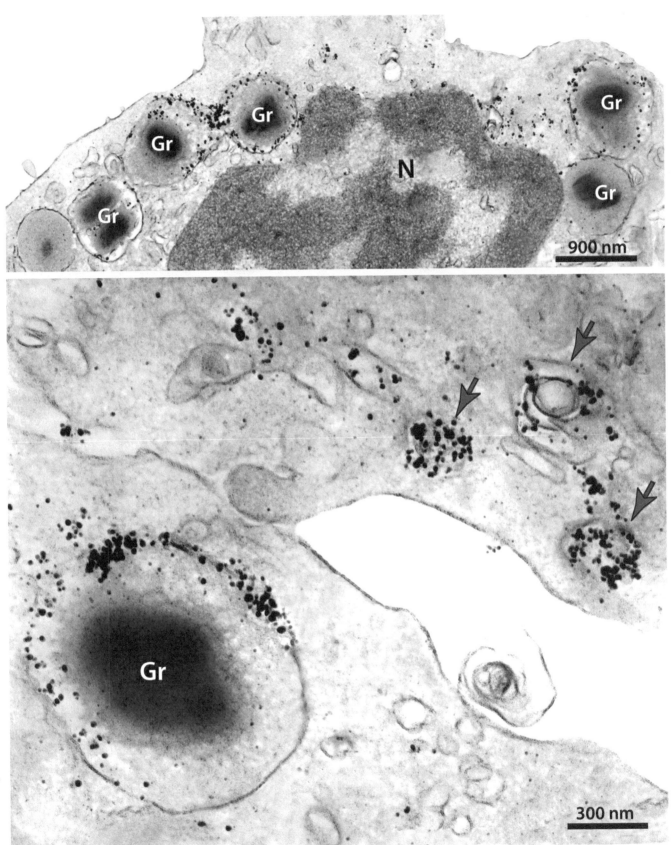

FIG. 5.18 Ultrastructural immunolabeling of CD63 in peripheral blood human eosinophils. CD63 is mostly seen on the limiting membranes of secretory granules (Gr) and scattered in the granule matrices. Note CD63-positive endosomal compartments *(arrows)*. Cells processed for immunonanogold EM.[23] *N*, nucleus.

A. The cell biology of human eosinophils

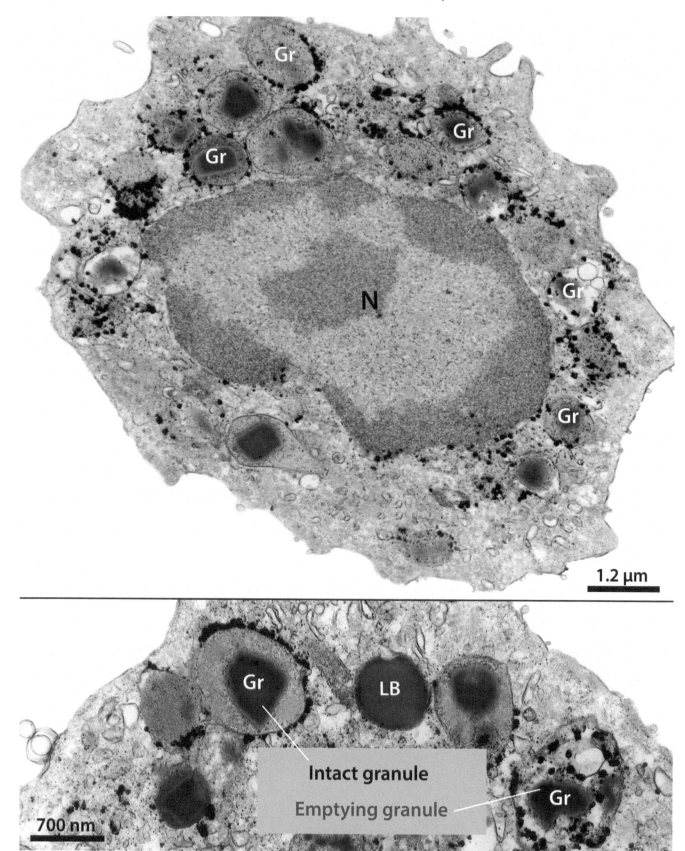

FIG. 5.19 CD63 immunolabeling in a CCL11-stimulated human eosinophil. CD63 associates with the limiting membranes of intact granules and accumulates within emptying granules. Peripheral blood eosinophils were stimulated and processed for immunonanogold EM.[23] *Gr*, specific granules; *LB*, lipid body; *N*, nucleus.

A. The cell biology of human eosinophils

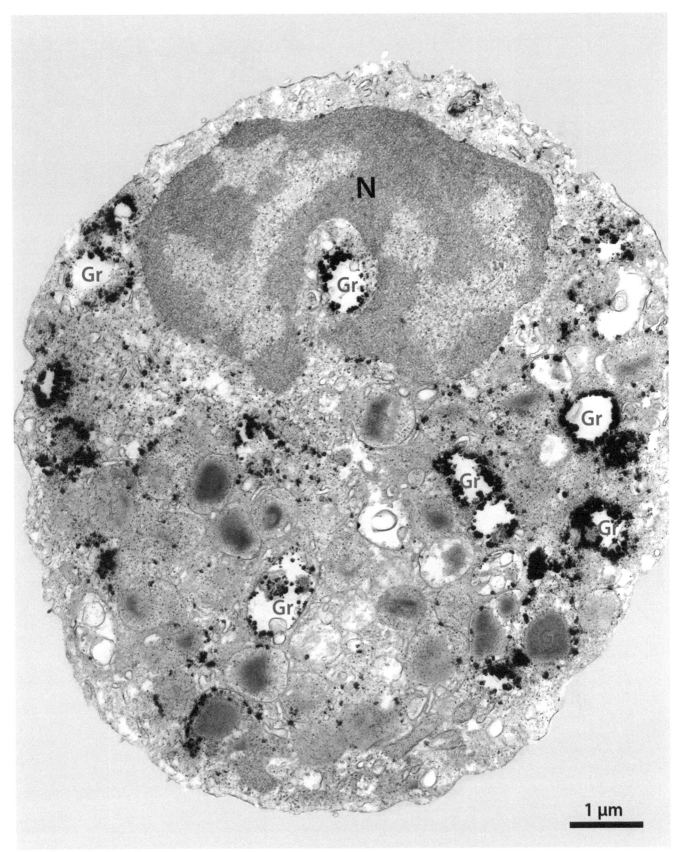

FIG. 5.20 CD63 is densely labeled in eosinophil granules undergoing PMD. CD63-positive secretory granules (Gr) show content losses. Cells were isolated from the peripheral blood, stimulated with CCL11, a well-known inducer of PMD, and processed for immunonanogold EM.[23] N, nucleus.

A. The cell biology of human eosinophils

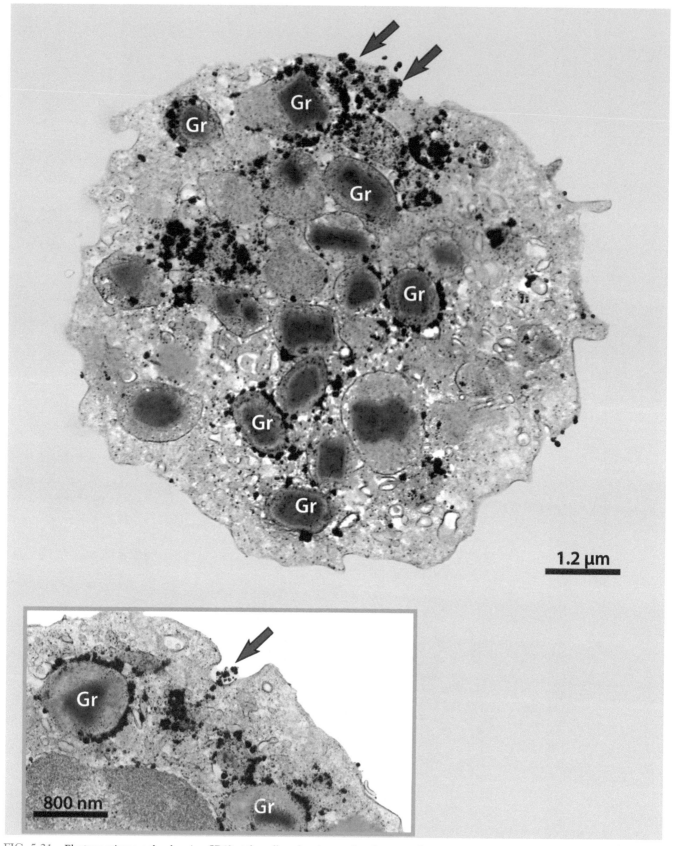

FIG. 5.21   Electron micrographs showing CD63 at the cell surface (arrows) and on cytoplasmic secretory granules (Gr) of human eosinophils. Note that the nucleus is not seen in the section plane (*top panel*). Eosinophils were isolated from the peripheral blood, stimulated with CCL11, and processed for immunonanogold EM.[23]

A.   The cell biology of human eosinophils

PMD

Compound exocytosis

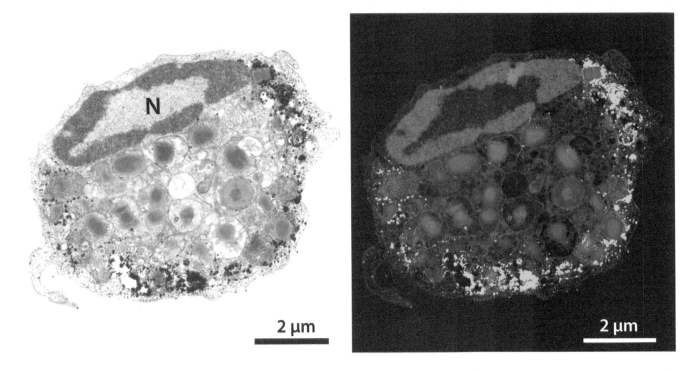

FIG. 5.22 Specific granules undergoing content release through PMD or compound exocytosis are intensely labeled for CD63. CD63-positive granules are distributed in the entire cytoplasm in eosinophils showing PMD while in eosinophils undergoing compound exocytosis, they appear fused and concentrated in the cell periphery. Identical panels highlight granules in *blue*. Peripheral blood eosinophils, stimulated with CCL11 or TNF-α to induce PMD or compound exocytosis, respectively, were processed for immunonanogold EM.[23] *N*, nucleus.

A. The cell biology of human eosinophils

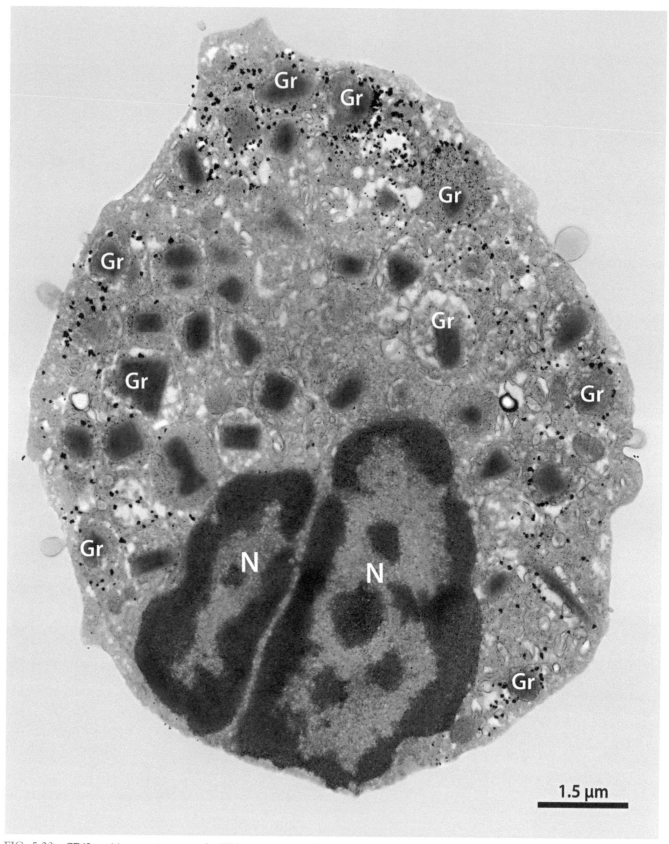

FIG. 5.23 CD63-positive secretory granules (Gr) concentrate in the peripheral cytoplasm of eosinophils undergoing compound exocytosis. Granules in deeper cytoplasmic areas are CD63-negative. TNF-α-stimulated peripheral blood eosinophils were processed for immunonanogold EM.[23] N, nucleus.

A. The cell biology of human eosinophils

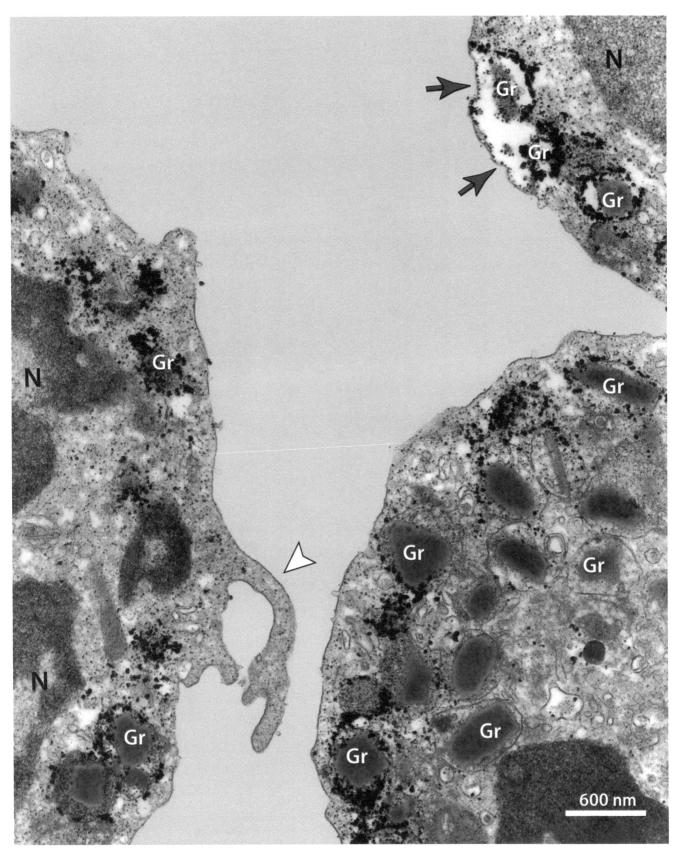

FIG. 5.24 CD63 in human eosinophils undergoing compound exocytosis. CD63-positive granules are seen in contact or fused *(arrows)* with the plasma membrane. Note a filopodium *(arrowhead)*. TNF-α-stimulated peripheral blood eosinophils were processed for immunonanogold EM.[23] *Gr*, secretory granules; *N*, nucleus.

A. The cell biology of human eosinophils

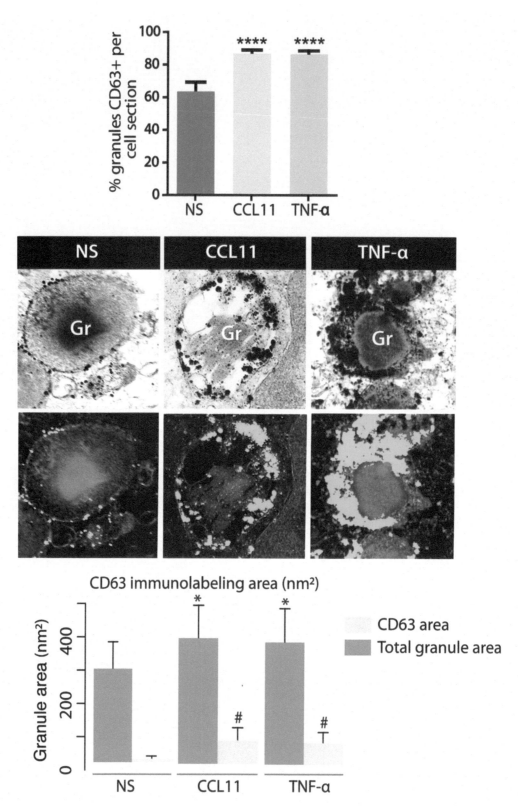

FIG. 5.25   Quantitative EM analyses of CD63 + secretory granules in activated human eosinophils. The number of CD63 + *(top panel)* significantly increases in response to CCL11 or TNF-α stimulation. ****$P < .0001$ vs. nonstimulated (NS) cells. Representative images of cytoplasmic granules (Gr) at high resolution *(middle panel)* show that CD63 concentrates within stimulated, emptying granules, while in NS granules (intact), the labeling is observed at the limiting membrane. The total granule area as well as the CD63-immunolabeled area *(highlighted in blue)* significantly increases after stimulation. *,*$P < .005$ vs. NS. Data represent mean±S.E.M. Analyses were derived from three experiments with 2005 granules counted in 54 electron micrographs randomly taken and showing the entire cell profile and nucleus. Cells were prepared as described.[23]

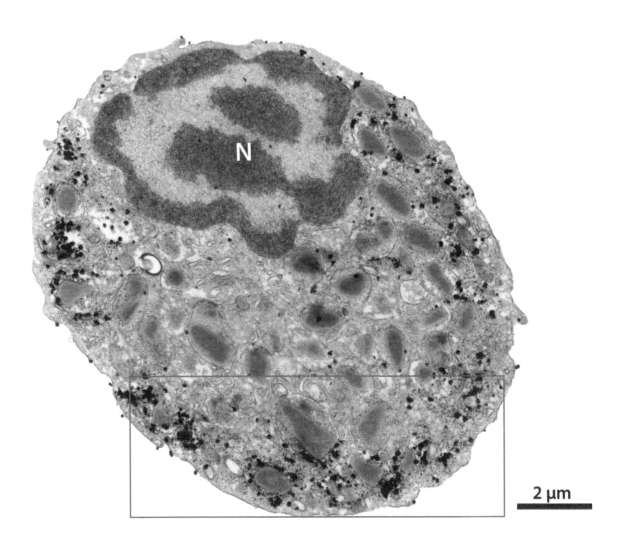

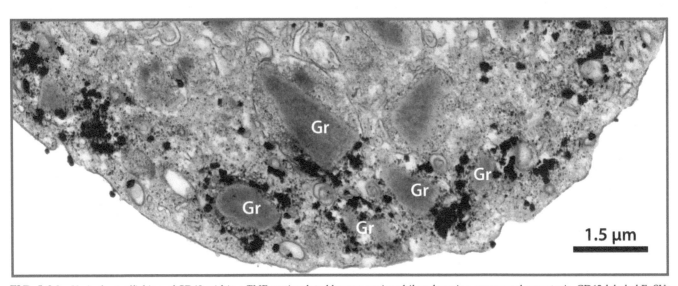

FIG. 5.26 Vesicular trafficking of CD63 within a TNF-α-stimulated human eosinophil undergoing compound exocytosis. CD63-labeled EoSVs *(highlighted in pink)* are predominantly at cell periphery, in association with CD63-positive, fused secretory granules (Gr). The *boxed area* is shown in higher magnification. Peripheral blood eosinophils were stimulated and processed for immunonanogold EM.[23] N, nucleus.

A. The cell biology of human eosinophils

CD63 immunolabeling

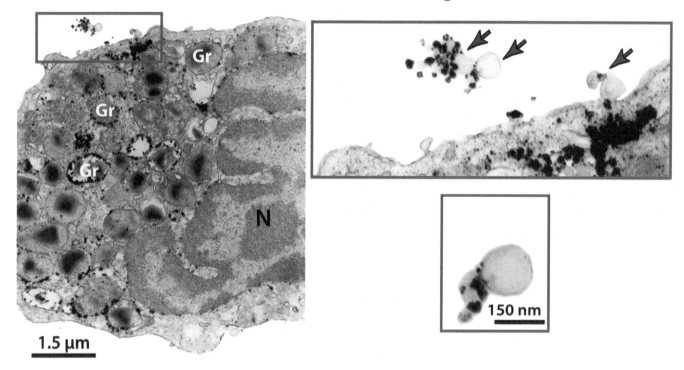

CD9 immunolabeling

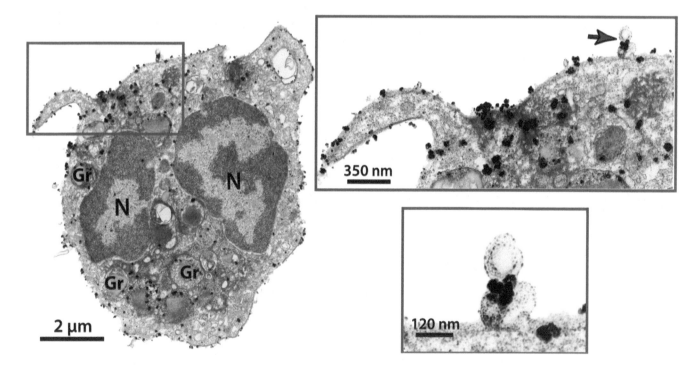

FIG. 5.27   CD63 and CD9 are associated with extracellular vesicles (EVs) at the surface of human eosinophils. CD63- and CD9-positive EVs, free and budding from the cell surface, are indicated *(arrows)*. Note that CD63 is consistently found intracellularly on secretory granules (Gr), while pools of CD9 are more detectable at the eosinophil surfaces. Eosinophils were isolated from the peripheral blood, stimulated with CCL11 *(top panels)* or TNF-α *(bottom panels)*, and processed for immunonanogold EM.[125] N, nucleus.

findings clearly showed CD63 and CD9 labeling associated with the delimiting membrane of EVs formed from the plasma membrane (microvesicles) in both resting and activated human eosinophils (Fig. 5.27).[125] However, in accordance with our nanoscale flow cytometry results, immuno-EM also revealed that not all microvesicles were labeled for these tetraspanins, regardless of whether cells were stimulated or not.[125] EM quantitative analysis demonstrated a higher number of these EVs labeled for CD9 than CD63-labeled EVs, and, therefore, CD9 might be a better marker than CD63 for microvesicles released by human eosinophils.[125]

5.6 Galectin-10 (Gal-10)

A predominant protein of human eosinophils is Gal-10, also known as Charcot-Leyden crystal protein (CLC-P) because of its remarkable ability to form CLCs, which are frequently found in tissues from patients with eosinophilic disorders.[33,36,272] Our group has formerly identified CLC-P as a protein with lysophospholipase activity.[273,274] Structural and genomic similarities between CLC-P and the carbohydrate-binding galectin superfamily have led to its reclassification as a galectin termed Gal-10.[275] CLC-P/Gal-10 can bind to mannose,[276] but other ligands for this protein including complex glycoconjugates remain to be determined.

Gal-10 is highly expressed in human eosinophils.[38] Comprehensive proteomic analyses found that Gal-10 is one of the most abundant proteins within eosinophils from healthy donors.[277,278] This protein has been receiving high attention as a biomarker for eosinophilic diseases[35,36,279] or even a potential target for drugs.[280]

The ability of human eosinophils to store and secrete Gal-10 has been addressed by our ultrastructural studies.[37] Our immunonanogold approach, in accordance with immunofluorescent data, revealed for the first time at the nanoscale level that Gal-10 accumulates in the eosinophil peripheral cytoplasm under the plasma membrane, and not within secretory granules (Figs. 5.28, 5.29, and 5.30).[37] Immunolabeling and EM procedures performed in situ on eosinophils kept attached on the surface of glass slides clearly show expanded, granule-poor areas of the peripheral cytoplasm densely labeled for Gal-10 (Fig. 5.29). The high concentration of Gal-10 in the peripheral eosinophil cytoplasm, prompt to be released, might explain the high levels of Gal-10 found in the sputum of patients with asthma and allergic bronchopulmonary aspergillosis (ABPA).[36] Moreover, the subplasmalemmal localization of Gal-10 may facilitate its roles in regulating cell surface secretion mediated by regulatory palmitoylation.[274]

While scattered labeling for Gal-10 is detected on the nucleus (Fig. 5.30) and associated with some cytoplasmic vacuoles, likely endosomes, quantitative imaging analysis demonstrated that the majority of Gal-10 (~ 80%) concentrates in an area of the eosinophil peripheral cytoplasm, measuring just ~ 250nm wide from the plasma membrane.[37] Moreover, Gal-10 not only concentrates under the plasma membrane but also interacts with it. Gal-10-immunoreactive microdomains with high immunolabeling density are distributed at the plasma membrane (Figs. 5.30 and 5.31), indicating that Gal-10 in human eosinophils is functionally linked to this membrane.[37] The association of galectin members with the plasma membrane has been reported. For example, galectins interact with membrane lipids and galectin-glycan clusters cause membrane bending (reviewed in[281]). However, the functional meaning of such association remains unknown.

Galectins can reenter different cells by endocytosis[282] and are able to populate endocytic and recycling compartments.[283,284] Cytosolic galectins rapidly accumulate in disrupted endocytic compartments due to their ability to bind to their ligands (glycans), which are normally present in the lumen of these endocytic vacuoles, but under disruption, are exposed to the cytosol (reviewed in[285]). We investigated this phenomenon in human eosinophils after treating the cells with PMA for 15min (10ng/mL), which causes rapid formation of plasma membrane-derived vacuoles in the cytoplasm of leukocytes and macrophages.[286–289] Newly formed endocytic vacuoles were consistently labeled for Gal-10 (Fig. 5.32), thus indicating that this protein may be interacting with the endocytic pathway in human eosinophils.

In contrast to other immune mediators, which are stored preformed within eosinophil-specific granules, Gal-10, as noted, does not have a granular localization. In earlier ultrastructural studies, we found that CLC-P was completely absent from specific granules of human eosinophils obtained from different origins (blood, tissue, and cultures) while cytoplasmic, granule-poor areas were consistently labeled.[146,147,290,291] On the other hand, we also reported immunolabeling on a minimal population of large vacuoles considered coreless granules ("primary granules").[247,290] However, these previous data are conflicting with the high concentration of Gal-10 existent in human eosinophils.[277,278] These cells in normal conditions contain only a single population of specific granules, and true core-free (immature) granules, which are rarely found in mature eosinophils, cannot account for such amount of Gal-10.[18] Additional former ultrastructural work using mature and immature eosinophils found that both core-containing granules and coreless granules were mainly negative for CLC-P.[292] This work also demonstrated that CLC-P is predominantly cytosolic and raised several concerns about the presence of CLC-P within few eosinophil granules.[292] However, subsequent to

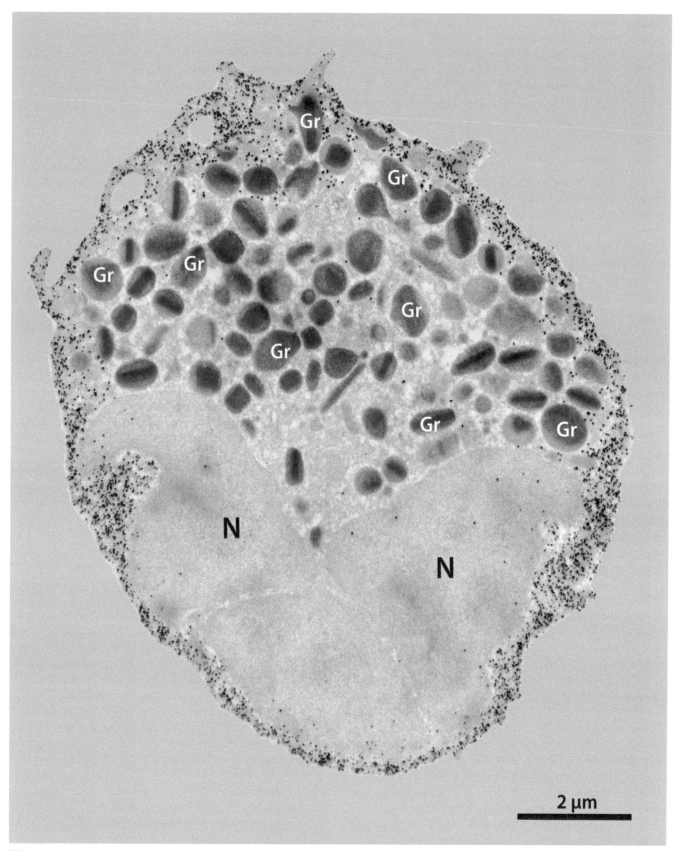

FIG. 5.28   Immunolabeling of galectin-10 (Gal-10) in human resting eosinophils. Gal-10 is densely stored in the peripheral cytoplasm under the plasma membrane. Specific granules (Gr) are Gal-10-negative. Peripheral blood eosinophils were processed for immunonanogold EM.[37] N, nucleus.

A. The cell biology of human eosinophils

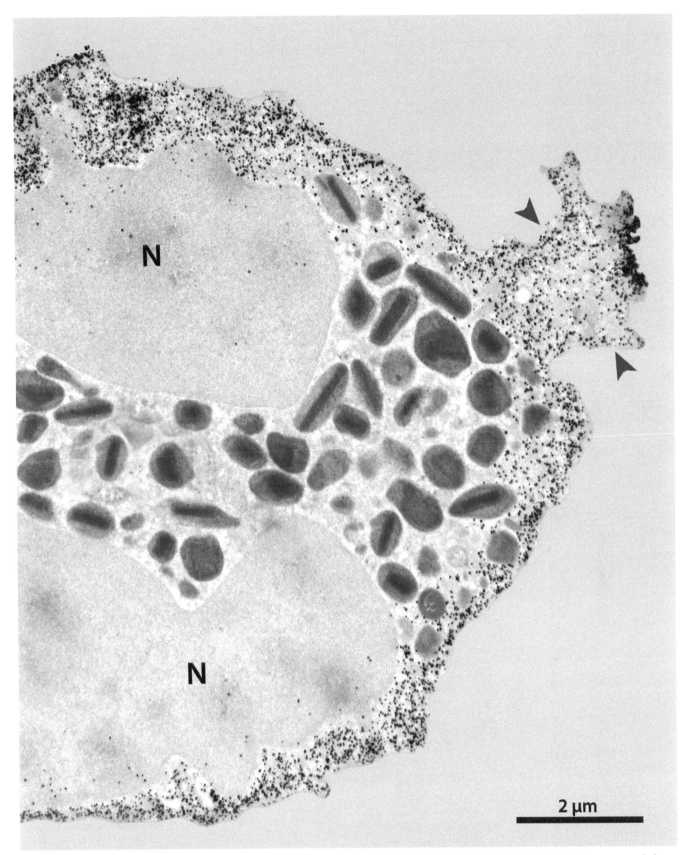

FIG. 5.29 Granule-poor cytoplasmic areas of the peripheral cytoplasm are densely labeled for Gal-10. *Arrowheads* indicate an expanded area of the cell surface rich in Gal-10. The population of specific granules is not labeled. Eosinophils were isolated from the peripheral blood and processed for immunonanogold EM.[37] *N*, nucleus.

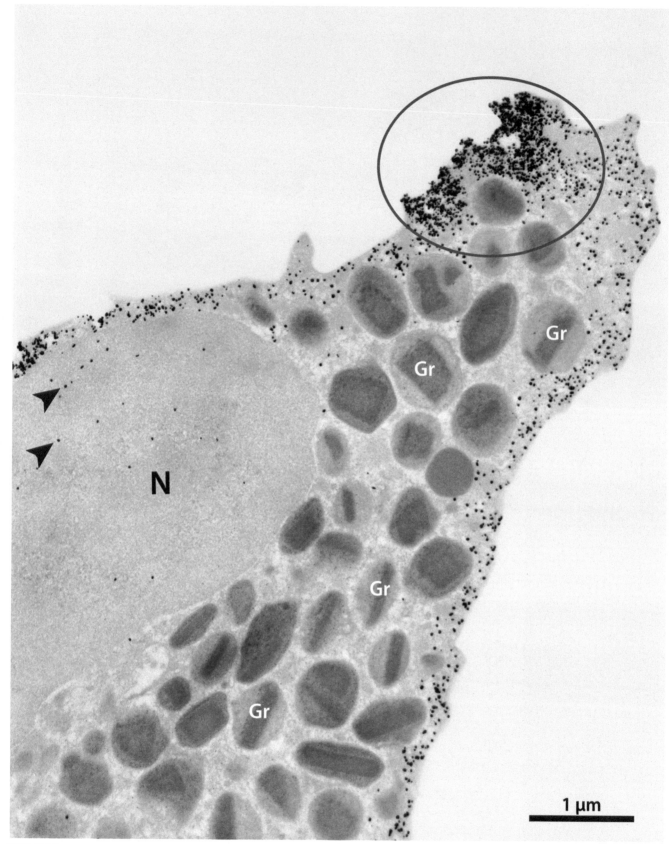

FIG. 5.30   Gal-10 accumulates in the eosinophil peripheral cytoplasm and interacts with the plasma membrane. Note the presence of Gal-10-immunoreactive microdomains with high immunolabeling density *(circle)* at the eosinophil plasma membrane while specific granules (Gr) are negative. A scattered labeling *(arrowheads)* on the nucleus (N) is also observed. Eosinophils were isolated from the peripheral blood and processed for immunonanogold EM.[37]

A. The cell biology of human eosinophils

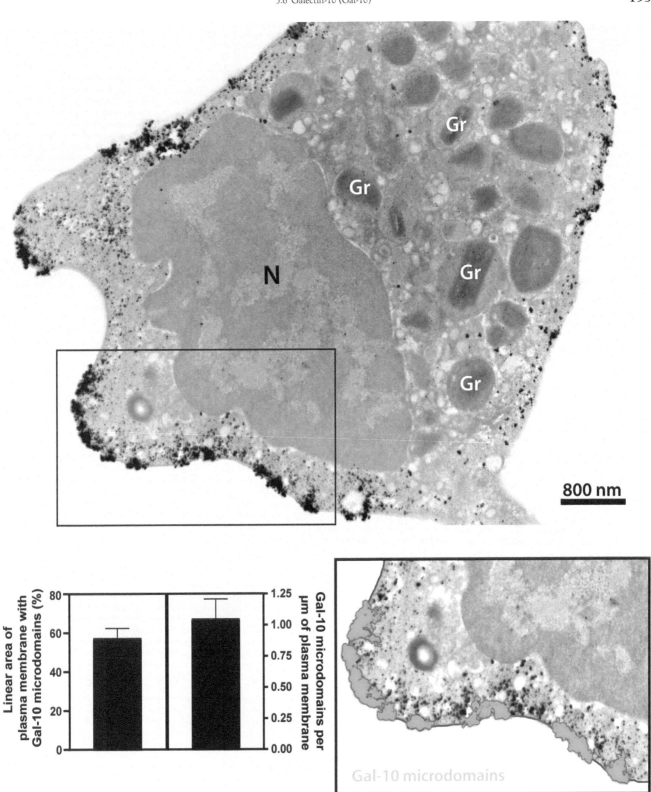

FIG. 5.31 Quantitative imaging analyses of Gal-10-positive microdomains distributed along the eosinophil plasma membrane (PM). Around 60% of the PM contains high-density-Gal-10 microdomains. Cells showing the entire cell profile and nucleus were analyzed using Fiji software after ultrastructural immunolabeling for Gal-10.[37] N, nucleus.

A. The cell biology of human eosinophils

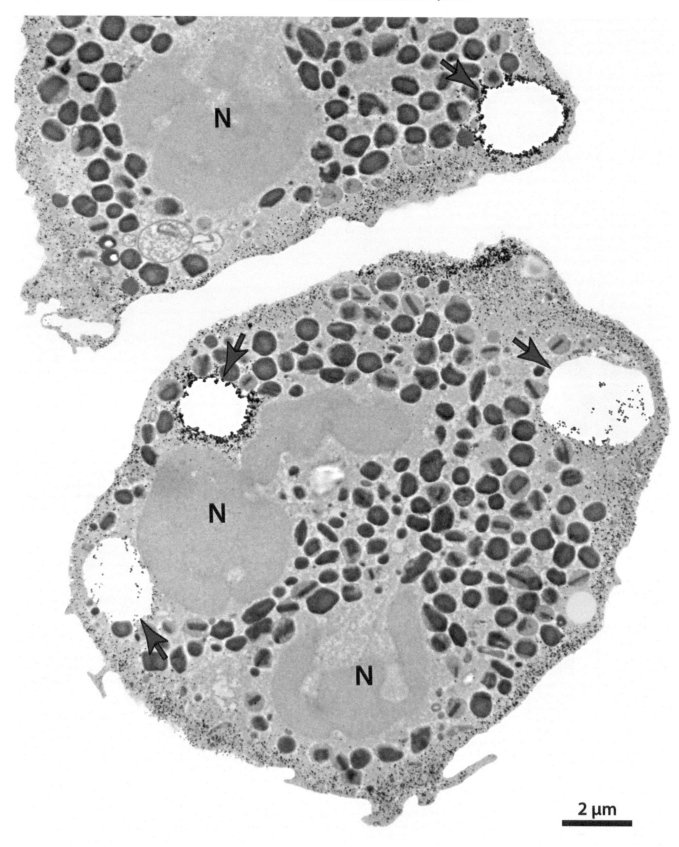

FIG. 5.32   Ultrastructural immunolabeling of Gal-10 within endocytic compartments from human eosinophils. PMA-induced endocytosis leads to the formation of Gal-10-positive vacuoles *(arrows)* in the eosinophil cytoplasm, mainly in the cell periphery. Eosinophils isolated from the blood were stimulated with PMA and processed for immunonanogold EM as described.[37] *N*, nucleus.

A. The cell biology of human eosinophils

these initial works during the 1990s, no attention was given to the subcellular localization of CLC-P within human eosinophils. With the contemporary recognitions that human eosinophils do not contain "primary granules" but just a single population of specific granules[18] and that Gal-10 is clearly not stored within granules,[37] it is conceivable that the "primary granules" originally positive for CLC-P represent in fact distinct membrane compartments from intracellular pathways, as demonstrated for the endocytic pathway.[283,284] Thus, we can assume that Gal-10-positive vacuoles reported in our former studies ("primary granules") are likely endocytic compartments (Fig. 5.32).[37]

It is interesting that degranulation observed after stimulation with CCL11 or TNF-α, which are known inducers of eosinophil secretion through PMD or compound exocytosis, respectively, do not affect the cytoplasmic localization of Gal-10 as observed by immunonanogold EM (Figs. 5.33 and 5.34).[37] This means that Gal-10 is neither stored within secretory granules nor exported through classical degranulation mechanisms. On the other hand, under cytolytic stimulation, Gal-10 is promptly released and deposited in the extracellular medium.[41] This mechanism is associated with the formation of extracellular traps and vesicles.[41] In eosinophils in the process of cytolysis with activated granules and plasma membrane still intact (Fig. 5.35), Gal-10 is densely observed at the cell periphery, but once the plasma membrane is disrupted, immunolabeling for Gal-10 is remarkably observed in extracellular vesicles (Fig. 5.36). High-resolution imaging of cytolytic eosinophils immunolabeled for Gal-10 revealed a large number of extracellular vesicles, produced as microvesicles from the plasma membrane, fully labeled for this protein (Figs. 5.36 and 5.37).

5.7 Enzymes involved in the formation of leukotrienes

As highlighted in Chapter 4, different stimuli trigger LB accumulation in eosinophils and these organelles act as platforms for the synthesis of inflammatory mediators (eicosanoids).[86] Eicosanoid production by newly formed LBs begins with the release of arachidonic acid (AA) by the action of cytosolic phospholipase A2 (cPLA2) localized on LBs. AA then enters cyclooxygenase (COX) or lipoxygenase (LOX) pathways to generating the eicosanoids prostaglandin D$_2$ (PGD$_2$) or leukotriene C4 (LTC4), respectively (Fig. 5.38).[86] These eicosanoids are potent inflammation mediators and immunity modulators.[205]

All major enzymes involved in the enzymatic conversion of AA into eicosanoids are localized specifically within LBs of activated eosinophils.[214,240,293,294] The enzymes demonstrated by immunogold EM include 5-LO (Fig. 5.39)[293] and COX (Figs. 5.40 and 5.41),[94,240] which localized abundantly within eosinophil LBs. Both naturally activated eosinophils from HES donors (Figs. 5.39 and 5.40) and pathogen-activated human eosinophils (Fig. 5.41) showed increased LB formation and concomitant expression of eicosanoid-generating enzymes in these organelles, not only at their peripheries but also in their cores. Because these enzymes are membrane proteins, this observation highlights the fact that LBs are intricate organelles and not just lipid-storage stations. The presence of internal membranes within eosinophil LBs, as demonstrated by 3D electron tomography (Chapter 2), explains how these enzymes are kept inside the LB cores.[94]

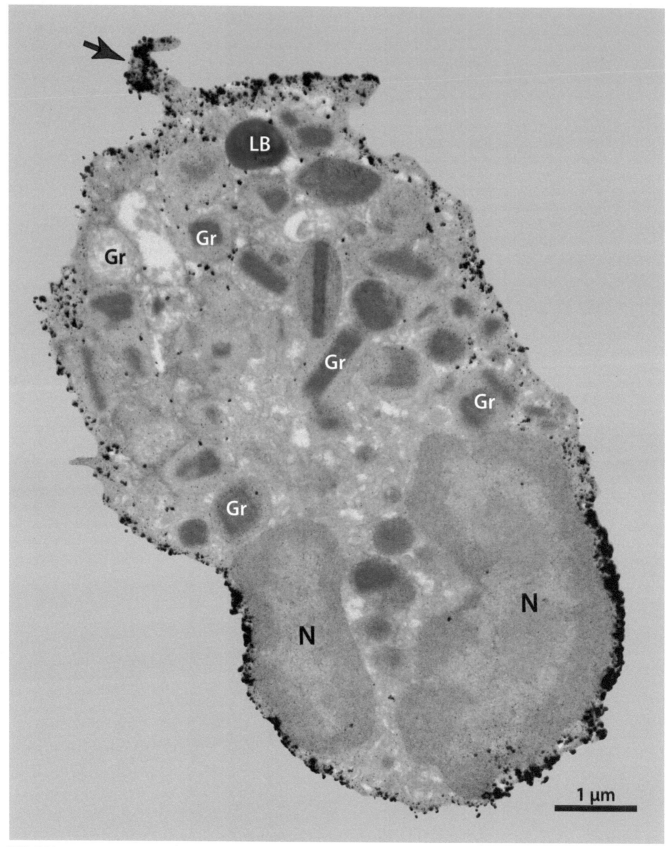

FIG. 5.33   Gal-10 in a CCL11-activated human eosinophil. Gal-10 is seen in the cell periphery, while secretory granules (Gr) are negative. Note a filopodium *(arrow)*. Peripheral blood eosinophils were stimulated and processed in situ for immunonanogold EM.[37] *N*, nucleus; *LB*, lipid body.

A. The cell biology of human eosinophils

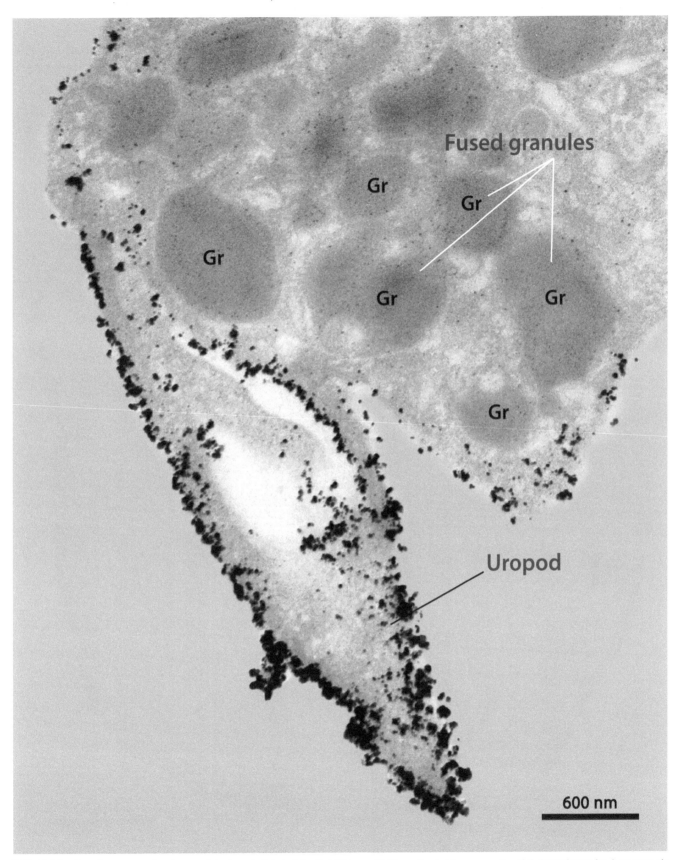

FIG. 5.34   A surface protrusion (uropod) is densely labeled for Gal-10 in an TNF-α-activated human eosinophil. Stimulation leads to granule fusion and shape changes. Gal-10 is distributed in the peripheral cytoplasm but not in secretory granules (Gr). Peripheral blood eosinophils were stimulated and processed in situ for immunonanogold EM.[37]

A. The cell biology of human eosinophils

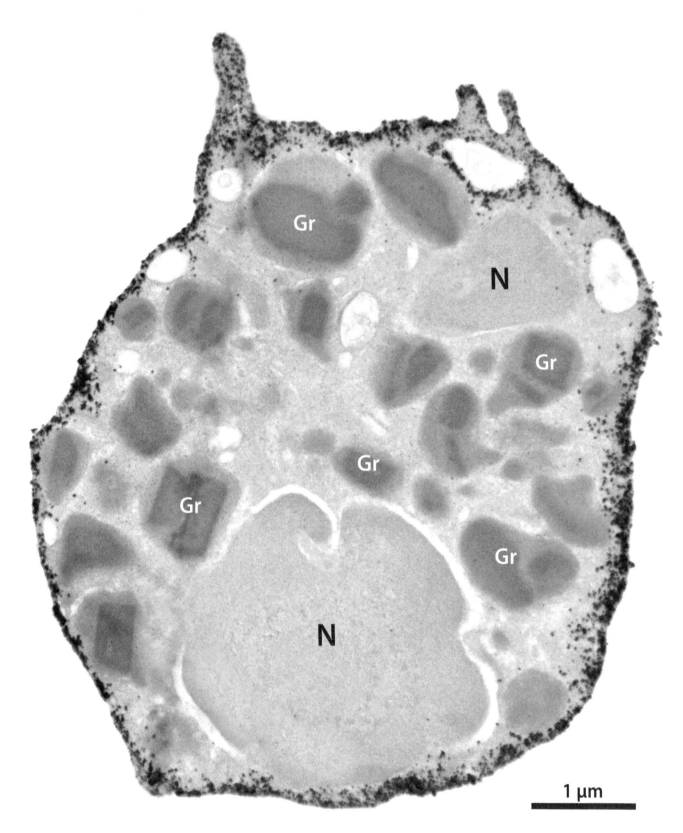

FIG. 5.35   A human eosinophil in the process of cytolysis showing labeling for Gal-10 in the peripheral cytoplasm. Note that enlarged, activated secretory granules (Gr) with disarranged cores are negative for Gal-10. Eosinophils isolated from the blood were stimulated with immobilized IgG to elicit lytic degranulation as described[46] and processed for immunonanogold EM. *N*, nucleus.

A. The cell biology of human eosinophils

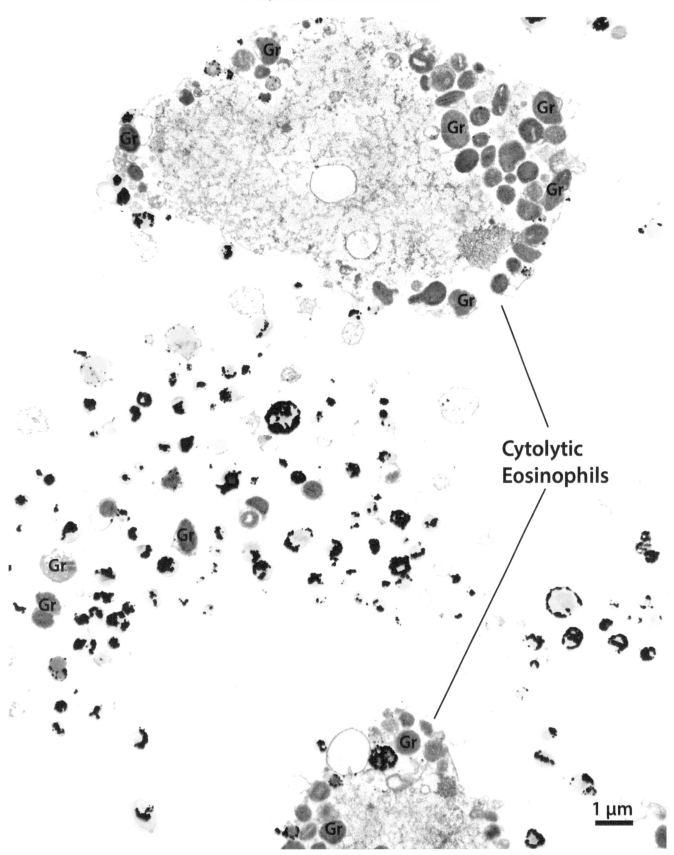

Cytolytic Eosinophils

1 μm

FIG. 5.36 Ultrastructural immunolabeling for Gal-10 in cytolytic human eosinophils. After cytolysis, Gal-10 is mostly observed extracellularly associated with extracellular vesicles *(highlighted in blue)*. Secretory granules (Gr) are not labeled. Eosinophils show disrupted plasma membranes and extensive chromatin decondensation and expansion *(highlighted in purple)*.

A. The cell biology of human eosinophils

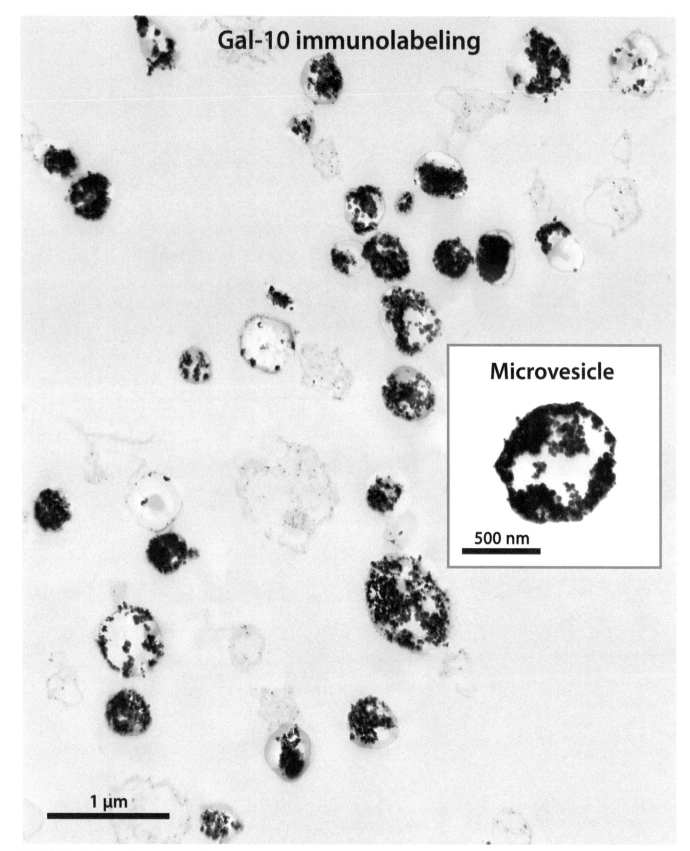

FIG. 5.37   High-resolution imaging of eosinophil extracellular vesicles (EVs) carrying Gal-10. EVs packed with Gal-10 *(highlighted in blue)* are generated as microvesicles with different sizes in response to eosinophil cytolysis. Eosinophils isolated from the blood were stimulated with immobilized IgG as before[46] and processed for immunonanogold EM.

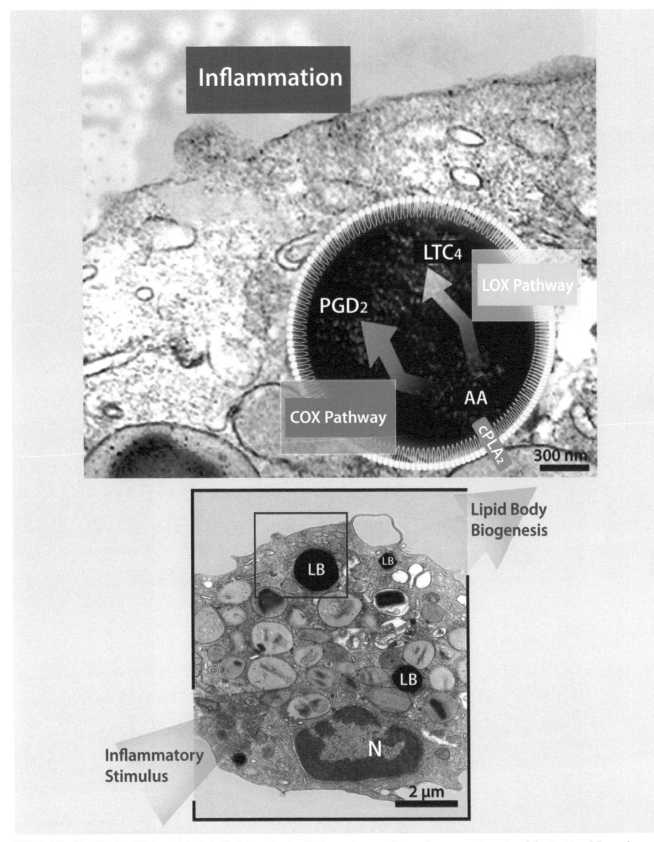

FIG. 5.38 Lipid bodies (LBs) participate in the biosynthesis of inflammatory mediators. In response to eosinophil activation, LBs are formed in the cytoplasm. Within LBs, arachidonic acid (AA) acts as a substrate for generation of eicosanoid lipid mediators. AA is released by phospholipase A_2 (cPLA2) and converted by cyclooxygenase (COX) or lipoxygenase (LOX) pathways to the eicosanoids prostaglandin D2 (PGD2) or leukotriene C4 (LTC4), respectively.[206] The electron micrograph is from a peripheral blood human eosinophil stimulated with CCL11 and processed for conventional TEM.[94] N, nucleus.

A. The cell biology of human eosinophils

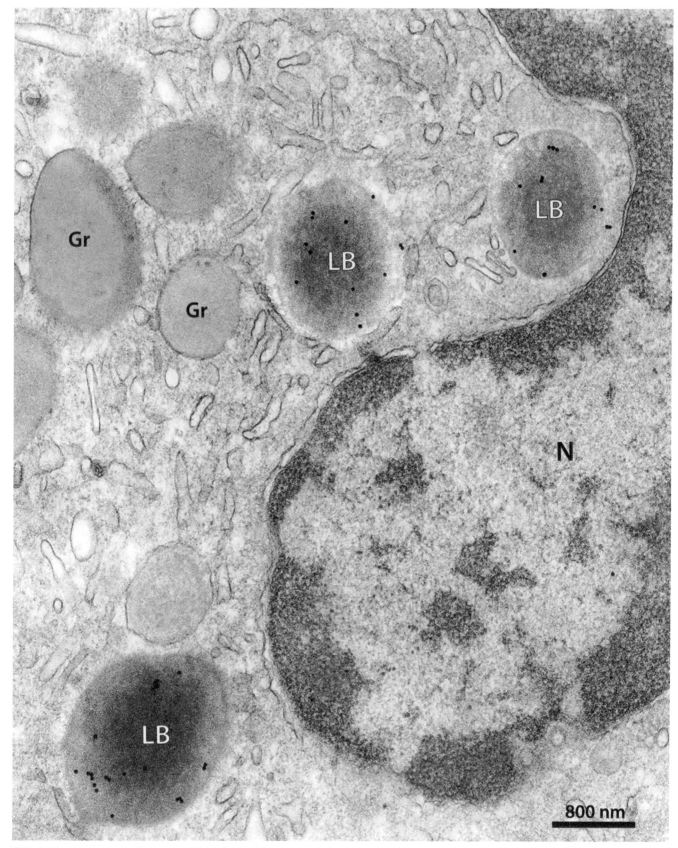

FIG. 5.39　Ultrastructural immunogold labeling of 5-lipoxygenase (5-LO) in eosinophil lipid bodies (LBs) from a patient with HES. Cells were processed as before.[293] *Gr*, specific granules; *N*, nucleus.

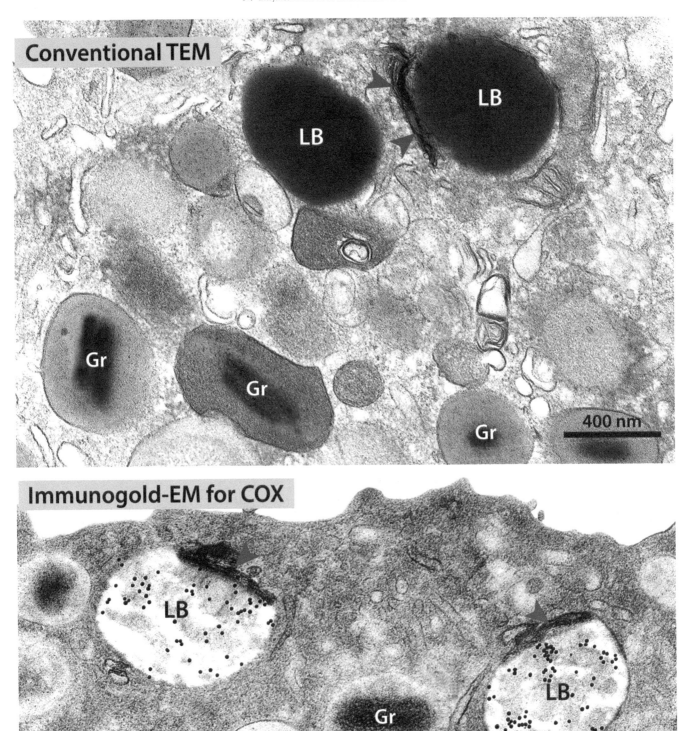

FIG. 5.40 LBs within HES peripheral blood eosinophils seen under conventional TEM *(top panel)* and after immunolabeling for cyclooxygenase (COX) *(bottom panel).* Labeling is observed within LBs. Note membranous structures *(arrowheads),* likely part of the ER, in close apposition to LBs. Cells were processed as before.[94]

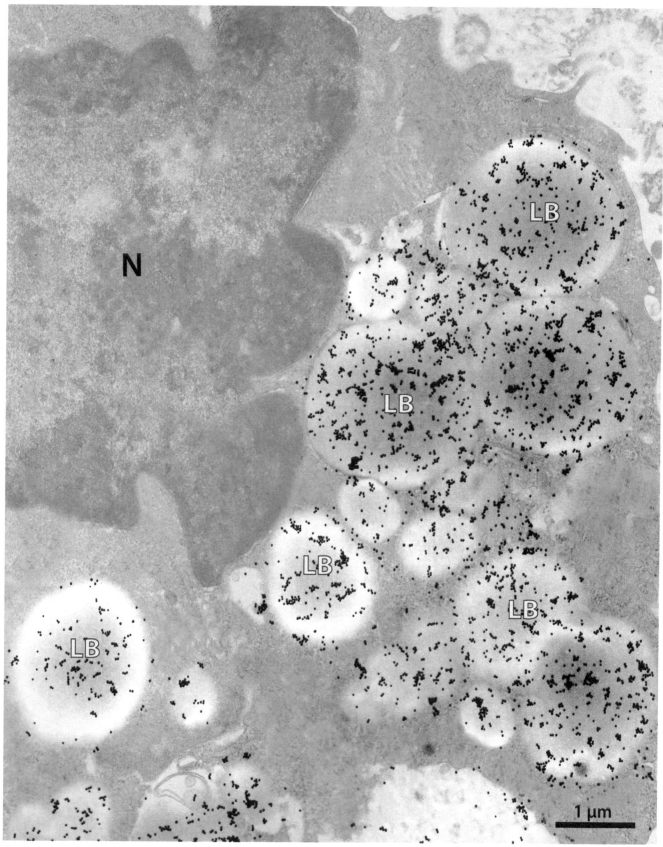

FIG. 5.41   LBs within a pathogen-activated eosinophil are heavily labeled for COX. Note the increased number of cytoplasmic LBs formed in response to the infection with HIV-1. Eosinophils cultured with HIV-1[204] were immunolabeled as before.[240] Nucleus (N) colored in *purple*.

6

Eosinophil cell death

6.1 Overview

Cell death can be defined as an "irreversible degeneration of vital cellular functions culminating in the loss of cellular integrity (permanent plasma membrane permeabilization or cellular fragmentation)."[295] The process of cell death is associated with multiple human diseases such as cancer, neurodegenerative, autoimmune, and infectious diseases and also considered an essential event for normal development and homeostasis.[296]

Cell death may occur in response to varied stresses. Basically, cells may die from accidental cell death or regulated cell death (reviewed in[296,297]). Accidental cell death is a biologically uncontrolled process; that is, it refers to the instantaneous and catastrophic death of cells exposed to severe insults of physical (high pressures, temperatures, or osmotic forces), chemical (extreme pH variations), or mechanical (for example, shear forces) nature. On the other hand, regulated cell death relies on the molecular machinery; that is, cell death occurs upon stimulation of specific signaling cascades (reviewed in[295,298]). Regulated cell death is also known as programmed cell death when it occurs in physiological conditions.[297]

There are different modalities of cell death under the umbrella of regulated cell death, which are orchestrated by distinct signaling pathways and have been considered from morphological, biochemical, and functional perspectives. Apoptosis is the most well-known form of regulated cell death, but during the past few decades, many other types of regulated cell death have been identified with a role in development, tissue homeostasis, inflammation, immunity, and varied pathophysiological situations.[296] Since 2005, the Nomenclature Committee on Cell Death (NCCD) has been formulating guidelines for the definition and interpretation of all aspects of cell death and, thus, to support the continued development of the field.[295,297,299–301] As presented in the 2018 NCCD position paper and based mainly on molecular machinery, there are 12 major cell death subroutines including intrinsic and extrinsic apoptosis, necroptosis, pyroptosis, autophagy-dependent cell death, lysosome-dependent cell death, NETotic cell death, and mitochondrial permeability transition-driven necrosis.[295] Thus, mammalian cells exposed to unrecoverable perturbations of the intracellular and extracellular microenvironment can activate one of many signal transduction cascades ultimately leading to their death.[295] Regulated cell death manifests with a large spectrum of immunomodulatory sequelae, ranging from anti-inflammatory and tolerogenic to pro-inflammatory and immunogenic.[295]

The molecular machinery associated with regulated cell death has been reviewed.[295,296] There is significant cross-talk between the molecular mechanisms that trigger different forms of regulated cell death, and therefore definition and interpretation of cell death modes can be complex.[295,297] For example, there are many cross-pathway signaling events between apoptosis, necroptosis, and pyroptosis.[298] To add to the complexity, it is now recognized that cell death is a more dynamic event than formerly reported with "gray zones" and "points of no return" between cell death modes.[302]

Cell death manifests with morphological changes. From a morphological point of view, cell death depicts fundamentally a cytolytic (also termed necrotic) or a noncytolytic (apoptotic) morphology. Both morphological profiles can be associated with regulated cell death. EM enables the unambiguous distinction of a cytolytic from a noncytolytic profile and reveals morphological hallmarks that have been associated with certain types of cell death. For example, cytolytic cells showing both nuclear decondensation and extrusion of extracellular DNA traps characterize a specific type of cell death (ETosis).[46,303]

Necrosis has long been considered an uncontrolled type of cell death (accidental cell death) classically linked to a cytolytic profile and the presence of cytoplasmic and organelle swelling, but regulated types of necrosis, such as necroptosis and mitochondrial permeability transition-driven necrosis, also occur in a controlled manner.[295,296] Therefore, while a precise definition depends on molecular delineation of pathways of necrosis,[295,304] a solely morphological distinction among all modalities of regulated cell death, as classified by molecular criteria, is not possible.

Eosinophils in the process of death have been ultrastructurally described in different conditions and diseases.[40,44,151,162,168,173,197,204] Considering the focus of this atlas, we will discuss in this chapter the ultrastructural features of the eosinophil cell death based on cytolytic and noncytolytic morphologies, in accordance with the NCCD.[295] Eosinophil cell death can manifest with an entire spectrum of morphological appearances as described below.

6.2 Cytolytic morphology: General aspects

A cytolytic morphology is characterized by the loss of plasma membrane integrity.[171] This is one important morphological indication that the cell is entering into a process of cell death with a cytolytic morphotype. Because the plasma membrane is well observed in morphologic detail only at high visual resolutions, a broken plasma membrane can be directly seen just by EM. In Figs. 6.1–6.3, eosinophils are observed under TEM in the initial stages of cytolysis with the plasma membrane partially disrupted and thus discontinuous. As the process progresses, the entire plasma membrane disappears, and the population of secretory granules is spilled into the extracellular medium (Figs. 6.4 and 6.5).

Cytolytic eosinophils observed under SEM show an "exploded" appearance due to the rupture of the plasma membrane while secretory granules are seen as intact organelles (Figs. 6.6 and 6.7). These granules, as detected by TEM, frequently preserve their delimiting membrane (Fig. 6.8) and can remain functionally active as free extracellular granules—FEGs also termed clusters of FEGs (cFEGs).[16,168] Cytolytic eosinophils and FEGs are common findings in tissue biopsies and secretions from patients with eosinophilic diseases (Figs. 6.9–6.11) (see also Chapter 8).[170,171,174,305] Therefore, for eosinophils, in many circumstances, a cytolytic cell death represents a degranulation (secretory) process as discussed in Chapter 3.

As noted, when eosinophil-specific granules are seen by TEM as FEGs scattered in the extracellular matrix (Fig. 6.8), the identity of their cellular origin, that is, eosinophils, is not in question since these granules have a unique internal ultrastructure, as explained in Chapter 2. Large vesicular carriers with a distinctive morphology (EoSVs) are also a particular feature of human eosinophils[21] (Chapter 2) and, intriguingly, intact EoSVs are often distributed around FEGs in biopsy sites from patients with varied inflammatory diseases (Fig. 6.10).[17] The meaning of these free EoSVs and their relationship with FEGS in the extracellular matrix awaits further investigation.

A cytolytic morphology is also characterized by nuclear alterations. The nuclear envelope undergoes morphological changes, disrupts, and disintegrates in parallel with chromatin decondensation and dissolution (chromatolysis) (Figs. 6.4 and 6.9–6.11). Nuclei in different stages of decondensation and degeneration can be observed in tissue sites characterized by eosinophil infiltration and lytic death, for example, in biopsies of patients with HES (Fig. 6.4), ECRS (Fig. 6.9), and ulcerative colitis (Fig. 6.11). Sometimes, eosinophils in the process of cytolysis with disrupted plasma membranes and dissolving nuclei are also seen in the peripheral blood, after cell isolation (Figs. 6.12 and 6.13), or even in vivo within blood vessels (see Fig. 3.32 in Chapter 3).[306] Thus, when manifesting a cytolytic death, rather than undergoing nuclear condensation and increased density, damaged eosinophil nuclei are generally less dense, and the distinction between euchromatin and heterochromatin, as seen in normal conditions (Figs. 6.12 and 6.13; see also Chapter 2), has disappeared (Figs. 6.4 and 6.9–6.11).

Cytolytic eosinophils in vivo are frequently associated with the formation of CLCs.[39,40] These structures, readily identifiable by TEM as bipyramidal, hexagonal, or amorphous crystals with diverse sizes and electron densities, are seen in the extracellular matrix in the proximity of eosinophils undergoing cytolysis during inflammatory responses (Fig. 6.14), as detailed in Chapter 8.

Eosinophil lytic death may denote varied forms of regulated cell death or even necrosis induced by accidental cell death.[46,162] Typically, necrosis is morphologically characterized by the gain in cell volume and the presence of swelling organelles in addition to plasma membrane disruption and nuclear alterations (Fig. 6.15). This profile can also be correlated with eosinophil necroptosis[162]; but this type of cell death, which is described as a cell-death receptor-induced, caspase-independent modality of regulated cell death with necrotic morphology[307] or merely regulated necrosis, is still poorly characterized in eosinophils. Nuclear dissolution and loss of recognizable organelles due to intense cell disintegration can also be seen in the late stages of necrosis (Fig. 6.15).[308]

As noted, while morphological distinction among all types of regulated cell death may not be feasible, a cytolytic profile combined with other morphological hallmarks can serve as a diagnosis for particular types of cell death at the EM level. This is the case of eosinophils undergoing cell death through ETosis, as described below.

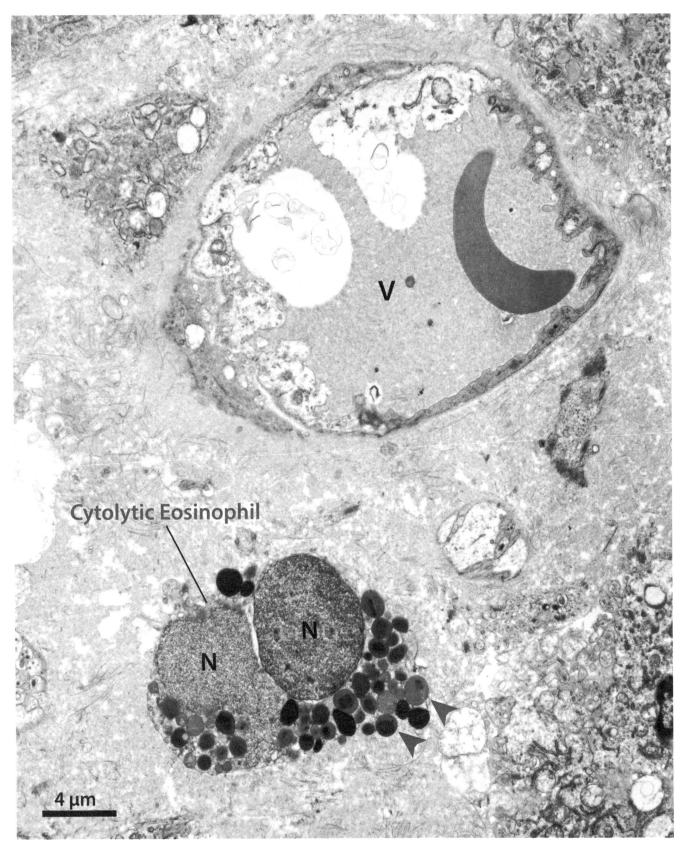

Cytolytic Eosinophil

4 μm

FIG. 6.1 TEM of a human tissue eosinophil undergoing cytolysis. Note the decondensed nucleus (N), plasma membrane partially disrupted, and secretory granules spilling into the extracellular medium *(arrowheads)*. A small blood vessel (V) is observed. Biopsy from a patient with ulcerative colitis. This eosinophil is shown at higher magnification in Fig. 6.20.

A. The cell biology of human eosinophils

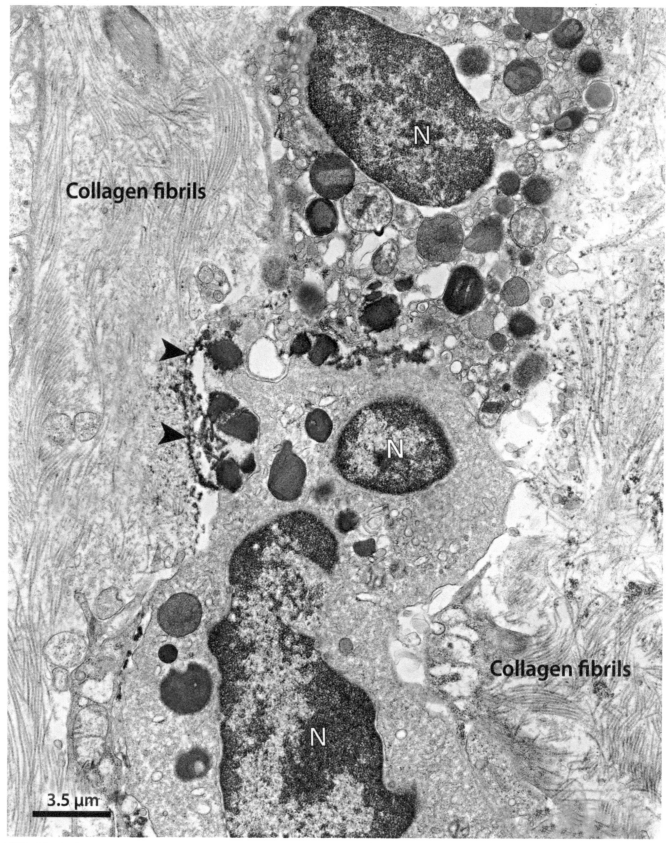

FIG. 6.2   Ileal tissue-infiltrating eosinophils from a patient with ulcerative colitis. Loss of the plasma membrane integrity is seen in one eosinophil in the initial process of cytolysis *(colored in purple)*, whereas the other one is intact *(colored in green)*. Note secreted material *(arrowheads)* at the
cell surface and collagen fibrils from the ECM. Sample prepared for TEM. *N*, nucleus.

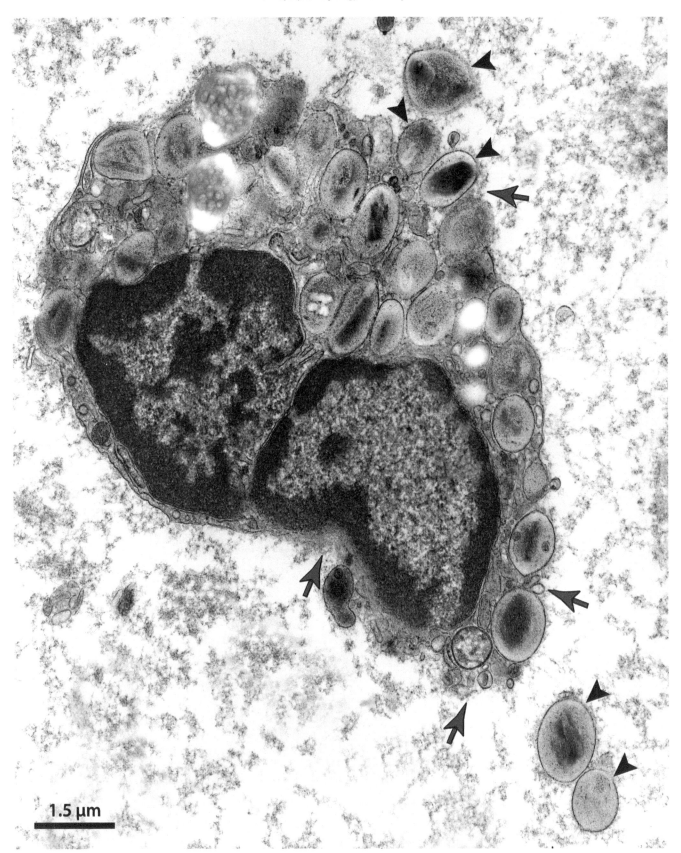

FIG. 6.3　A human eosinophil *(pseudocolored)* with the plasma membrane partially disrupted *(arrows)* observed under TEM. Some secretory granules *(arrowheads)* were released. Cytoplasm *(pink)*; granules *(yellow)*; nucleus *(purple)*.

A. The cell biology of human eosinophils

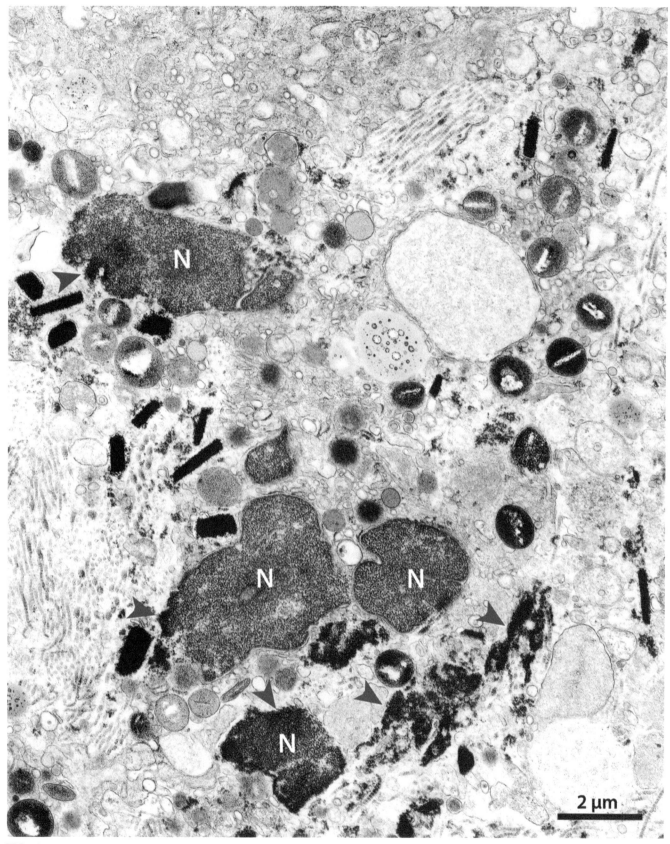

FIG. 6.4 Eosinophils infiltrated in a tissue site (skin) from a patient with HES. Note free extracellular granules *(colored in yellow)* and decondensed nuclei (N) with disrupted nuclear envelope and chromatolysis *(arrowheads)*. Biopsy prepared for TEM.

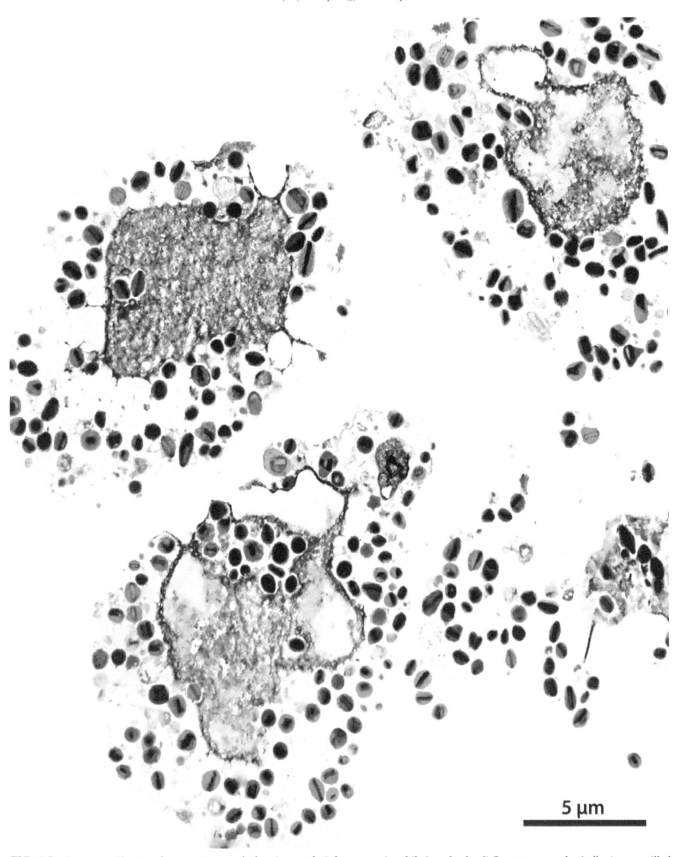

FIG. 6.5   Low-magnification electron micrograph showing cytolytic human eosinophils *(pseudocolored)*. Secretory granules *(yellow)* were spilled into the extracellular medium following plasma membrane rupture. Note decondensed nuclei *(purple)*. Peripheral blood eosinophils were stimulated with PMA for two hours and processed in situ for TEM.[46]

A. The cell biology of human eosinophils

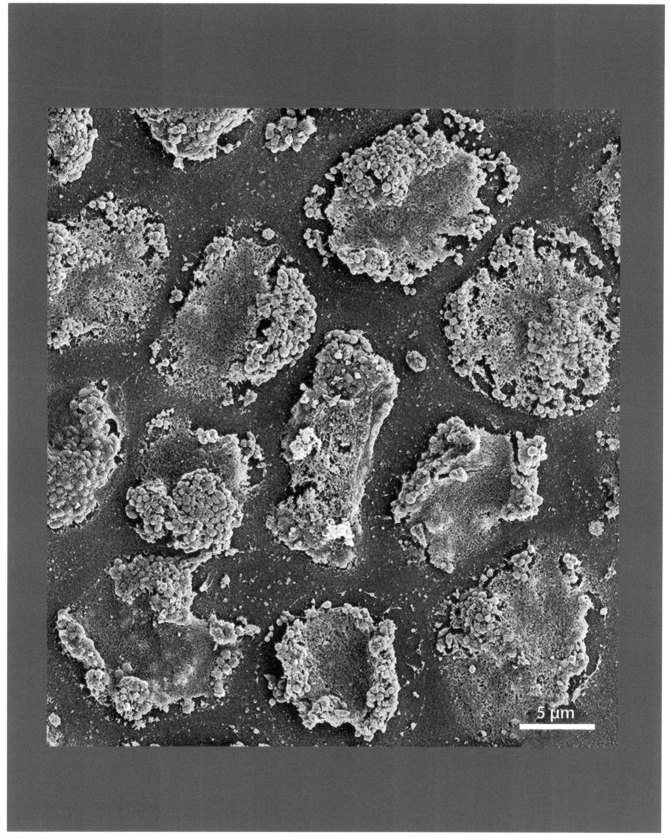

FIG. 6.6 Cytolytic human eosinophils *(pseudocolored)* seen by SEM at low magnification. Note the "exploded" 3D aspect of the cells with released secretory granules colored in *orange* and remaining cytoplasm in *blue*. Peripheral blood eosinophils were stimulated with PMA for two hours and processed for SEM as described.[46]

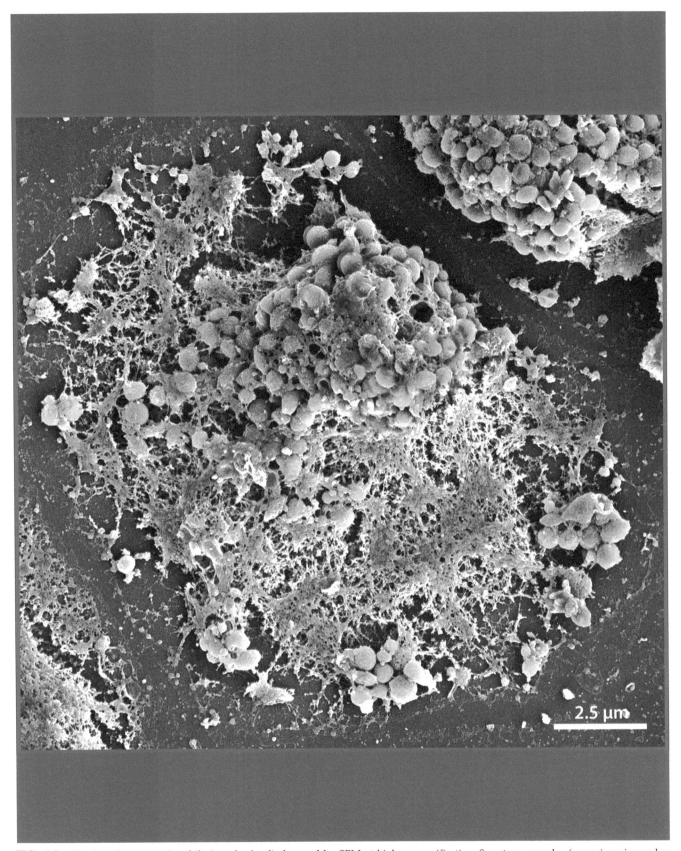

FIG. 6.7 Cytolytic human eosinophils *(pseudocolored)* observed by SEM at higher magnification. Secretory granules *(orange)* are imaged as round/elliptical, apparently intact organelles. Eosinophils isolated from the peripheral blood were stimulated for two hours with PMA and processed for SEM as described.[46] Remaining cytoplasm *(blue)*.

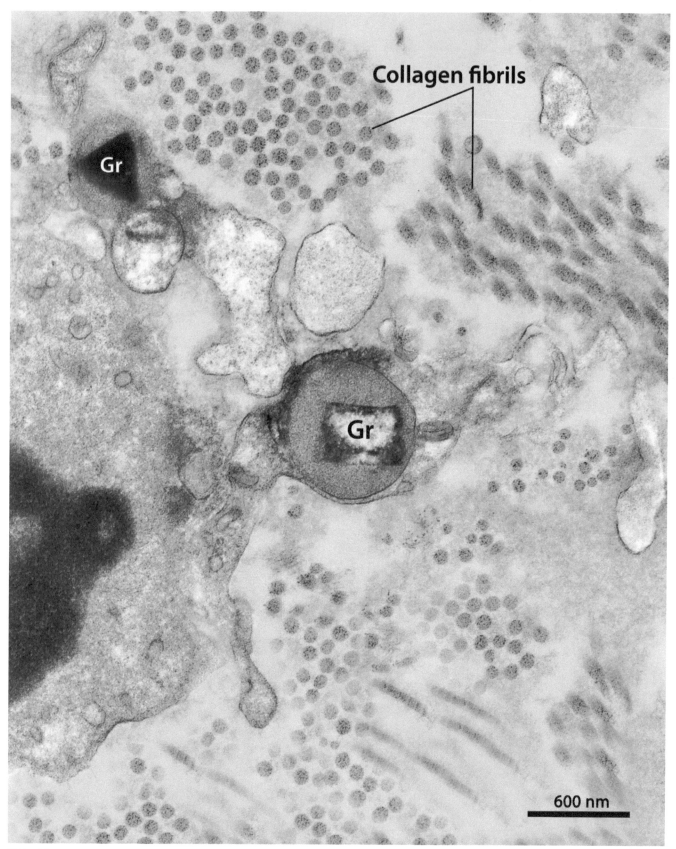

FIG. 6.8   High-magnification electron micrograph showing released eosinophil secretory granules (Gr) in a human skin site. One granule surrounded by EoSVs *(colored in pink)* has intact delimiting membrane and partial loss of its core. Note collagen fibrils in the ECM *(green)*. Skin biopsy prepared for TEM.

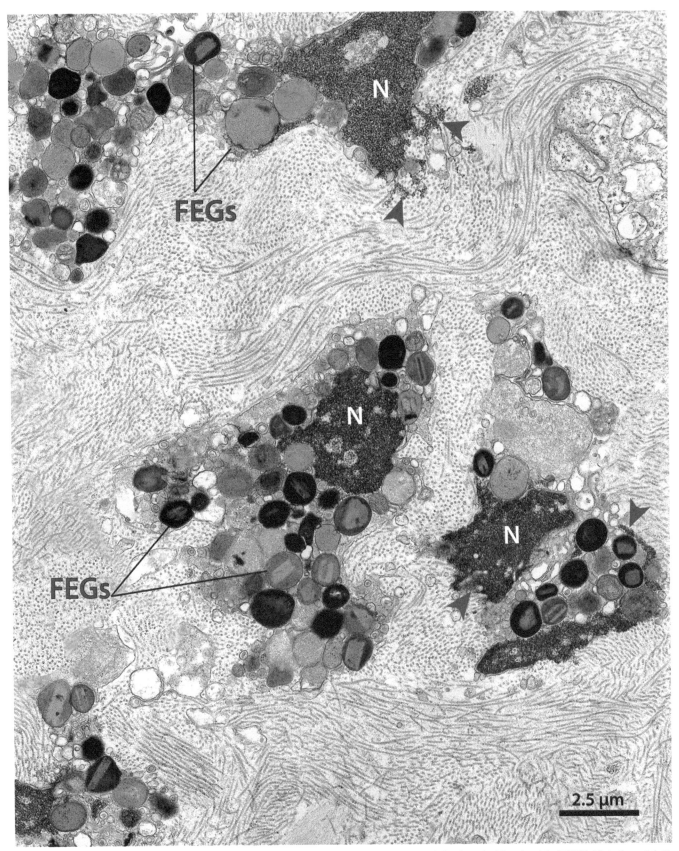

FIG. 6.9 TEM showing eosinophils in the frontal sinus tissue from a patient with eosinophilic chronic rhinosinusitis (ECRS). Note the cytolytic morphology with disruption of both plasma membrane and nuclear envelope and released granules (*FEGs*) and chromatin *(arrowheads)*. The nuclei (N) lost their distinction between heterochromatin and euchromatin. *FEGs*, free extracellular granules.

A. The cell biology of human eosinophils

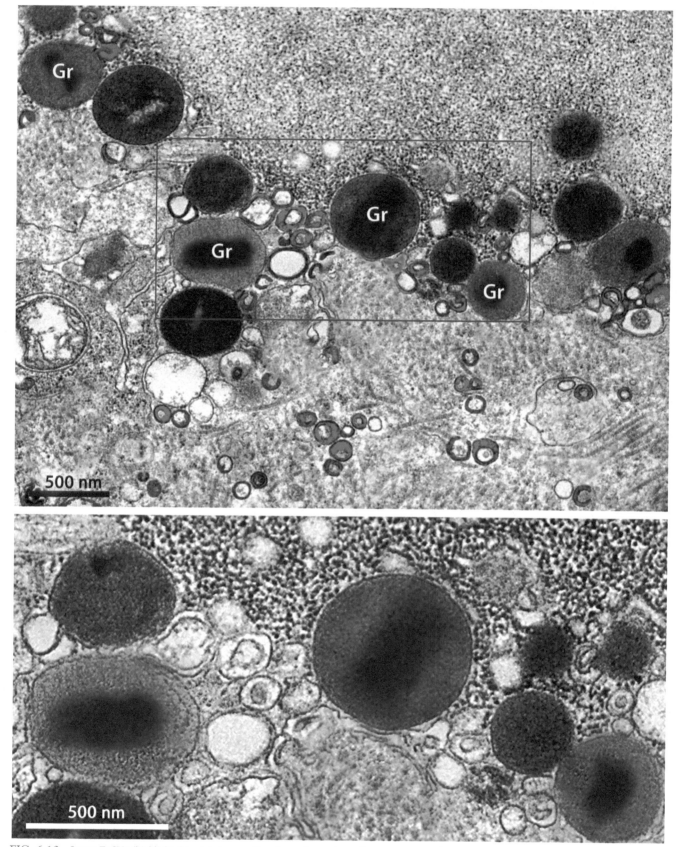

FIG. 6.10 Intact EoSVs *(highlighted in pink, top panel)* and membrane-bound secretory granules (Gr) released in the extracellular medium by a cytolytic tissue human eosinophil. The boxed area is shown in higher magnification. Note the decondensed chromatin *(colored in blue, top panel)* in direct contact with some granules and EoSVs. Biopsy (frontal sinus, ECRS) prepared for TEM.

A. The cell biology of human eosinophils

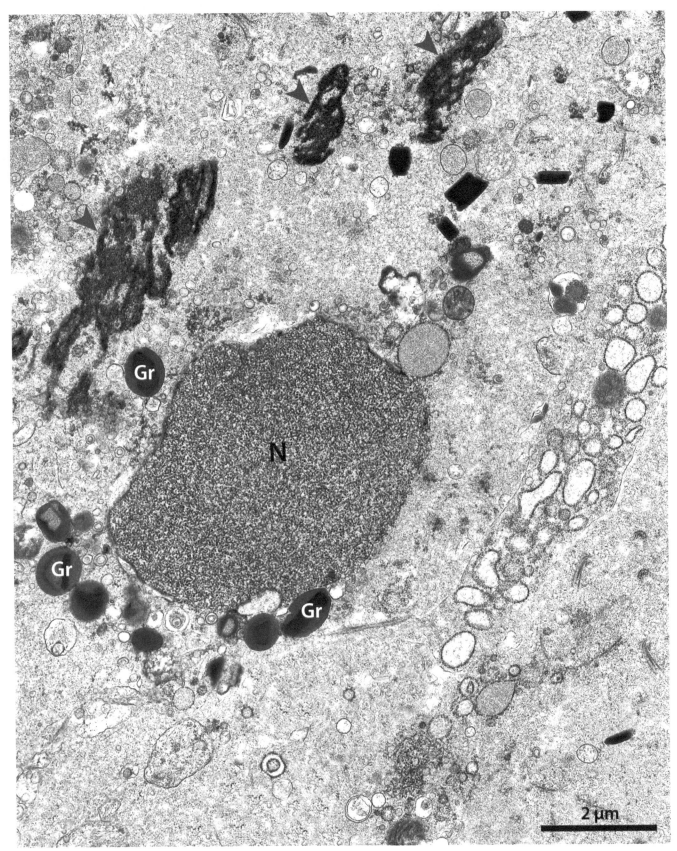

FIG. 6.11   A tissue human eosinophil in the late stage of cytolysis observed in the intestinal mucosa. A decondensed nucleus (N) and pieces of chromatin in advanced chromatolysis *(arrowheads)* are seen together with intact secretory granules (Gr). Biopsy (ulcerative colitis) prepared for TEM.

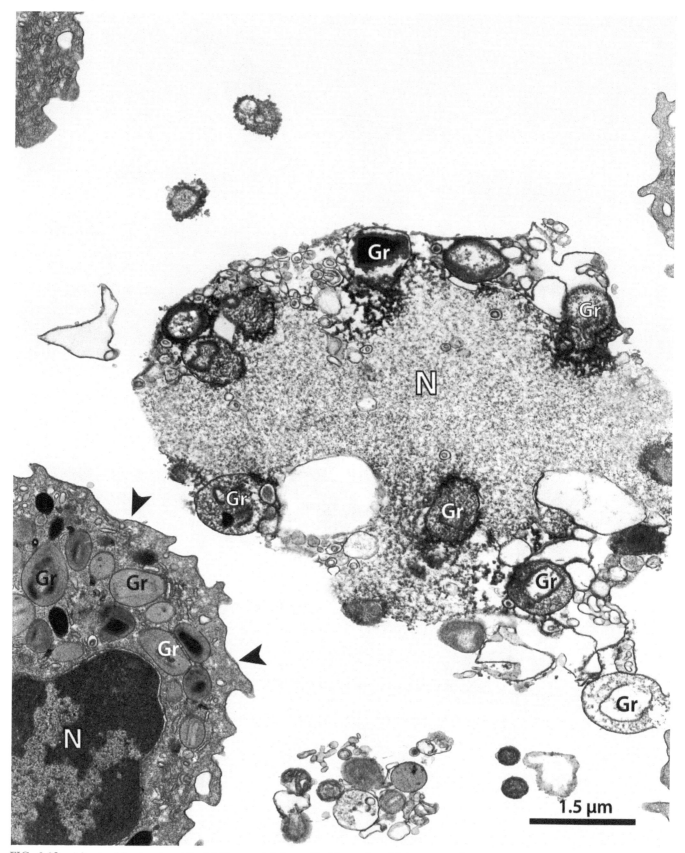

FIG. 6.12   A human blood eosinophil exhibiting a cytolytic profile. The nucleus (N) lost its envelope and appears as decondensed, dissolving chromatin. A normal eosinophil with typical marginal heterochromatin is partially observed (*arrowheads*). Eosinophils were isolated from the peripheral blood using negative selection and processed for TEM.[20] *Gr*, secretory granules.

A.   The cell biology of human eosinophils

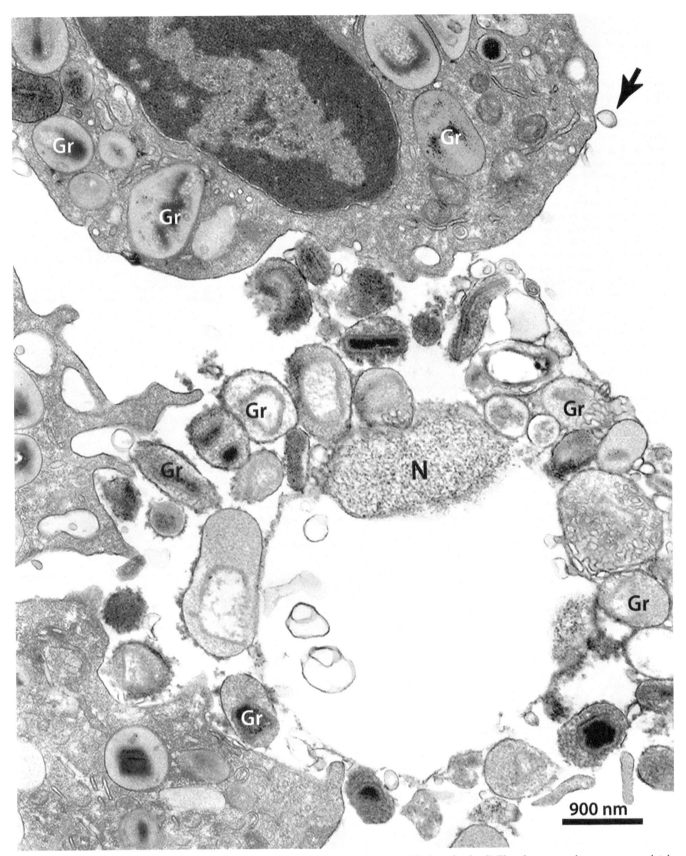

FIG. 6.13   TEM showing a cytolytic human eosinophil nearby two normal eosinophils *(pseudocolored)*. The plasma membrane was completely disrupted and the nucleus (N) is decondensed while the normal nucleus *(purple)* shows areas with heterochromatin and euchromatin. An extra-cellular vesicle is being formed at the cell surface *(arrow)*. Cell were isolated from the peripheral blood. *Gr*, secretory granules.

A. The cell biology of human eosinophils

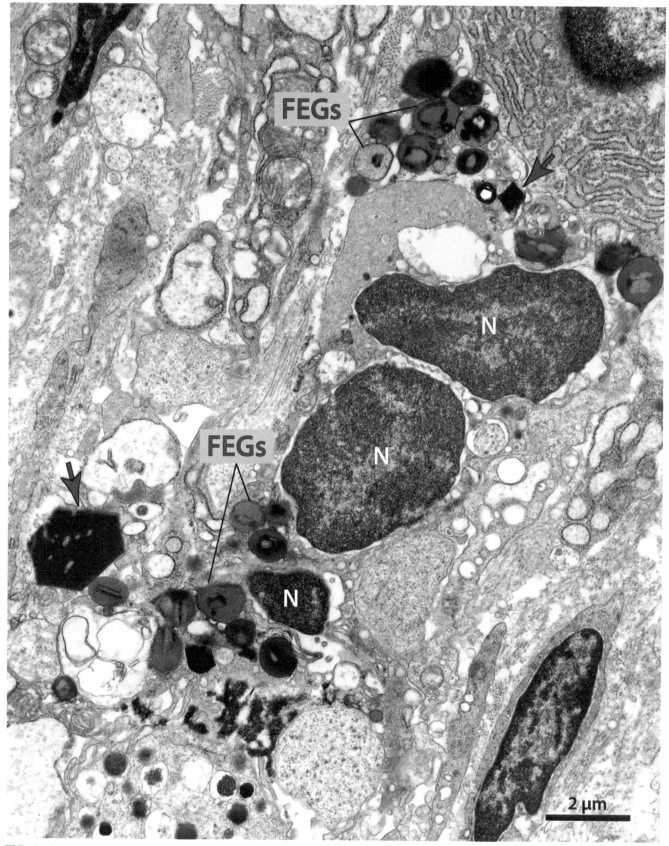

FIG. 6.14   Electron micrograph showing two tissue CLCs *(arrows)*, one of them very small, in the proximity of FEGs generated from cytolytic human eosinophils. Intestinal biopsy (ulcerative colitis) prepared for TEM. *N*, nucleus.

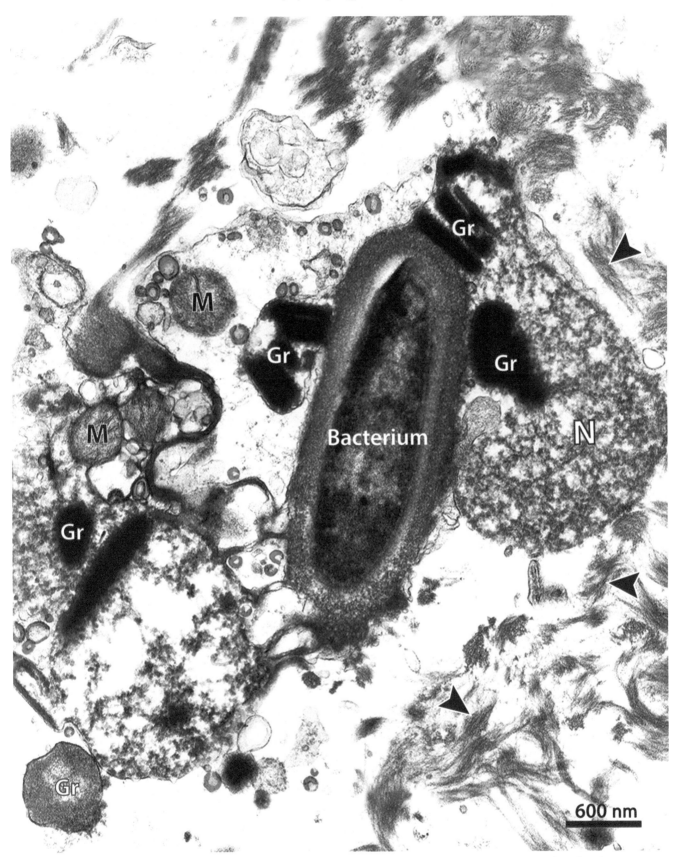

FIG. 6.15 A human eosinophil undergoing necrosis in a tissue site (bacterially infected ileum, ulcerative colitis). The cell underwent intense disintegration. Note nucleus (N) dissolution, swollen and disrupted secretory granules (Gr), and remaining mitochondria (M) and EoSVs *(colored in pink)*. Fibrin strands *(arrowheads)* are seen nearby. Biopsy processed for TEM.

A. The cell biology of human eosinophils

6.3 Cytolytic morphology: ETosis

ETosis is a form of regulated cell death characterized by the release of filamentous chromatin structures, termed extracellular traps (ETs), in response to different stresses. Using EM, these extracellular net-like DNA-protein structures were initially recognized in neutrophils and named neutrophil extracellular traps (NETs).[303] The term NETosis was then coined to refer to the process of NET formation and release[309]; but, because the ability to form NET-like structures had been subsequently identified in mast cells and considered relevant for additional cell types, the term ETosis was also proposed.[310] The terms NETotic[260,261] and ETotic[165] cell death have also been used to refer to cell death associated with extrusion of a meshwork of chromatin and histone-containing fibers. It is now clear that ETosis is a relevant process, not only for neutrophils but also for diverse cell types, including monocytes, macrophages, basophils, and eosinophils.[296]

In human eosinophils, ETosis was described for the first time in 2013, when Ueki and colleagues introduced the concept of eosinophil ETosis (EETosis).[46] By using different stimuli such as calcium ionophore A23187, PMA, immobilized immunoglobulins, and PAF in combination with cytokines, this study demonstrated molecular and morphological eosinophil features similar to those described for neutrophils undergoing this process of death.[46,311,312] EETosis and NETosis are active processes that appear to rely on multiple signaling pathways, including NADPH oxidase-mediated reactive oxygen species (ROS) production.[46,296,311] However, the signals and steps that drive cytolytic cell death in eosinophils associated with the extrusion of extracellular traps are not fully understood, and EETosis, in some circumstances, can occur independently of the major systems critical for ROS production in leukocytes.[313]

Human eosinophils in the process of ETosis undergo dramatic structural alterations in the nucleus, as observed by TEM both in vitro (Figs. 6.16, 6.17, and 6.18) and in vivo (Figs. 6.19–6.22). The typically bilobed nucleus progressively loses its shape (delobulation) to coalesce into single round nuclear structures in parallel with chromatin decondensation with loss of the euchromatin/heterochromatin distinction (Figs. 6.17 and 6.19–6.22).[46,47]

Studies that applied TEM to understand the process of ET release during active cell death of neutrophils in vitro documented that one of the most distinctive features of NETosis is the rupture of the nuclear envelope, which can also depict structural changes characterized by a separation between the inner and outer membranes and vesiculation.[311,314] We detected, for the first time, equivalent morphological alterations in eosinophils undergoing ETosis in vivo during human diseases, for example, in HES (Fig. 6.21) and ECRS (Fig. 6.22). Tissue-infiltrated eosinophils with round, decondensed nuclei showed clear enlargement of the nuclear space, thus leading the two membranes that compose this envelope to become apart and to form vesicles (Figs. 6.21 and 6.22). However, it should be noted that the presence of a swollen nuclear envelope and vesiculation associated with this envelope is not an exclusive characteristic of ETosis. We have occasionally been observing these features in vivo in eosinophils with a cytolytic profile without evidence for ETosis or in vitro in apoptotic eosinophils (Section 6.4).

During the process of ETosis, the breakdown of the nuclear envelope allows further expansion and swelling of the chromatin, which, in consequence, fills a great part of the cell (Figs. 6.16–6.18 and 6.20). Therefore, nuclear changes usually shown by human eosinophils undergoing EETosis are decondensation, delobulation/rounding, alterations/disruption of the nuclear envelope, and chromatin expansion (Figs. 6.16–6.22)[46] As the plasma membrane breakdowns, the decondensed/expanded chromatin is then released into the extracellular space as extracellular fibers of DNA. These structures, which are frequently referred to as eosinophil extracellular traps (EETs), thus emerge from dying eosinophils and can cover large extracellular areas. Figures 6.17, 6.18, and 6.23 show the two-dimensional appearance of EETs visualized in situ by TEM of adhered cells.[46,47] By light microscopy, these extruded EETs are usually observed in vitro as aggregated of net-like structures after staining with cell-impermeable DNA or immunolabeling for the detection of histones, which bind nuclear DNA and form nucleosomes, the smallest constituent units of chromatin fibers.[312] Of note, although the plasma membrane breakdown, as a pattern, is considered the final step of ETosis, this event can occur earlier, preceding disruption of the nuclear envelope (Fig. 6.22).

Interestingly, eosinophil secretory granules released during the process of ETosis retain their morphology and contents or show little evidence of loss of their protein compounds (Figs. 6.17, 6.23, and 6.24).[46,163] Hence, in contrast to NETosis, whereby neutrophil granules lose their limiting membranes and release their components, which are mixed with decondensed chromatin,[311,315] EETosis is a process that can release intact eosinophil secretory granules.[46,163]

While TEM provides visualization of the intracellular morphological events of cells releasing ETs, such as nuclear changes and aspect of the secretory granules, SEM has been used to observe extruded ETs in three dimensions under different conditions (Figs. 6.25 and 6.26). SEM showed that these structures, observed after stimulation with calcium ionophore A23187 (two hours), consist of 25- to 35-nm diameter chromatin fibers aggregated into larger fibers and forming a weblike structure (Figs. 6.27 and 6.28).[46] Of note, because fibrin can also show a weblike appearance when

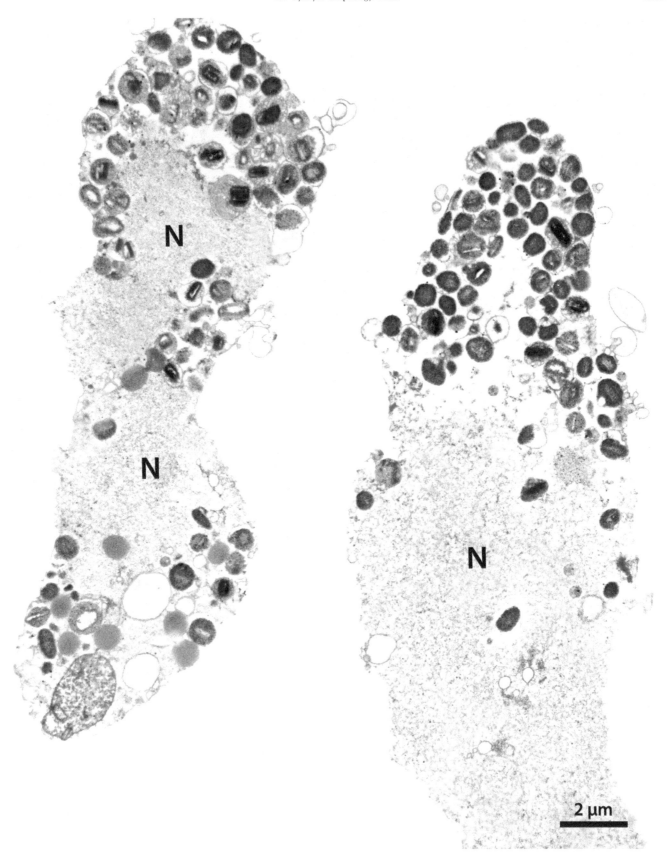

FIG. 6.16   Human eosinophils undergoing extracellular trap cell death (ETosis). The nuclear envelope disintegrated, and the cell is filled with decondensed nuclear (N) chromatin *(colored in purple)* while secretory granules *(yellow)* are mostly intact. Eosinophils isolated from the peripheral blood were activated for two hours with immobilized IgG and processed in situ for TEM.[46]

A. The cell biology of human eosinophils

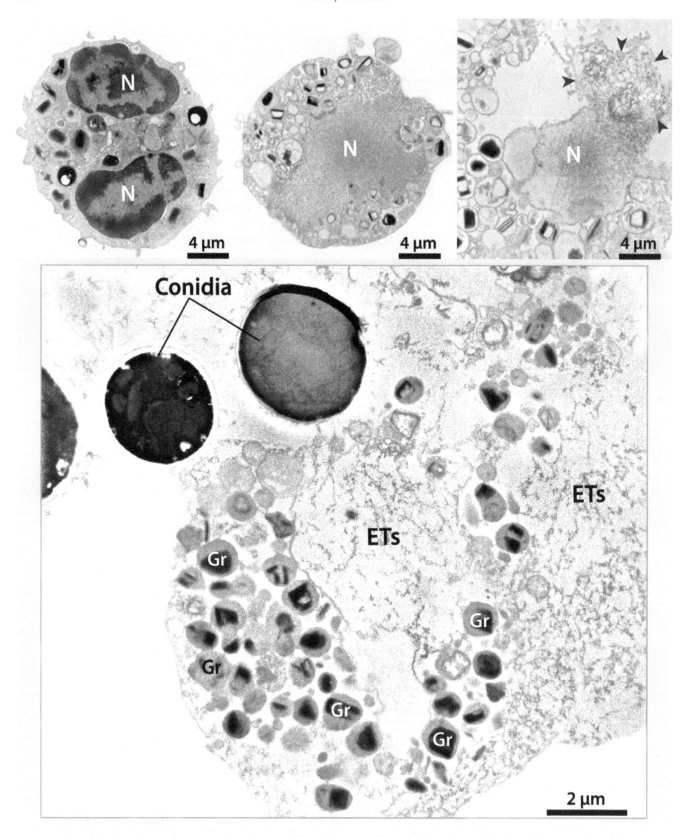

FIG. 6.17 Human eosinophils exposed to *Aspergillus fumigatus* conidia undergo ETosis. A representative unstimulated eosinophil *(top panel, left)* shows a typical bilobed nucleus (N) with well-defined euchromatin and heterochromatin. After six hours of interaction with the fungal cells, eosinophils show nuclear delobulation, disintegration of the nuclear envelope, chromatin *(colored in purple)* decondensation/expansion, and release *(arrowheads)* as DNA extracellular traps (ETs). Note ETs covering a large area outside the cell and entrapping conidia *(bottom panel)*. Most secretory granules (Gr) keep their morphology. Cells were processed in situ for TEM.[47]

A. The cell biology of human eosinophils

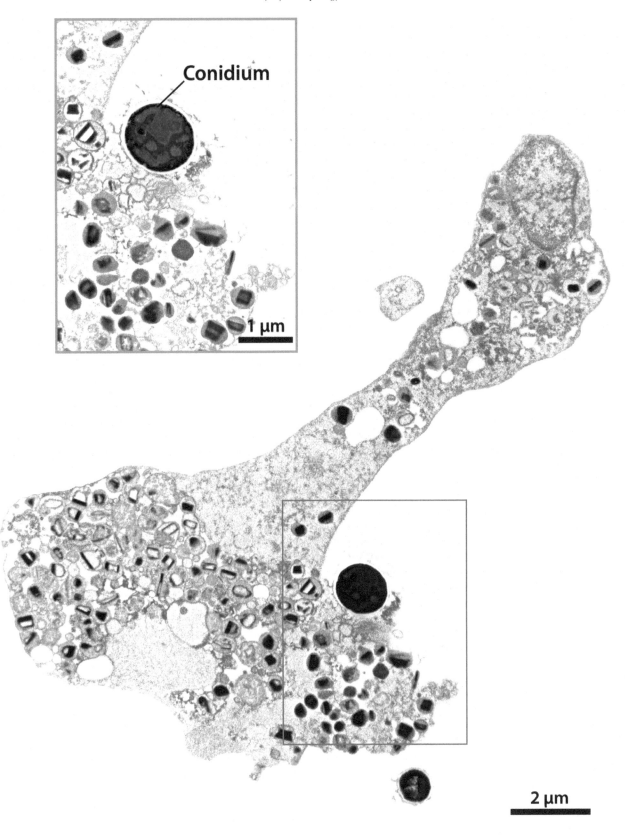

FIG. 6.18   A human eosinophil showing dramatic morphological changes associated with ETosis under interaction with *Aspergillus fumigatus*. The decondensed chromatin *(colored in purple)* is expanded, mixed with cytoplasmic contents, and released as extracellular traps, which surrounds a conidium and secretory granules *(yellow)*. Cells were processed for TEM directly on the slide surface after six hours of incubation with *A. fumigatus* conidia.[47]

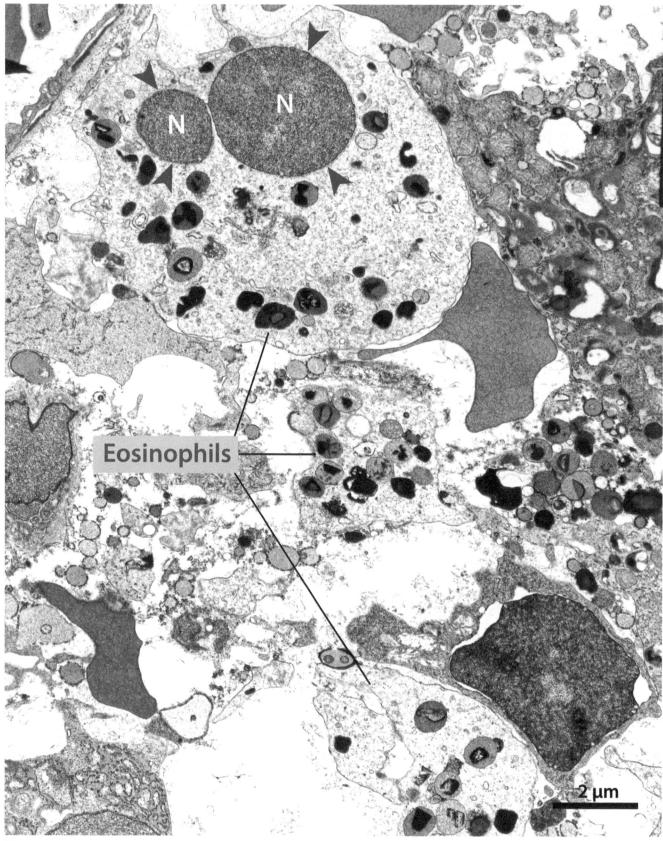

FIG. 6.19   Tissue eosinophils in a biopsy from a patient with ulcerative colitis. One eosinophil shows nuclear delobulation and rounding (*arrowheads*), morphological features indicative of ETosis. Sample processed for TEM. *N*, nucleus.

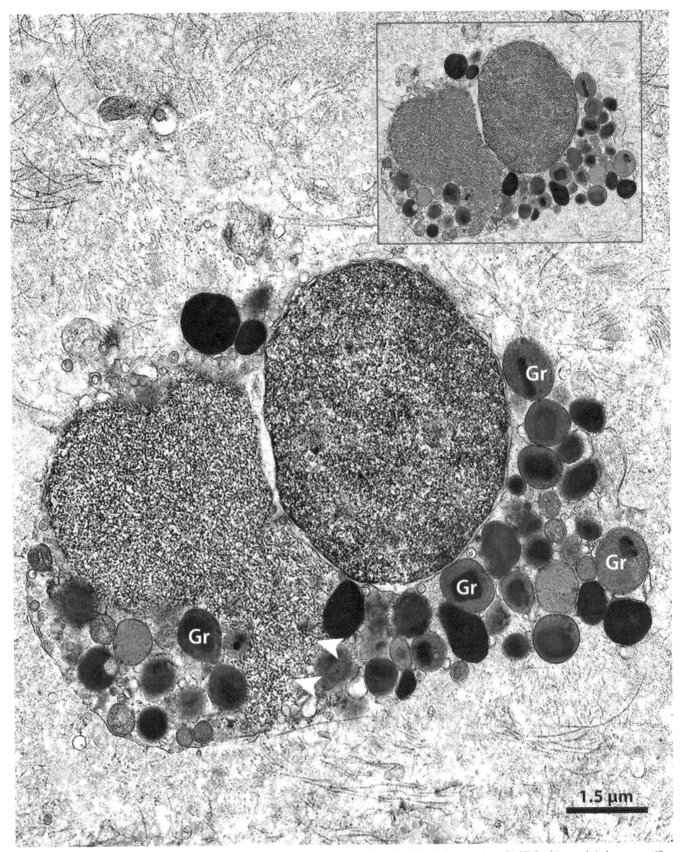

FIG. 6.20 Ultrastructural nuclear changes in a human tissue eosinophil undergoing ETosis. The nucleus *(highlighted in purple in lower magnification)* becomes round, and the chromatin decondenses. Note that the nuclear envelope *(light blue in lower magnification)* is seen around one nuclear lobe but has disintegrated around the other lobe, and the chromatin is expanding *(arrowheads)* into the cytoplasm. Most part of the plasma membrane is disrupted, and intact secretory granules (Gr) are being released into the ECM. Biopsy (ulcerative colitis) processed for TEM.

A. The cell biology of human eosinophils

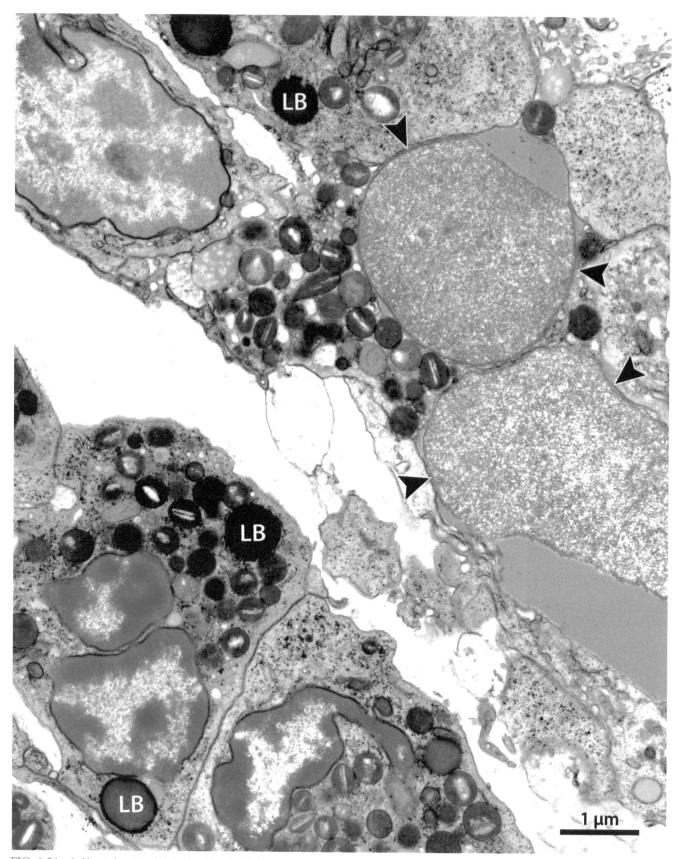

FIG. 6.21   Infiltrated eosinophils in the skin of a HES patient. While two eosinophils show nuclei with well-defined euchromatin and heterochromatin regions, one eosinophil display expanded nuclear lobes *(arrowheads)* with decondensed chromatin and a nuclear envelope with swollen areas (pink). Chromatin *(purple)*; secretory granules *(yellow)*. Sample processed for TEM. *LB*, lipid bodies.

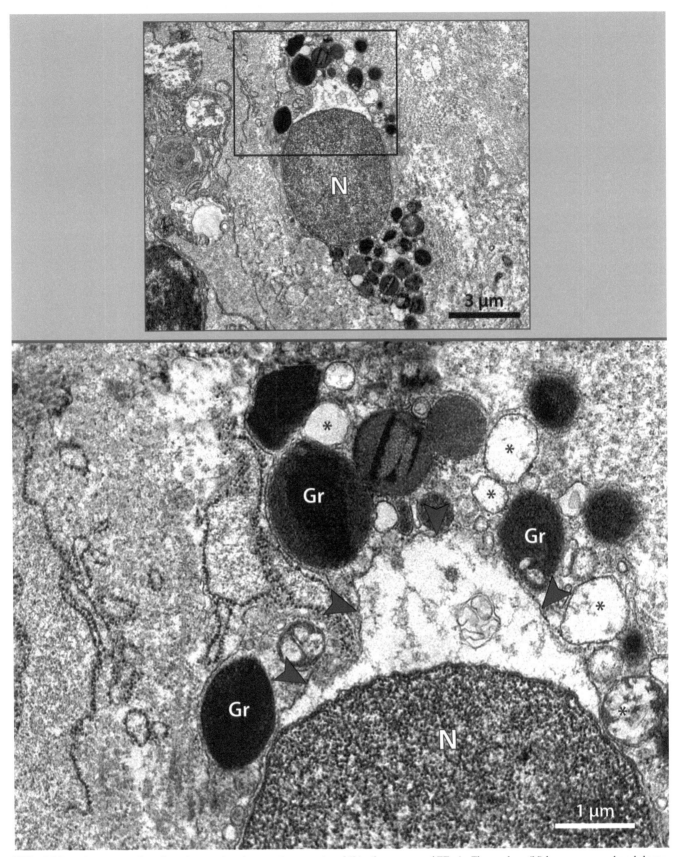

FIG. 6.22 Ultrastructural nuclear changes in a human tissue eosinophil in the process of ETosis. The nucleus (N) becomes round and decondensed while the nuclear envelope undergoes separation of the inner and outer membranes and vesiculation (*). *Arrowheads* indicate the outer membrane of the nuclear envelope. Note the enlargement of the nuclear space between the two membranes and intact secretory granules (Gr). The plasma membrane was lost. ECRS biopsy prepared for TEM.

A. The cell biology of human eosinophils

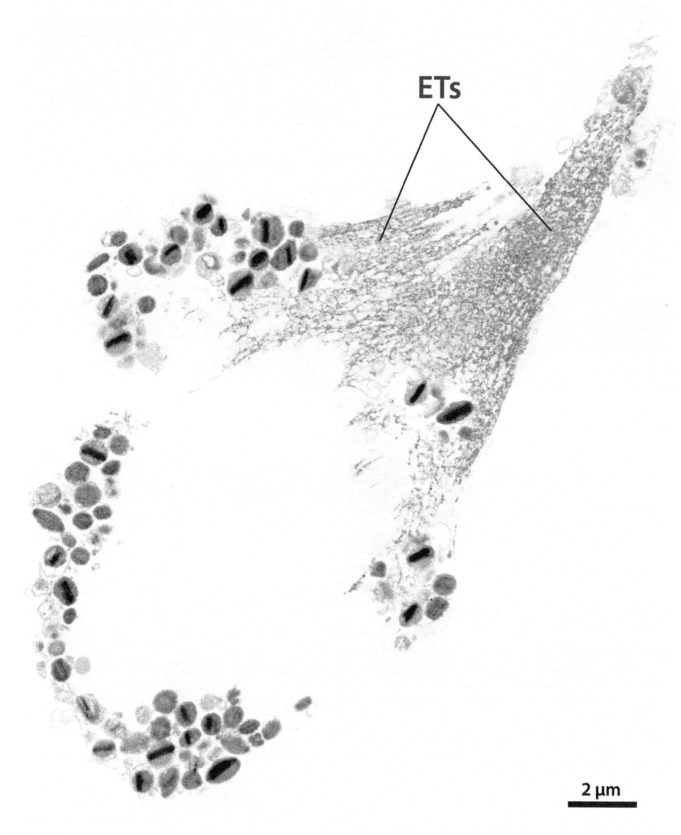

FIG. 6.23 Extracellular traps (ETs), composed of nuclear DNA fibers *(colored in purple)*, emerge from a cytolytic human eosinophil. Note released secretory granules with typical morphology. Peripheral blood eosinophils were stimulated with PMA for two hours and processed in situ for conventional TEM.[46]

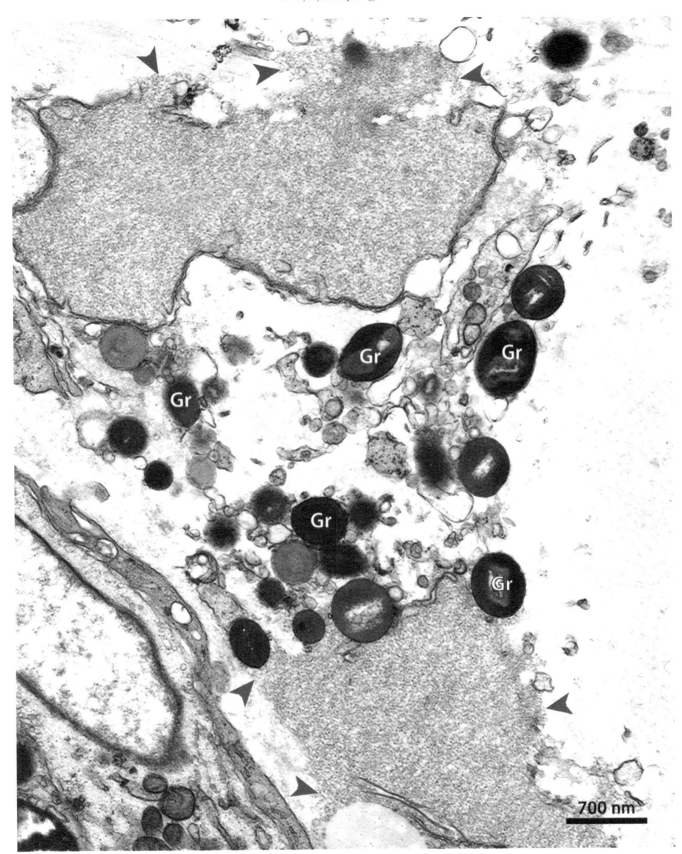

FIG. 6.24 A human tissue eosinophil with a cytolytic profile showing free secretory granules (Gr) and extruded chromatin *(arrowheads)*. Granules exhibit intact delimiting membranes and content losses in the cores. HES biopsy (skin) prepared for TEM.[46]

A. The cell biology of human eosinophils

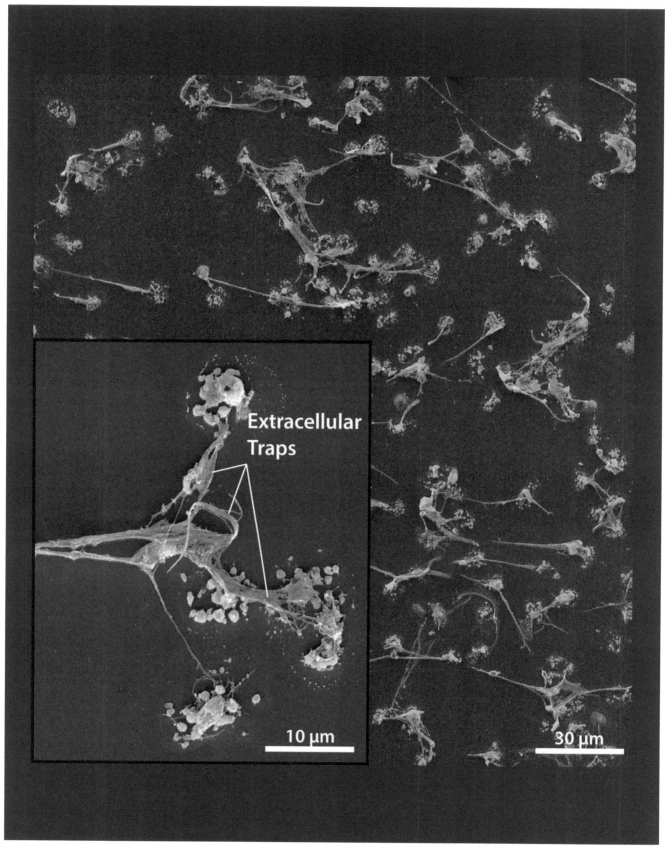

FIG. 6.25   SEM of human eosinophils releasing extracellular traps (ETs). DNA-composed ETs *(colored in orange)* are seen at low magnification after being released from lytic eosinophils *(colored in green)*. Eosinophils isolated from the peripheral blood were stimulated with calcium ionophore A23187 for one hour and processed for SEM.[46]

A. The cell biology of human eosinophils

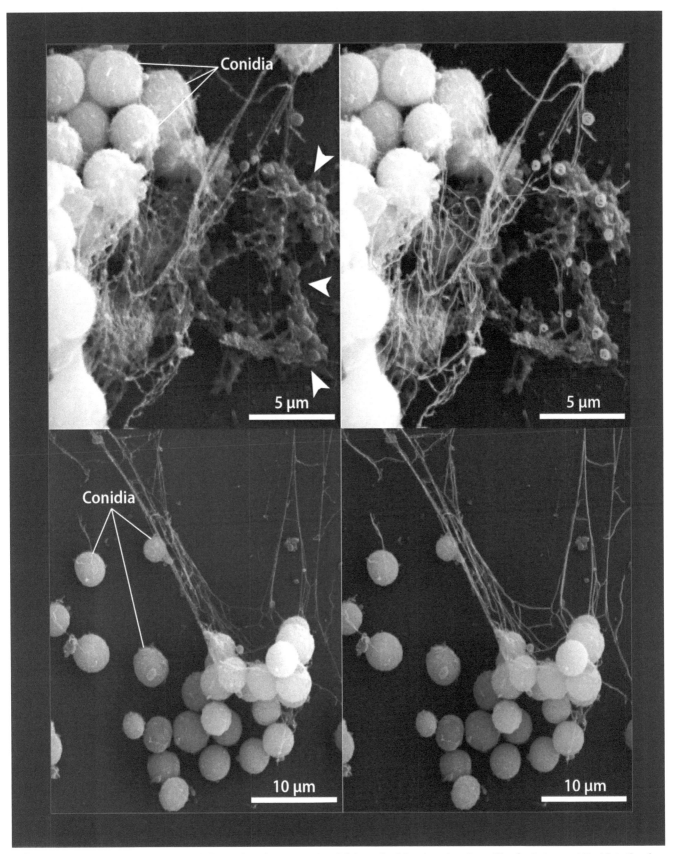

FIG. 6.26 Weblike DNA traps released from a lytic human eosinophil *(arrowheads)* entrapping *Aspergillus fumigatus* conidia. Note the close association of extracellular DNA traps *(highlighted in orange)* with conidia. Released secretory granules were colored in *green*. Human eosinophils were incubated with *A. fumigatus* conidia for six hours and processed for SEM.[49]

A. The cell biology of human eosinophils

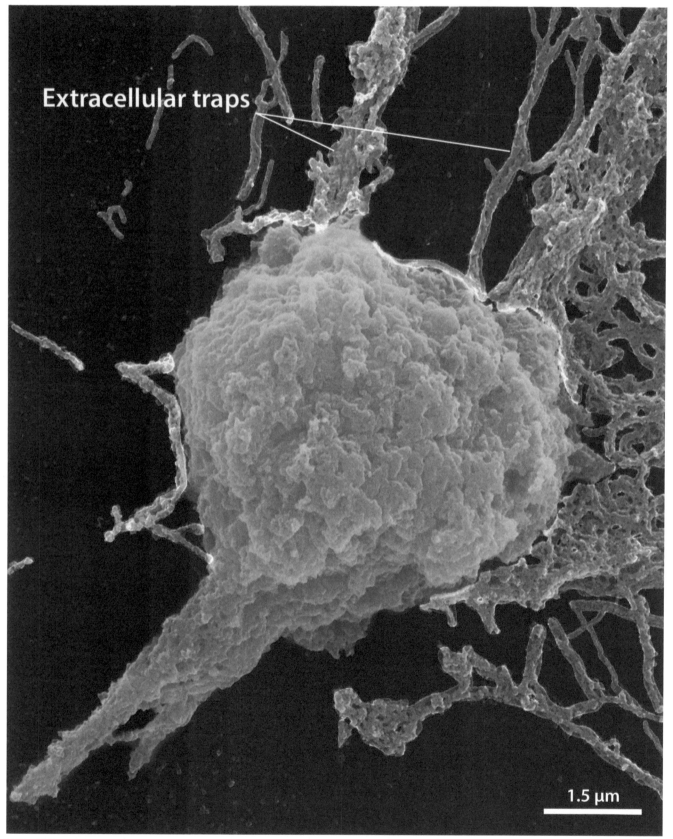

Extracellular traps

1.5 μm

FIG. 6.27   Weblike aspect of extracellular traps *(colored in orange)* released by a human eosinophil *(green)*. Eosinophils isolated from the peripheral blood were stimulated with calcium ionophore A23187 for one hour and processed for SEM.[46]

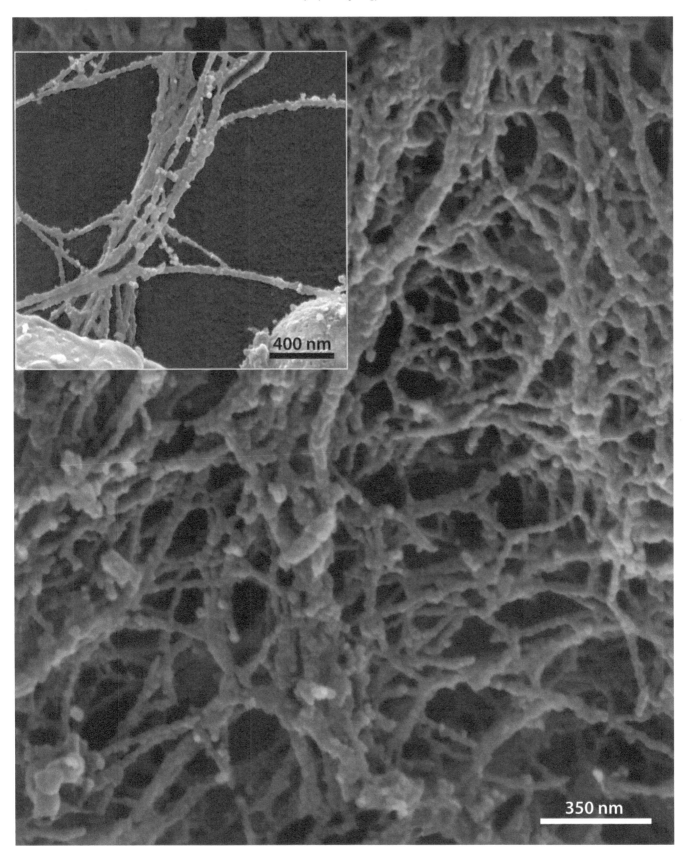

FIG. 6.28 High-resolution SEM showing eosinophil extracellular traps (EETs). EETs consist of 25- to 35-nm diameter chromatin fibers *(high-lighted in orange in lower magnification)* aggregated into larger fibers. Note the weblike aspect of EETs. Eosinophils isolated from the human peripheral blood were stimulated with calcium ionophore A23187 and processed for SEM.[46]

A. The cell biology of human eosinophils

seen by SEM and therefore might be misinterpreted as ETs, samples from inflammatory exudates should be analyzed with additional approaches using appropriate markers.[316]

A common observation at sites of eosinophilic inflammation, in tissues, body fluids, and secretions, is the presence of CLCs. As noted above, these structures are often concomitantly accompanying lytic eosinophils and have been associated with EETosis.[41] TEM has demonstrated that lytic eosinophils with nuclear changes characteristic of ETosis are frequently seen in conjunction with typical CLCs in biopsies of eosinophilic diseases (Figs. 6.29–6.31).

EETosis occurs in vivo during eosinophilic and other diseases such as ulcerative colitis (Figs. 6.19, 6.20, and 6.31), HES (Figs. 6.21, 6.24, and 6.29),[46] ECRS (Figs. 6.22 and 6.30),[46,163] otitis media,[163] chronic obstructive pulmonary disease,[317] and eosinophilic granulomatosis with polyangiitis (EGPA).[279] EETs can also form in response to diverse stimuli derived from pathogens and their products (reviewed in[165]). For example, *Aspergillus fumigatus*, the opportunistic fungus that dominates in allergic bronchopulmonary aspergillosis (ABPA), a disease characterized by massive infiltration of eosinophils,[48] leads to EET formation from cytolytic eosinophils[313] as observed in vitro by both TEM (Figs. 6.17 and 6.18)[47] and SEM (Fig. 6.26; see also Fig. 3.36 in Chapter 3).[47,49,318] DNA traps were also demonstrated by immunofluorescence in clinical samples (bronchial secretions) of patients with ABPA in association with eosinophils.[49,319]

Bacteria, such as *Escherichia coli*, likewise induce the formation of EETs, but the resultant responses were associated with noncytolytic eosinophils that were releasing mitochondria-originated (not nuclear) DNA to form ETs.[320] Although the microbicidal properties of varied DNA EETs are not fully established, these structures have been considered potentially relevant for the host defense to invasive pathogens, as already demonstrated for neutrophils (reviewed in[165]).

6.4 Apoptotic morphology: General aspects

TEM was the tool used by John Kerr, Andrew Wyllie, and Alastair Currie to identify and coin the term "apoptosis" in 1972 as a form of programmed cell death morphologically different from necrosis.[321] These authors described apoptosis as a "distinctly different mode of cellular death with ultrastructural features that are consistent with an active, inherently controlled phenomenon," which plays an important role in the regulation of cell numbers in a variety of tissues under both physiological and pathological conditions.[321]

By using TEM, Kerr and colleagues studied the evolution of apoptosis in different tissues and situations, thus demonstrating that (i) apoptosis characteristically affects scattered single cells, that is, cells that have separated from their neighbors; (ii) the initial morphological events in apoptosis comprise cell shrinkage, chromatin condensation (pyknosis), and aggregation of the nuclear chromatin in dense masses beneath the nuclear envelope; (iii) condensed cells fragment in apoptotic bodies, large membrane-bound vacuoles containing packed organelles; and (iv) apoptotic bodies are then phagocytosed and degraded by other surrounding cells.[321] Therefore, in contrast to cells exhibiting a cytolytic profile, in cells undergoing apoptosis, the plasma membrane remains intact. Interestingly, in every case evaluated by these authors, in both humans and experimental models, in health and disease, including malignant neoplasms, the ultrastructural features were essentially the same.[321] As anticipated by Kerr and colleagues, apoptosis is extensively recognized as a widespread modality of cell death being initiated and inhibited by a variety of environmental stimuli, and TEM is still considered a "gold standard" for the identification of apoptotic cells.[322]

Eosinophils undergo canonical apoptosis with ultrastructural alterations comparable to those of other cells dying by this process.[204,323,324] Eosinophils in the process of apoptosis initially become rounded and show one of the most noticeable features of apoptosis: condensation of the nucleus (Figs. 6.32–6.36) and its fragmentation (karyorrhexis) into smaller pieces (Fig. 6.37).[325] Margination of compacted chromatin, a particular aspect of the nucleus during apoptosis, can be observed in apoptotic eosinophils (Figs. 6.36 and 6.37). Chromatin margination is first limited to clusters underlying the nuclear envelope (Fig. 6.36) and then organized in cup-shaped structures (Fig. 6.37), as seen in other cells.[308,326] These electron-dense areas of rearranged chromatin are due to DNA fragmentation.[326] An approach in which the TUNEL technique was applied in situ at the ultrastructural level revealed the specific localization of DNA breaking points in the electron-dense apoptotic chromatin clumped at the nuclear envelope periphery.[327] In human eosinophils, in addition to chromatin alterations, vesiculation from the external membrane of the nuclear envelope can be noted at times (Fig. 6.38). Nuclear ultrastructural changes, typical of apoptosis, have been described, for example, in eosinophils' aging in cultures (Figs. 6.33 and 6.35), under effects of glucocorticoids,[324] or in response to HIV infection[204] (Figs. 6.34 and 6.38).

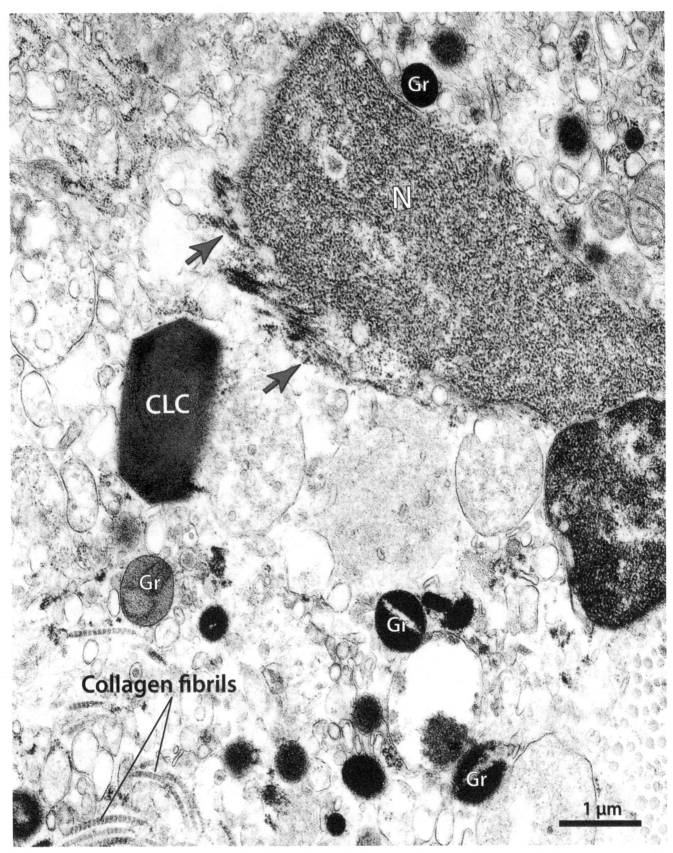

FIG. 6.29 Eosinophil cytolysis in the papillary dermis from a HES patient. Free extracellular granules (Gr) and a typical CLC are observed in the ECM. Note released decondensed chromatin *(arrows)* and collagen fibrils. Biopsy processed for TEM.[41] *N*, nucleus.

A. The cell biology of human eosinophils

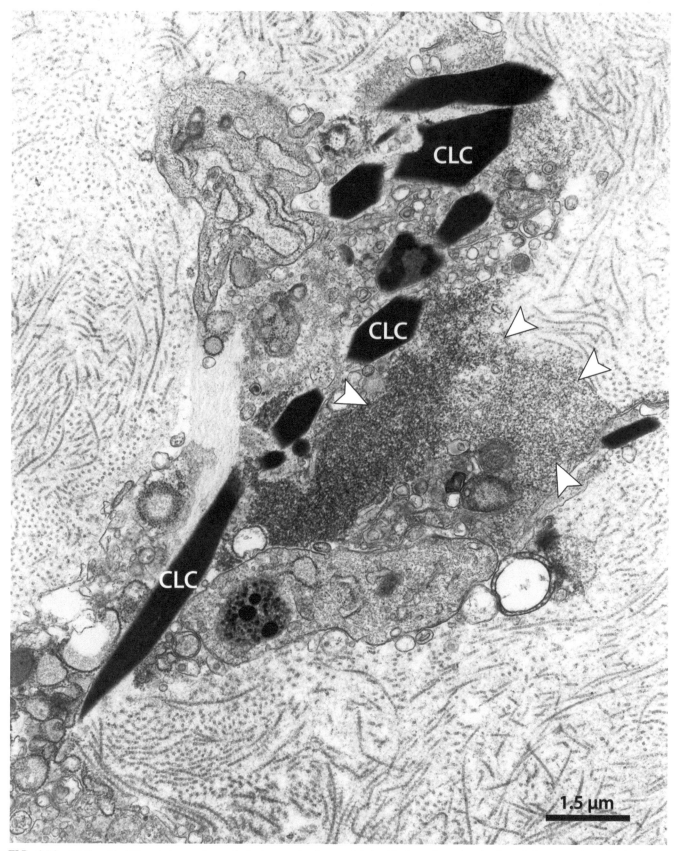

FIG. 6.30   CLCs with different sizes and morphologies are closely associated with decondensed and released chromatin *(arrowheads)* from a cytolytic human eosinophil. Note free EoSVs *(colored in pink)* intermixed in the chromatin. Biopsy (frontal sinus, ECRS) processed for TEM.

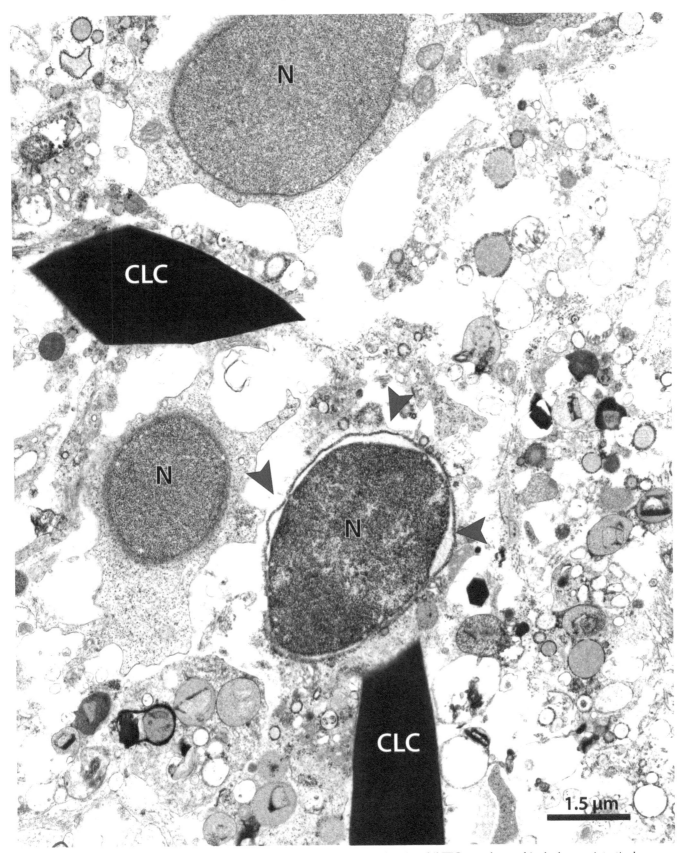

FIG. 6.31   Tissue CLCs, nuclei showing ultrastructural features of ETosis, and eosinophil FEGs are observed in the human intestinal mucosa. Nuclei (N) are decondensed and round. Note a swollen nuclear envelope *(arrowheads)*. Biopsy (ulcerative colitis) processed for TEM.

A. The cell biology of human eosinophils

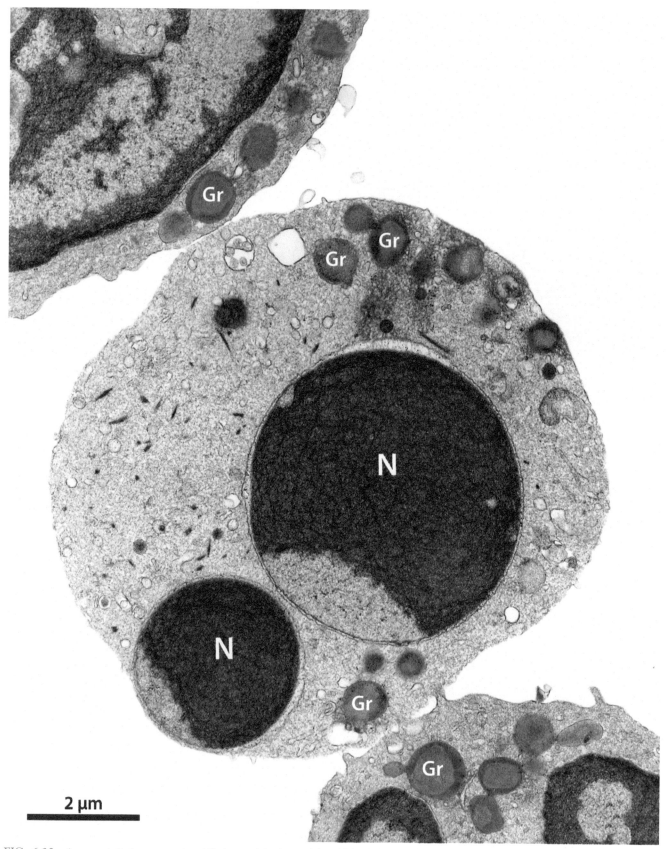

FIG. 6.32   An apoptotic human eosinophil observed in the peripheral blood. Note cell rounding and nuclear (N) condensation while the plasma membrane is intact. Eosinophils were isolated from the peripheral blood and processed for TEM. *Gr*, secretory granules.

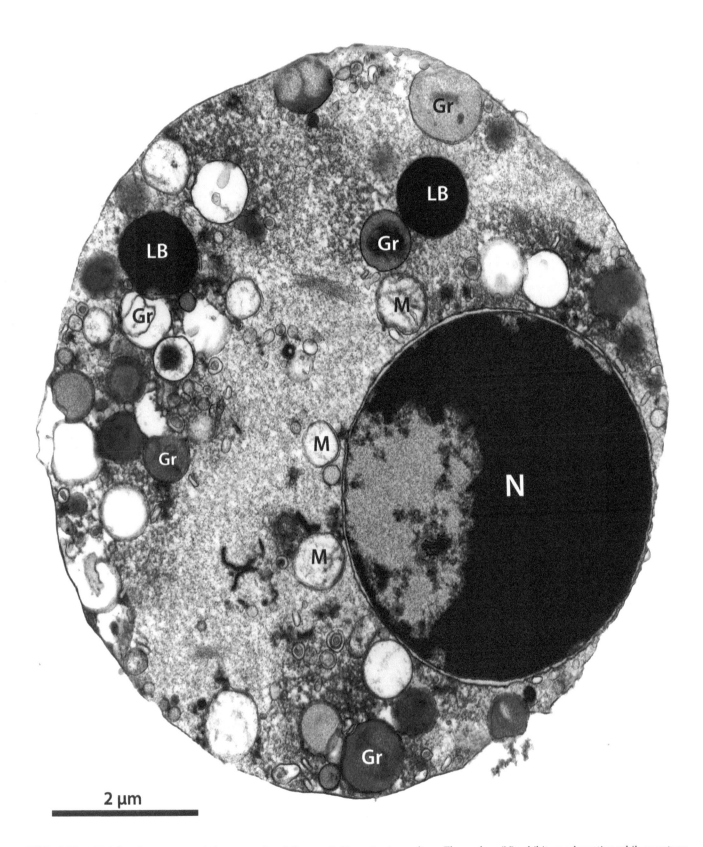

FIG. 6.33 TEM showing an apoptotic human eosinophil generated by aging in a culture. The nucleus (N) exhibits condensation while secretory granules (Gr), lipid bodies (LB), and mitochondria (M) are compacted in the cytoplasm.

A. The cell biology of human eosinophils

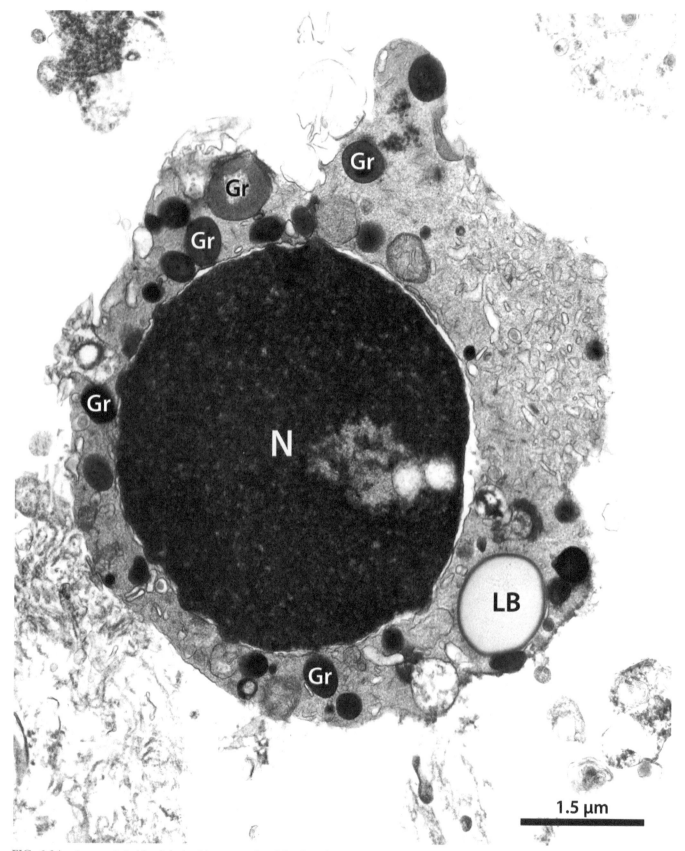

FIG. 6.34   An apoptotic blood-derived human eosinophil cultured with HIV-1. The nucleus (N) is highly condensed, and the cytoplasm has shrunken. Note an electron-lucent lipid body (LB) and specific granules (Gr). Culture processed for TEM.[204]

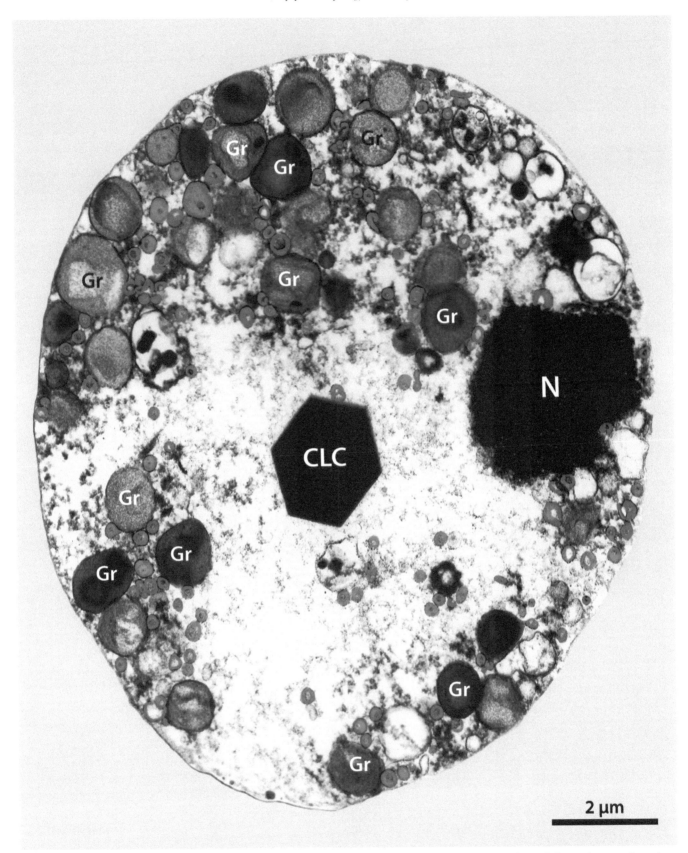

FIG. 6.35 A rare image of an aging human eosinophil showing apoptosis features and a CLC in the cytoplasm. Cell contains a condensed nucleus (N), compacted EoSVs *(colored in pink)* and granules (Gr), and intact plasma membrane. Cultured cells were processed for TEM.

A. The cell biology of human eosinophils

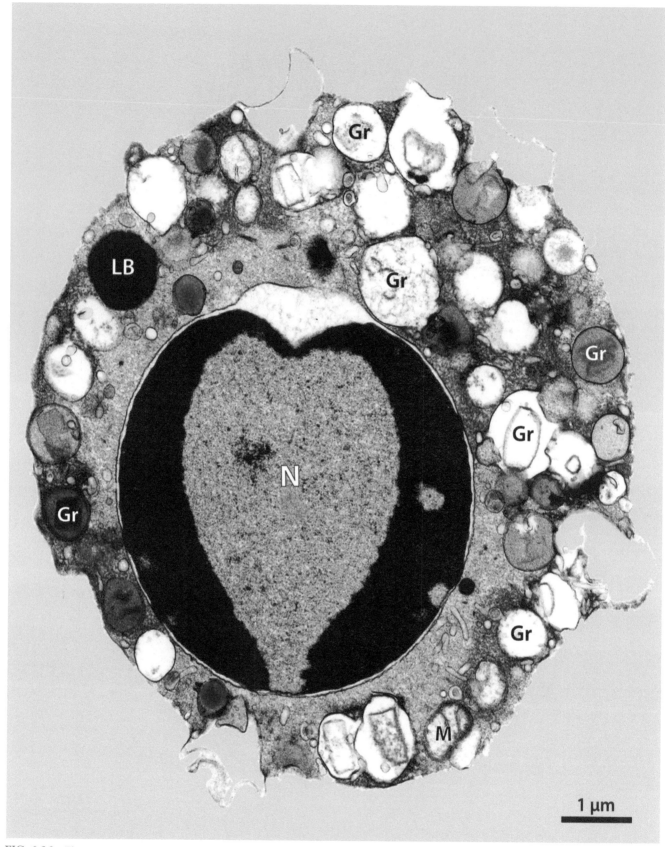

FIG. 6.36   Electron micrograph of an apoptotic human eosinophil. Note the strongly electron-dense marginal chromatin within the nucleus (N), secretory granules (Gr), and a single lipid body (LB). Cultured eosinophils were processed for TEM.[522] *M*, mitochondrion.

A. The cell biology of human eosinophils

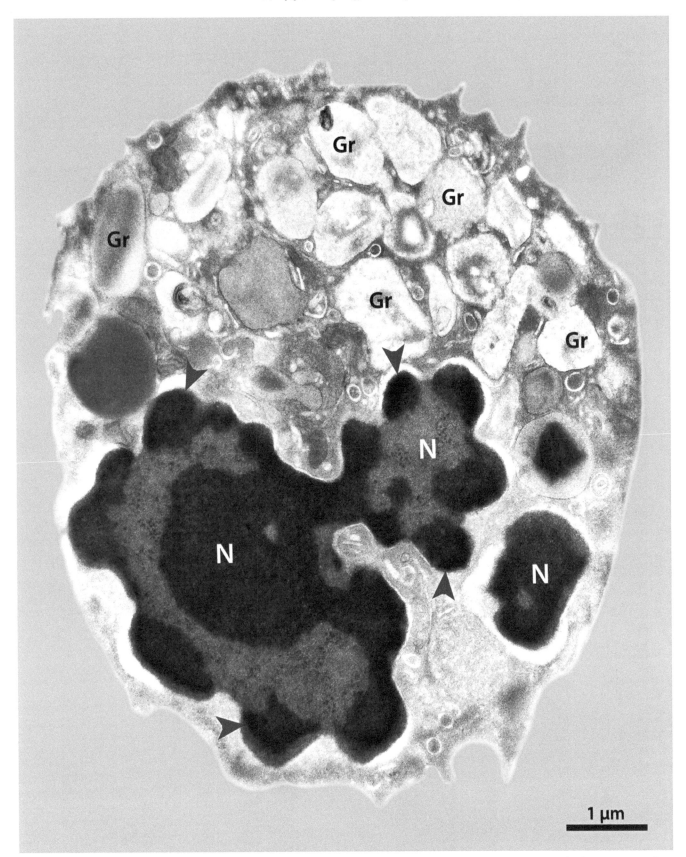

FIG. 6.37   An apoptotic human eosinophil showing condensed chromatin organized as cup-shaped structures *(arrowheads)* due to DNA fragmentation. Most secretory granules (Gr) are enlarged and show mobilized content. Cultured eosinophils were prepared for TEM.

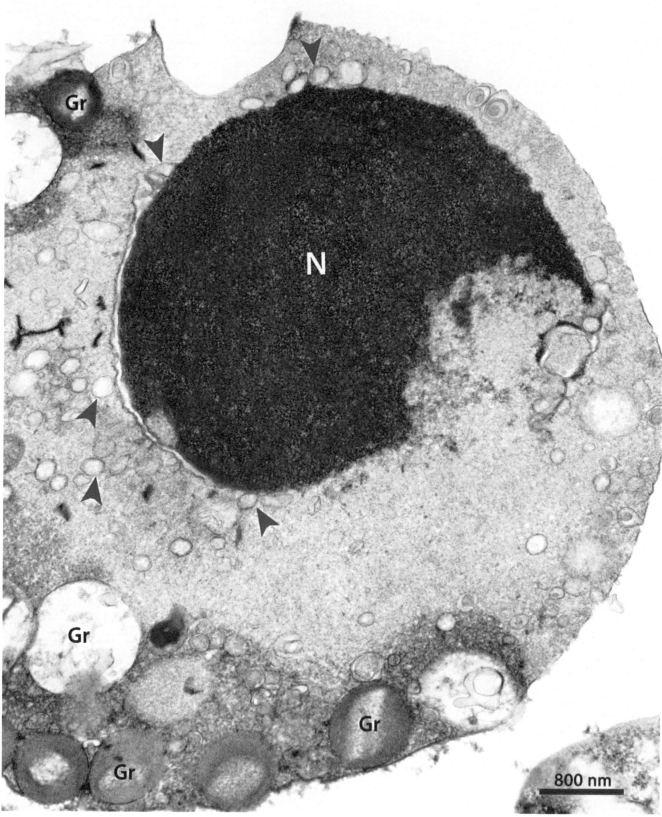

FIG. 6.38 Electron micrograph of an infected eosinophil showing alterations of the nuclear envelope at high magnification. Note the presence of vesiculation *(arrowheads)* from the outer membrane and a highly electron-dense nucleus (N). Human eosinophils cultured with HIV-1 were processed for TEM.[204]

A. The cell biology of human eosinophils

Cells in apoptosis also undergo changes in the cytoplasm, such as cytoplasmic contraction, and the cytoplasmic matrix usually becomes more electron-dense and associated with aggregated organelles. Eosinophils with such characteristics can be occasionally observed in body fluids, cultures (Fig. 6.33), and tissues (Fig. 6.39).

Cells that undergo apoptosis are dismantled in a controlled manner that minimizes damage. Apoptotic cells show the initial formation of membrane blebs (circular protuberances at the plasma membrane) and subsequent separation of blebs to generate apoptotic bodies, which are recognized and removed by phagocytes.[328] The process of phagocytic clearance of apoptotic bodies is named efferocytosis and occurs without an apparent inflammatory or immune reaction.[296,325,329,330] The collapse of apoptotic cells into numerous apoptotic bodies, which allows cells to be removed with their plasma membranes intact, prevents the potentially damaging release of cellular constituents into the surrounding milieu and likely makes this task easier for phagocytes.[325] Apoptotic bodies have also been deemed a subtype of extracellular vesicle (EV) and, in conjunction with other types of EVs released from dying cells, broadly known as apoptotic cell-derived EVs, are becoming recognized as structures with immune regulatory roles (reviewed in[179,328,331]). It is now recognized that multiple signs anchored on apoptotic bodies promote their uptake by phagocytes.[328] Therefore, apoptotic bodies, as other EVs produced by viable cells, seem to have the ability to communicate with surrounding cells and even activate or dampen immune responses, but this concept remains under investigation.[328]

The content of an apoptotic body depends on the cellular constituents that are present in the cytoplasmic protuberance that gave rise to it.[321] Thus, apoptotic bodies can consist of cytoplasm with tightly packed organelles with or without a nuclear fragment or can consist almost entirely of condensed nuclear chromatin.[321] Regardless of their contents, apoptotic bodies are always surrounded by the plasma membrane and exhibit large sizes with a range usually between 1 and 5 μm.[179,329]

In eosinophils, blebbing and formation of apoptotic bodies can be observed in vitro (Fig. 6.40), and uptake of entire apoptotic eosinophils by professional and nonprofessional phagocytes, such as epithelial cells, is demonstrated both in vitro and in vivo by TEM. For example, cultures of peripheral blood–derived eosinophils, in which apoptotic eosinophils were generated by aging in cultures and examined by TEM, showed apoptotic eosinophils being ingested by bronchial epithelial cells after one hour of interaction.[332] The authors suggested a nonpassive role of the airway epithelium in the resolution of eosinophilic inflammation in asthma.[332] On the other hand, failure or inhibition of clearance mechanisms permits apoptotic cells to enter into secondary necrosis. In this phase of necrosis, cellular disintegration is identifiable ultrastructurally as cells showing some features of apoptosis and necrosis, such as condensed nuclei and rupture of the plasma membrane (see Fig. 8.75 in Chapter 8).[173,333] These changes might result in an inflammatory response.[331,334] Progression of apoptosis to secondary necrosis can be observed in situations where clearance by scavengers does not operate after a complete apoptotic program.[333,335,336]

In the absence of exogenous stimuli, eosinophils die spontaneously by apoptosis briefly after terminal differentiation (life span of 8–18h in circulation).[334,337,338] Apoptotic eosinophils can be sporadically found in the peripheral blood (Fig. 6.32), cultures (Figs. 6.33–6.38), and tissues (Fig. 6.39). However, apoptotic eosinophils are not a common finding in vivo in the presence of inflammatory signaling. For example, when the frequency of apoptotic eosinophils was investigated in inflamed upper airway tissues from patients with nasal polyposis[174] and seasonal allergic rhinitis[153] by TEM and quantitative analyses, no eosinophils displaying classical ultrastructural signs of apoptosis were detected in contrast with the presence of eosinophils undergoing cytolysis and PMD. Another study from the same group evaluated eosinophils in human nasal polyps in vivo from nonselected patients.[173] By using TEM and molecular markers such as TUNEL to detect DNA fragmentation occurring at an advanced stage of apoptosis, it was also demonstrated that apoptotic eosinophils are rare in human nasal polyps in vivo, suggesting that the turnover of human airway tissue eosinophils in vivo, irrespective of steroid treatment, largely involves mechanisms other than apoptosis.[173] On the other hand, the maintenance of eosinophils from polyp tissues or isolated from the peripheral blood in culture without growth factors leads to eosinophil apoptosis.[173,339] Indeed, multiple cytokines and extracellular matrix components have been shown, both in vivo and in vitro, to antagonize eosinophil programmed cell death and promote cell longevity.[334,337] Thus, eosinophil apoptosis can result from the elimination of survival factors such as IL-3, IL-5, IL-13, and granulocyte-macrophage colony-stimulating factor (GM-CSF), which significantly prolong eosinophil survival under nonpathological conditions (reviewed in[338,340,341]). Other cytokines that support eosinophil survival include leptin, TNF-α, IL-33, and IFN-γ.[334,338,340,341] Contrarily, glucocorticoids and varied other stimuli are capable of accelerating eosinophil apoptosis (reviewed in[337]).

From the time when apoptosis was described, the molecular mechanisms regulating this type of cell death have been extensively investigated in multiple organisms. It is now established that apoptosis occurs through two primary pathways, namely extrinsic and intrinsic (reviewed in[295,296]).

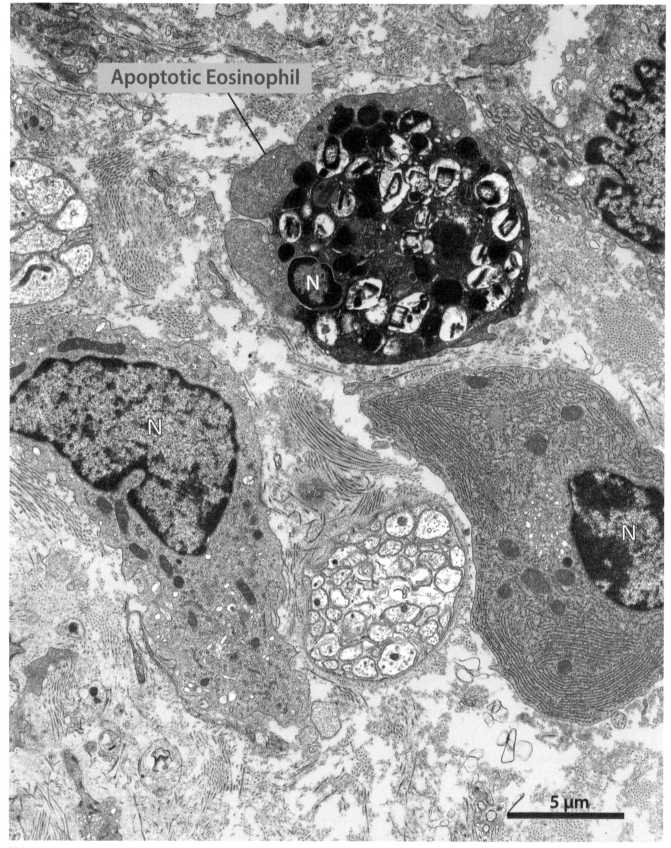

Apoptotic Eosinophil

FIG. 6.39 Tissue human eosinophil in the process of apoptosis. Cell rounding and cytoplasmic changes (cell shrinkage, aggregation of organelles, and electron-dense matrix) are observed. Note other cell types with normal morphology. Intestinal biopsy (Crohn's disease) processed for TEM. The ECM was colored in *green*. *N*, nucleus.

A. The cell biology of human eosinophils

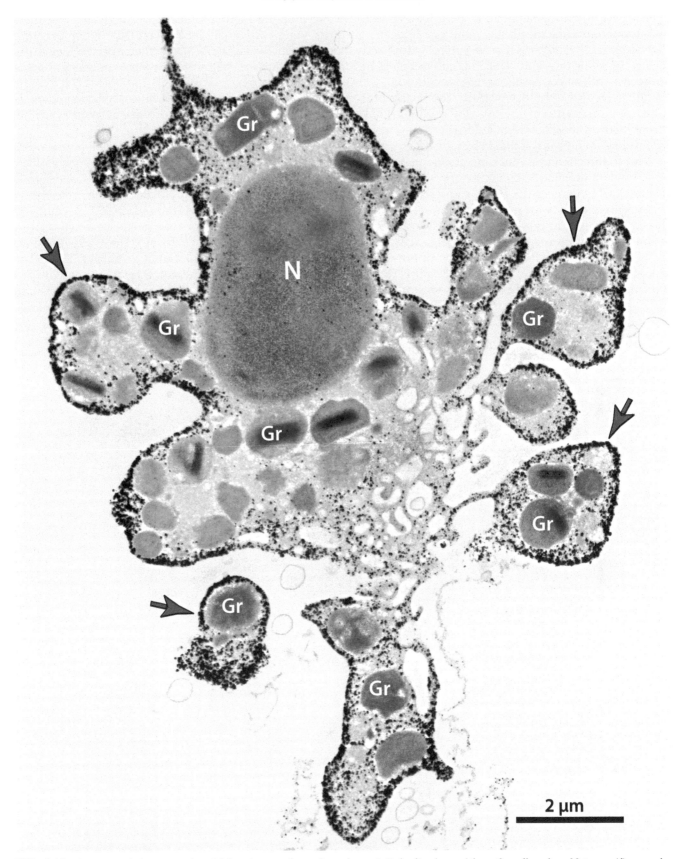

FIG. 6.40   An apoptotic human eosinophil forming membrane-bound apoptotic bodies *(arrows)* from the cell surface. Note specific granules (Gr) in the cytoplasm and within apoptotic bodies. Immunolabeling for Gal-10 shows the typical localization of this protein underneath the plasma membrane. Immunonanogold EM was performed as before.[37] *N*, nucleus.

A. The cell biology of human eosinophils

Extrinsic apoptosis can occur by perturbations of the extracellular microenvironment and is activated by either of two types of plasma membrane receptors: (i) death receptors, whose activation depends on the binding of cognate ligand(s); and (ii) dependence receptors, whose activation happens when the levels of their specific ligand drop below a specific threshold level.[295] Death receptors include Fas cell surface death receptor [also known as apoptosis antigen 1 (APO-1) or cluster of differentiation 95 (CD95)] and other members of the TNF-α receptor (TNFR) super-family such as TNFRSF1A/TNFR1. The family of dependence receptor ligands comprises several members that promote cell survival in physiological conditions but can activate lethal signaling cascades when ligand availability drops below a specific threshold level.[295,296]

Intrinsic apoptosis is initiated by varied perturbations of the extracellular or intracellular microenvironment trig-gered by mitochondrial outer membrane permeabilization, which is tightly controlled by the pro-survival B-cell lymphoma 2 protein (Bcl-2) family of proteins.[296] These perturbations include, for example, withdrawal of growth factor(s), DNA damage, ER stress, and reactive oxygen overload.[295] Caspase (CASP) proteins, cysteine proteases typ-ically with specificity for aspartate, act as apoptotic initiators or effectors. CASP3, CASP6, and CASP7 are considered the common effector caspases for both extrinsic and intrinsic apoptosis.[296]

Eosinophil apoptosis can result from the activation of either extrinsic or intrinsic pathways. The extrinsic pathway is initiated in eosinophils by activation of pro-death receptors such as TNFR superfamily members, TNF-related apoptosis-inducing ligand receptors 1, 3, and 4 (TRAIL-R1, R3, and R4), and Fas.[337] Human resting eosinophils express both the Fas receptor and Fas ligand.[341] After treatment with Fas antibody, TEM of cultured eosinophils demonstrates changes consistent with apoptosis.[342,343] Phagocytosis of anti-Fas antibody-treated eosinophils by monocyte-derived macrophages can be noted in cocultures.[343] Moreover, mitochondria have been considered as another pathway with a role in facilitating eosinophil apoptosis.[129]

7

Immature eosinophils

7.1 Overview

The life cycle of the eosinophil is divided into bone marrow, blood, and tissue phases. Like other hematopoietic lineages, eosinophils are derived from common hematopoietic stem cells and multilineage progenitors in the bone marrow. Eosinophil differentiation from multipotent progenitor cells leads to the formation of eosinophil progenitors (EoPs) committed to the eosinophil lineage (reviewed in[344,345]). Once committed, further maturation and proliferation of eosinophils are directed mainly through extrinsic signals from the eosinophil's tissue microenvironment, where each component has the potential to impact the eosinophil phenotype and function.[345,346] Thus, hematopoietic cells depend on a finely balanced network of signaling pathways to progress from multipotent progenitors to terminally differentiated cells, and cellular identity is established through activating and repressing specific gene sets.[347]

GM-CSF, IL-3, and IL-5 are cytokines with an important role in eosinophil maturation and effector functions. IL-3 and GM-CSF are relatively nonspecific and stimulate the proliferation of neutrophils, basophils, and eosinophils. In contrast, IL-5 is eosinophil-selective, acting as a key cytokine in the regulation of eosinophil proliferation and terminal differentiation in the bone marrow as well as in the survival and activation of eosinophils in blood and tissues.[345]

The specific IL-5R consists of the IL-5R α chain (IL-5Rα) and the beta common (βc) subunit.[348] The expression of IL-5Rα in human common myeloid progenitors (hCMPs) drives proliferation and differentiation of human EoPs. These IL-5Rα-positive human EoPs give rise exclusively to eosinophils but not neutrophils or basophils. The early human EoP is now considered to be generated from the hCMP or its upstream multipotent progenitor independent of the human granulocyte/macrophage progenitor and megakaryocyte/erythrocyte progenitor, which originate neutrophils/monocytes/macrophages and megakaryocyte/erythroid cells, respectively, without developing into the eosinophil lineage.[349] Therefore, the identification of surface markers, notably IL-5Rα and CD34, in conjunction with characteristic stem cell markers, is used to define EoPs and to distinguish lineage-committed EoPs from other progenitors.[345]

Although eosinophilopoiesis occurs primarily in the bone marrow, emerging evidence suggests that eosinophils in different states along their differentiation continuum can be found at extramedullary sites, particularly during inflammatory diseases (reviewed in[345,346,350]). It is recognized that EoPs are equipped with the appropriate chemokine receptors as well as adhesion and homing molecules for trafficking to inflammatory sites and for responding to local signals, which would lead to eosinophil differentiation.

In fact, CD34 + progenitors can be detected in the blood. Moreover, other cytokines such as IL-33 have been considered with a role in the eosinophil development within the bone marrow, as well as in the activation of mature cells and/or activation of EoPs within the tissue.[351] Thus, eosinophils can traffic to tissues in immature states, or even develop locally in the tissue via in situ hematopoiesis amplifying the source of effector cells (reviewed in[345,346,350]).

Eosinophil expansion is a hallmark of several allergic and inflammatory diseases, and different scenarios may explain this amplification: increased output from the bone marrow, increased trafficking of blood eosinophils into tissues, increased survival, and local maturation of EoPs in tissues or even proliferation of resident eosinophils. Because mature eosinophils are terminally differentiated cells without a robust proliferative capacity, more research focused on the eosinophil developmental processes outside the bone marrow is needed.[351] By our ultrastructural observations, eosinophils showing immature specific granules (lacking a crystalline core) can be at times observed in the peripheral blood and tissues in situations in which there is an overproduction of these cells, for example, in HESs, which may be reflecting cell release from the bone marrow with morphological features of cellular immaturity.[18] In animal models (Chapter 9), we noticed a small population of morphologically immature eosinophils in the liver of mice infected with *Schistosoma mansoni*, a parasite that induces striking recruitment of eosinophils into the circulation, peritoneal cavity, and target tissues.[352]

While the initial stages of eosinophil development can be defined by the expression of surface markers, morphological features observed at the EM level are also useful to characterize lineage-committed EoPs. Our group has been studying the ultrastructure of developing eosinophils derived from umbilical cord blood or bone marrow cells in cultures, as well as in bone marrow biopsies.[40,54,56,353,354] The ultrastructural features of the human eosinophil developmental pathway are discussed below.

7.2 About immature and mature eosinophil granules

As pointed out in Chapter 2, human eosinophils contain a single population of secretory granules termed specific granules and not two populations ("primary" and "secondary") as considered in the past (reviewed in[18]). Several lines of evidence have indicated that "primary" granules do not represent a separate granule population in eosinophils.[18] For example, it is well documented that during granule formation and granule protein genesis, cationic proteins, notably MBP, undergo progressive processing passing from precursor forms, such as proMBP, to finally MBP that forms the signature crystalline core in eosinophil-specific granules.[355,356] We provided the first ultrastructural localization of MBP to condensing immature granules from eosinophilic myelocytes.[246] In fact, both proMBP and MBP are expressed in immature granules within precursors of the eosinophilic lineage that are actively forming cored granules.[356] Other cationic proteins such as ECP and EPO were demonstrated in both coreless and cored granules in human EoPs of the bone marrow.[357] Moreover, studies in mice have shown that granule formation is closely associated with the maturation/terminal differentiation of eosinophils from bone marrow progenitors.[358,359] For example, combined loss of both MBP-1 and EPX gene expression caused the disruption of eosinophilopoiesis, with rare eosinophils being found in the peripheral blood.[358] Interestingly, these cells showed specific granule-like structures with limiting membranes but devoid of electron-dense cores.[358] Accordingly, it was demonstrated that a requisite proteolytic processing of granule cationic proteins, including MBP and EPX, during eosinophil cell maturation is required for both specific granule formation and eosinophil survival.[359] Altogether, accumulated data provide evidence that "primary" granules are indeed early "secondary" cored granules. Thus, there are no distinct populations of "primary" and "secondary" granules—there are only immature and mature specific granules.[18]

As the eosinophil matures in the bone marrow, immature specific granules (coreless granules) undergo remarkable structural changes to develop into core-containing granules (Fig. 7.1).[18] Immature specific granules are large, membrane-bound organelles that undergo condensation and crystallization of their contents and size reduction during maturation.

Granules undergoing structural events by which their contents are being accumulated and condensed are considered the earliest ultrastructurally recognizable form of eosinophil-specific granules (Fig. 7.1).[360] These immature specific granules are observed by TEM as large organelles with irregular contour and partially filled with variable amounts of material moderately granular and/or homogeneously electron-dense (Figs. 7.1–7.5).

Tiny vesicles can be noted within these granules and are seen below the limiting granule membrane, compacting at the granule center and/or aggregating into the already deposited, electron-dense material (Figs. 7.6 and 7.7). These very small vesicles can be clearly seen in partially filled granules until the granule is completely occupied with accumulating materials (Fig. 7.8). Therefore, the event of granule filling seems related to an intense deposition of vesicles and condensation of their material within membrane-bound containers (Fig. 7.8). Interestingly, large vesiculotubular structures are also observed in association with very immature eosinophil granules (Fig. 7.7). Whether these vesicles are being incorporated into or released from these granules in these initial steps of eosinophil development is unclear, but they likely represent the first noticeable formation of EoSVs.

Membranes are also distinguished within immature specific granules in the early process of formation (Fig. 7.6) in accordance with findings in mature eosinophils. As noted, membranous structures have been reported by our group and others in blood and tissue human eosinophils during different situations, including

Developmental morphology of specific granules in human eosinophils

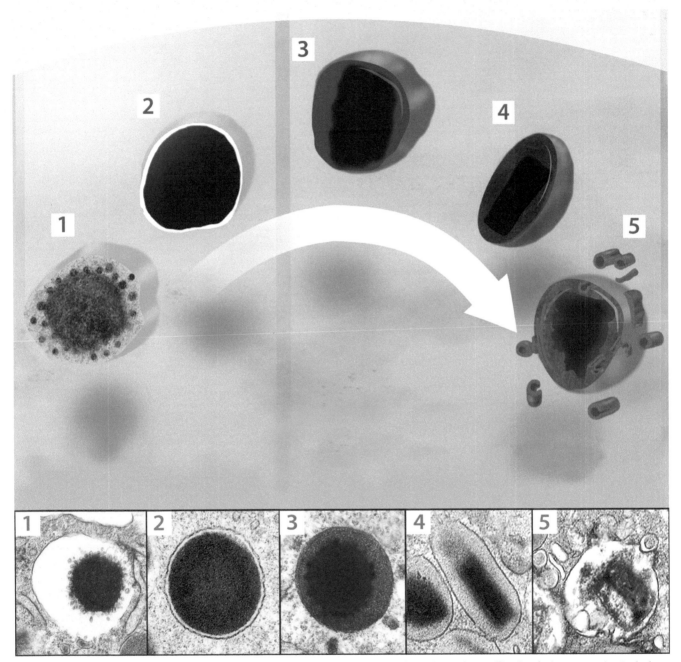

FIG. 7.1 Immature specific granules from human eosinophils undergo content condensation and crystallization during maturation in the bone marrow. (1) Immature granule in the process of filling and condensation showing intragranular vesicles surrounding the internal dense material; (2) round immature granule with homogeneously dense content; (3) spherical core-containing granule seen with an irregular electron-dense central area surrounded by a less dense region; (4) resting, elliptical granule with a well-defined electron-dense core and an electron-lucent matrix; (5) an activated granule shows disassembled core and matrix and formation of EoSVs. All granules are delimited by a phospholipid bilayer membrane. A mixture of granules (1), (2), and (3) is observed in progenitor cells from the eosinophilic lineage. The amount of these granules is variable. Initially, coreless granules (1) and (2) predominate and the number of granule (3) progressively increases. Granules (4) and (5) are typical of mature eosinophils from the peripheral blood and tissues.

A. The cell biology of human eosinophils

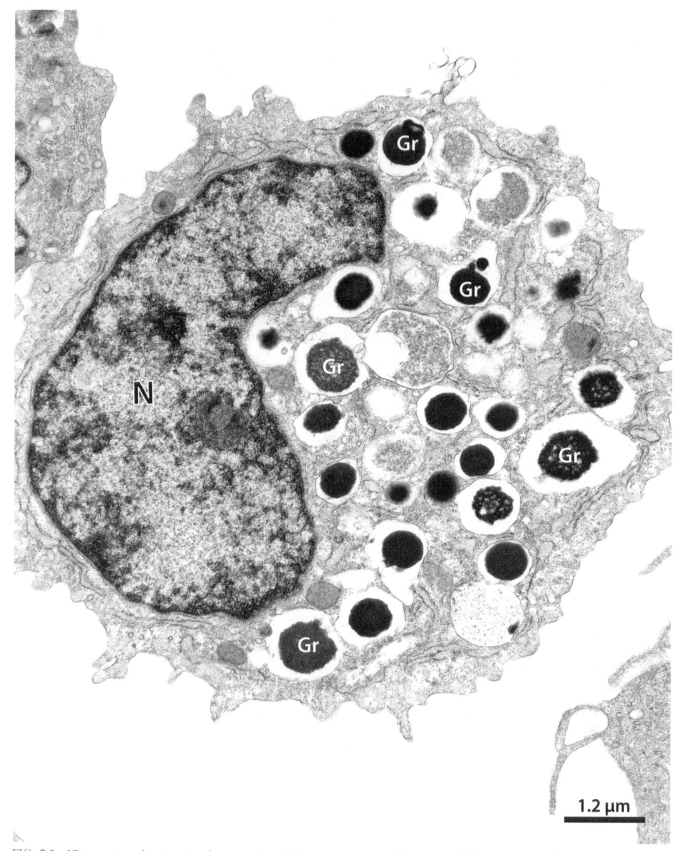

FIG. 7.2 Ultrastructure of an immature human eosinophil. Large immature specific granules (Gr) in the process of formation/condensation are seen in the cytoplasm. Note the variable electron density of the intragranular material and the large, monolobed, and eccentric nucleus (N). Human cord blood mononuclear cells were cultured in the presence of cytokines and prepared for TEM at 3-week culture.[353,354]

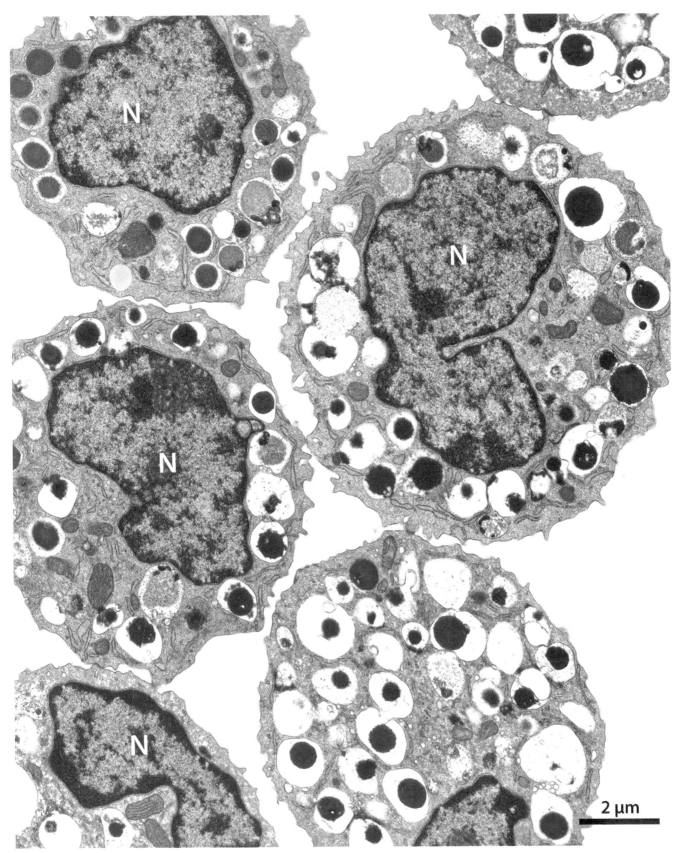

FIG. 7.3   Immature human eosinophils in the process of granulogenesis. The cells show the cytoplasm packed with granules partially filled with coarse and/or electron-dense material. Cultured cord blood mononuclear cells were processed for TEM.[353] *N*, nucleus.

A.   The cell biology of human eosinophils

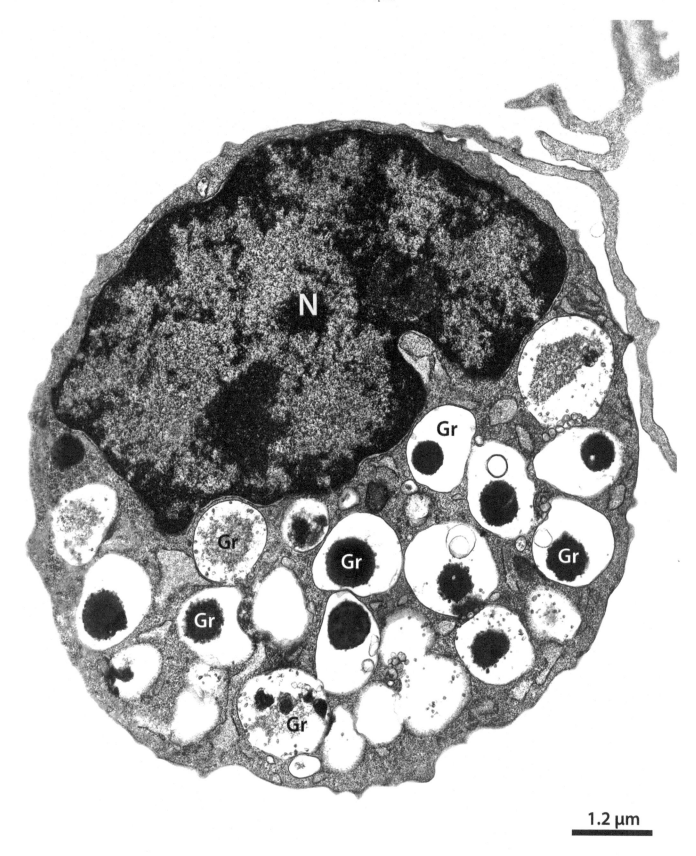

1.2 μm

FIG. 7.4 A human eosinophil in the early stage of development. Heterogeneous materials are being accumulated and condensed within immature specific granules (Gr). A large monolobed and eccentric nucleus (N) occupies a great part of the cytoplasm. Cultured cord blood mononuclear
cells were processed for TEM.[353]

A. The cell biology of human eosinophils

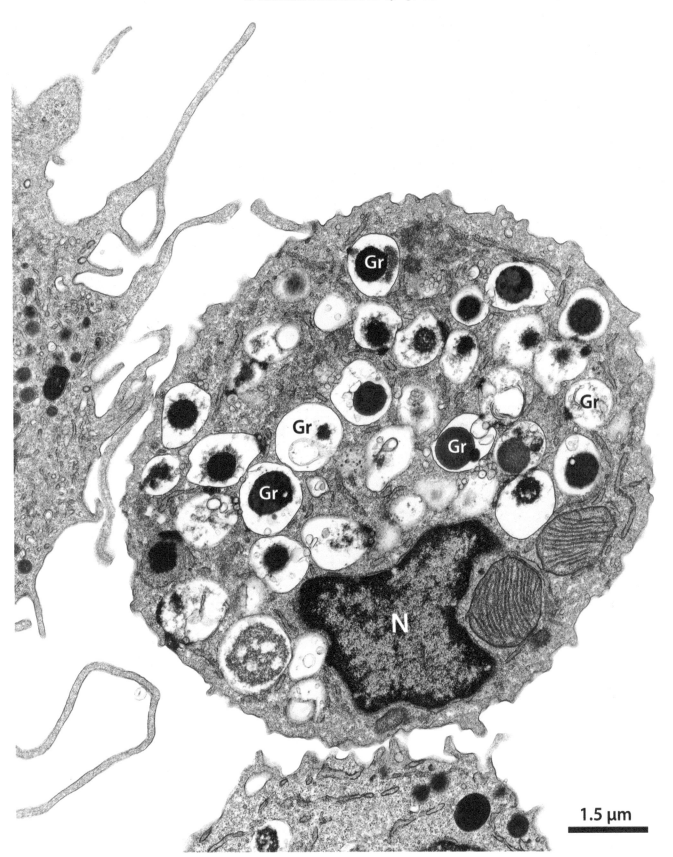

FIG. 7.5 Electron micrograph showing an immature human eosinophil with partially filled secretory granules (Gr) and large mitochondria *(colored in green)*. Cultured cord blood mononuclear cells were processed for TEM.[353] *N*, nucleus.

A. The cell biology of human eosinophils

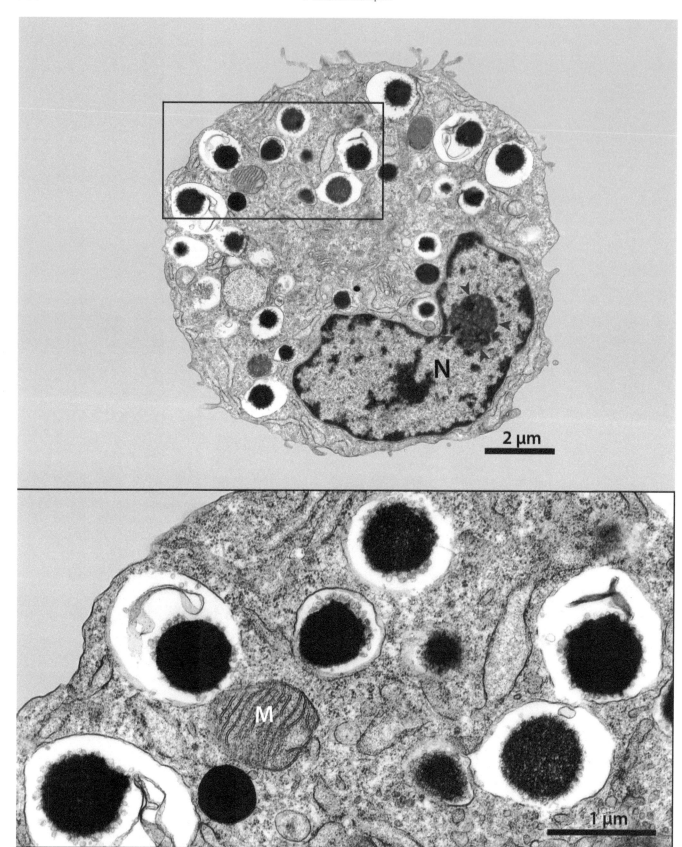

FIG. 7.6   Ultrastructural features of an immature human eosinophil. The boxed area shows in high magnification the presence of tiny vesicles *(colored in green)* being deposited within granules in the process of formation. Intragranular membranes *(highlighted in purple)* are also observed. A prominent nucleolus *(arrowheads)* is seen within the eccentric nucleus (N). Cultured cord blood mononuclear cells were prepared for TEM.[353] M, mitochondrion.

A. The cell biology of human eosinophils

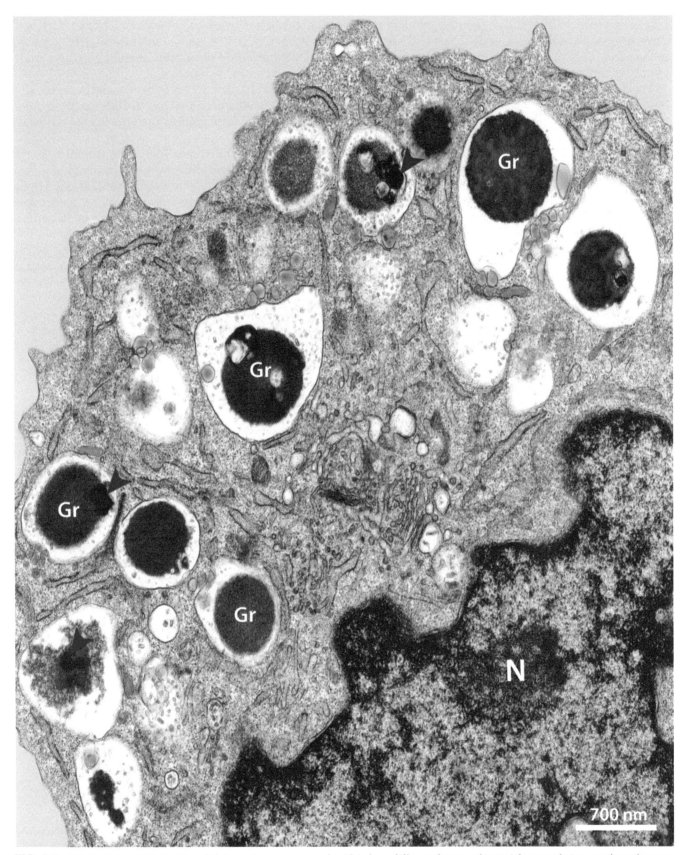

FIG. 7.7 A human developing eosinophil. Large immature granules (Gr) show different degrees of material accumulation, condensed masses *(arrowheads)*, and deposition of tiny vesicles *(highlighted in green)*. Large vesiculotubular structures *(colored in pink)* are also associated with forming granules. Cultured cord blood mononuclear cells were processed for TEM.[353] *N*, nucleus.

A. The cell biology of human eosinophils

several diseases.[20,189,195–201] Intragranular membranes within mature eosinophils are shown in Chapter 4 by both conventional TEM (Figs. 4.2 and 4.8) and 3D electron tomography (Fig. 4.10). Thus, immature specific granules in the early stage of development exhibit heterogeneous materials composed of vesicles, membranes, and amorphous contents with variable electron density (Fig. 7.8).

The prominence of the Golgi region in developing eosinophils (Fig. 7.9) has implicated this organelle in the packing of secretory material. However, the mechanism of granulogenesis is not well understood; it is still lacking information on how the earliest granules are elaborated in eosinophils, and the origin of the granule material is not well established yet.[360]

With the progressive accumulation and condensation of materials within very immature granules as described above, they are transformed into organelles filled with homogeneously electron-dense contents (Fig. 7.8). We, therefore, consider that these immature granules full of dense contents, which are frequently seen as perfect round organelles, correspond to the sequential stage of maturation from granules showing heterogeneous contents (Figs. 7.1 and 7.8). As the granule matures, round granules full of homogeneous dense contents mature to core-containing granules.[360] This transition is clearly captured by TEM, and spherical granules with an electron-dense central area surrounded by a less dense region are observed. In this stage, the central cores do not show well-defined limits, as seen in completely mature eosinophils (Fig. 7.1). These early cores appear as distinct electron-dense, expanded intragranular areas that are crystalizing to become, in a more advanced stage, an ellipsoid, crystalloid-containing mature granule (Fig. 7.1).

Because the granule development is a continuum of gradual morphological changes, the number of the coreless and cored granules in the cytoplasm is variable, and a mixture of these granules can be observed in different proportions within precursors of the eosinophilic lineage, even in a single cell section.

7.3 Maturation stages of human eosinophils characterized by TEM

Human eosinophils undergo a spectacular series of ultrastructural changes during maturation. The sequential morphological alterations, which accompany the human eosinophil development in the bone marrow, are related not only to the process of granulogenesis but also to modifications in other organelles, notably in the nucleus, ER, and Golgi complex. The eosinophil nucleus, which is large, eccentric, and mostly euchromatic or with finely dispersed heterochromatin in early EoPs committed to the eosinophil lineage, undergoes size reduction, lobulation, and progressive distinction of the heterochromatin and euchromatin regions. Moreover, very immature eosinophils can show one or more nucleoli, which are conspicuously observed within the nucleus in contrast to mature cells. The ER and Golgi complex are also plentiful organelles in the initial stages of the human eosinophil but undergo significant reduction during cell maturation[40,56,72] and, as discussed in Chapter 2, are not frequently seen in electron micrographs of mature cells. Mitochondria are also considered more numerous and with higher size in earlier stages of the eosinophil development (Figs. 7.5 and 7.6) compared with mature eosinophils.

The classical nomenclature used for EoPs from the bone marrow during the process of hematopoiesis (promyelocyte, myelocyte, metamyelocyte, and band form),[361] which is largely based on the light microscopic appearance of these cells, cannot always be applied when thin sections of these cells are observed under TEM. However, considering the granule morphology correlated with other cytoplasmic features, it is possible to distinguish sequential stages of eosinophil development and unique structural aspects of this cell by TEM. Thus, we define the differentiation and maturation morphologies of the eosinophilic lineage in three main stages (early, middle, and late stages of development) as described below. Of note, immature eosinophils in the process of differentiation have also been referred to in the literature as merely eosinophil myelocytes or early, middle, or late eosinophil myelocytes.[40,362]

7.3.1 Early stage of eosinophil development

Eosinophils in the early stage of development are characterized by the presence of a large, eccentric, and single-lobed nucleus containing a prominent nucleolus.[52,109] The euchromatin predominates in the nucleus while the heterochromatin is finely dispersed and begins to aggregate at the marginal nuclear regions (Figs. 7.2, 7.3, and 7.4). Another morphological feature found in more immature eosinophils is the presence, as noted, of a prominent Golgi (Fig. 7.9) and peripheral RER (Figs. 7.10 and 7.11), which can appear as dilated cisternae (Fig. 7.11).

Eosinophil granulogenesis

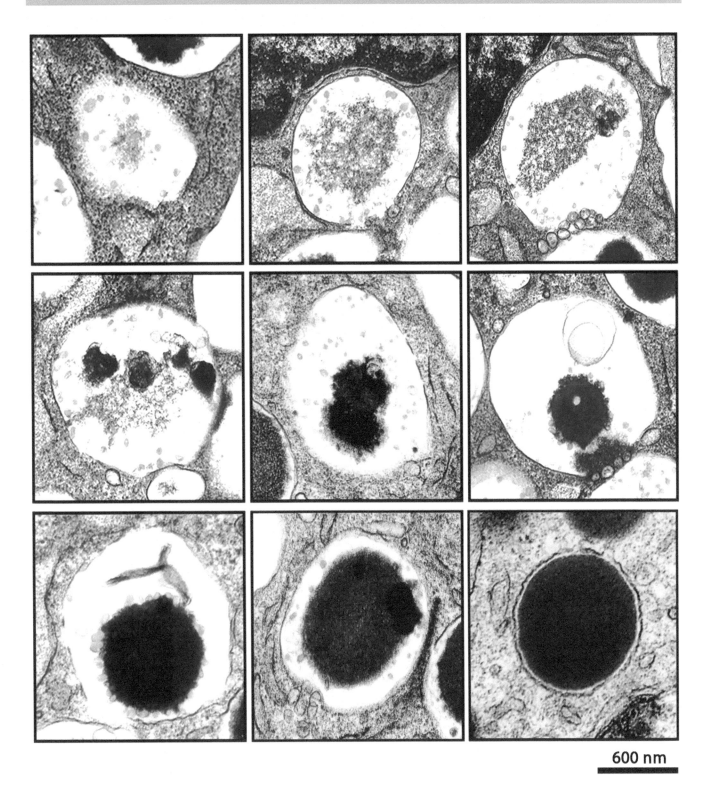

600 nm

FIG. 7.8 Granulogenesis observed in early immature human eosinophils. The sequence of images shows progressive filling of very immature granules. The developmental events include deposition of diminutive vesicles *(colored in green)* and content condensation resulting in spherical granules completely filled with a very electron-dense material. No crystalline cores are identified in these stages. Granules are representative of immature eosinophils collected from cultures or bone marrow biopsies and processed for TEM.

A. The cell biology of human eosinophils

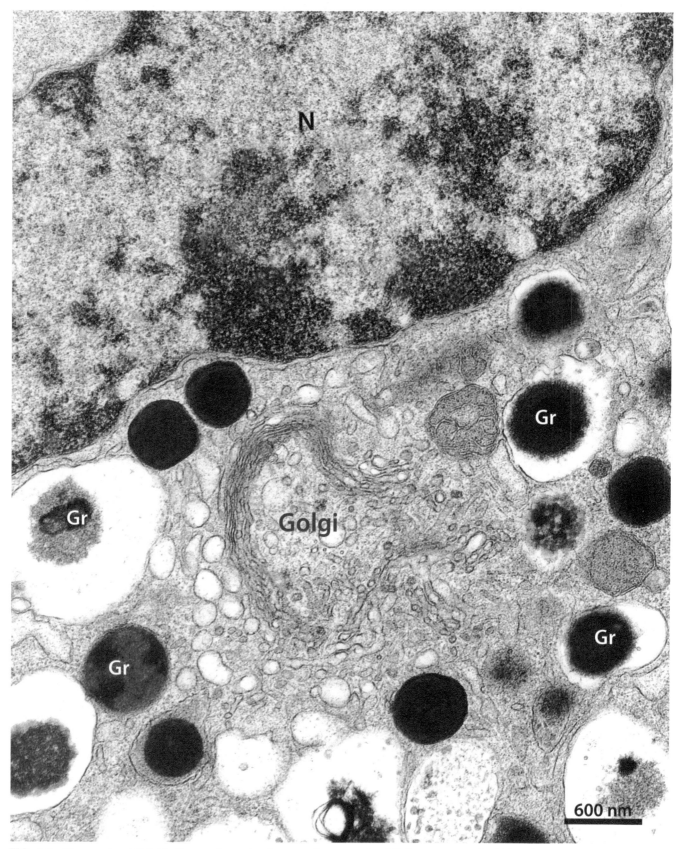

FIG. 7.9   The prominent Golgi in the cytoplasm of an immature human eosinophil. Secretory granules (Gr) in the process of formation and a large nucleus (N) are also observed. Cultured cord blood mononuclear cells were processed for TEM.[353]

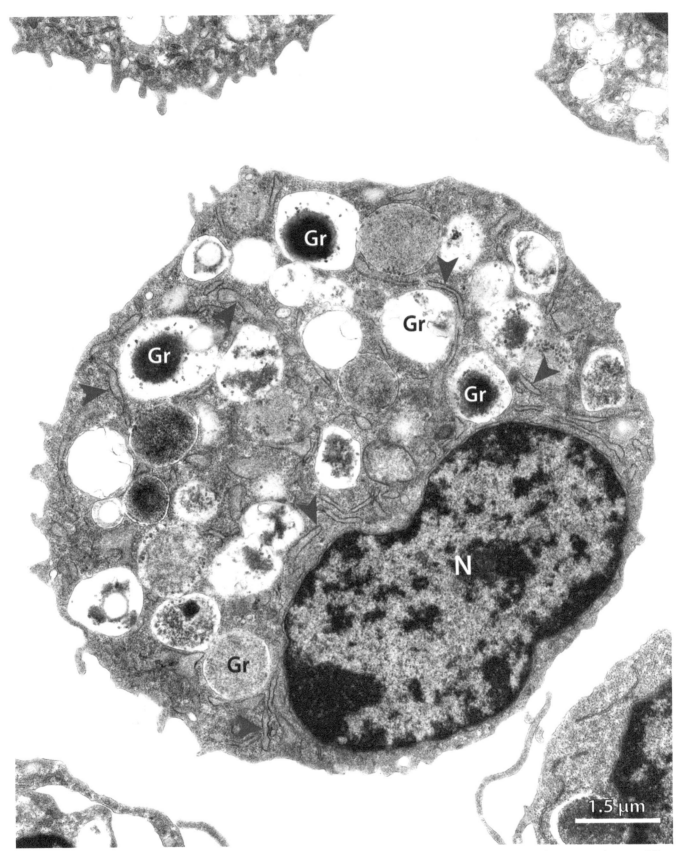

FIG. 7.10 TEM showing a bone marrow-derived, human developing eosinophil with secretory granules (Gr) in the process of content filling. Note RER cisternae *(arrowheads)* distributed in the cytoplasm and the eccentric monolobed nucleus (N), typical of progenitors from the eosinophilic lineage.

A. The cell biology of human eosinophils

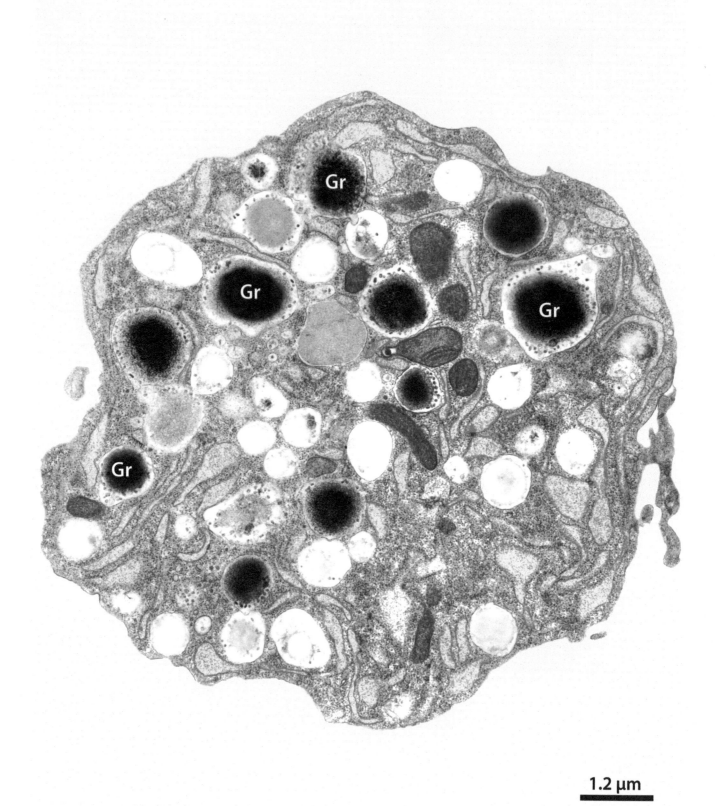

1.2 μm

FIG. 7.11 Immature human eosinophil showing abundant RER. Dilated RER cisternae *(colored in blue)* and large immature granules (Gr) are seen in the cytoplasm. The nucleus is not visible in the section plane. Bone marrow-derived cells were prepared for TEM.

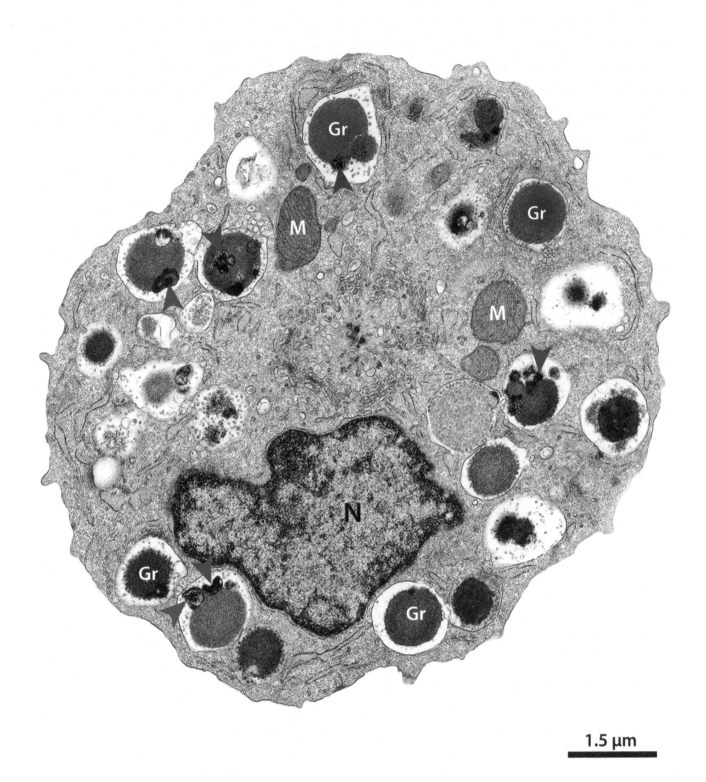

FIG. 7.12 An immature human eosinophil showing granules in the process of condensation. Electron-dense irregular areas *(arrowheads)* are seen within immature specific granules. Cultured cord blood mononuclear cells were prepared for TEM.[353] *M*, mitochondria; *N*, nucleus.

A. The cell biology of human eosinophils

RER strands are clearly distinguishable in the cytoplasm, an observation that is not possible to make in mature eosinophils, in which RER are considered minor organelles (Chapter 2).[52] Mitochondria are large and frequently observed (Figs. 7.5 and 7.6).

Early eosinophils show numerous large granules in the process of filling and thus exhibiting variable aggregates of heterogeneous contents associated or not with intragranular tiny vesicles (Figs. 7.4, 7.5, and 7.6). The accumulation of materials within these granules occurs in parallel to their processes of condensation, and TEM clearly captures these events by revealing the presence of aggregates within the granule with different electron densities (Fig. 7.12). Of note, as a general characteristic of the granulogenesis in eosinophils, not all granules are in the same stage of development, and granules almost or totally full of electron-dense contents are interspersed with granules in processes of filling and condensation (Figs. 7.12 and 7.13).[109] As the cell matures, the numbers of these homogeneously electron-dense granules increase while the numbers of granules showing heterogeneous materials/vesicles decrease and tend to disappear (Fig. 7.8). Eosinophils undergoing mitosis (Fig. 7.14) and showing the nucleus with initial indentation (Fig. 7.15) can be occasionally observed at this stage.

7.3.2 Middle stage of eosinophil development

We consider that immature eosinophils are in an intermediate stage of eosinophilopoiesis when they show a predominance of spherical granules full of electron-dense contents and a large monolobed nucleus (Figs. 7.16–7.23). The RER (Fig. 7.18) and Golgi (Fig. 7.19) membranes continue to be prominent at this stage of development but will undergo progressive diminution in parallel to cell maturation. Moreover, the mitochondrial population (Figs. 7.17, 7.18, 7.22, and 7.24) is more evident than in mature eosinophils.

The gradual accumulation and condensation of materials within more immature granules lead them to be transformed into granules completely filled with a compact, homogeneous, and very electron-dense material (Fig. 7.8). Secretory granules at this point of maturation are generally seen as regularly spherical granules intermixed with few granules containing heterogeneous contents and/or showing the initial formation of the central cores (Figs. 7.20 and 7.21). The emergence of the crystalloids within the granules is thus seen in this maturational stage (Fig. 7.23). The process of granule crystallization is revealed by the presence of a centrally localized, electron-dense, and irregularly defined region surrounded by a narrow less-dense (matrix) area present initially in a limited number of granules (Fig. 7.23) and increasing gradually within cytoplasmic granules (Fig. 7.25).

7.3.3 Late stage of eosinophil development

Eosinophils in the late stage of development, identified in cultures and bone marrow biopsies, show nuclear lobulation and an increased number of cored granules, which now predominate in the cytoplasm (Figs. 7.26-7.29). These granules show progressive crystallization seen as irregular electron-dense areas (Figs. 7.27) and are intermingled with more immature granules, which can also be present in lower numbers in the late stages (Fig. 7.27).

The production of eosinophils is a continuum of general and gradual morphologic changes. The amount of heterochromatin increases within the nucleus as the cell becomes more mature and its typical marginal localization and demarcation from the euchromatin are observed in more advanced stages of maturation (Figs. 7.28 and 7.29). Thus, lobed eosinophils with many cored granules in the process of crystallization can still exhibit a considerable level of euchromatin and dilated strands of peripheral RER (Fig. 7.26), which are features seen in more immature eosinophils. Progressive maturation leads to a reduction of the RER volume in parallel with the formation of more heterochromatin (Figs. 7.27–7.29). The remarkable alterations of the eosinophil nuclei during cell development are shown in Fig. 7.30. These changes include segmentation, reduction in size, eccentric to a more central localization, increased chromatin condensation, and nucleoli reduction in volume.

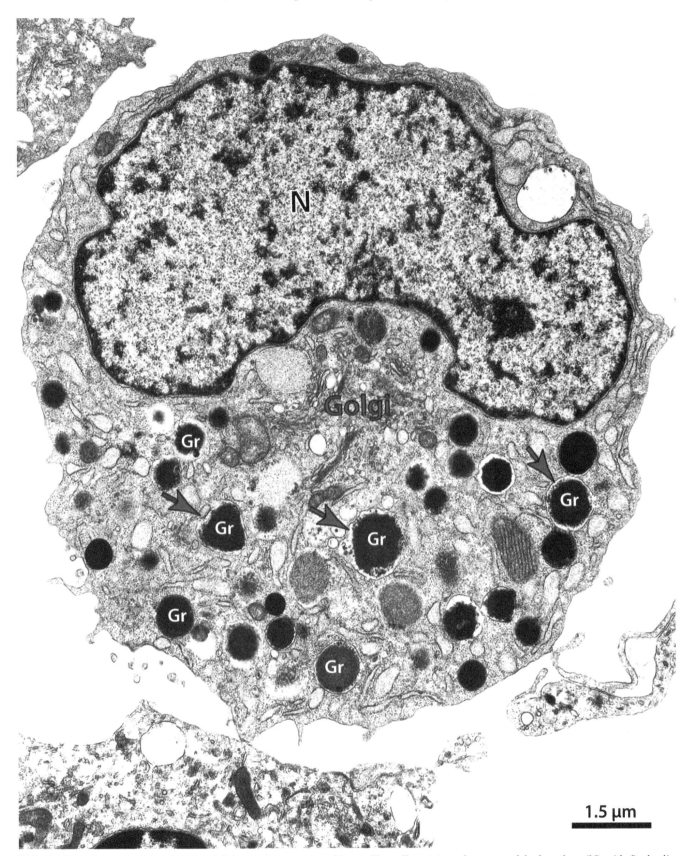

FIG. 7.13 A human immature eosinophil from a bone marrow biopsy. The cell contains a large monolobed nucleus (N) with finely dispersed heterochromatin, prominent Golgi and RER, and round immature granules (Gr). *Arrows* indicate granules with almost completely filled electron-dense material. Sample prepared for TEM.

A. The cell biology of human eosinophils

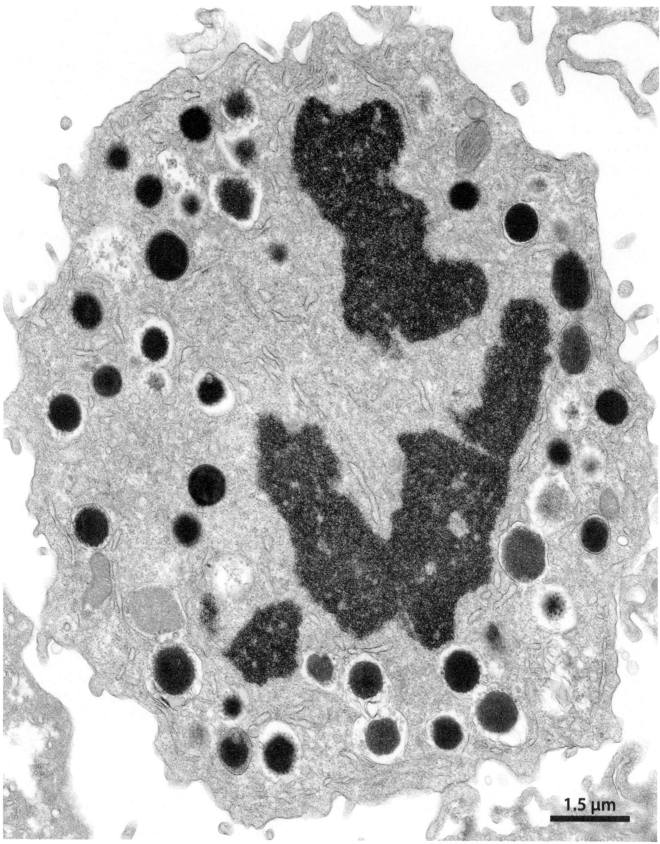

FIG. 7.14 Electron micrograph of a dividing immature human eosinophil. Immature specific granules containing mostly electron-dense material and condensed chromatin *(colored in blue)*, typical of mitosis, are observed in the cytoplasm. Cultured cord blood mononuclear cells were prepared for TEM.[353]

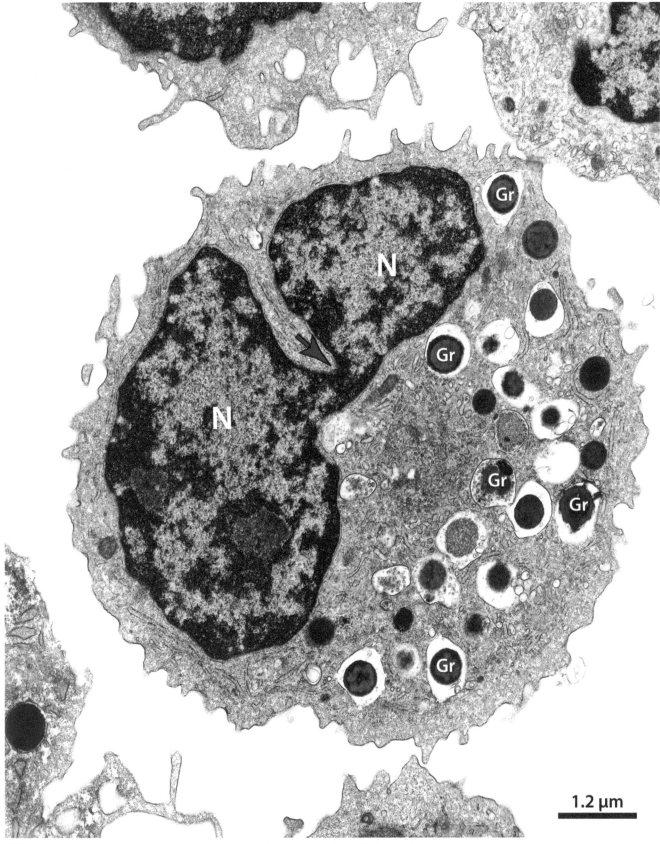

FIG. 7.15 An immature human eosinophil showing a large nucleus (N) with deep indentation *(arrow)* and a mixture of developing secretory granules (Gr). Cultured cord blood mononuclear cells were prepared for TEM.[353]

A. The cell biology of human eosinophils

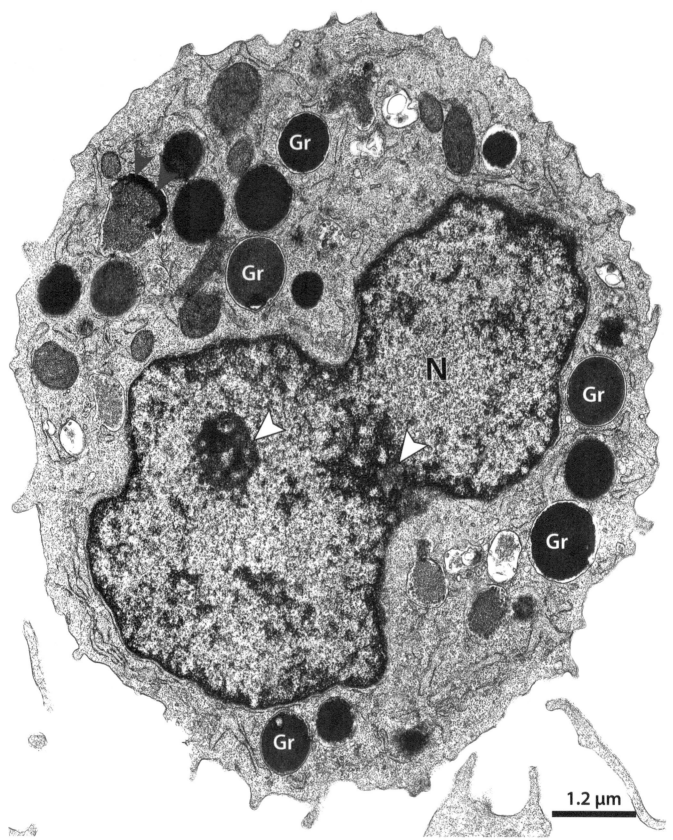

FIG. 7.16   An immature human eosinophil in the middle stage of development displaying secretory granules (Gr) full of contents. Note the large nucleus (N) with evident nucleolar areas *(white arrowheads)* and a dense aggregate in a granule *(pink arrowheads)*. Cultured cord blood mononuclear cells were prepared for TEM.[353]

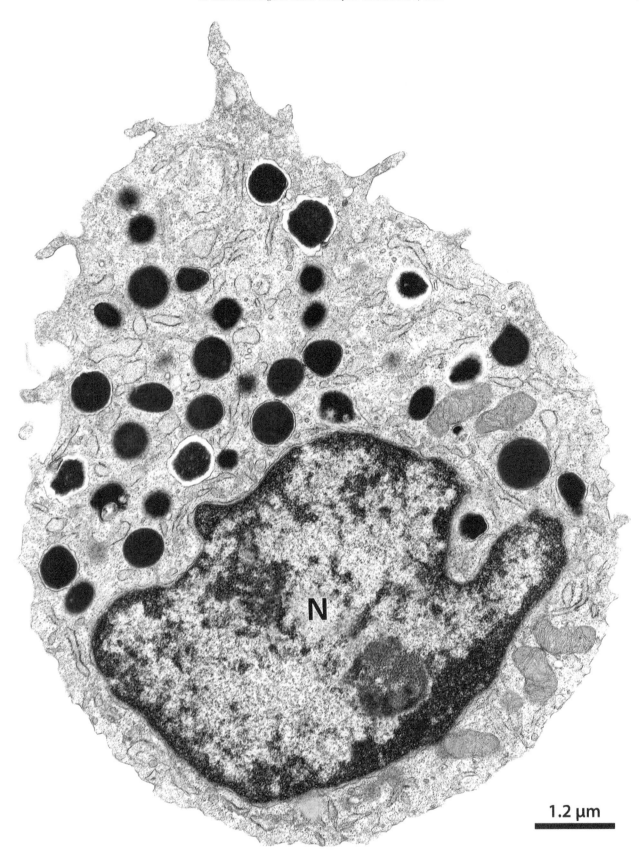

FIG. 7.17 Immature human eosinophil containing predominance of spherical granules full of electron-dense material. Cultured cord blood mononuclear cells were prepared for TEM.[353] Mitochondria (colored in green); N, nucleus.

A. The cell biology of human eosinophils

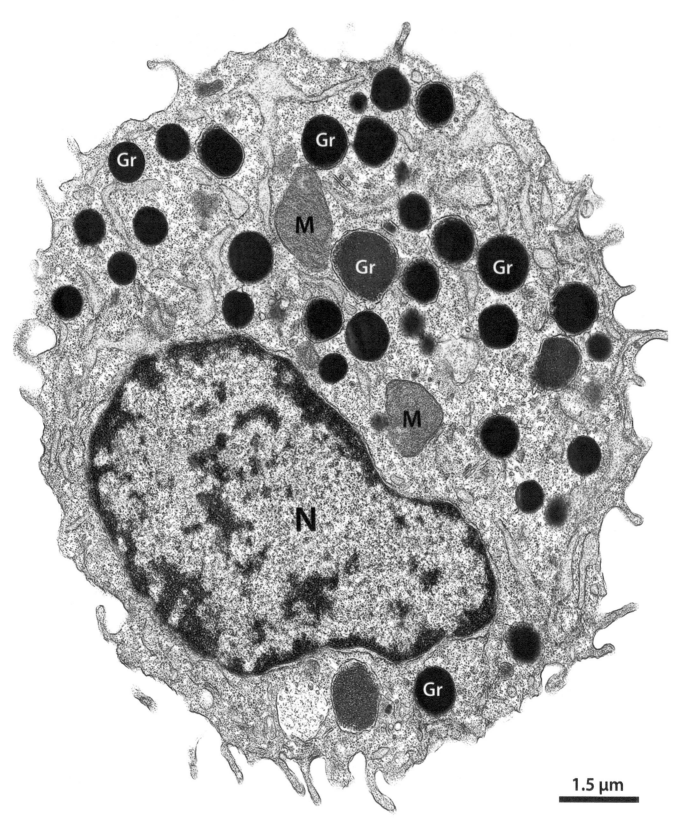

FIG. 7.18 Electron micrograph of a human immature eosinophil. The cytoplasm contains numerous strands of dilated RER *(colored in blue)*, large mitochondria (M), and immature, full of contents, specific granules (Gr). Cultured cord blood mononuclear cells were prepared for TEM.[353] *N*, nucleus.

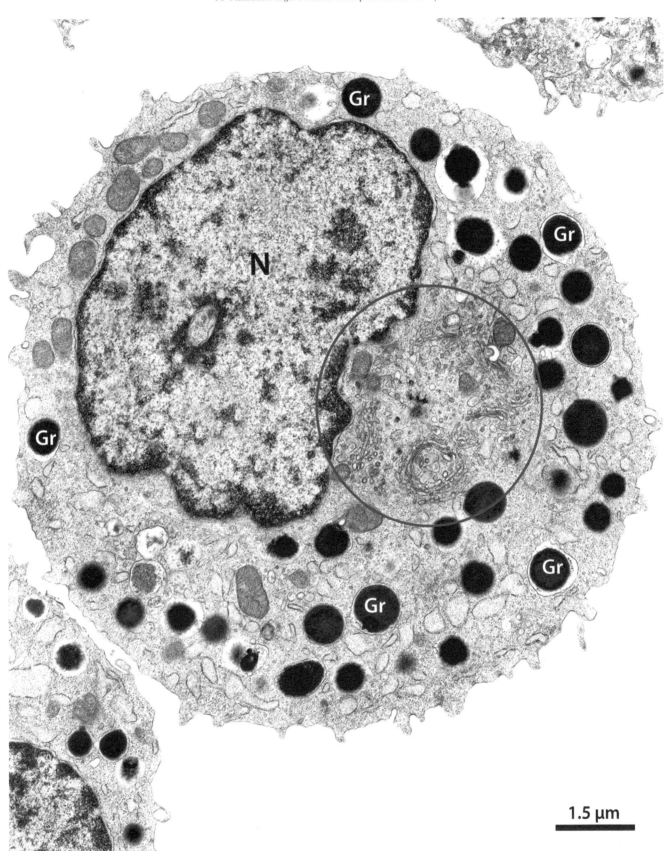

FIG. 7.19 Human immature eosinophil. The cytoplasm displays round immature granules (Gr) filled with homogeneously dense contents. Note the predominantly euchromatic monolobed nucleus (N) and the active Golgi region *(circle)*. Cultured cord blood mononuclear cells were processed for TEM.[353]

A. The cell biology of human eosinophils

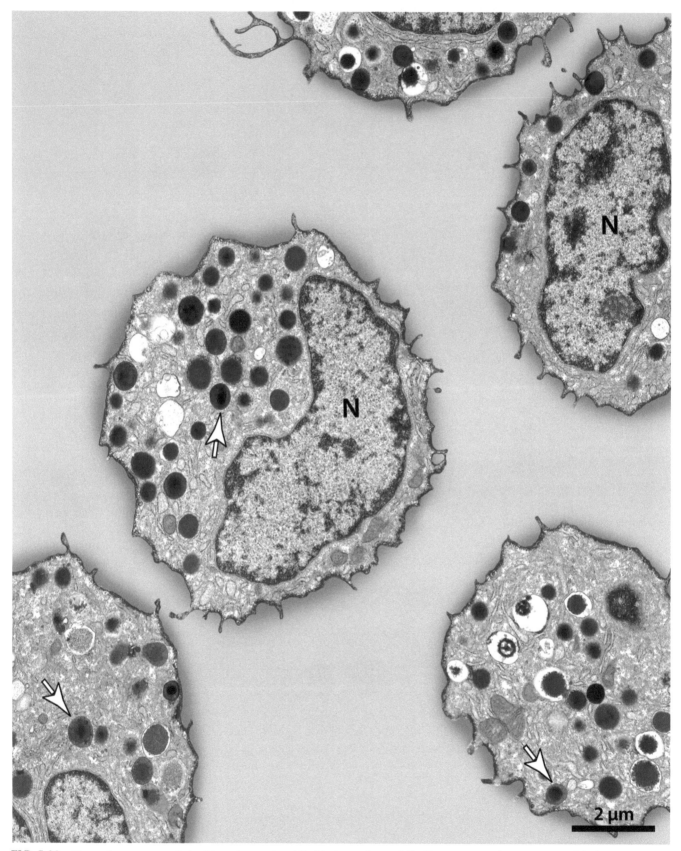

FIG. 7.20   Human bone marrow-derived immature eosinophils. Most granules are full of homogeneous dense contents while few granules *(arrows)* show a more electron-dense central area reflecting the ongoing crystallization process. The irregular cell surface was highlighted in purple. Samples prepared for TEM. *N,* nucleus.

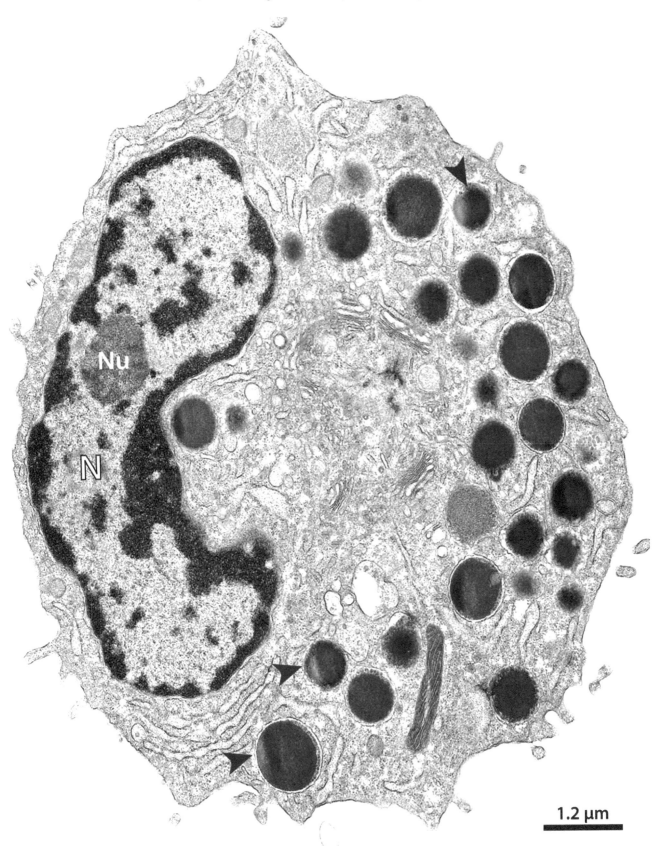

FIG. 7.21 Immature human eosinophil at the intermediate stage of development. Some granules in the process of crystallization with ill-defined cores and matrices *(arrowheads)* are intermingled with granules containing homogeneous dense contents. Cultured cord blood mononuclear cells were processed for TEM.[353] *N*, nucleus; *Nu*, nucleolus.

A. The cell biology of human eosinophils

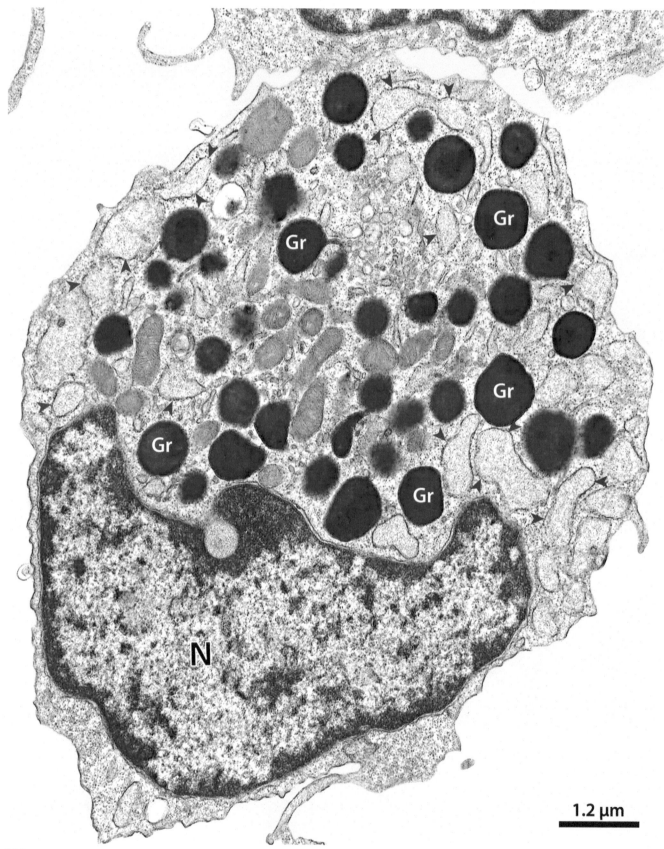

FIG. 7.22   Immature human eosinophil showing numerous mitochondria *(colored in green)* in the cytoplasm. Note dilated cisternae of RER *(arrowheads)* and immature secretory granules (Gr). Cultured cord blood mononuclear cells were processed for TEM.[353] *N*, nucleus.

A. The cell biology of human eosinophils

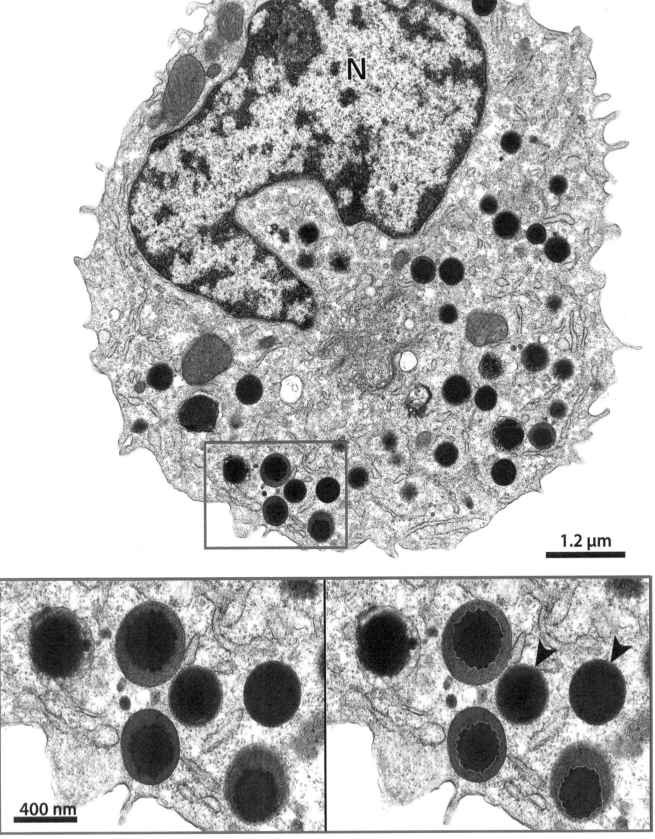

FIG. 7.23 TEM showing granules at different developmental stages in a human bone marrow-derived immature eosinophil. In higher magnification *(boxed area)*, note granules full of dense contents *(arrowheads)* and with irregularly defined cores *(outlined in orange)*. One granule is almost full and shows tiny vesicles *(colored in green)*. N, nucleus.

A. The cell biology of human eosinophils

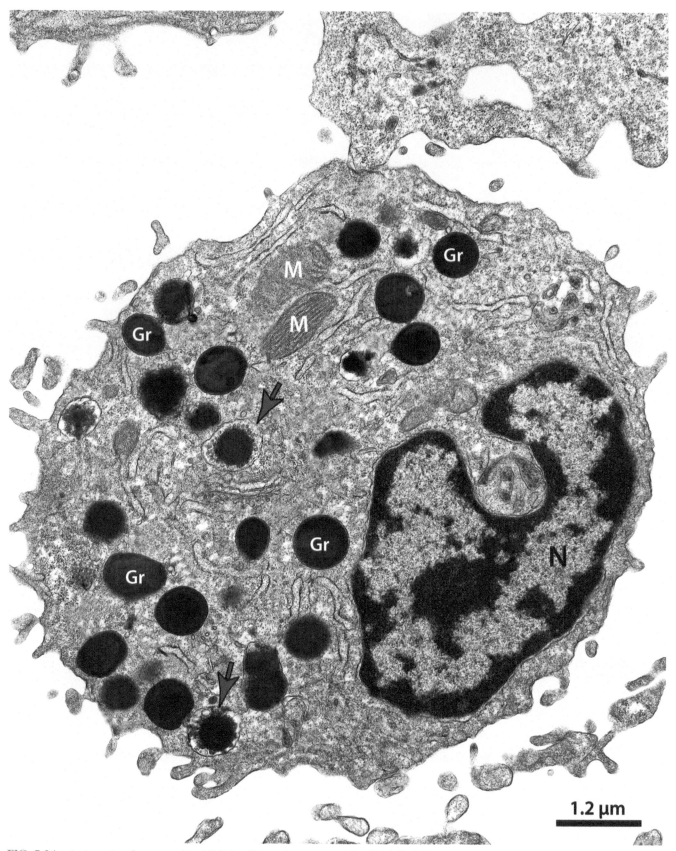

FIG. 7.24   An immature human eosinophil. This cell contains most granules (Gr) full of dense materials while few granules are still in the process of filling *(arrows)*. Note the indented nucleus (N) with marginal well-characterized heterochromatin. Cultured cord blood mononuclear cells were processed for TEM.[353] *M*, mitochondria.

A. The cell biology of human eosinophils

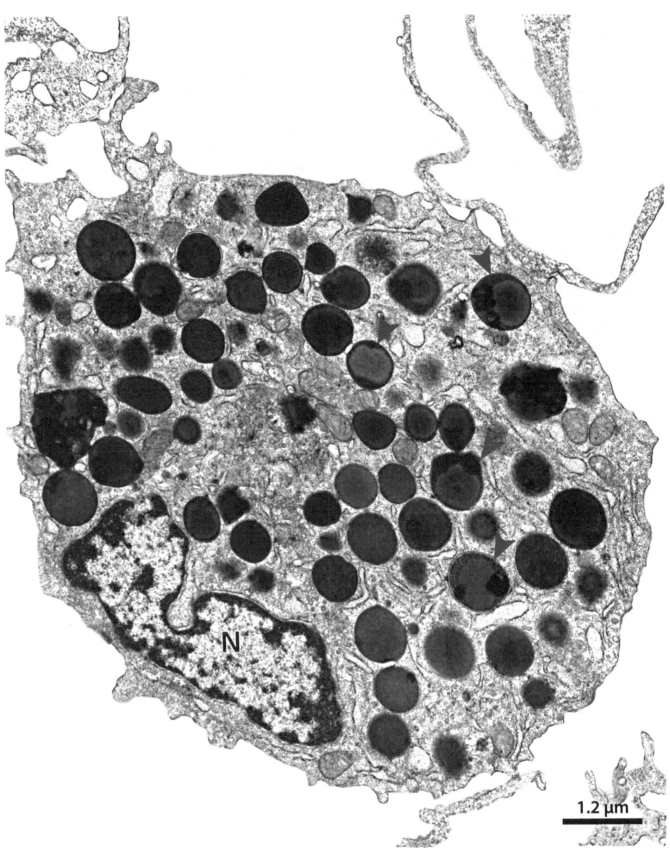

FIG. 7.25   An immature human eosinophil showing granules undergoing events of condensation and crystallization. Several granules are in the process of core formation *(arrowheads)*. Cultured cord blood mononuclear cells were prepared for TEM.[353] N, nucleus.

A. The cell biology of human eosinophils

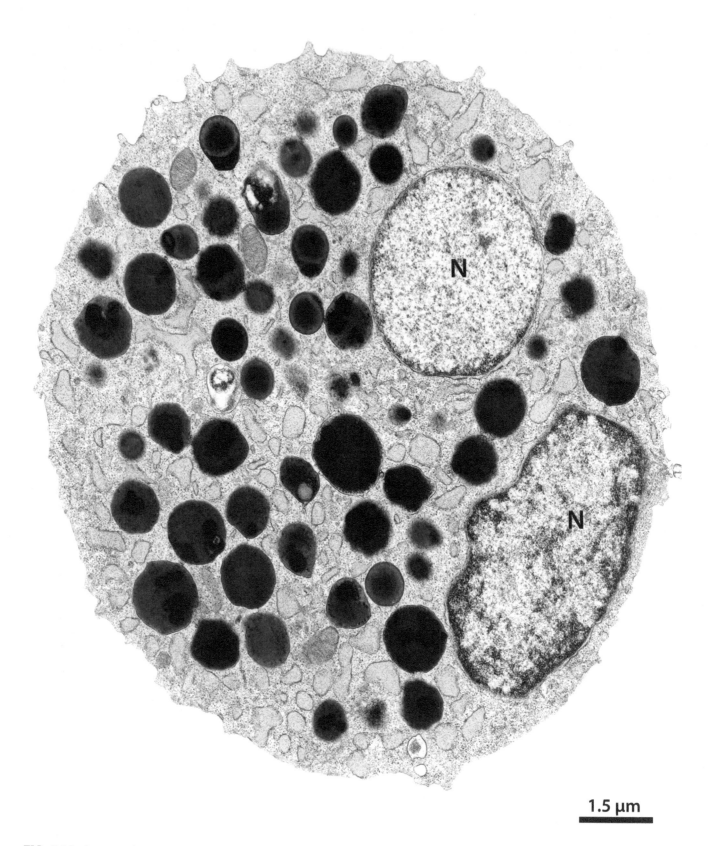

FIG. 7.26   Immature human eosinophil at the late stage of development. The nucleus (N) is bilobed, and the cytoplasm contains many cored granules. A rich peripheral RER with dilated strands *(colored in blue)* is still observed. Cultured cord blood mononuclear cells were prepared for TEM.[353]

A. The cell biology of human eosinophils

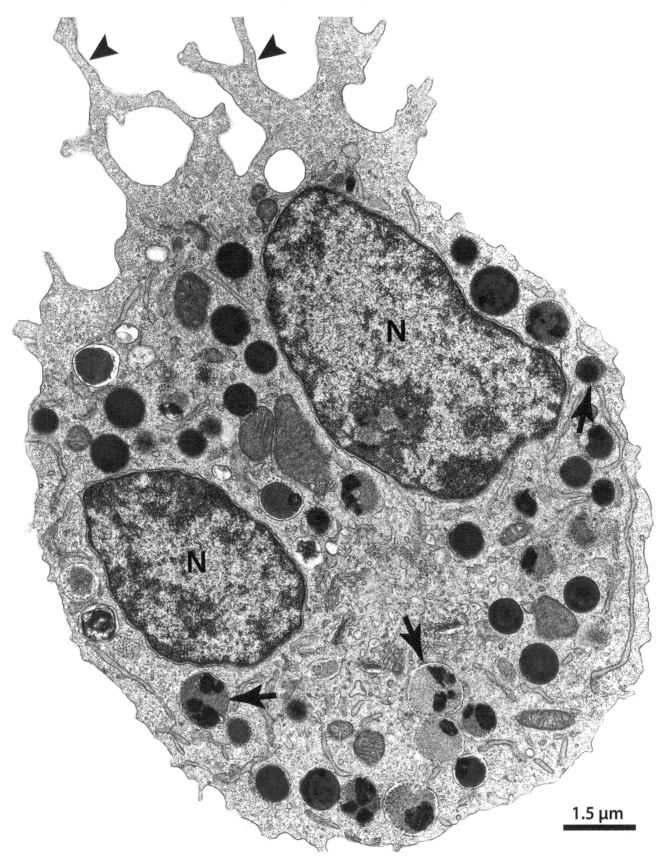

FIG. 7.27 Immature human eosinophil at the late stage of development. Note the moderate amount of RER *(colored in blue)*, bilobed nucleus (N), and most granules in the process of core formation *(arrows)*. *Arrowheads* indicate surface projections. Cultured cord blood mononuclear cells were prepared for TEM.[353]

A. The cell biology of human eosinophils

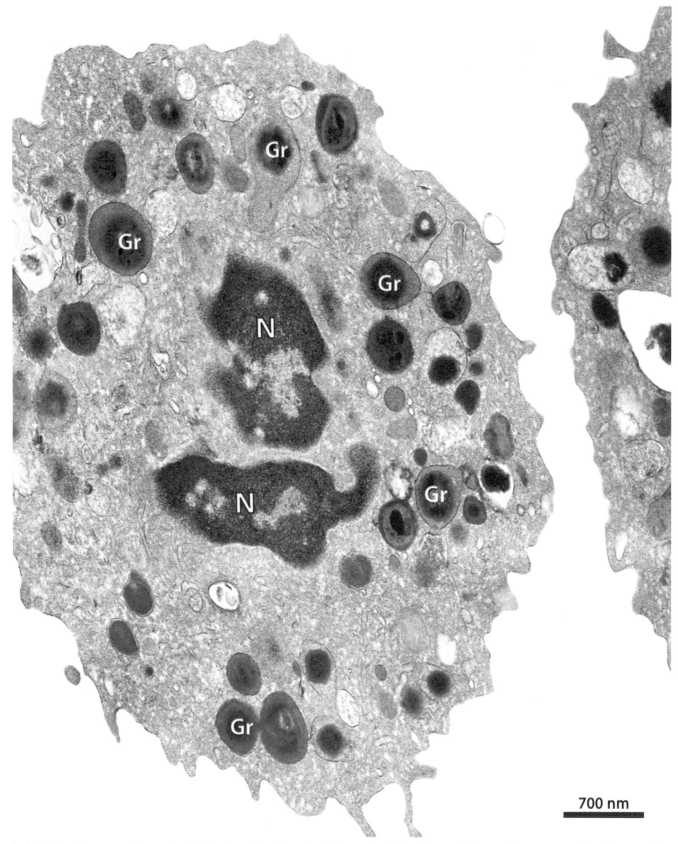

700 nm

FIG. 7.28   A human mature eosinophil developed in culture. Specific granules (Gr) with central electron-dense cores and less dense matrices and nucleus (N) showing typical marginal heterochromatin are observed. Cultured cord blood mononuclear cells were prepared for TEM.[353]

A. The cell biology of human eosinophils

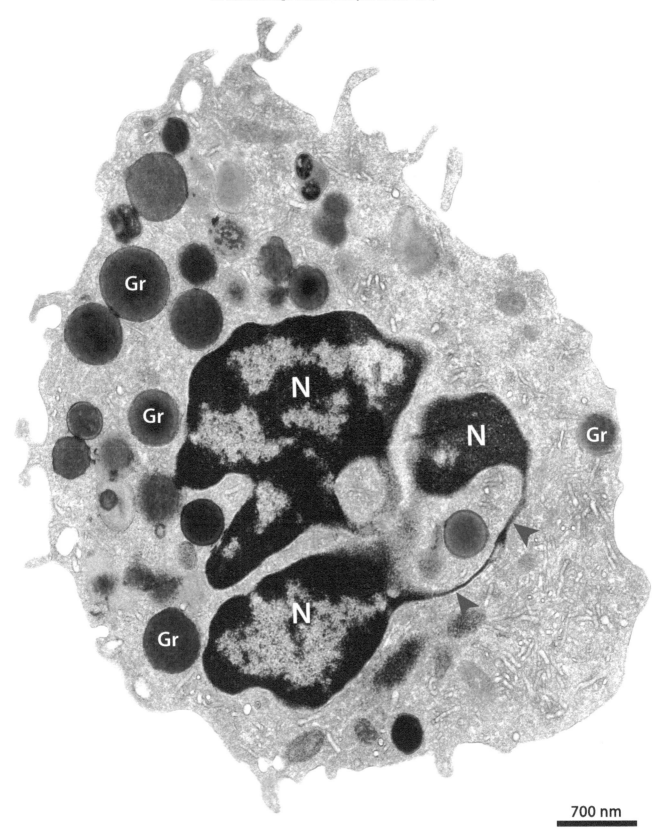

FIG. 7.29 A human mature eosinophil developed in culture. Note the polylobed nucleus (N) with clumped heterochromatin and cored specific granules (Gr). Arrowheads indicate a thin bridge of chromatin connecting two nuclear lobes. Cultured cord blood mononuclear cells were prepared for TEM.[353]

A. The cell biology of human eosinophils

Nuclear changes during eosinophilopoiesis

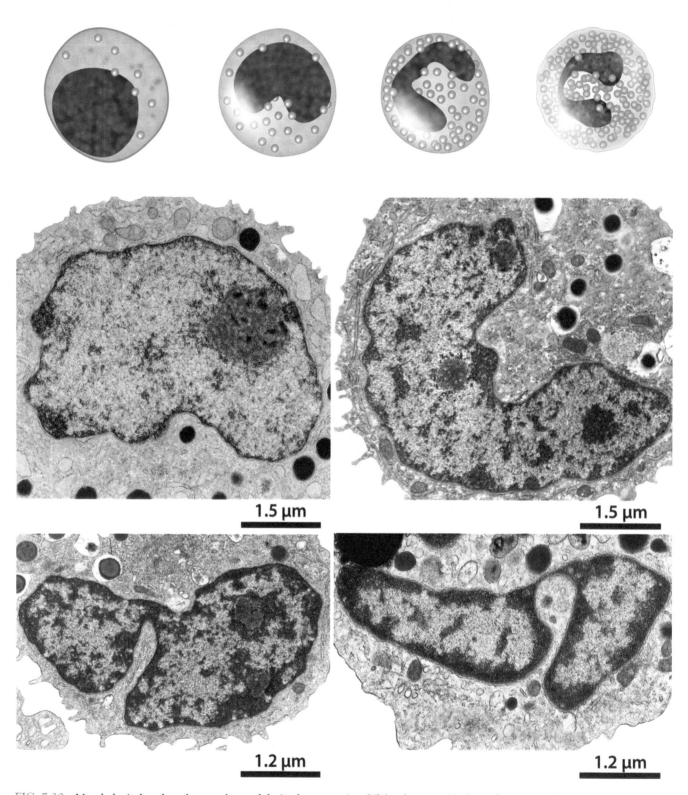

FIG. 7.30 Morphological nuclear changes observed during human eosinophil development. Nuclei undergo size reduction and segmentation. The amount of heterochromatin increases, and marked distinction from the euchromatin is observed. The nucleoli *(colored in purple)* are larger and clearly seen in the early stages but not always noticeable in mature cells. Electron micrographs are representative of eosinophils derived from bone marrow or cord blood cultures and processed for TEM. Drawing courtesy of Zlatibor Anđelković.

Eosinophils in human diseases

Eosinophil-associated diseases (EADs)

8.1 Overview

Although the eosinophil is a formed element of the peripheral circulation, it is primarily a tissue-dwelling cell. In the blood, eosinophils are a minor component of circulating leukocytes (less than 5%), but larger numbers are present outside the vasculature.[3] Substantial numbers of eosinophils are normally found in tissues from the gastrointestinal tract (with the exception of the esophagus), mammary gland, uterus, thymus, and bone marrow, but these cells are just occasionally seen in other human tissues and organs. Eosinophils residing within specific tissue niches function to maintain tissue, metabolic, and immune homeostasis in the steady state (reviewed in[3,5,6]). On the other hand, eosinophils can be prominent in virtually any tissue during inflammatory responses, increasing in number as a constituent of mixed inflammation or being the predominant or even the only cell found in inflammatory infiltrates.[3,363]

Peripheral blood and tissue eosinophilia can occur in patients with varied allergic and inflammatory diseases, as well as diverse hematologic malignancies (reviewed in[364]). Although the criteria to define eosinophilic disorders involve several aspects and have been refined, the terms EADs or eosinophilic diseases are generally used to describe diseases characterized by the presence of eosinophil-rich inflammatory infiltrates and/or extensive extracellular deposition of eosinophil-derived proteins resulting in clinically relevant organ pathology.[364,365] EADs encompass a broad spectrum of diseases of different etiologies including allergies, gastrointestinal disorders, dermatoses, HESs, infectious diseases, and neoplastic disorders.[364] The biology of resident and recruited eosinophils, as well as the precise role of these cells in the pathogenesis of EADs, remains mostly unclear.

Methods for evaluation of eosinophil involvement in tissue human pathology include quantification of eosinophil numbers and detection of eosinophil granule proteins in biopsies with the use of histological and immunohistochemical approaches, respectively.[365] Because TEM can unambiguously identify eosinophils and their tissue-deposited FEGs without the use of a specific marker and is also a premier technique to evaluate eosinophil activation, the application of EM has helped to detect these cells and understand their degranulation/secretory activity in a variety of inflammatory conditions.

In general, ultrastructural studies demonstrate that PMD and cytolysis are the most common degranulation patterns found in human biopsies from patients with EADs while exocytosis is occasionally observed in vivo, and, by our experience, when present, is associated with other eosinophil secretory processes. Other cellular events such as cell polarization and migration, production of extracellular vesicles, and morphological profiles of cell death (apoptotic and cytolytic) can be fully identified by TEM with critical insights into the functional activity of eosinophils in both human health and disease. The ultrastructural signature of a particular eosinophil cytolytic cell death (EETosis) has also been disclosed by TEM studies both in vitro, as a result of eosinophil activation,[41,47,312] and in vivo during eosinophilic diseases,[46,279,312,317] thus helping to identify EETosis as a mechanism increasingly associated with eosinophil degranulation.

EM uniquely enables imaging of the object of interest at high resolution in its structural context.[45] In the study of eosinophil immunobiology, EM is the only technique to reveal the ultrastructure of infiltrating eosinophils and their microenvironment. The identification of CLCs in their varied forms and sizes by TEM in the extracellular matrix, adjacent to lytic eosinophils and FEGs, provided important clues about the functional activity of these cells related to inflammation. Ultrastructural studies showing eosinophils interacting with other cells present in the tissue microenvironment such as mast cells,[197,366] plasma cells,[42] and tumor cells[367] are also insightful in situations of diseases and need to be better explored as they point out to more complex roles for eosinophils.[250] Therefore, despite the progress achieved so far into eosinophil biology, there is still much to learn regarding eosinophils and their response to tissue components.[365] The application of TEM can be beneficial in this direction.

The purpose of this chapter is to illustrate the ultrastructure of eosinophils participating in a wide variety of human diseases. The electron micrographs shown in this chapter are by no means exhaustive in terms of displaying all ultrastructural features of a particular disease. We included electron microscopic findings of eosinophils identified in numerous human biopsy specimens collected during decades of pathological studies. Therefore, our goal here is to show the ultrastructural characteristics exhibited by human eosinophils in selected EADs affecting several tissues and organ systems. The identification of eosinophils participating in diseases, as well as their ultrastructural alterations, is important for diagnostic purposes and understanding the immunobiology of these enigmatic cells.

8.2 Allergic airway diseases

Eosinophils are remarkably increased in numbers during allergic airway diseases such as ECRS, allergic rhinitis, asthma, and nasal polyposis (reviewed in[368]). Allergic inflammation is a complex multifactorial response, and the pathogenesis of these diseases, as well as the role played by eosinophils, remains speculative.

ECRS is an inflammatory and persistent disease of the nose and paranasal sinuses in which accumulation and secretion of activated eosinophils and mucus production are considered to have an important role in its pathogenesis, but the mechanisms involved are not well understood (reviewed in[369]). We analyzed in detail the ultrastructure of eosinophils in biopsies from ERCS patients. Significant tissue eosinophilia, with both intact (Figs. 8.1 and 8.2) and cytolytic (Fig. 8.3) eosinophils, was observed in the lamina propria of nasal sinuses. Infiltrating intact eosinophils were distributed in the extracellular matrix as isolated or grouped cells showing frequently tight interactions with each other (Figs. 8.4 and 8.5) and/or with plasma cells (Figs. 8.5 and 8.6).

PMD (Fig. 8.2) and cytolysis (Figs. 8.3, 8.6, and 8.7) were the two major modes of eosinophil degranulation observed in ECRS biopsies. Quantitative EM analyses of 80 eosinophils, in randomly taken electron micrographs, showed that 52.5% of the cells had morphological features of PMD while cytolysis, in different levels, was noted in 47.5% of the eosinophils. Resting, as well as eosinophils with morphological features of exocytosis, were not observed. FEGs (Figs. 8.6–8.8), typical CLCs (Figs. 8.8–8.10), and free EoSVs (Figs. 8.6, 8.7, and 8.10) deposited in the extracellular matrix were recurrent ultrastructural findings.

Interestingly, cytolytic eosinophils showing features of ETosis, a cytolytic cell death process involving the release of DNA extracellular traps from the nuclear origin (detailed in Chapter 6), also associated with CLC formation,[41] were identified in ECRS biopsies. By TEM, ETosis was characteristically observed in these samples by the presence of nuclear alterations such as decondensation and roundness (see Fig. 6.22 in Chapter 6) and release of chromatin, which appeared spread in the extracellular matrix in association with FEGs (Figs. 8.6, 8.9, and 8.10) and CLCs (Figs. 8.9 and 8.10; see also Fig. 6.30 in Chapter 6). Quantitative analyses detected that 40% of all cytolytic eosinophils with identifiable chromatin had ultrastructural features typical of ETosis. Our TEM studies demonstrating the ultrastructural EETosis signature confirm previous findings of EETs in secretions[163] and histological biopsies[41] from patients with ECRS.

Ultrastructural quantitative studies of airway biopsy specimens from patients with allergic rhinitis, asthma, and nasal polyposis also reported cytolysis and PMD as important degranulation patterns of eosinophils during these diseases.[151,174] By analyzing the eosinophil ultrastructure in seasonal allergic rhinitis, these authors observed that at nonsymptomatic baseline conditions, eosinophils from nasal biopsies already exhibited a moderate level of degranulation through PMD while pollen exposure led to extensive PMD with severe-to-complete loss of granule contents observed exclusively during seasonal allergen exposure.[153] Cytolysis, but not classical exocytosis, was also detected during the pollen season.[153]

As highlighted in Chapters 3 and 4, for human eosinophils, cytolysis is not considered just a degenerating process, but a critical degranulation mode, an "ultimate activation" mechanism culminating with the release of intact FEGs,[160,168,171] which are considered to remain active in the tissue microenvironment through the local release of preformed cytokines and other proteins.[16,167] As noted, our comprehensive EM analyses also revealed that, together with FEGs, EoSVs (Figs. 8.6, 8.7, and 8.10) persist intact after eosinophil cytolysis, indicating that these vesiculotubular structures might have a role in the tissue microenvironment. Furthermore, recent works highlight that in human allergic diseases, local cytolysis of eosinophils not only releases FEGs but also generates nuclear-derived DNA traps that are major extracellular structural components within eosinophil-rich secretions.[163]

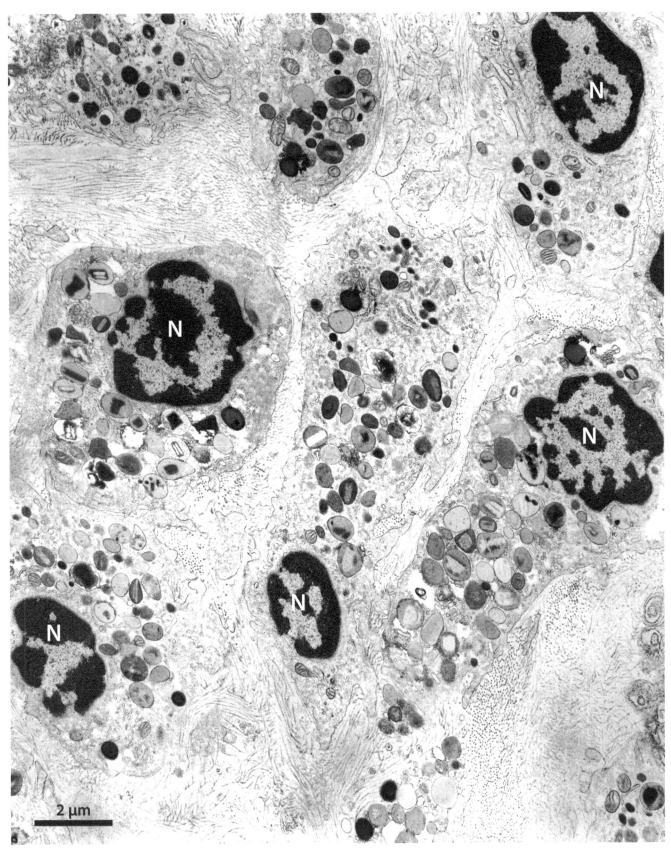

FIG. 8.1   Tissue eosinophilia observed in the frontal sinus from a patient with eosinophilic chronic rhinosinusitis (ECRS). A collection of eosinophils is seen in the lamina propria by TEM. Secretory granules with varied morphology fill the cytoplasm. *N*, nucleus.

B. Eosinophils in human diseases

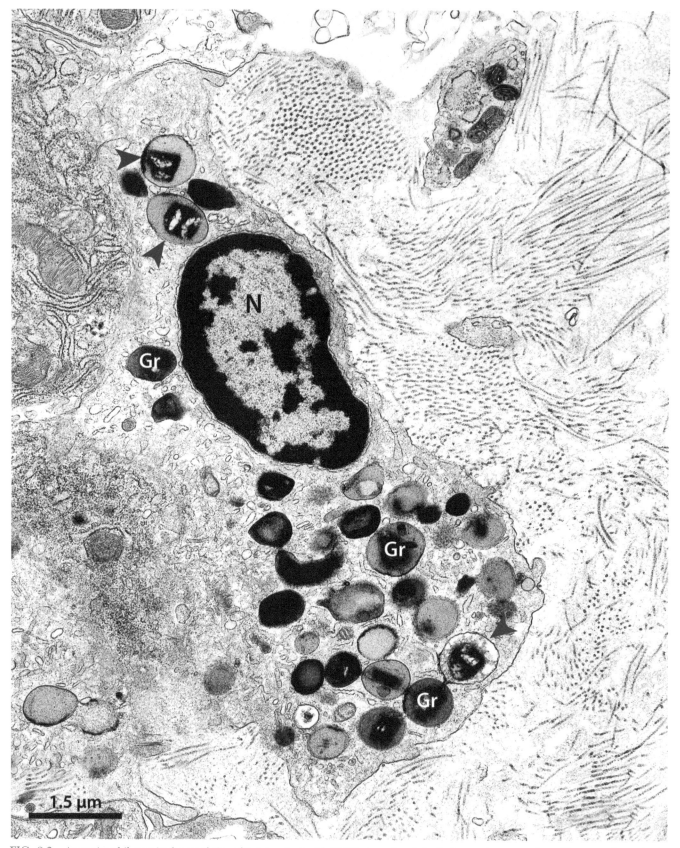

FIG. 8.2   An eosinophil seen in the nasal sinus from a patient with ECRS. Arrowheads indicate secretory granules with content losses in the cores and matrices indicative of PMD. Samples were prepared for TEM. *Gr*, secretory granules; *N*, nucleus.

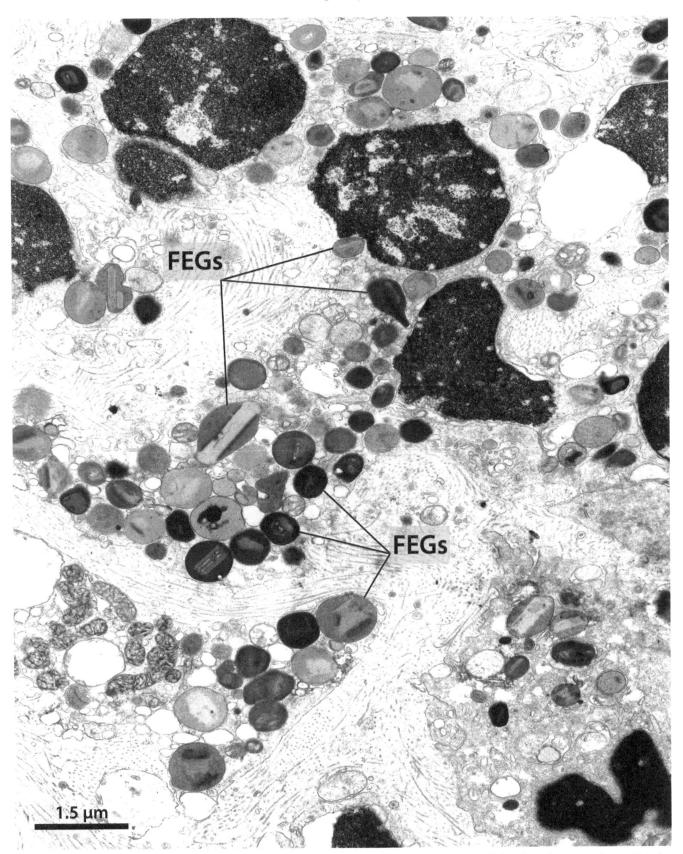

FIG. 8.3 Cytolytic eosinophils observed by TEM in a nasal biopsy from a patient with ECRS. Most cells lost their plasma membrane. Note free extracellular granules (FEGs) and eosinophil nuclei *(colored in purple)*.

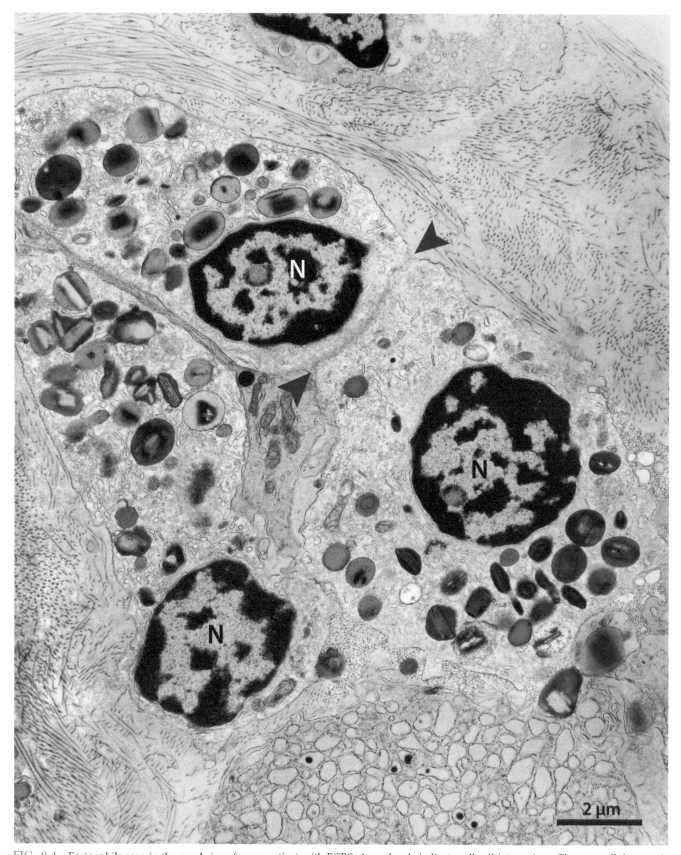

FIG. 8.4   Eosinophils seen in the nasal sinus from a patient with ECRS. Arrowheads indicate cell-cell interactions. The extracellular matrix (ECM) and surrounding cells were colored in *green*. Samples were prepared for TEM. *N*, nucleus.

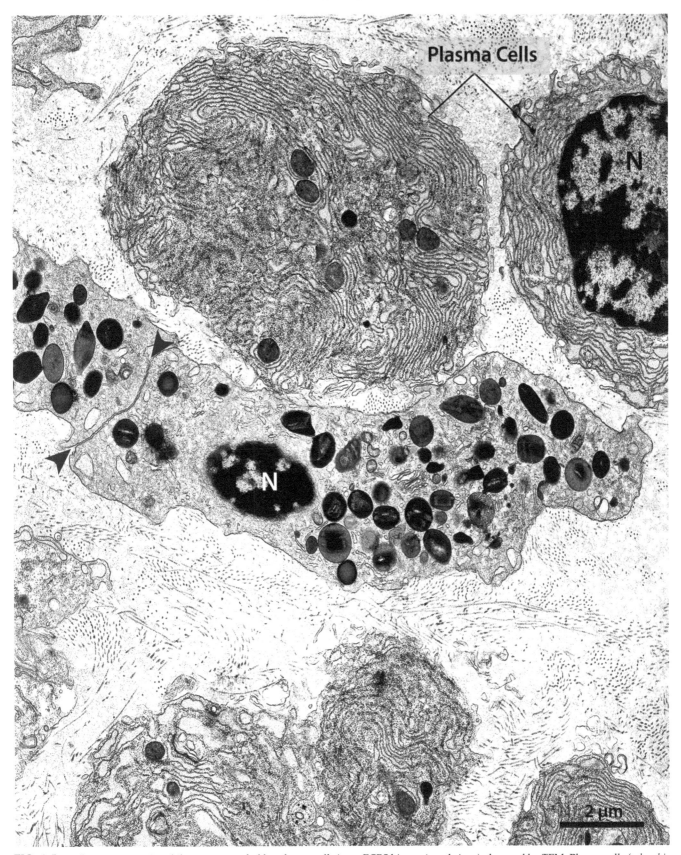

FIG. 8.5   Inflammatory eosinophils are surrounded by plasma cells in an ECRS biopsy (nasal sinus) observed by TEM. Plasma cells *(colored in purple)* show typical morphology with abundant RER cisternae *(light blue)*. Arrowheads indicate eosinophil-eosinophil interaction. Mitochondria *(green)*. N, nucleus.

B. Eosinophils in human diseases

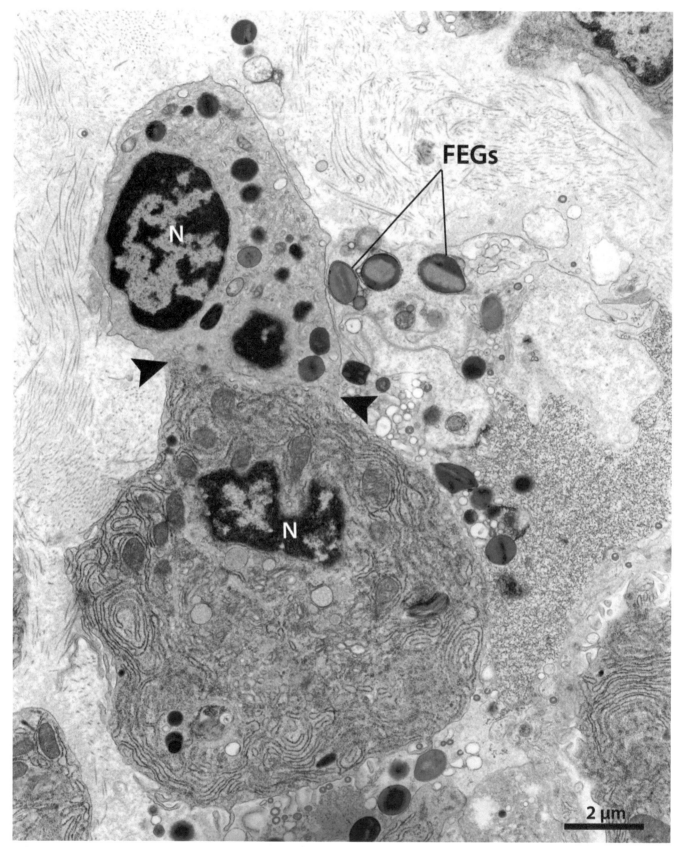

FIG. 8.6   TEM of tissue eosinophils in an ECRS biopsy. An intact eosinophil interacts with a plasma cell *(arrowheads)* while a neighboring eosinophil exhibits cytolytic features of ETosis [very decondensed and spread chromatin *(highlighted in blue)*] in association with FEGs (colored in *purple*) and EoSVs *(pink)*. *N*, nucleus.

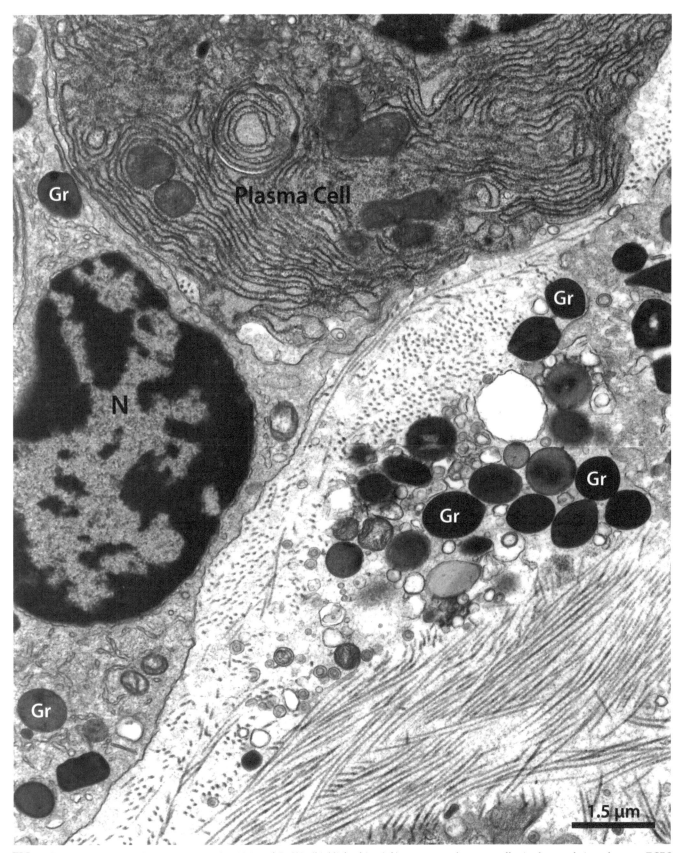

FIG. 8.7 Intact eosinophil secretory granules (Gr) and EoSVs *(highlighted in pink)* are seen as free organelles in the nasal sinus from an ECRS patient. An intact eosinophil interacts with a plasma cell. Biopsy prepared for TEM. *N*, nucleus.

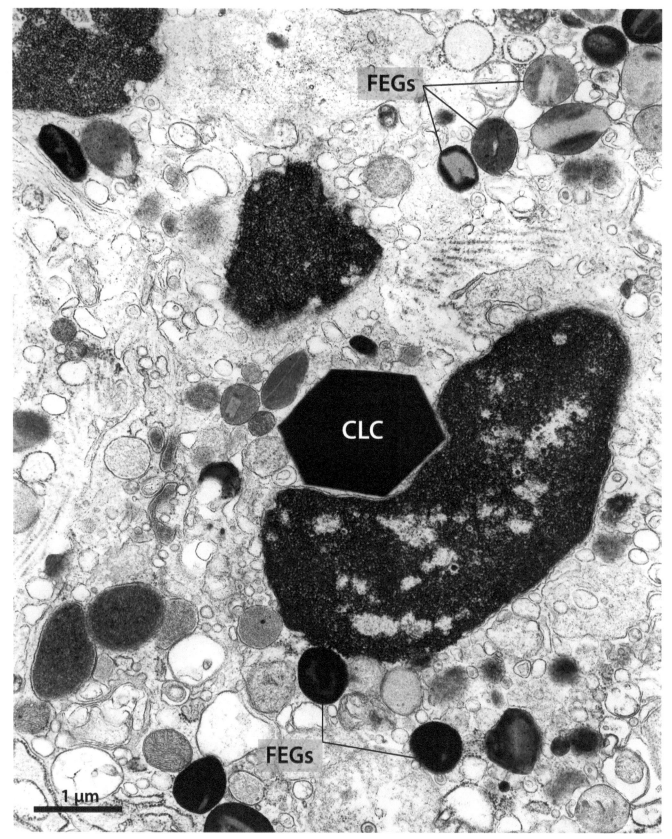

FIG. 8.8 Cytolytic eosinophils seen in an ECRS biopsy (nasal sinus) by TEM. Note membrane-bound FEGs and a hexagonal Charcot-Leyden crystal (CLC) in close proximity to a nucleus. Nuclei were colored in *purple*.

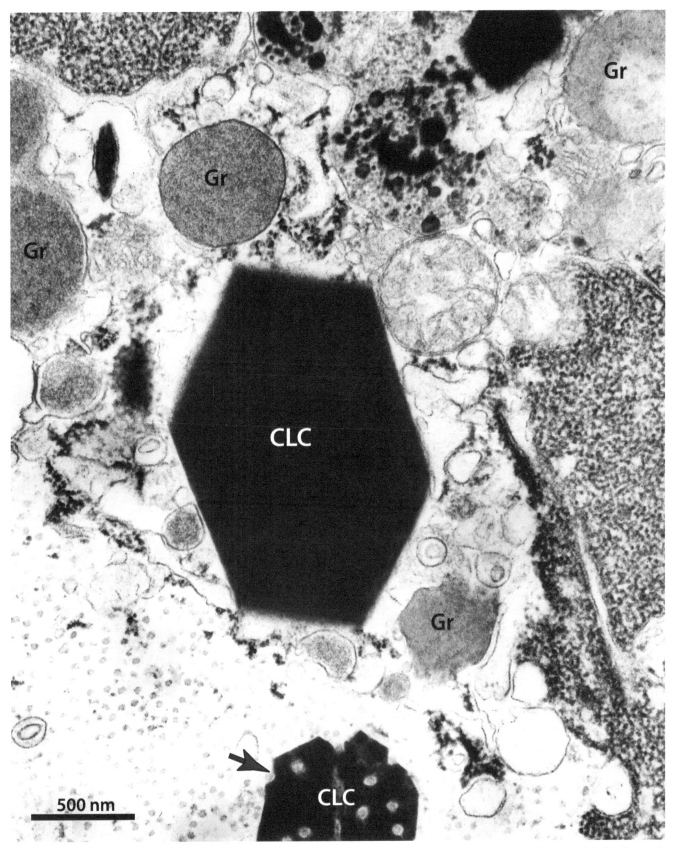

FIG. 8.9 Charcot-Leyden crystals (CLC) are observed in high magnification in the nasal sinus from a patient with ECRS. Note the chromatin filaments *(highlighted in purple)* spread around CLCs and FEGs (Gr). One of the CLCs shows focal electron-lucent areas *(arrow)*. Biopsy prepared for TEM.

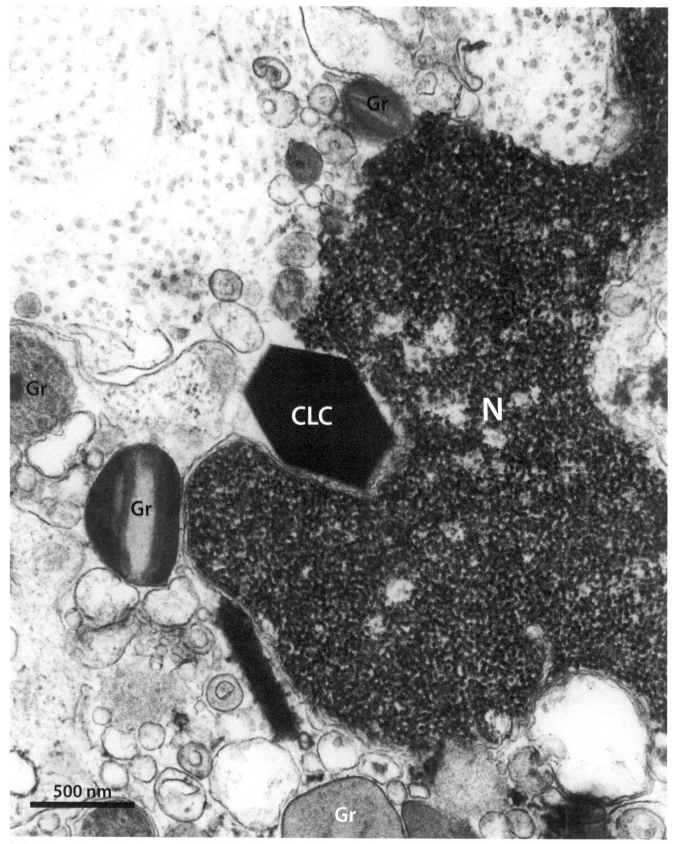

FIG. 8.10   A Charcot-Leyden crystal (CLC), FEGs (Gr), and free EoSVs *(colored in pink)* are seen in close association with a disrupted eosinophil in the nasal sinus from a patient with ECRS. nucleus (N). Sample prepared for TEM.

8.3 Skin disorders

Eosinophilic skin disorders, generally referred to as eosinophilic dermatoses, represent a broad spectrum of diseases of different etiologies characterized by skin and/or mucosal infiltration of eosinophils and/or eosinophil degranulation, with or without blood eosinophilia (reviewed in[368,370-372]). Most of these disorders, for example, atopic dermatitis, urticarias, and drug reactions are also included under the umbrella of allergic diseases. Other common eosinophilic skin disorders are linked with parasitic infestations and autoimmune blistering diseases such as bullous pemphigoid. Types of eosinophilic dermatoses also comprise rare diseases, for example, eosinophilic cellulitis, eosinophilic pustular folliculitis, granuloma faciale, eosinophilic fasciitis, and eosinophilic vasculitis, in which eosinophil infiltration can affect specific tissue layers or skin structures (dermis, subcutaneous fat, fascia, follicles, and cutaneous vessels).[371,373] Eosinophilic skin disorders can also be associated with carcinomas, bacterial/viral infections, and hematological malignancies.[372]

A categorized list of eosinophilic skin disorders, including systemic diseases with cutaneous eosinophilia as well as their clinical and pathological properties, has been presented in different reviews.[371-373] Although eosinophils are considered a key cell implicated in the pathogenic mechanisms of eosinophilic dermatoses, their roles in these diseases are not well defined yet and their ultrastructural aspects remain to be addressed in more detail in many of these diseases.

8.3.1 Atopic dermatitis

In histopathological sections, eosinophilic infiltration is identified by the acidophilic staining properties of the specific granules, usually after application of hematoxylin and eosin. However, when most tissue eosinophils underwent activation and cytolytic degranulation, these cells are poorly visualized by routine histological techniques. Thus, the number of intact eosinophils cannot be the only parameter to be considered to evaluate the contribution of these cells to inflammatory infiltrates. This applies to several skin disorders, particularly atopic dermatitis, in which intact eosinophils are not a prominent component of the cellular infiltration while their granule-derived products are remarkably deposited in the tissue as well documented by immunohistochemistry and quantitative analyses.[372,374]

In an ultrastructural study of skin biopsy specimens from 10 patients diagnosed with chronic recurrent atopic dermatitis and with active disease at the time of biopsy, cytolysis was the main eosinophil degranulation process found in the skin lesions.[154] By performing an extensive quantitative analysis of approximately 600 electron micrographs, the authors found dermal eosinophils with varied levels of degeneration, including loss of the plasma membrane and/or nuclear envelope and the presence of FEGs (Fig. 8.11). Moreover, when intact eosinophils were observed, they had ultrastructural signs of PMD combined or not with uropod formation,[154] both morphological appearances of eosinophil activation as addressed in Chapter 4. The authors did not find evidence of classical exocytotic degranulation or normal-appearing eosinophils.[154] In another ultrastructural study comparing the ultrastructure of human eosinophils in tissue (skin biopsies) and peripheral blood from patients with atopic dermatitis, cytolysis and PMD were detected in tissue eosinophils but not when these cells were in the circulation,[190] indicating that eosinophil activation is more prominent in the tissue microenvironment.

8.3.2 Bullous pemphigoid

Among eosinophilic dermatoses, there are diseases classified as autoimmune blistering skin disorders. Bullous pemphigoid is the most common of them, affecting mainly elderly individuals and, sometimes, children. Bullous pemphigoid is characterized by dermal-epidermal junction separation with inflammatory infiltrate composed of lymphocytes, neutrophils, and mainly eosinophils (Figs. 8.12 and 8.13) (reviewed in[371,375-378]). High levels of Th2 cytokines such as IL-4, IL-5, and IL-13, as well as other cytokines and chemokines affecting eosinophil recruitment, activation, and, survival, have been found in lesional skin, blister fluids, and peripheral blood of patients with bullous pemphigoid.[375,378]

The diagnosis of bullous pemphigoid relies not only on histopathological evaluation demonstrating a subepidermal blister with an accumulation of eosinophils (Fig. 8.12) or eosinophilic spongiosis (intraepidermal eosinophils in spongiotic regions), but also on the deposition of IgG and/or complement component 3 (C3) (Fig. 8.12) in a linear band at the dermal-epidermal junction, and quantification of circulating autoantibodies against the hemidesmosome proteins BP180 (mainly) and BP230.[375,377] BP180 (type XVII collagen) is a transmembrane glycoprotein with an extracellular C-terminus that mediates adhesion between the epidermis and the basement membrane, while BP230 is an intracellular plakin-like protein of the hemidesmosomal plaque. The formation of these autoantibodies, primarily of the IgG and IgE classes, is considered a leading event in blister formation.[375]

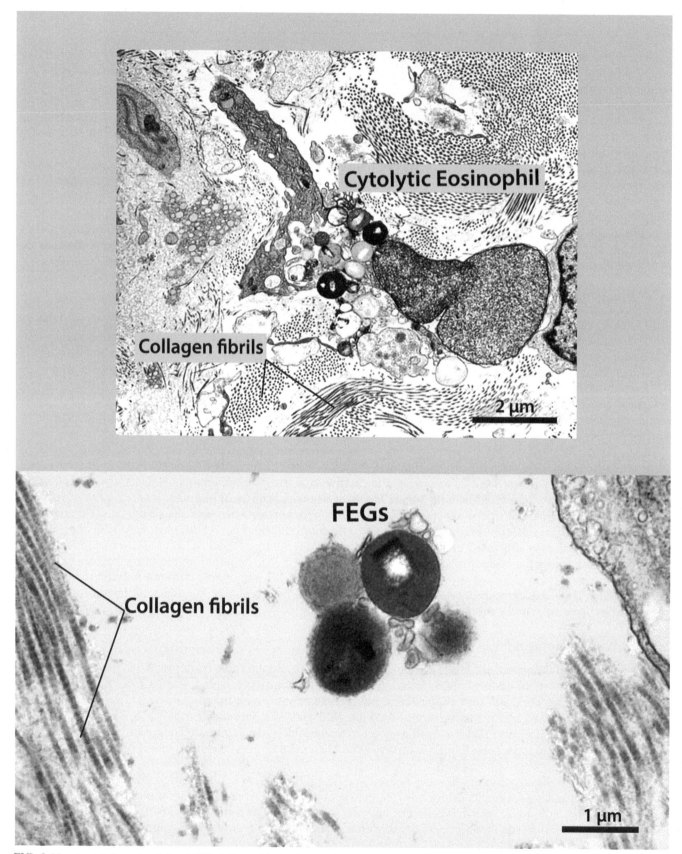

FIG. 8.11   A cytolytic eosinophil and FEGs infiltrated in the dermis of a patient with atopic dermatitis. A decondensed nucleus *(colored in purple)* undergoing chromatolysis is seen in the top panel. The plasma membrane has disintegrated, and membrane-bound FEGs *(yellow)* and EoSVs *(pink)* are spread in the ECM. Note bundles of collagen fibrils sectioned in both transversal and longitudinal planes. Skin biopsies prepared for TEM.[154]

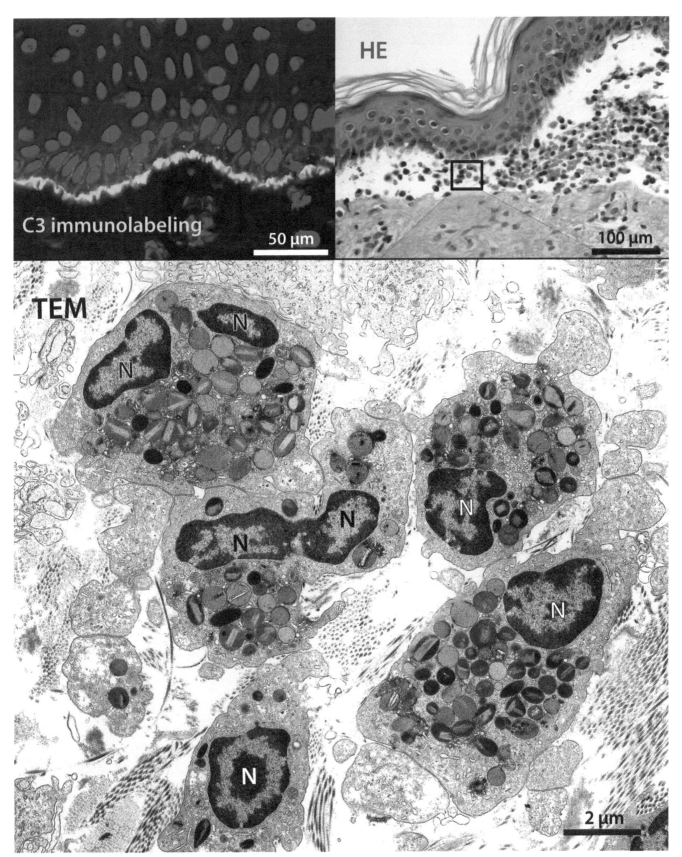

FIG. 8.12 Eosinophils in an erythematous plaque lesion from a patient with bullous pemphigoid. Note dermis-epidermis separation with deposition of complement component 3 (C3) in the basement membrane zone (fluorescent line) and accumulation of eosinophils. Biopsies were prepared for C3 immunolabeling,[523] histology (section stained with HE), and TEM.[379] N, nucleus.

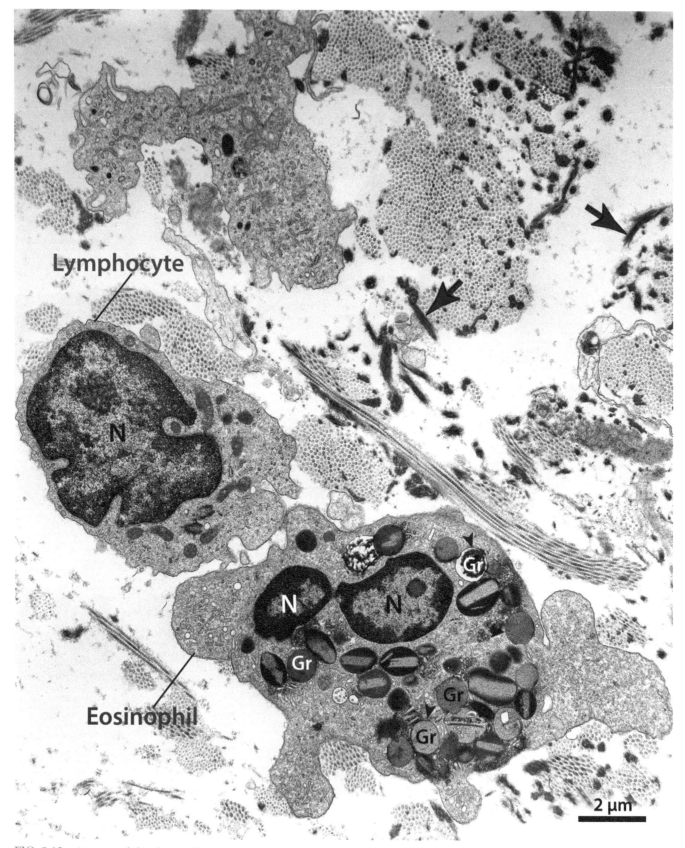

FIG. 8.13   An eosinophil in the papillary dermis of a bullous pemphigoid erythematous plaque lesion. Specific granules (Gr) show PMD signs *(arrowheads)*. Note a lymphocyte close to the eosinophil and fibrin *(arrows)* in the ECM. Skin biopsy prepared for TEM.[379] *N,* nucleus.

We studied histopathological events associated with eosinophil infiltration and eosinophil ultrastructure in biopsy specimens obtained from bullous pemphigoid lesions at different stages of clinical development.[379] Semiquantitative histological analyses showed that eosinophils changed in number and appearance as lesions developed. When seen in adjacent normal skin and erythematous maculae, the earliest detectable lesions, eosinophils appeared mostly normal with some scattered free granules observed. At the later stages of lesions (erythematous plaques and bullae), eosinophils became the predominant infiltrating cell and numerous FEGS were detected throughout the dermis. The formation of a blister was associated with more marked eosinophil infiltration.[379]

Ultrastructural examination showed both intact (Figs. 8.12–8.14) and cytolytic eosinophils (Figs. 8.15 and 8.16) in the erythematous plaques and bullae. Intact eosinophils showed signs of activation with frequent morphological changes of the secretory granules, including content losses and a marked reduction in the density of the entire granule (Fig. 8.13). Ultrastructural alterations also included the occurrence of clouds of amorphous material around the granule limiting membranes together with granule-associated EoSVs, events reflecting vesicle formation and content release through PMD (Fig. 8.14).

A large number of eosinophils were undergoing cytolysis with the breakup of both plasma membrane and nuclear envelope and release of membrane-bound granules (Figs. 8.15 and 8.16). Nuclear chromatin particles, FEGs, which displayed partial or complete lucent areas in the crystalline core, and EoSVs were scattered widely within erythematous plaques and bullous lesions (Figs. 8.15 and 8.16). Eosinophils were not observed to degranulate through exocytosis, which, as noted, requires granule-granule fusion and/or granule fusion with the plasma membrane.[379] Fibrin-gel, which is composed of a loose interlacing mesh of fibrin and entrapped edema fluid and irregular elastic fibers, were also observed by TEM (Fig. 8.13).

Although the exact role of eosinophils in the pathogenesis of bullous pemphigoid has not been completely elucidated, these cells have been implicated in the disease development through different pathways. Eosinophils are a source of a large number of mediators putatively associated with blister formation after eosinophil activation and degranulation, including matrix metalloproteinase-9 (MMP9), which is able to cleave BP180, granule-stored cationic proteins (MBP, ECP) with a potentially toxic effect on basal keratinocytes, and IL-31, a cytokine playing a significant role in itch-related inflammation.[375,378] Moreover, the interplay between the high-affinity IgE receptor (FcεRI) expressed on eosinophils in bullous pemphigoid patients and anti-BP180 IgE autoantibodies may contribute to eosinophil degranulation in bullous pemphigoid lesions and consequent blister formation.[375,378] Eosinophils also produce extracellular traps and ROS, additional components considered to be involved in the pathogenic mechanisms of bullous pemphigoid.[375]

8.4 Hypereosinophilic syndromes (HESs)

HESs encompass a diversity of eosinophil-related diseases with varied clinical manifestations that may affect one isolated organ or several systems. A cardinal early study, based on eosinophilic subjects referred to the US National Institutes of Health (NIH), provided guidelines for what was then termed the idiopathic HES.[380] For study purposes, the diagnosis required sustained blood eosinophilia $> 1500/mm^3$ for 6 months, absence of other etiologies for eosinophilia, and finally signs and symptoms of organ involvement.[210,381] Subsequently, definitions of HES have been modified. No longer must eosinophilia be documented for 6 months duration, and repeated eosinophilia over shorter periods can be sufficient.[382] Moreover, the scope of HESs has been expanded to include and incorporate the diversity of eosinophil-related diseases.[383] In this broader expanded definition of HESs, multiple diseases, including the eosinophil-associated vasculitis, EGPA, and a variety of organ-focused diseases (e.g., eosinophilic pneumonias and eosinophilic gastroenteropathies), can be included under the umbrella classifications of HESs.

While many forms of HES remain idiopathic, etiologies of two categories of HES have been elucidated, and these are referred to as lymphocytic variant HES and "myeloproliferative" HES. Lymphocytic variant HESs are due to aberrant clones of T cells that can overproduce IL-5 that in turn drives the hypereosinophilia.[384] "Myeloproliferative" variants of HES are forms of chronic eosinophilic leukemia and are due to one of many possible chromosomal translocations. The most prevalent of these chromosomal translocations involve the gene encoding the tyrosine kinase receptor, platelet-derived growth factor (PDGF) alpha, which may be implicated in translocations with various fusion partners. Among these, the most common translocation yields a fusion protein combining a FIP1L1-like domain with a PDGF alpha domain that yields a FIP1L1-PDGFRA protein, a constitutively active tyrosine receptor kinase.[385] Here, we discuss the ultrastructure of eosinophils in the context of an undefined form of HES, with negative results for the FIP1L1-PDGFRA mutation. Our studies are based on numerous samples collected from the bone marrow (Fig. 8.17), peripheral blood (Figs. 8.18–8.27), and tissues (Figs. 8.28–8.40) from a patient diagnosed with longstanding HES.

As noted in Chapter 4, morphologic evidence of eosinophil activation in vivo becomes more apparent in tissues after eosinophil migration from the vascular compartment as observed by us[40] and other groups.[190] However, in

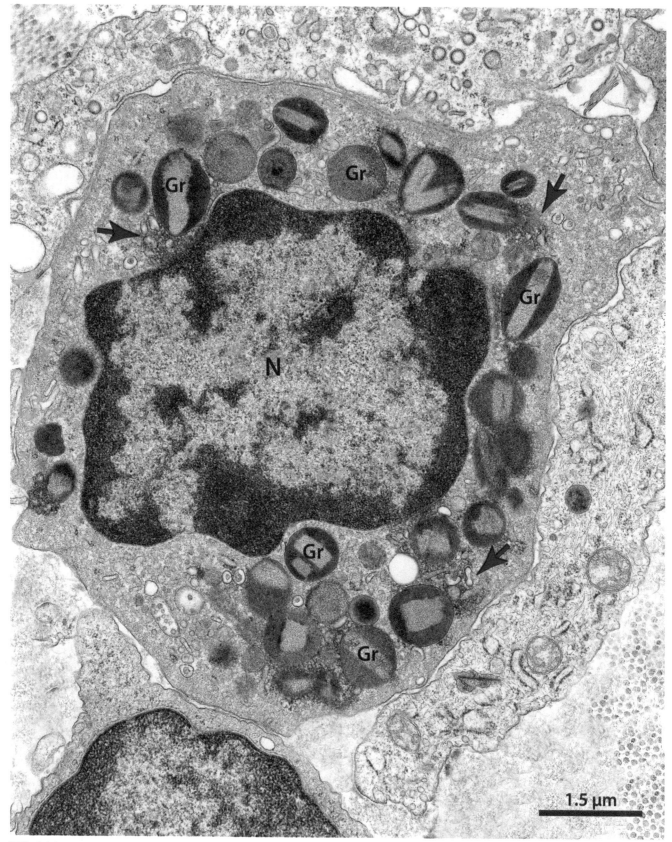

FIG. 8.14   A dermal eosinophil in a bullous pemphigoid erythematous plaque lesion. Arrows indicate clouds of dense material associated with granule vesiculation. Several EoSVs *(colored in pink)* are observed. Granules (Gr) show reversal of the electron densities. Skin biopsy prepared for TEM.[379] *N*, nucleus.

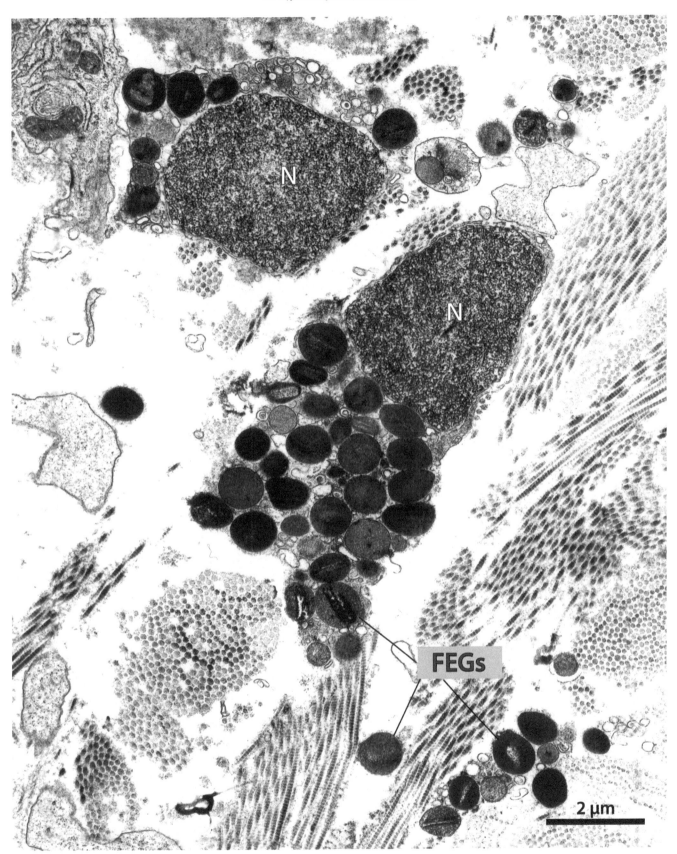

FIG. 8.15   Cytolytic eosinophils in the reticular dermis of a bullous pemphigoid erythematous plaque lesion. Ruptured plasma membranes, decondensed nuclei (N), and membrane-bound FEGs are observed. Note intact EoSVs around FEGs. Skin biopsy prepared for TEM.[379]

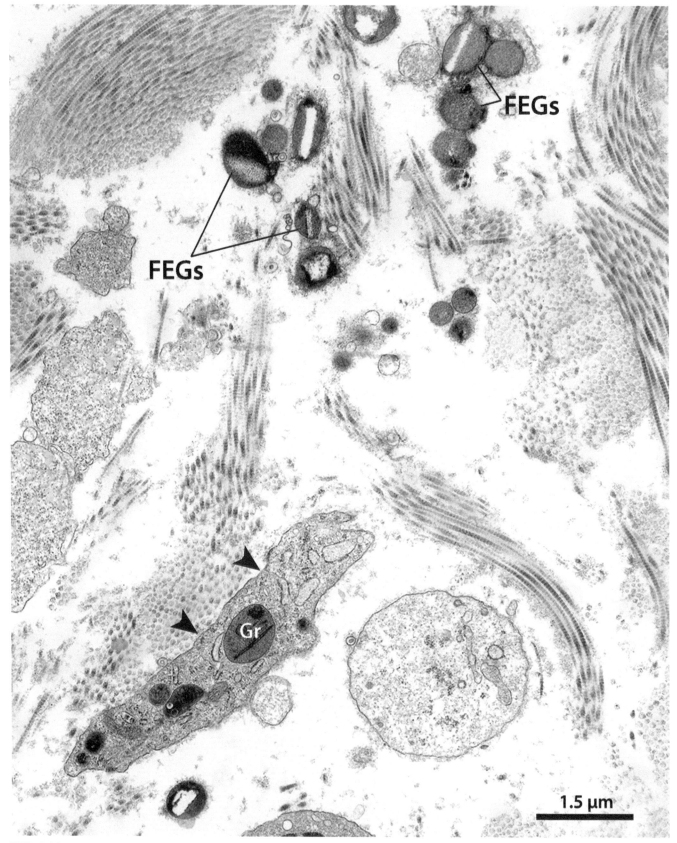

FIG. 8.16   FEGs and part of an eosinophil cytoplasm *(arrowheads)* are dispersed between bundles of collagen fibrils *(highlighted in green)* in the skin (erythematous plaque lesion) from a patient with bullous pemphigoid. Note losses in the granule (Gr) central core. Biopsy prepared for TEM.[379]

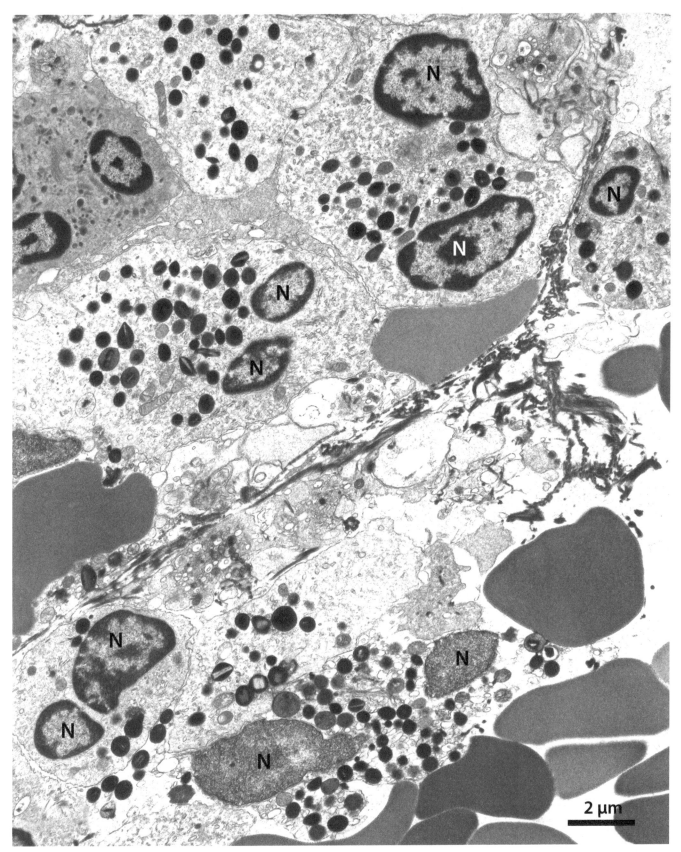

FIG. 8.17 Electron micrograph showing a high number of eosinophils in the bone marrow from a patient with HES. One neutrophil *(colored in green)* is also observed. Bone marrow aspirate prepared for TEM. *N*, nucleus.

B. Eosinophils in human diseases

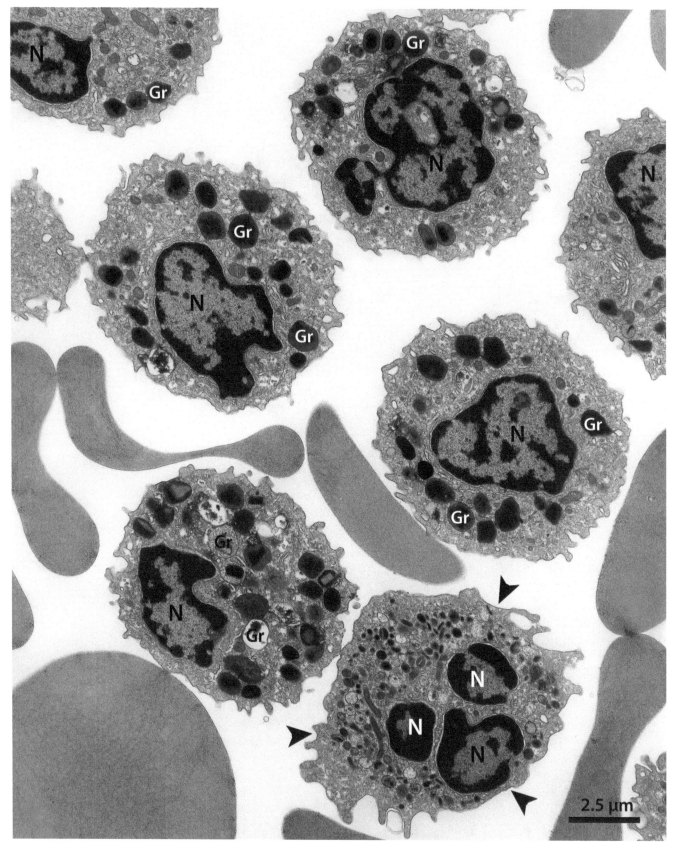

FIG. 8.18 Granulocytes in the peripheral blood from a HES patient. Many eosinophils and one neutrophil *(arrowheads)* are observed among red blood cells *(highlighted in red)*. Samples prepared for TEM. *Gr*, secretory granules; *N*, nucleus.

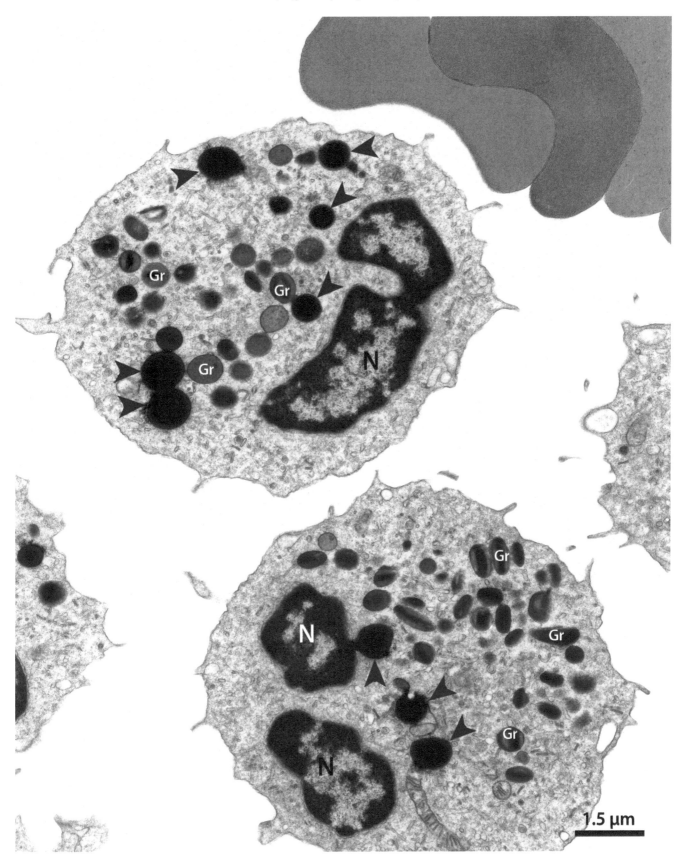

FIG. 8.19 Eosinophils in the HES peripheral buffy coat blood. Note the high numbers of LBs *(arrowheads)*, which typically appear as highly osmiophilic, nonmembrane-bound organelles. Samples prepared for TEM. Red blood cells *(colored in red)*; *Gr*, secretory granules; *N*, nucleus.

B. Eosinophils in human diseases

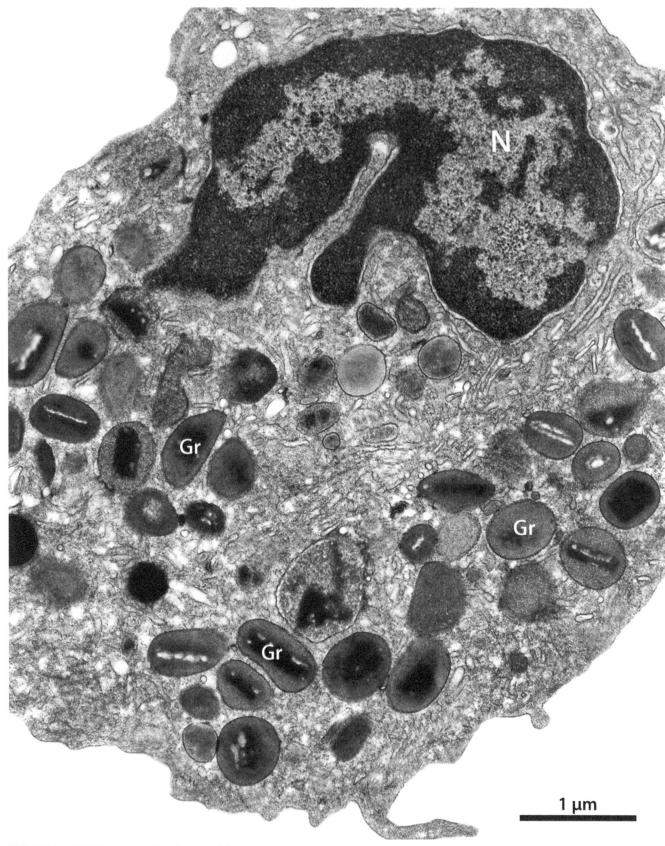

FIG. 8.20   A HES peripheral blood eosinophil. Most specific granules (Gr) show ragged losses of the cores associated with PMD. The nucleus (N) is not typically lobulated and RER cisternae are evident *(colored in blue)*, indicating a certain degree of cell immaturity. Cells were isolated and prepared for TEM.

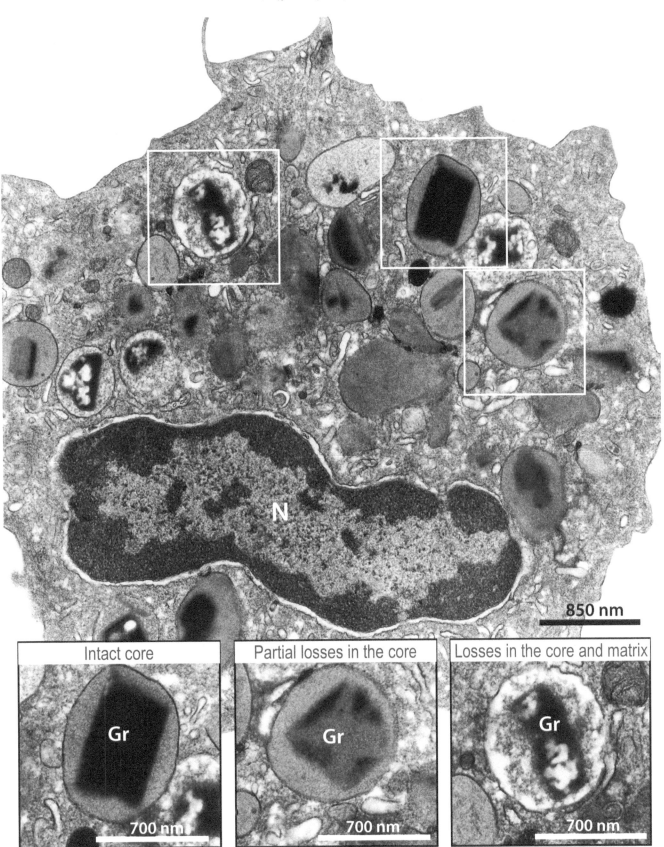

FIG. 8.21 Ultrastructure of a HES blood eosinophil. Specific granules (Gr) exhibit different degrees of content losses characteristic of PMD. Cells were isolated from the peripheral blood and processed for TEM. *N*, nucleus.

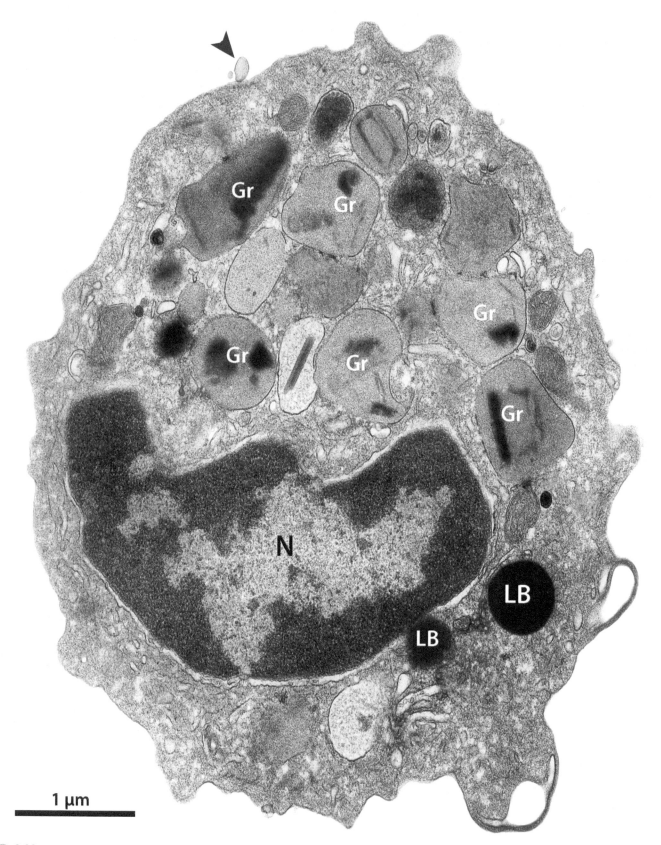

FIG. 8.22 Ultrastructure of a HES peripheral blood eosinophil. Note the presence of enlarged specific granules (Gr) with content disarrangement and losses. An extracellular vesicle is observed at the cell surface *(arrowhead)*. Cells were processed for TEM. *N*, nucleus; *LB*, lipid body.

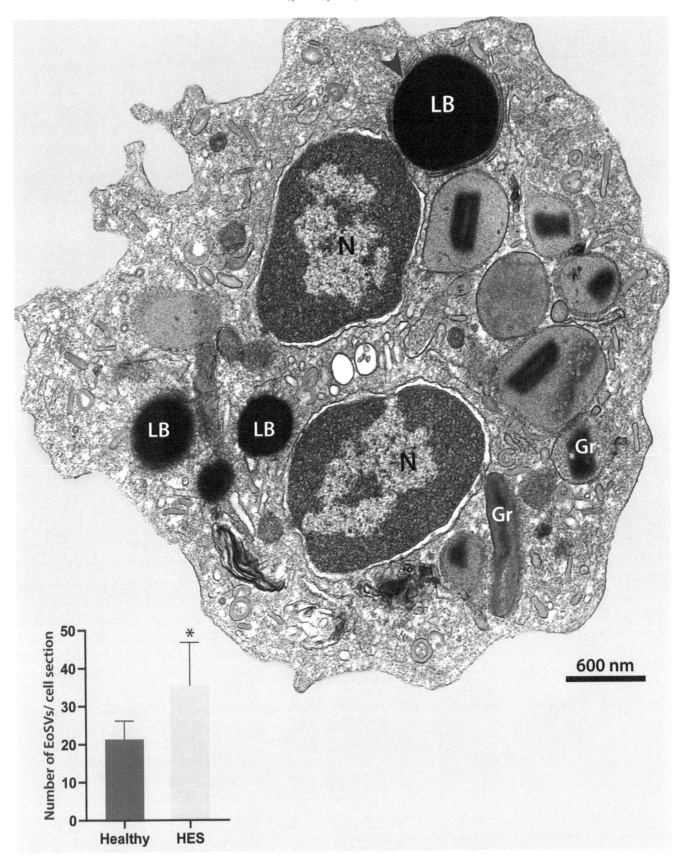

FIG. 8.23 Electron micrograph of a HES blood eosinophil. Increased numbers of EoSVs *(colored in pink)* are seen in the cytoplasm when compared with a healthy donor (*P < .001). Specific granules (Gr) and lipid bodies (LB) show different sizes. A large LB *(arrowhead)* is partially surrounded by ER cisternae. Cells were isolated from the peripheral blood and prepared for TEM. *N*, nucleus.

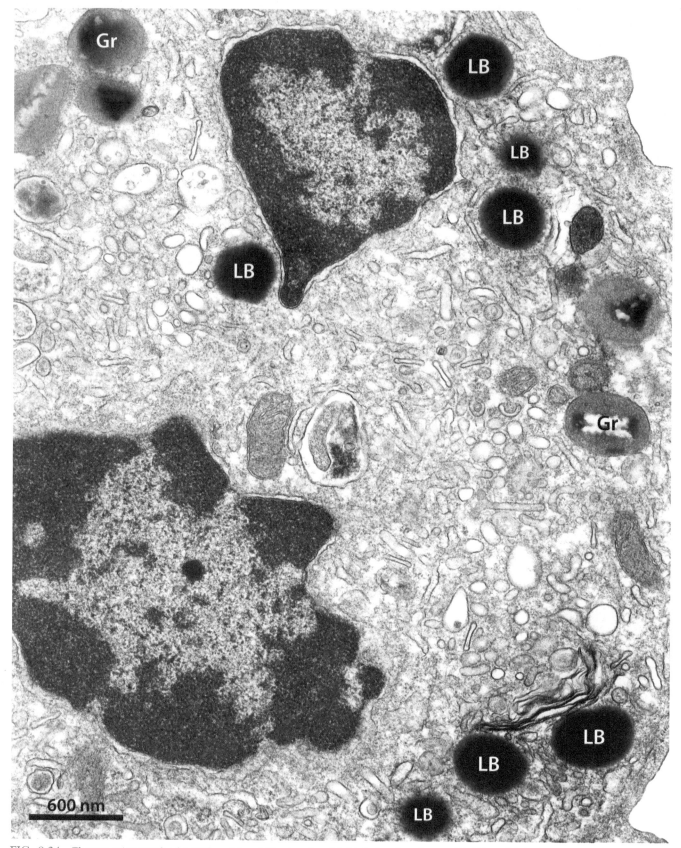

FIG. 8.24   Electron micrograph of a HES peripheral blood eosinophil. Note the abundance of EoSVs and LBs distributed in the cytoplasm. Nuclear lobes were colored in *purple*. Cells were prepared for TEM. *Gr*, specific granules.

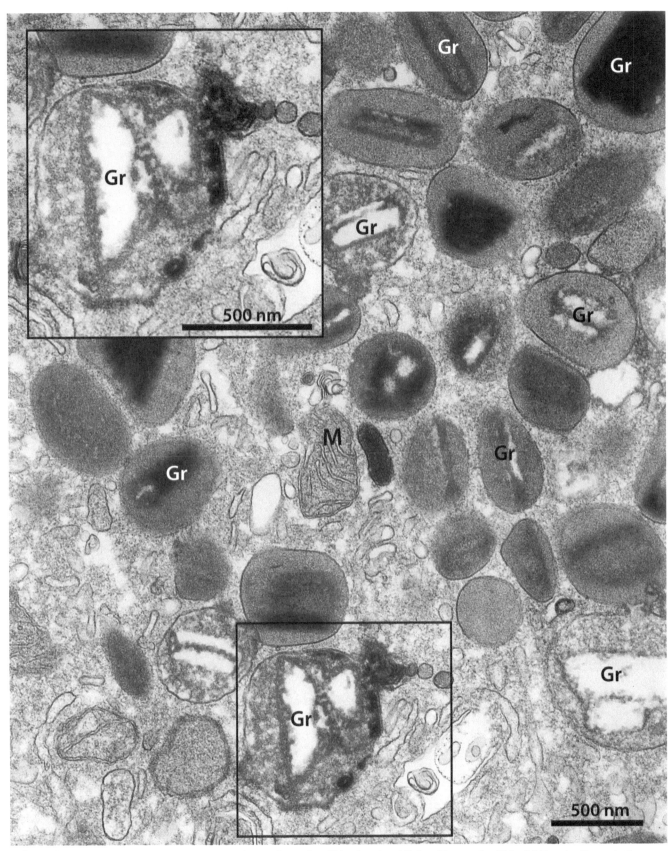

FIG. 8.25 TEM showing part of the cytoplasm of an HES peripheral blood eosinophil. Specific granules (Gr) exhibit content losses in the cores. Vesicle formation was captured from a granule surface (*boxed area*, also shown pseudocolored in higher magnification). Note amorphous material at the vesiculation site. *M*, mitochondrion.

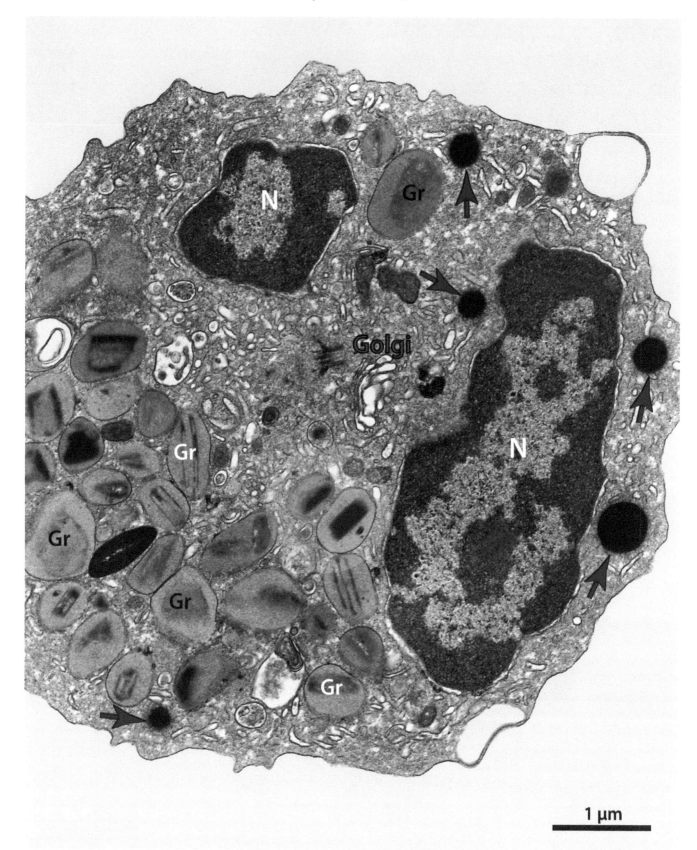

FIG. 8.26   Electron micrograph of a peripheral blood eosinophil from a patient with HES. While only two specific granules *(highlighted in yellow)* exhibit intact electron-dense cores, the remaining granules (Gr) show ragged losses to these structures. Note a dilated Golgi and LBs *(arrows)*. Cells were prepared for TEM. *N*, nucleus.

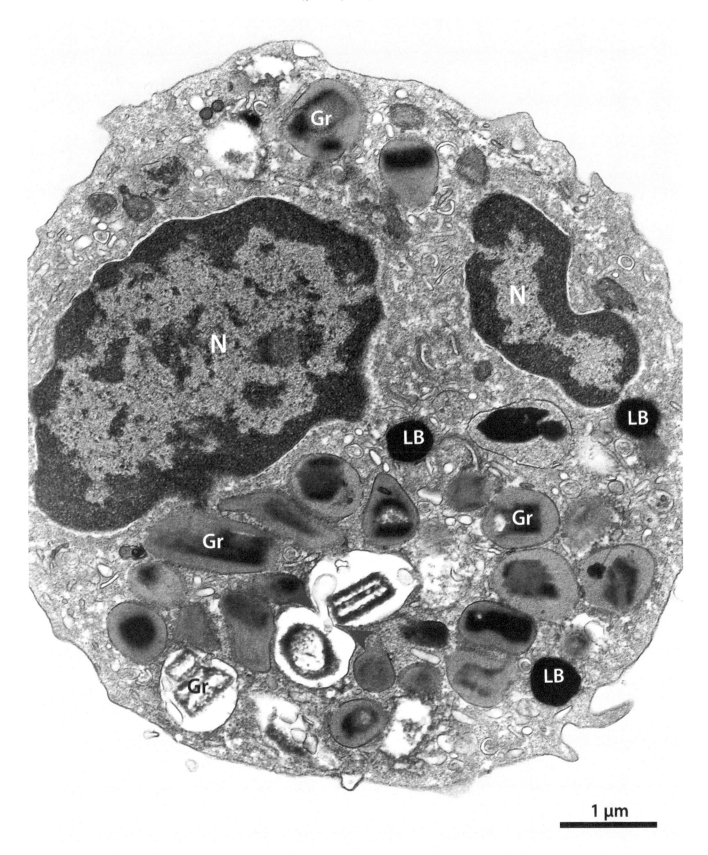

FIG. 8.27 Electron micrograph of a peripheral blood eosinophil from a patient with HES. Specific granules (Gr) show different electron densities indicating progressive losses of their products. Two of these granules are fused *(arrowheads)*. Sample prepared for TEM. *LB*, lipid bodies; *N*, nucleus.

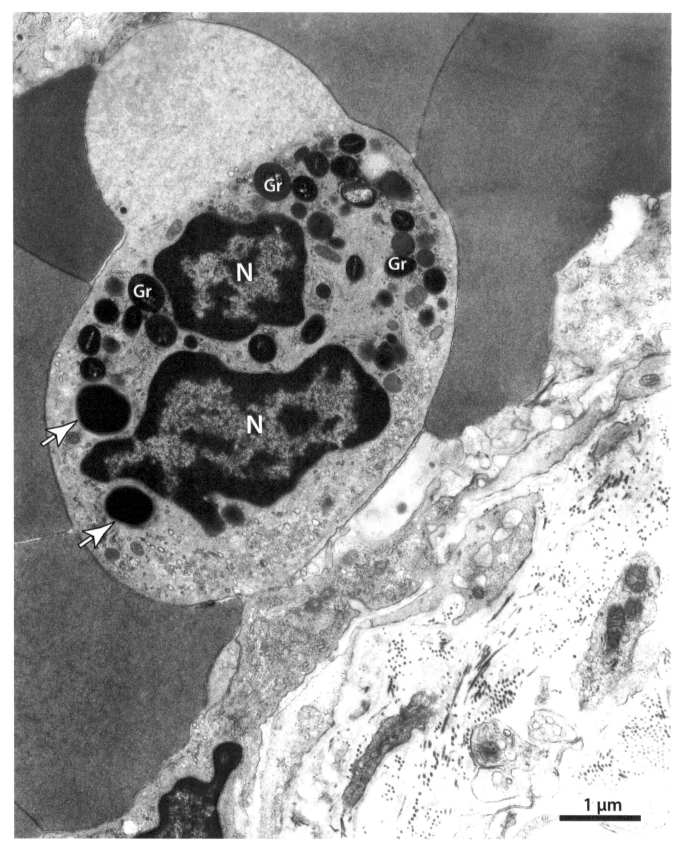

FIG. 8.28   An eosinophil within a skin blood vessel of a HES patient. Note the bilobed nucleus (N), specific granules (Gr) with ragged losses to the cores, and lipid bodies *(arrows)*. Red blood cells were colored in *red* and the endothelium in *orange*. The nucleus *(purple)* of one endothelial cell is partially seen. Biopsy processed for TEM.

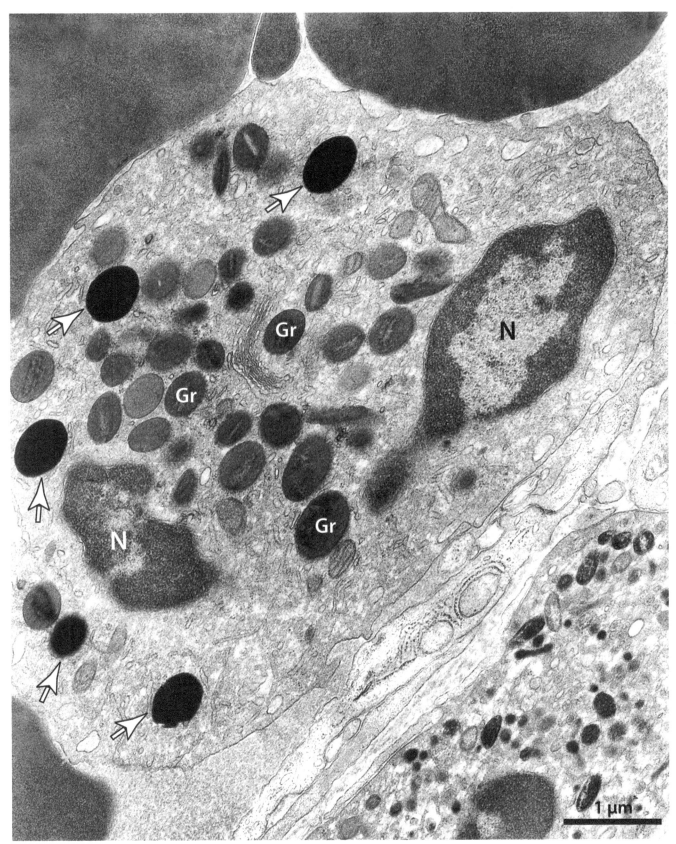

FIG. 8.29   An eosinophil within a small blood vessel in the skin of a HES patient. Several lipid bodies *(arrows)* and specific granules (Gr) are observed. A neutrophil is partially seen inside another blood vessel. Red blood cells *(red)*; endothelium *(yellow)*. Biopsy processed for TEM. *N*, nucleus.

B. Eosinophils in human diseases

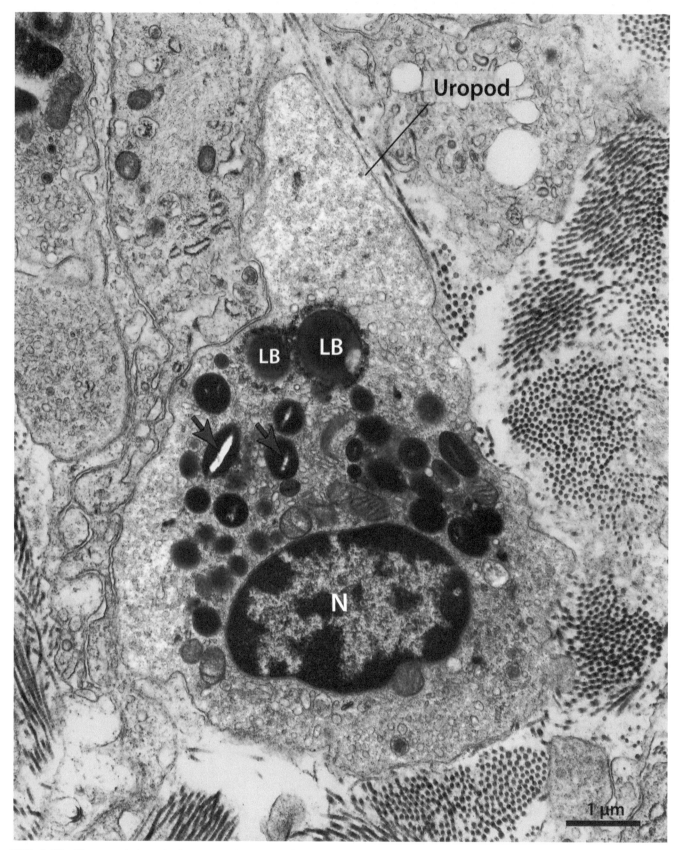

FIG. 8.30 Electron micrograph of an eosinophil infiltrated in the skin of a HES patient. Lipid bodies (LB) are observed in the cytoplasm. Note content losses *(arrows)* in the cores of nonfused secretory granules and uropod formation. The ECM and surrounding cells were colored in *green*. Biopsy prepared for TEM. *N*, nucleus.

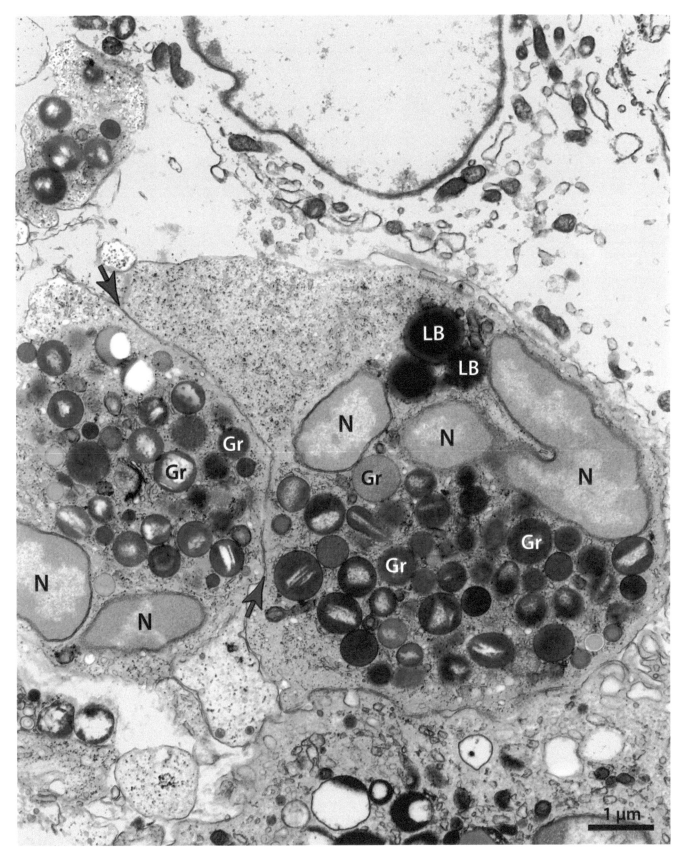

FIG. 8.31 An interaction *(arrows)* between two eosinophils in the skin of a HES patient. Note nonfused emptying secretory granules (Gr), indicative of PMD, the multilobed nuclei (N), and lipid bodies (LB). The ECM and surrounding cells were colored in green. Biopsy processed for TEM with ferrocyanide-reduced osmium postfixation.

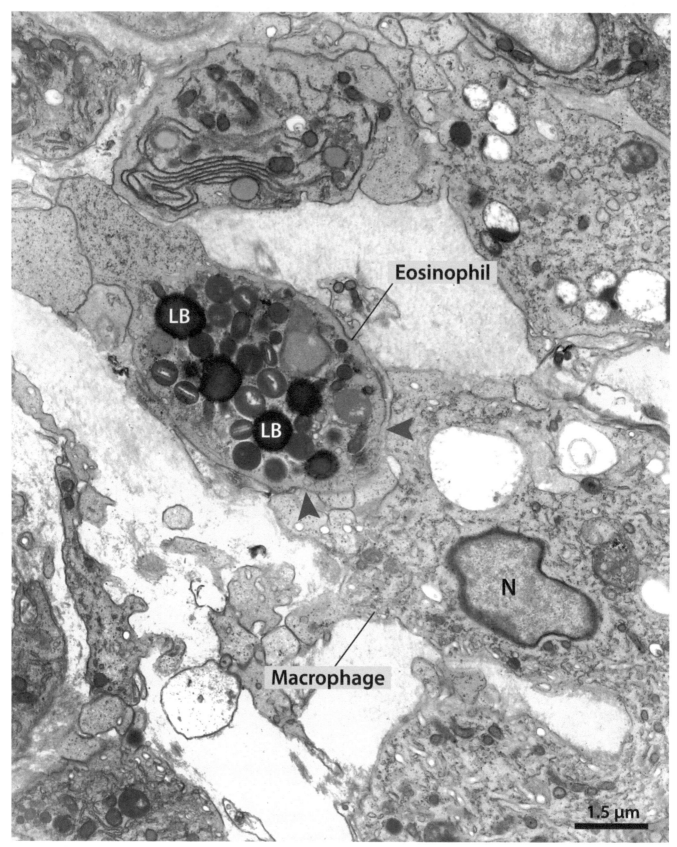

FIG. 8.32   An interaction *(arrowheads)* between an eosinophil and a macrophage in the skin of a HES patient. Granules with content losses denoting PMD are interspersed with lipid bodies (LBs). The Macrophage cytoplasm and the ECM were colored in beige and green, respectively. Biopsy processed for TEM with ferrocyanide-reduced osmium postfixation.

B. Eosinophils in human diseases

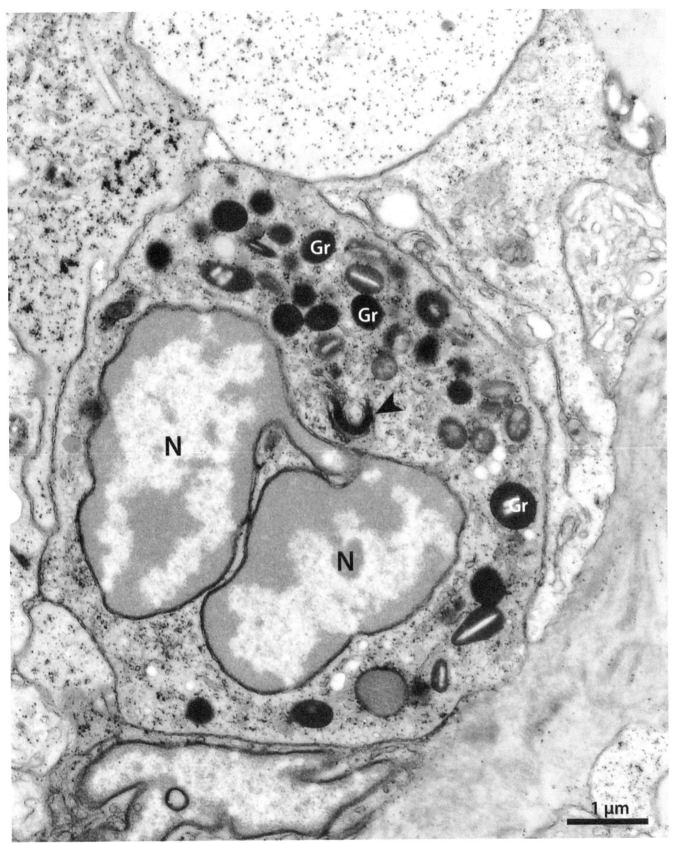

FIG. 8.33   Electron micrograph of an eosinophil in a HES skin biopsy. Membranous structures such as Golgi *(arrowhead)* and nuclear envelope are very contrasted while nucleus (N) appears less dense after postfixation with a TEM technique using ferrocyanide-reduced osmium. Secretory granules (Gr) show losses in the cores. ECM colored in *green*.

B. Eosinophils in human diseases

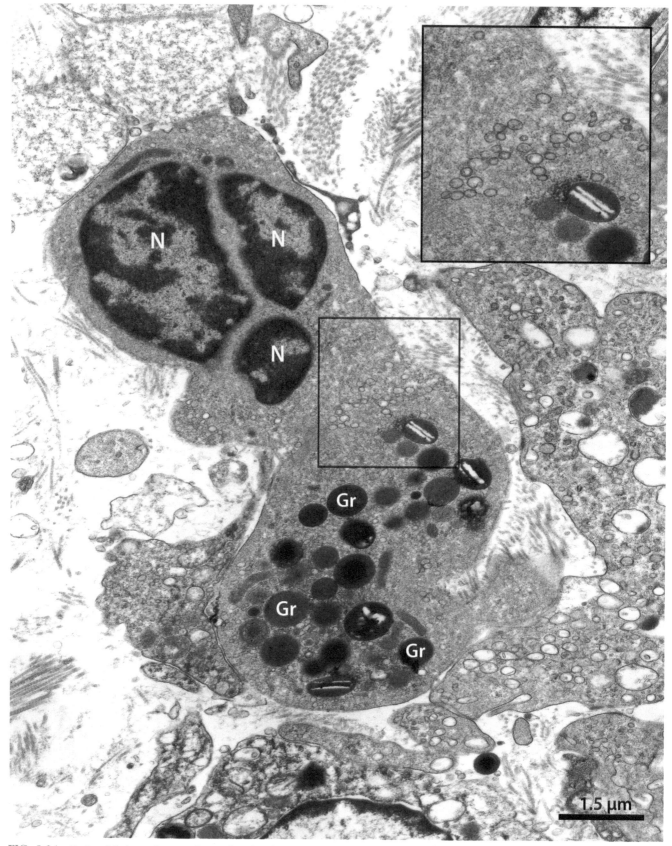

FIG. 8.34 Eosinophil shape changes observed at the ultrastructural level in the skin of a HES patient. Cell is elongated and polarized with nucleus (N) and granules (Gr) in opposite sides. Note accumulation of EoSVs *(highlighted in pink in higher magnification)* in the cell periphery. Granules exhibit content losses. Biopsy prepared for TEM.

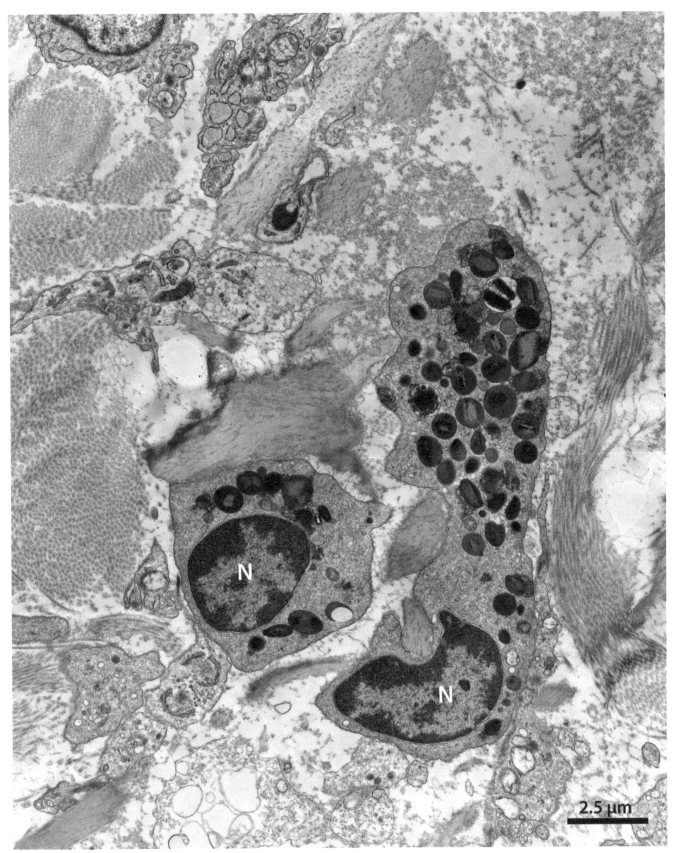

FIG. 8.35   Eosinophil shape changes observed at the ultrastructural level in the skin of a HES patient. One eosinophil shows clear cell polarization with nucleus (N) and the population of specific granules located in opposite sides. Biopsy prepared for TEM.

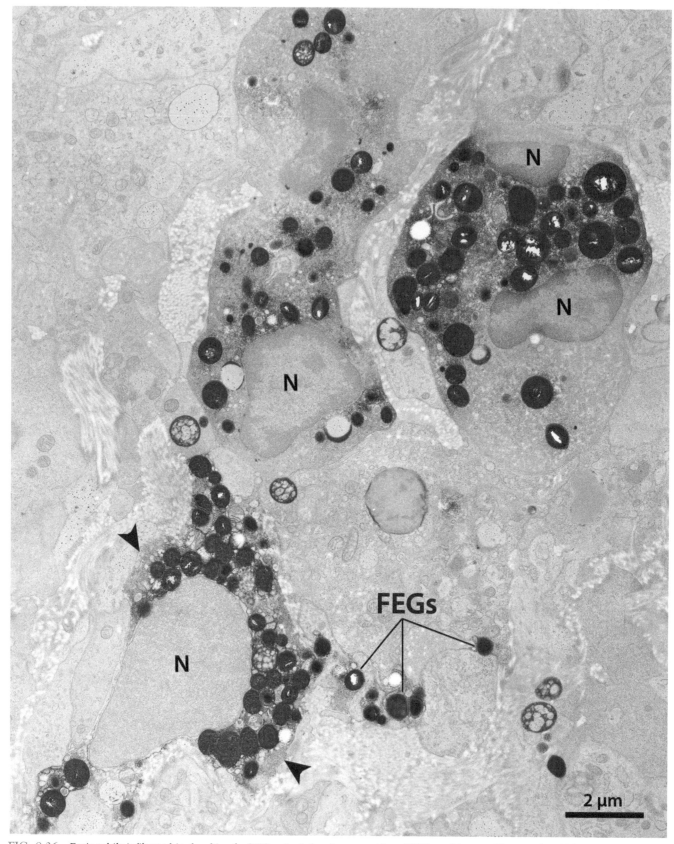

FIG. 8.36 Eosinophils infiltrated in the skin of a HES patient showing peroxidase (EPO)-positive specific granules. EPO is imaged by TEM as a dense reaction product following a cytochemistry procedure.[108,147] Note a cytolytic eosinophil *(arrowheads)* and FEGs. Nuclei (N) and other organelles as well as the ECM appear less dense.

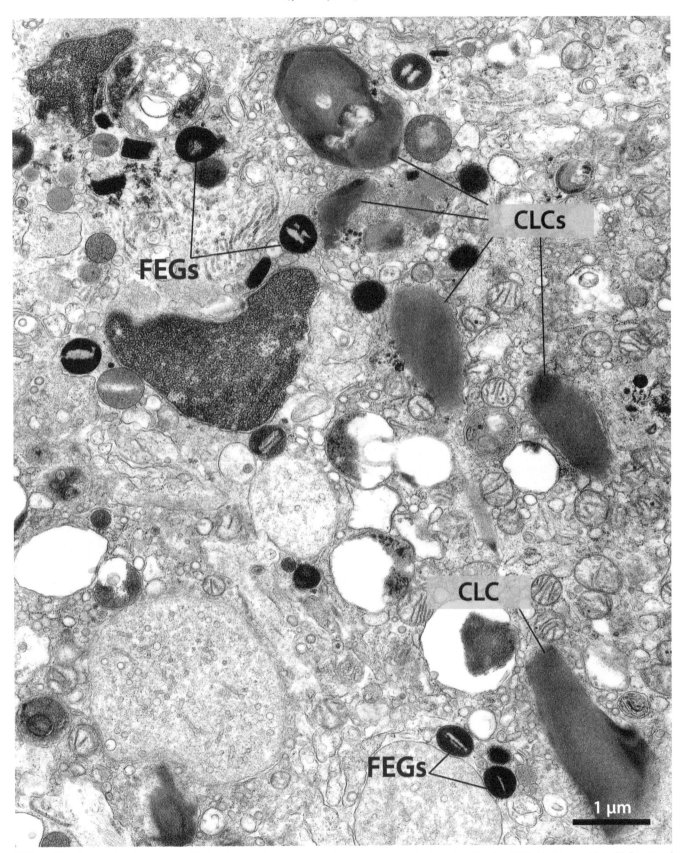

FIG. 8.37 Charcot-Leyden crystals (CLCs) showing heterogeneous forms, sizes, and electron densities in a HES skin biopsy. Note CLCs around decondensed eosinophil nuclei *(colored in purple)* and FEGs. Sample prepared for TEM.

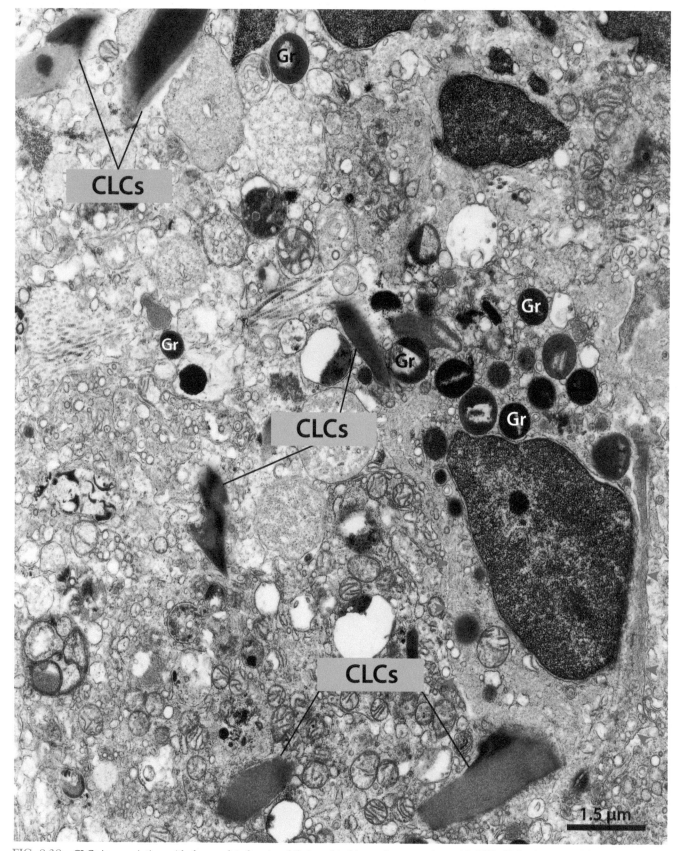

FIG. 8.38   CLCs in association with degranulated eosinophils in a skin biopsy from a patient with HES. CLCs show variable forms and electron densities. Note a cytolytic eosinophil with partially preserved plasma membrane *(arrowheads)* and granules (Gr) being spilled into the ECM. Nuclei *(purple)* lost their distinction between euchromatin and heterochromatin. Sample prepared for TEM.

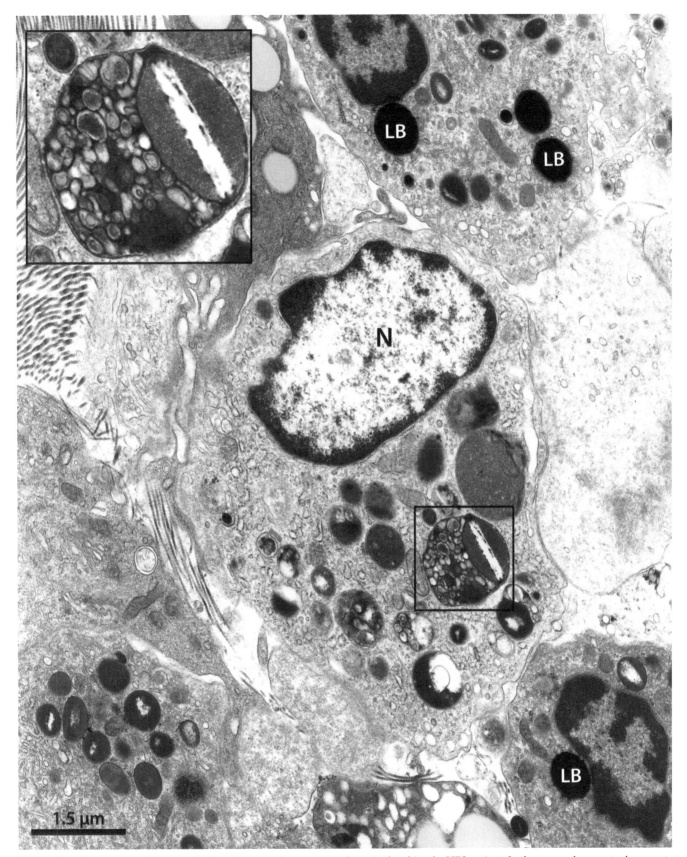

FIG. 8.39 Tissue eosinophils *(pseudocolored)* surrounding a macrophage in the skin of a HES patient. In the macrophage cytoplasm, note a phagosome *(boxed area seen in higher magnification)* containing an eosinophil specific granule *(yellow)* and EoSVs. Eosinophil nuclei were colored in *purple* and cytoplasm in *pink*. Biopsy processed for TEM. *LB*, lipid bodies; *N*, macrophage nucleus.

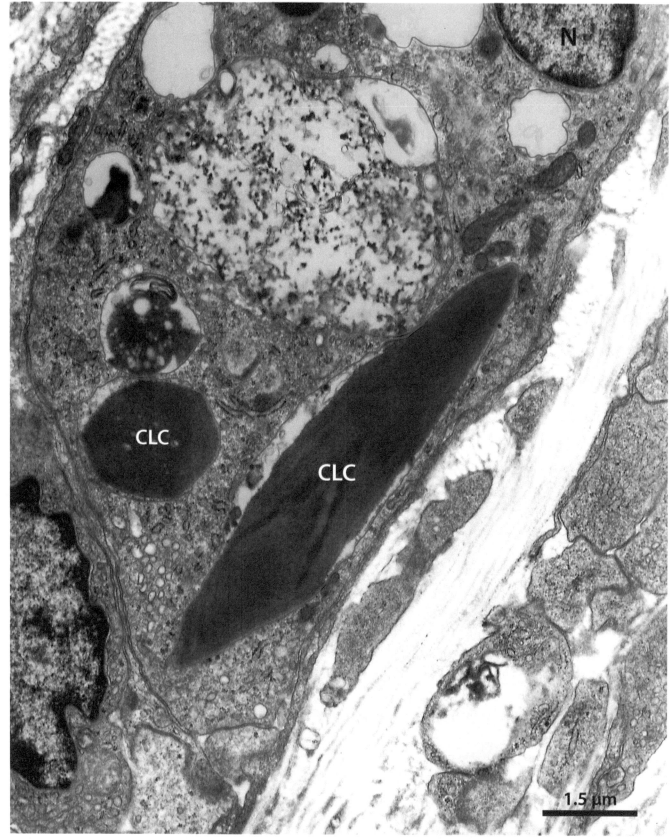

FIG. 8.40   A macrophage with phagocytosed CLCs in the cytoplasm seen in the skin of a HES patient. Note many membrane-bound phago-somes *(colored in purple)* with digesting material. Two phagosomes contain a large pyramidal and a hexagonal CLC. CLCs show heterogeneous electron densities. Sample prepared for TEM. *N*, nucleus.

HESs, eosinophils can show ultrastructural signs of activation when they are still circulating in the blood.[40] Specific granules, LBs, and the vesiculotubular system (EoSVs) of human eosinophils are organelles/structures that respond to stimuli/combinatorial signals with alterations in their number, size, and contents. In HES blood eosinophils, all of these parameters can be altered as detected ultrastructurally by the presence of (i) PMD features, (ii) increased numbers/sizes of LBs,[94] and (iii) increased formation of EoSVs.[22] Ultrastructural changes captured by TEM in HES circulating eosinophils are supported by many of our analyses in which eosinophils were purified from the blood using negative selection or evaluated directly in the whole blood without any previous sorting (blood buffy coats). We also noted that the Golgi complex, a nonprominent organelle in mature eosinophils, as pointed out in Chapter 2, appears at times with dilated cisternae (Fig. 8.26), a morphological feature that can indicate a certain degree of cell activation.

Altogether, our EM studies support a natural activation status of blood HES eosinophils, but it is important to highlight that we also noticed fluctuations in the level/frequency of the morphological changes, which is expected considering the intricate in vivo environment that impacts eosinophil responses. Moreover, HES blood eosinophils can show at times some morphological features of immaturity (Chapter 7) represented by the presence of a few immature specific granules, nuclei not completely lobulated, and/or RER cisternae in a degree not usually seen in mature eosinophils (Fig. 8.20). The occurrence of not entirely differentiated eosinophil may be consequential from the overproduction of these cells in the bone marrow.[18]

The PMD features found in specific granules of blood HES eosinophils are represented by partial to complete ragged losses of the cores (Figs. 8.20 and 8.21), progressive granule swelling (Figs. 8.22), granule disarrangement with content losses in both cores and matrices (Figs. 8.21 and 8.22), and loss of density to both compartments (Fig. 8.21). Of note, the images showing ragged losses differ entirely from the well-known reversible density expressed frequently by specific granules when variable staining methods are used, as discussed in Chapter 2. In this case, the cores are retained but appear less dense compared with the matrix. In HES eosinophils, the lucencies in the granule cores and matrices reflect the partial release of granule-derived products, likely MBP and cytokines. However, the circumstances of release and if it is happening in the blood or other compartment are not understood. High levels of MBP were found in sera from patients with HES,[386] but, clinically, the utility of measuring circulating levels of eosinophil-granule products to monitor disease activity in HESs is still a matter of discussion.[387]

Activated eosinophils show a remarkable ability to form EoSVs, which increase in terms of numbers and interactive events with mobilized granules (reviewed in[18,21]). Collections of EoSVs in their circular, elongated, or C-shaped forms populate the cytoplasm of HES eosinophils in numbers significantly higher compared with resting eosinophils[22] (Figs. 8.23 and 8.24). These vesicles interact with and bud off from secretory granules, events clearly captured by TEM and associated with the vesicular transport of granule products.[18] The presence of vesiculotubular structures and clouds of dense amorphous material associated with the granule limiting membrane reflect these events (Fig. 8.25).

LB formation is a hallmark of eosinophil activation. Increased LB numbers arise in response to a plethora of inflammatory diseases and stimuli (reviewed in[86]), as addressed in Chapter 4. In HES patients, amplified numbers of LBs can be demonstrated by quantitative analyses by both light microscopy[214] and TEM[94] of eosinophils freshly isolated from the blood, and these LBs are sites of localization of 5-LO (see Fig. 5.39 in Chapter 5), 15-LO, and COX (see Fig. 5.40 in Chapter 5),[94,214,240] all enzymes involved in the synthesis of inflammatory mediators in activated (but not resting) eosinophils.[86] Moreover, LBs can enhance in size in HES eosinophils, and these large LBs can be together with small LBs in the same thin section (Fig. 8.23). By our analyses, the in vivo formation of LBs observed in HES peripheral blood (Figs. 8.19, 8.23, 8.24, 8.26, and 8.27) is additionally increased when the cells reach the tissue microenvironment. For example, in HES skin biopsies, we found a remarkable number of LBs in infiltrating eosinophils (Figs. 8.28–8.32), which can show up to 15 LBs in a single thin cell section.

Comprehensive EM analyses showed that a large number of secretory granules observed within the cytoplasm of HES tissue eosinophils had morphological signs indicative of PMD (Figs. 8.30–8.33), with rare granule-granule fusion events. Thus, exocytosis is not a degranulation process associated with HES eosinophils. HES tissue eosinophils can also exhibit cell-cell interactions (Fig. 8.31) and other signs of activation such as shape changes and cell polarization (Figs. 8.34 and 8.35).

In addition to PMD, tissue HES eosinophils degranulate through cytolysis. Cytolytic eosinophils with disrupted plasma membranes, decondensed nuclei, and FEGs and EoSVs released to the extracellular matrix were frequently observed in HES skin biopsies (Figs. 8.36–8.38). Quantitative analyses of intact and cytolytic eosinophils and tissue regions with FEGs revealed that 46% and 54% of the eosinophils were undergoing cytolysis or PMD, respectively. Therefore, these two mechanisms underlie eosinophil degranulation with a deposit of its products in situ reported for different forms of HESs.[387] The release of these mediators together with reactive oxygen species and lipid mediators has been associated with tissue damage or dysfunction and can lead to end-organ manifestations of HES.[387]

The presence of CLCs in HES skin biopsies drew our attention. A large number of CLCs were found adjacent to cytolytic eosinophils (Figs. 8.37 and 8.38) and occasionally within the eosinophil cytoplasm. Moreover, our quantitative evaluations revealed that most eosinophils undergoing cytolysis in the HES skin had ultrastructural features of eosinophil extracellular trap cell death (EETosis) (see Figs. 6.21, 6.24, and 6.29 in Chapter 6) as also observed in the spleen and lymph nodes of HES patients in association with CLC formation.[388] Of note, as previously described,[40] tissue eosinophils found in the HES skin specimens established at times close interactions with inflammatory macrophages (Fig. 8.39). Phagocytosed single granules and CLCs were occasionally seen in the cytoplasm of these macrophages (Figs. 8.39 and 8.40).

8.5 Eosinophilic granulomatosis with polyangiitis (EGPA)

EGPA, formerly Churg-Strauss syndrome, is a prototypical eosinophilia-associated vasculitis. EGPA is a rare systemic form of vasculitis characterized by necrotizing vasculitis with eosinophilic infiltration of small vessels associated with a history of asthma, marked blood eosinophilia, and antineutrophil cytoplasmic antibodies (ANCAs) (reviewed in[389–392]).

EGPA affects diverse organ systems, including the airways, peripheral nerves, heart, kidney, paranasal sinus, gastrointestinal tract, and skin. Patients with EGPA may present cutaneous manifestations, which include palpable purpura, petechiae, hemorrhagic blisters, urticarial lesions, and papulonodular lesions on the extremities and scalp.[373] Eosinophil-rich granulomatous inflammation and small-to-medium-sized vessel vasculitis characterize the pathological findings of EGPA.[390,391]

Eosinophils are considered to have a critical role in the clinical manifestations of EGPA with a clinical benefit of eosinophil-targeted anti-IL-5 antibody therapy.[393] However, the precise mechanisms of eosinophil-mediated inflammation, as well as the interplay between the eosinophilic and vasculitis processes, remain to be fully elucidated.[389] Possible mechanisms underlying eosinophilic inflammation and ANCA-mediated vasculitis in EGPA have been summarized.[391]

Eosinophils undergoing degranulation through cytolysis and intact FEGs were observed by TEM in biopsies from EGPA patients (Fig. 8.41) and associated with tissue damage[279,394] since eosinophil granule-derived proteins have marked potential toxicity for host tissues and significant functions relevant to the EGPA pathogenesis (reviewed in[391]). By applying several methodological strategies, Fukuchi and colleagues showed that the cytolytic profile of eosinophils in tissues such as lung, skin, and gastric mucosa was associated with the release of nucleus-derived extracellular DNA traps, galectin-10, and intact secretory granules, thus identifying EETosis (Chapter 6) as a mechanism of eosinophil degranulation in EGPA.[279] Galectin-10, a protein that is highly abundant in the peripheral cytoplasm of human (but not animal) eosinophils,[37] was detected in the serum of EGPA patients and correlated with disease activity. The elevated levels of this protein in the serum might reflect the systemic occurrence of cytolytic EETosis and act as a novel biomarker for patients with EGPA.[279]

8.6 Gastrointestinal disorders

Eosinophils are normal residents of the gastrointestinal tract, where they are found in significant numbers in the lamina propria and to a lesser extent in the epithelium (reviewed in[363,395]), but their distribution is not homogeneous and the esophagus is considered the only segment without tissue-dwelling eosinophils.[396,397]

Eosinophils are important cells of the gastrointestinal immune system, a unique environment that demands robust protective immunity against pathogens while maintaining tolerance to dietary proteins or commensal bacteria. Several lines of evidence point out that different phenotypic subpopulations of eosinophils with activated profiles reside in the gastrointestinal tract at baseline conditions (reviewed in[395,398]). The heterogeneity of tissue eosinophils is likely shaped by intestinal environmental factors including cytokines, which promote eosinophil longevity and activation.[395] Accordingly, our ultrastructural quantitative analyses of resident intestinal eosinophils in murine models showed that these cells display morphological signs of activation represented by considerable features of PMD, thus reflecting a higher secretory activity.[352]

As effector cells, intestinal eosinophils provide immunity to pathogens, but they are also implicated in both homeostasis of the gastrointestinal tract and disease exacerbation. Eosinophils that reside in the gastrointestinal tract are required for the homeostatic intestinal functions such as maintenance of epithelial barrier function, regeneration, and monitoring and modulation of complex immune responses throughout the vast gastrointestinal tract.[395,399,400]

An assortment of inflammatory conditions related to the gastrointestinal tract can show mucosal eosinophilia with eosinophils being part (e.g., inflammatory bowel disease, drug-related esophagitis, autoimmune gastritis, radiation enteritis) or a major component of the inflammatory infiltrates (e.g., eosinophilic gastroenteritis, gastroesophageal reflux disease, eosinophilic esophagitis, eosinophilic gastritis).[363] Primary eosinophilic gastrointestinal disorders (EGIDs) are defined as disorders that selectively affect the gastrointestinal tract with eosinophil-rich inflammation in the absence of known causes for eosinophilia (e.g., drug reactions, parasitic infections, and malignancy).[401] The

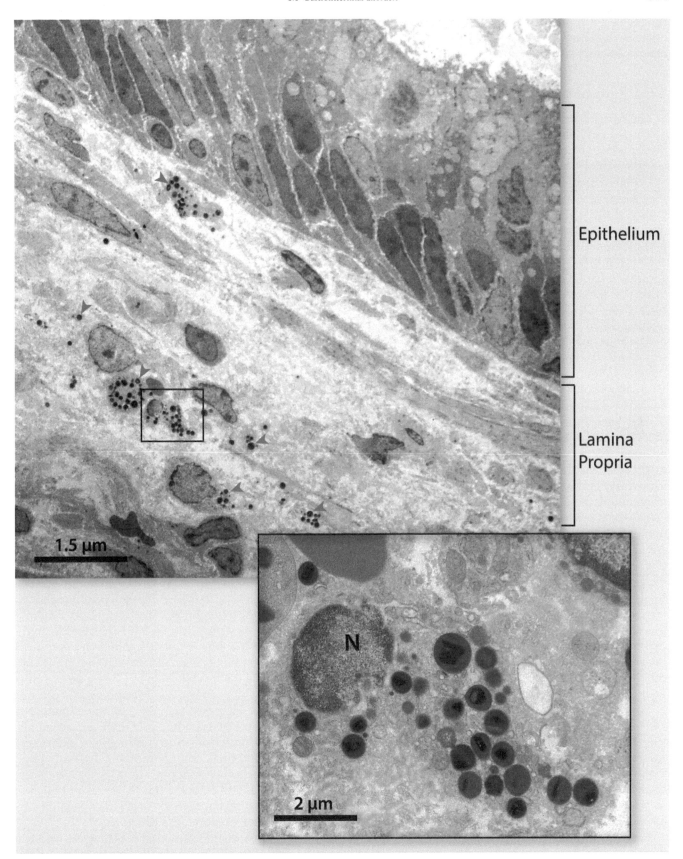

FIG. 8.41 Eosinophils in the lamina propria of the intestinal mucosa in a colon biopsy of a patient with eosinophilic granulomatosis with polyangiitis (EGPA). One cytolytic eosinophil *(boxed area)* with decondensing nucleus (N) and disrupted plasma membrane is seen in higher magnification. *Arrowheads* indicate eosinophils and scattered FEGs in the ECM. Sample processed for TEM. *Images courtesy of Shigeharu Ueki.*

functional contributions of eosinophils in each of these diseases are complex as the presence of these cells may be correlated with ameliorating and/or exacerbating tissue inflammation.[6,395] Here, we discuss the ultrastructural aspects of eosinophils participating in representative diseases of the spectrum of gastrointestinal disorders.

8.6.1 Inflammatory bowel disease

The term inflammatory bowel disease (IBD) refers mainly to two major categories of chronic inflammatory disorders associated with the gastrointestinal tract: Crohn's disease and ulcerative colitis (reviewed in[402,403]). The pathogenesis of these diseases is considered multifactorial and likely results from a complex interplay between genetic predisposition, environmental factors, and altered microbiota, leading to dysregulated innate and adaptive immune responses.[404,405]

Crohn's disease is typically characterized by transmural inflammation of the intestine and can affect any part of the gastrointestinal tract.[6,403] Ulcerative colitis affects the colon, and patients with this disease have mucosal inflammation starting in the rectum that can extend continuously to proximal segments of the colon.[406] The key features for diagnosing Crohn's disease and ulcerative colitis comprise a combination of clinical, radiographic, endoscopic, and pathological findings.[403,406]

Our group has performed comprehensive ultrastructural studies of biopsy specimens from patients with the diagnosis of Crohn's disease and ulcerative colitis.[42,145,197,407–411] Samples selected for these studies covered both longitudinal and horizontal axes of bowel specimens, including the full bowel thickness from mucosal epithelium to the serosa, thus providing a depth ultrastructural analysis. These biopsies were obtained from jejunum, ileum, colon, rectum, conventional ileostomies, or continent pouches, and the analyses involved hundreds of samples.[197,408,410,411]

The intestines are covered by a single layer of columnar epithelial cells comprising undifferentiated (intestinal stem cells) and differentiated cell types with functions associated with absorption (enterocytes) and secretion (mucus-secreting goblet, antimicrobial-secreting Paneth, hormone-secreting enteroendocrine, and chemosensing/immunomodulatory cytokine-secreting tuft cells).[412] Several ultrastructural changes associated with IBDs were revealed by TEM in the epithelium. We detected epithelial barrier defects characterized by alterations of both absorptive (glycocalyx absent or markedly reduced and microvilli with significant reduction to complete loss)[407] and Paneth (decreased granule numbers, vacuolization, and lysosomal inclusions) cells.[407,409] Moreover, extensive axonal necrosis of autonomic nerves was also identified in ileal biopsies of Crohn's patients, indicating that autonomic nerve damage might be one of the earliest pathologic changes in this disease.[408,410,411] While these former TEM studies offered significant insights into the understanding of the pathogenesis of IBD, here we will focus on the inflammatory response triggered by these diseases with emphasis on eosinophils.

Intestinal biopsies from patients with Crohn's disease and ulcerative colitis may show significant mucosal eosinophilia in areas of either active or quiescent disease as documented by histopathological quantitative evaluations.[363] Accordingly, our TEM studies detected numerous eosinophils in the lamina propria (Figs. 8.42–8.47) and submucosa (Fig. 8.48) of intestinal biopsies from patients with IBD.[42,197] Eosinophils dispersed singly or in clusters were seen in the extracellular matrix primarily interspersed among other immune cells, mainly plasma cells, macrophages, and mast cells, which were likewise prominent in numbers (Figs. 8.43, 8.44, 8.46, and 8.47).[42,197] Basophils, lymphocytes (Fig. 8.43; see also Fig. 4.17 in Chapter 4), and neutrophils (Fig. 8.45) were also present in the inflammatory infiltrates. Moreover, eosinophils were observed within small blood vessels of the lamina propria (Fig. 8.49). In these EM studies of intestinal biopsies from IBD patients, we have observed that tissue eosinophils frequently contain numerous LBs (Fig. 8.44 and Fig. 4.17 in Chapter 4), a sign of cell activation and that LBs can occasionally be seen free in the extracellular matrix (Fig. 8.47). This ultrastructural finding is interesting since the mechanisms of LB release or content secretion from eosinophils remain to be established.

Eosinophil-mast cell (Figs. 8.44 and 8.47) and eosinophil-plasma cell (Figs. 8.44 and 8.46) intimate interactions were remarkably observed. The number of plasma cells in the lamina propria was particularly high, and eosinophils very frequently kept physical interaction with several of them (Figs. 8.44 and 8.46). Such finding is of interest considering that plasma cell survival both in the bone marrow and in the intestinal lamina propria is dependent upon the presence of eosinophils (reviewed in[413]). In addition to cell-cell interactions, eosinophils in IBDs also established considerable interactions with nerve cells. Both intact (Figs. 8.50 and 8.51) and cytolytic (Fig. 8.52) eosinophils were in proximity or direct contact with nerve terminals, which showed structural damage represented by swelling, rarefaction, or absence of organelles, including microtubules, and presence of membranous structures.[408]

Eosinophils migrating across the intestinal epithelium (Figs. 8.53–8.56) were observed in IBD biopsies and were more recurrent in specimens from patients with ulcerative colitis. Intraepithelial eosinophils were noted in different levels of the intestinal epithelium from the basal (Fig. 8.54) to the apical cytoplasm close to the lumen, where the microvilli from the surface of the absorptive cells can be observed (Fig. 8.55). Eosinophil extensions denoting cell

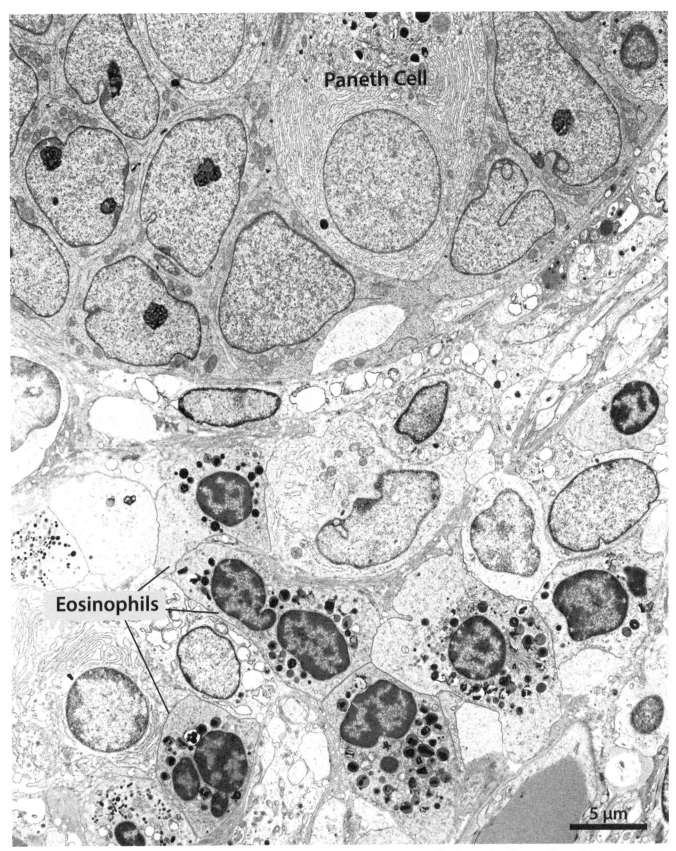

Paneth Cell

Eosinophils

5 μm

FIG. 8.42 TEM showing a cluster of eosinophils infiltrated in the intestinal lamina propria (ileal crypt) in a Crohn's disease biopsy. Eosinophils are intermingled with other inflammatory cells. Most epithelial cells are enterocytes (euchromatic nuclei, evident nucleoli, and high number of mitochondria). A Paneth cell (round nucleus, prominent RER, and supranuclear dense granules) is partially seen. Cells were pseudocolored.

B. Eosinophils in human diseases

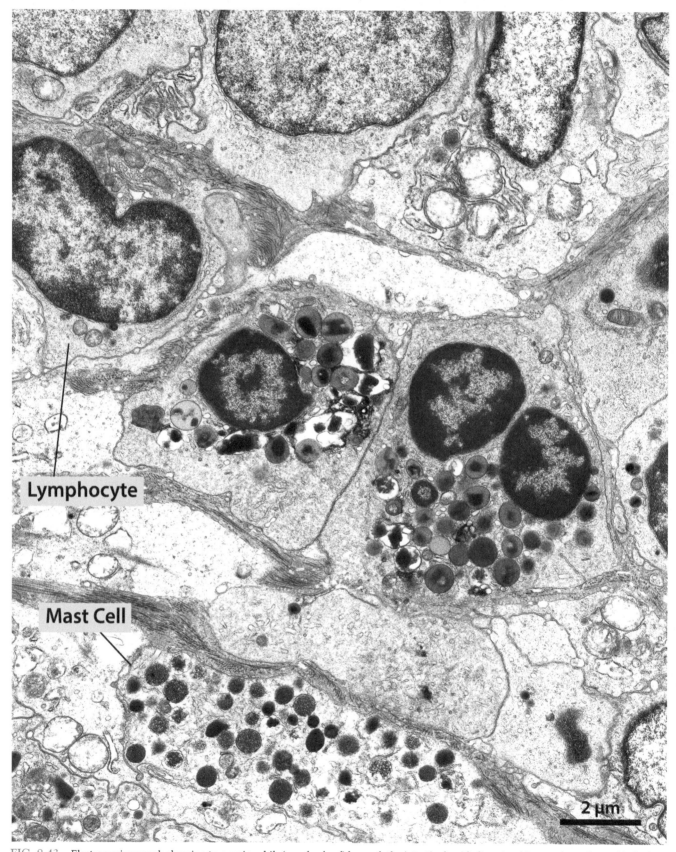

Lymphocyte

Mast Cell

2 µm

FIG. 8.43   Electron micrograph showing two eosinophils (*pseudocolored*) beneath the intestinal epithelium in a biopsy of a patient with Crohn's disease. Secretory granules *(yellow)* show content losses. Note a mast cell and a lymphocyte. The eosinophil cytoplasm was colored in *pink* and nucleus in *purple*. Sample processed for TEM.

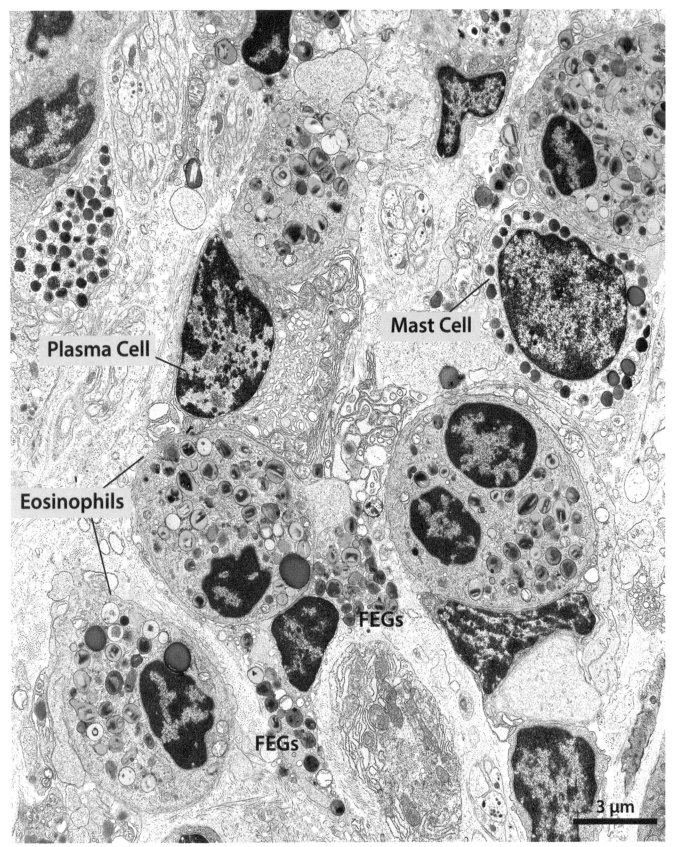

FIG. 8.44 Inflammatory infiltrate with many activated eosinophils in the human intestinal lamina propria (ulcerative colitis). Eosinophils show secretory granules *(colored in yellow) with content* losses and large electron-dense LBs in the cytoplasm. Note the intimate interaction of eosinophils with plasma cells and with a mast cell and the presence of FEGs. Cells were pseudocolored. Biopsy processed for TEM.

B. Eosinophils in human diseases

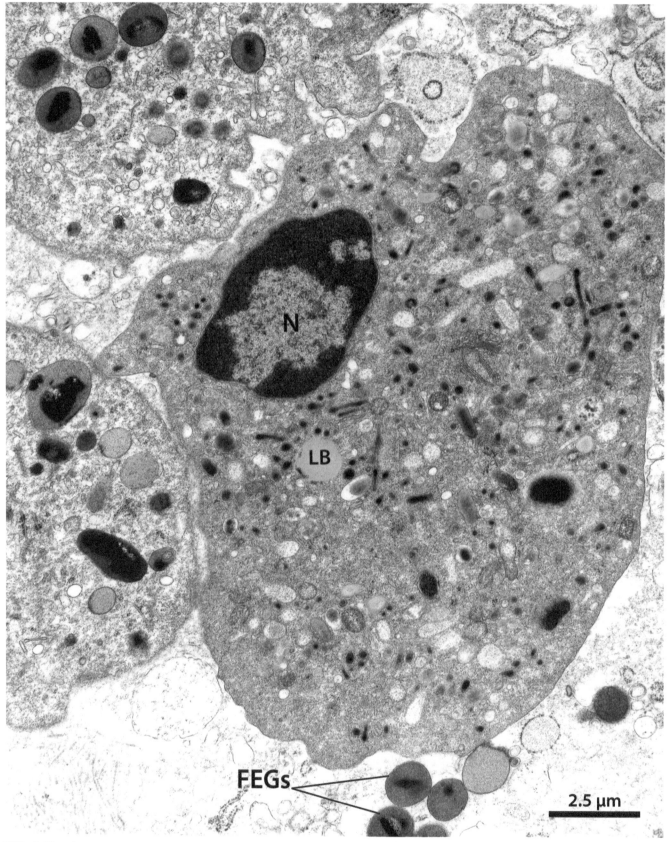

FIG. 8.45   Electron micrograph showing partial views of two eosinophils *(pseudocolored)* around a neutrophil in the intestinal lamina propria (jejunum) of a patient with a Crohn's disease. Specific granules *(colored in yellow)* are seen in the eosinophil cytoplasm *(pink)* and also appear as FEGs. Note an electron-lucent LB, typical of neutrophils. Biopsy processed for TEM. *N*, nucleus.

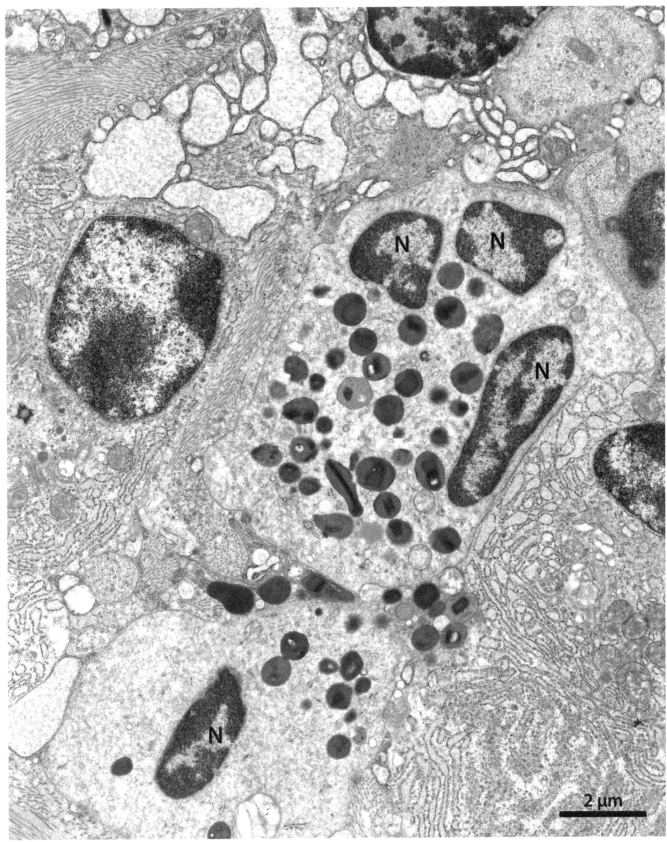

FIG. 8.46 Eosinophils interacting with plasma cells in the human intestinal lamina propria (ulcerative colitis). Plasma cells *(pseudocolored)* are recognized by their abundant dilated RER *(light blue)* and round eccentric nuclei *(purple)*. Biopsy processed for TEM. Mitochondria *(green)*; N, eosinophil nucleus.

B. Eosinophils in human diseases

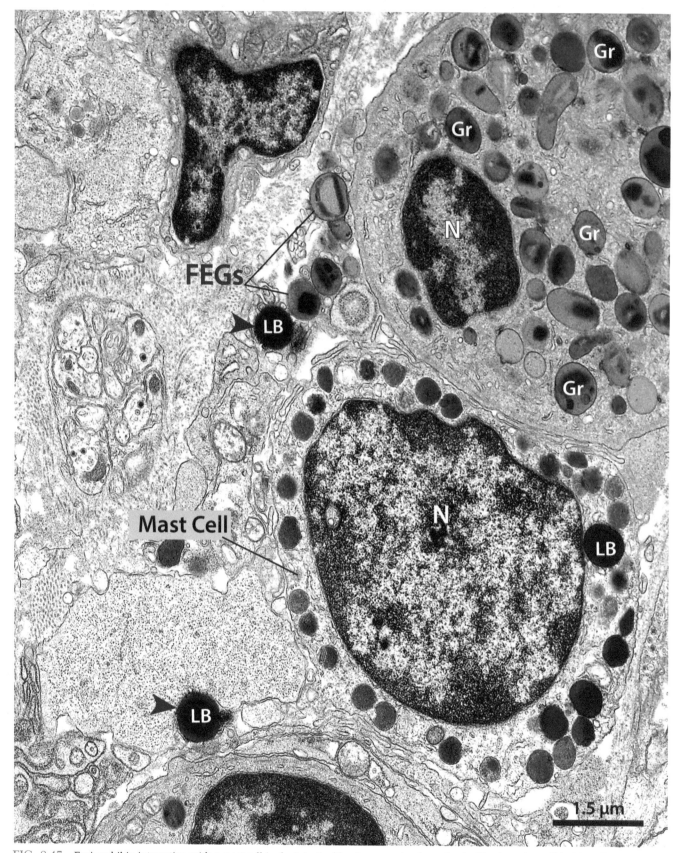

FIG. 8.47   Eosinophil in interaction with a mast cell in the intestinal lamina propria of a patient with ulcerative colitis. Note the presence of scattered FEGs and free LBs *(arrowheads)*. Biopsy processed for TEM. *Gr*, specific granules; *N*, nucleus.

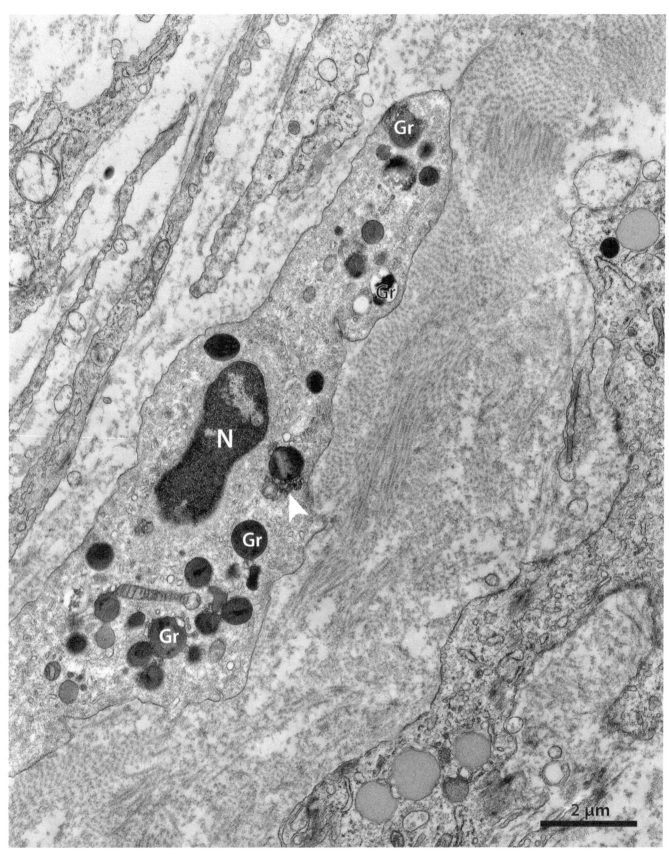

FIG. 8.48   Activated eosinophil in the intestinal submucosa observed in a Crohn's disease biopsy. Note mobilized nonfused specific granules (Gr). Arrowhead indicates a cloud of dense material associated with granule vesiculation. The ECM and surrounding cells were colored in *green*. Sample processed for TEM. *N*, nucleus.

B. Eosinophils in human diseases

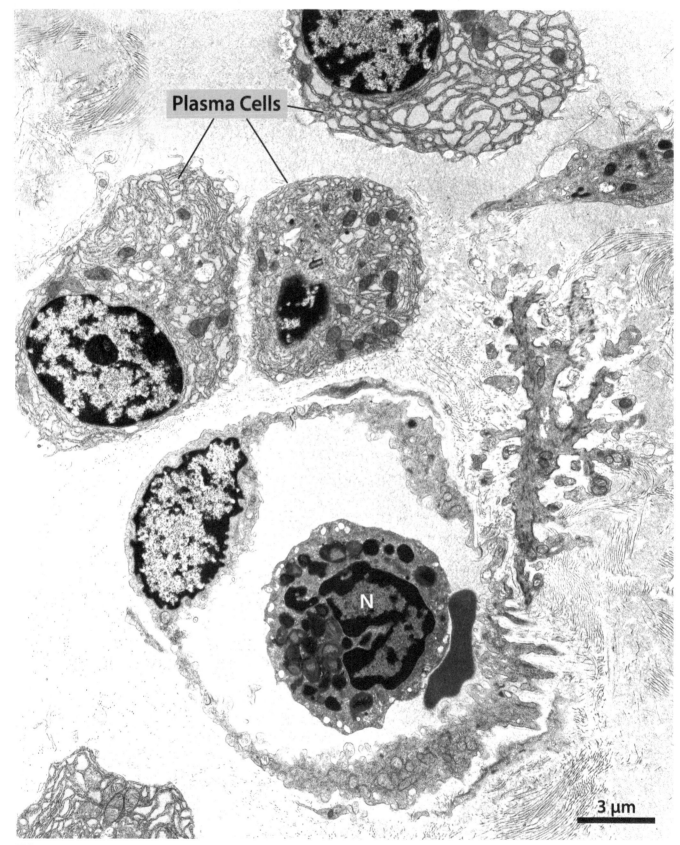

FIG. 8.49   An eosinophil within a small blood vessel in a Crohn's disease biopsy. The endothelium was colored in *yellow* and a single red blood cell in *red*. Note surrounding plasma cells *(purple)* with typical morphology. Sample processed for TEM. *N*, nucleus.

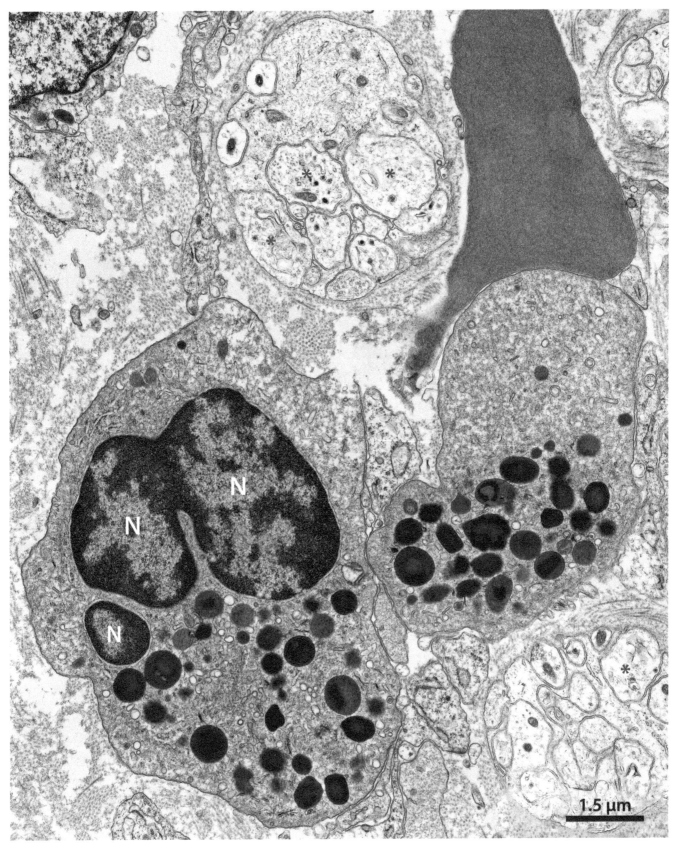

FIG. 8.50 Electron micrograph showing eosinophils in the intestinal lamina propria of a patient with ulcerative colitis. Note nerve terminals (*colored in light yellow*) with damaged swollen axons (*). Biopsy processed for TEM.[408] N, nucleus.

B. Eosinophils in human diseases

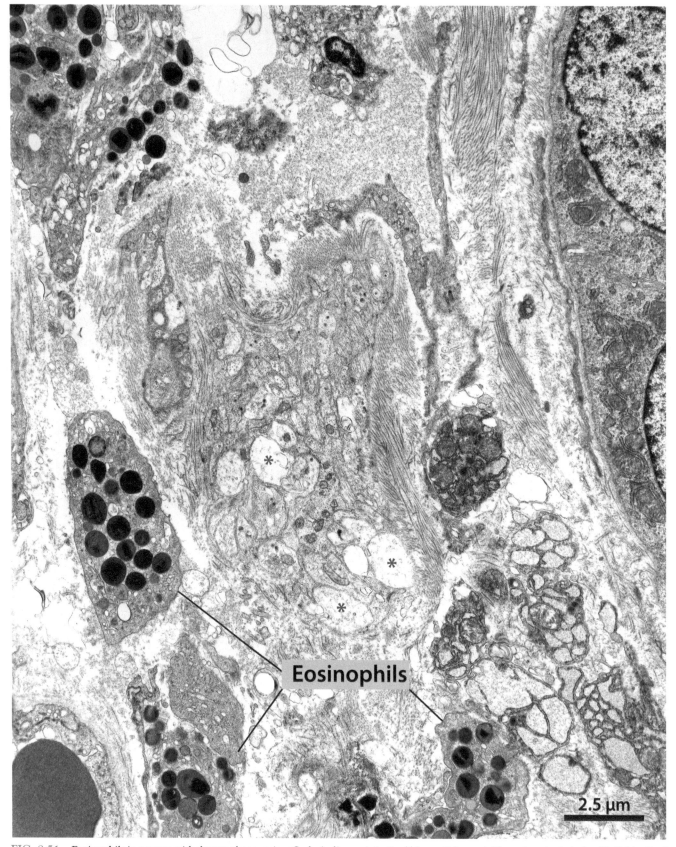

FIG. 8.51   Eosinophils in an area with damaged nerves in a Crohn's disease intestinal biopsy (jejunum). Nerve terminals *(colored in light yellow)* exhibit swollen axons (*). Sample processed for TEM.[408] *N*, nucleus.

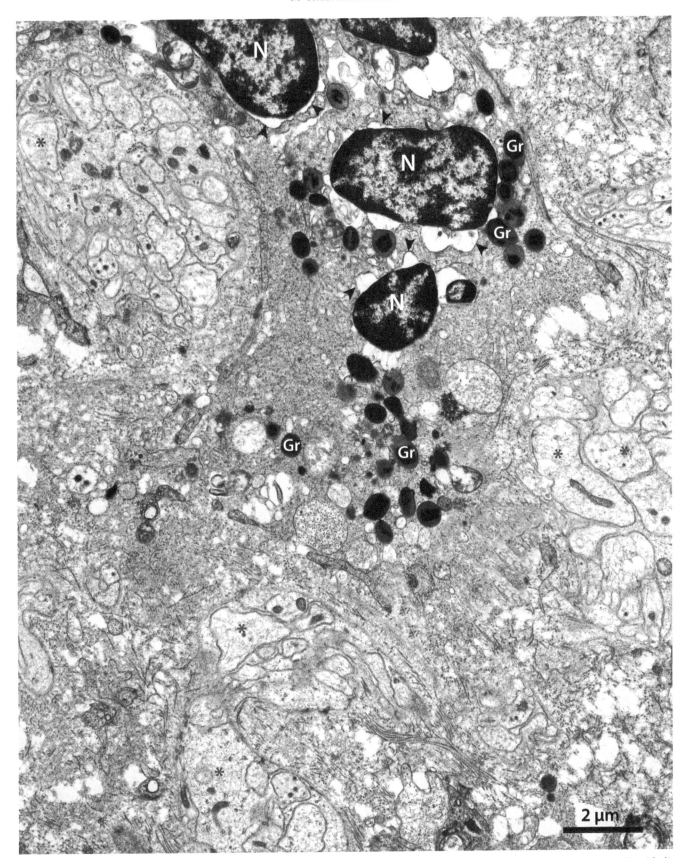

FIG. 8.52   A cytolytic eosinophil observed in an ulcerative colitis biopsy. Nerve terminals *(colored in yellow)* show enlarged axons (*) with diminished microtubules. Arrowheads indicate nuclear envelope alterations, characteristic of cytolysis. Sample processed for TEM.[408] *Gr*, specific granules; *N*, nucleus.

B. Eosinophils in human diseases

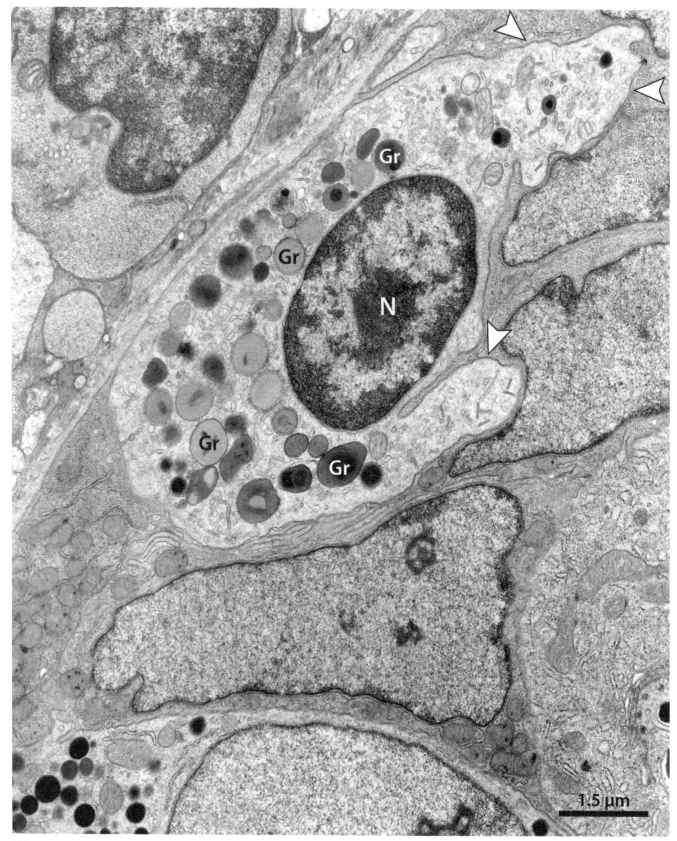

FIG. 8.53 Electron micrograph showing an eosinophil migrating across the intestinal epithelium in a biopsy of a patient with ulcerative colitis. The eosinophil cellular processes *(arrowheads)* extend among enterocytes. An enteroendocrine cell with dense granules is partially seen at the bottom. Sample processed for TEM. *Gr*, specific granules; *N*, nucleus.

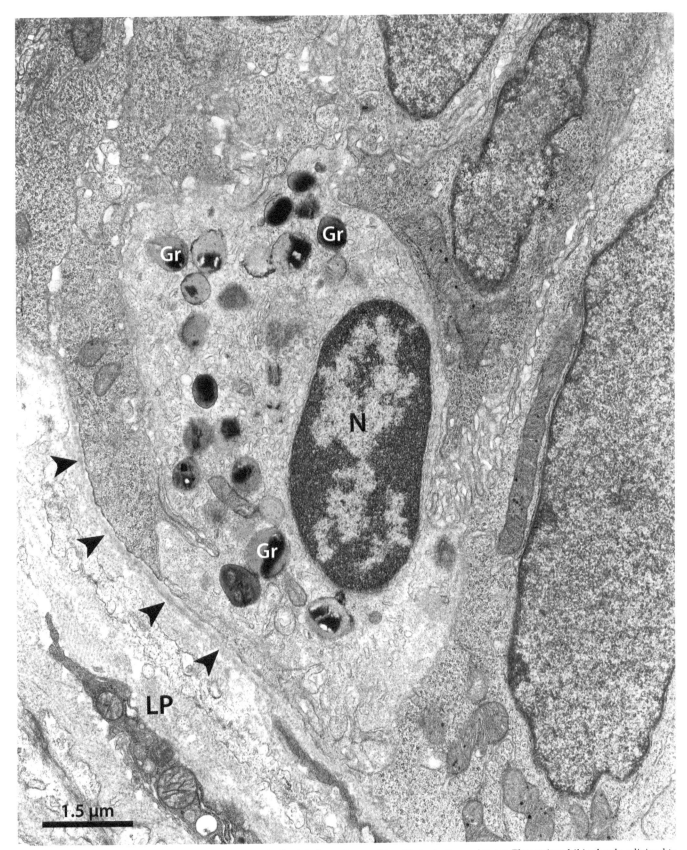

FIG. 8.54 An eosinophil migrating across the intestinal epithelium observed in a Crohn's disease biopsy. The eosinophil is closely adjoined to enterocytes (*pseudocolored pink* with mitochondria in *green* and nuclei in *purple*). Note the basement membrane *(arrowheads)* separating the basal epithelial cells from the lamina propria (LP). Sample prepared for TEM. *N*, eosinophil nucleus.

B. Eosinophils in human diseases

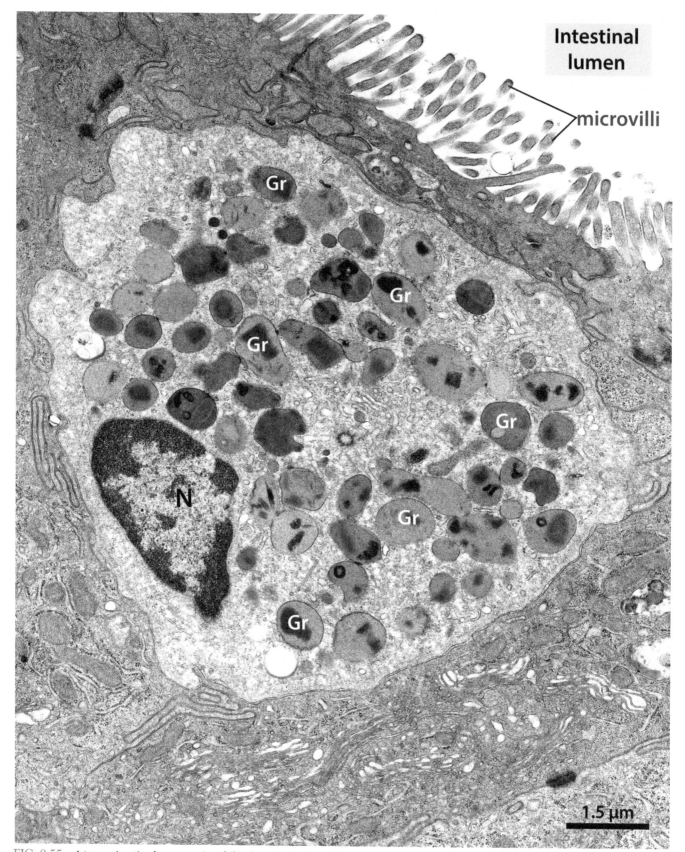

FIG. 8.55   A transmigrating human eosinophil in the apical region of the intestinal epithelium (ulcerative colitis). Note the expanded cytoplasm with secretory granules (Gr) undergoing PMD (structural disarrangement and content losses). Typical microvilli are seen at the surface of entero-cytes (*pseudocolored pink* with mitochondria in *green* and RER in *light blue*). Biopsy prepared for TEM. *N*, nucleus.

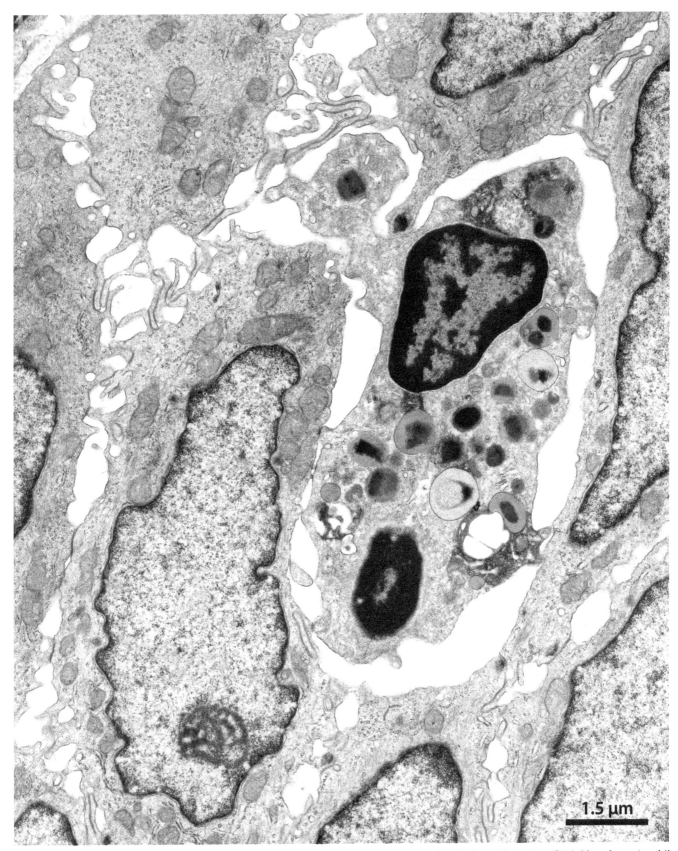

FIG. 8.56 TEM showing a human eosinophil *(pseudocolored)* migrating across the intestinal epithelium (ulcerative colitis). Note the eosinophil elongated morphology and emptying cytoplasmic granules *(colored in yellow)*. Cell-cell cohesion is deficient, with many blank areas indicating loss of contact.

B. Eosinophils in human diseases

migration were captured in electron micrographs (Figs. 8.53 and 8.56). In general, transmigrating eosinophils were intimately connected with epithelial cells (Figs. 8.53–8.55) but, at times, a defective cohesion was noted by the presence of intercellular spaces between eosinophils and epithelial cells and also between epithelial cells with each other (Fig. 8.56), which can impact the epithelial barrier function. Most intraepithelial eosinophils were not resting and exhibited evidence of PMD (Figs. 8.55 and 8.56).

In patients with IBD, eosinophil activation and increased concentrations of eosinophil granule proteins in the intestinal lumen are reported (reviewed in[399]). TEM unveiled diverse ultrastructural changes underlying the degranulation of mucosal eosinophils from IBD patients.[151,197] The following degranulation patterns were described: (i) PMD, characterized by the predominance of non-fused secretory granules with different degrees of core disarrangement and content losses (Figs. 8.55–8.57), and accumulation of clouds of dense material associated with granule vesiculation (Fig. 8.48)[151,197]; (ii) cytolysis with identification of disrupted eosinophils and/or FEGs and CLCs in the extracellular matrix (Figs. 8.44, 8.45, 8.52, and 8.58–8.61)[151]; and, to a smaller extent (iii) compound exocytosis characterized by the fusion of secretory granules and extruded granule products at the cell surface (Figs. 8.62–8.64; see also Figs. 3.8 and 3.9 in Chapter 3).[142]

As highlighted in Chapter 3, PMD and cytolysis comprise the most frequent degranulation processes of human eosinophils found in vivo during many diseases while compound exocytosis is rarely detected in vivo in response to eosinophil activation. However, compound exocytosis, as discussed in Chapter 3, can at times coexist with PMD. Both secretory processes were recorded for eosinophils participating in immune responses during Crohn's disease and ulcerative colitis, with compound exocytosis found in 22% of the intestinal biopsies in one quantitative study involving 117 biopsy samples.[142,197] Interestingly, compound exocytosis was more frequently observed in eosinophils from intestinal samples positive for bacterial cultures.[142]

By revisiting ultrastructural aspects of cytolytic eosinophils found in biopsies from patients with ulcerative colitis, we observed that they often exhibited features of ETosis (Fig. 8.60). Thus, in addition to classical changes of cytolysis (cell disruption, loss of the plasma membrane, and release of membrane-bound granules), these eosinophils had additional ultrastructural signs indicative of ETosis (nuclear delobulation/rounding as shown in Figs. 6.19 and 6.20 in Chapter 6) and chromatin decondensation/expansion and released as extracellular traps (Fig. 8.60).

In IBD biopsies, CLCs were found in the vicinity of cytolytic eosinophils and FEGs (Figs. 8.58 and 8.59; see also Figs. 6.14 and 6.31 in Chapter 6). These crystals were also occasionally seen as phagocytosed material within inflammatory macrophages (Fig. 8.61) or even within intact eosinophils (Fig. 8.65).

Although most eosinophils showing ultrastructural aspects of cell death were associated with cytolysis, our comprehensive EM analysis detected, in the intestinal submucous of patients with Crohn's disease, some eosinophils in the process of death through apoptosis (see Fig. 6.39 in Chapter 6), a finding not commonly reported for IBD.

8.6.2 Eosinophilic esophagitis

Eosinophilic esophagitis (EoE) is a chronic immunoallergic disease characterized clinically by symptoms of esophageal dysfunction and histologically by findings of eosinophil-predominant inflammation. EoE has emerged as an important contributor to upper gastrointestinal morbidity and can affect both pediatric and adult patients (reviewed in[414–416]).

The diagnosis of EoE is based on a combination of clinical manifestations, endoscopic data, and histological findings in esophageal mucosa biopsies.[416] Patients may have symptoms that mimic those of gastroesophageal reflux disease, but the two diseases are quite distinct in terms of histopathology and response to therapy.[414]

The most characteristic histological feature of EoE is the infiltration of eosinophils into esophageal epithelium (>15/high-power field). Eosinophils are more abundant in the superficial epithelium, where they tend to cluster and form microabscesses (defined as ≥ 4 eosinophils clustered together) (Fig. 8.66). Other histological alterations include dilated intercellular spaces, a thickened mucosa with basal layer hyperplasia and papillary elongation, and eosinophil degranulation (reviewed in[363,417]).

Marked extracellular deposition of eosinophil granule-derived proteins such as MBP,[418,419] EPX,[420] and EDN[421] has been reported in EoE biopsies, thus indicating that eosinophils degranulate during active disease and that their products may carry functional consequences to EoE pathogenesis.[422] These products have been considered important markers of the EoE activity, particularly in individuals in which eosinophils lose their morphological identity as intact cells and, therefore, are not enumerated on microscopic examination.[419,421]

In accordance with histological observations (Fig. 8.66), ultrastructural studies of EoE biopsies identified eosinophils migrating and degranulating (Figs. 8.66 and 8.67) between cells of the esophageal epithelium, which is composed of nonkeratinized, stratified squamous cells.[17] A significant decrease in the number of desmosomes on the surface of this epithelium of pediatric subjects with active EoE was also observed, supporting evidence of altered barrier function in EoE.[178]

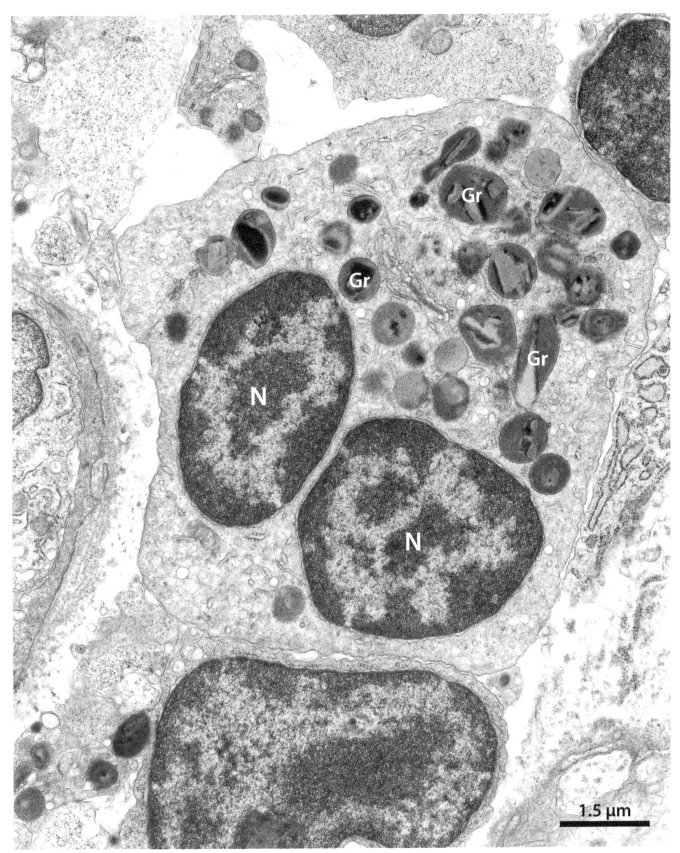

FIG. 8.57 Tissue human eosinophil (ileum, ulcerative colitis) showing secretory granules (Gr) undergoing content losses in the cores. Note the typical bilobed appearance of the nucleus (N). Biopsy processed for TEM.

B. Eosinophils in human diseases

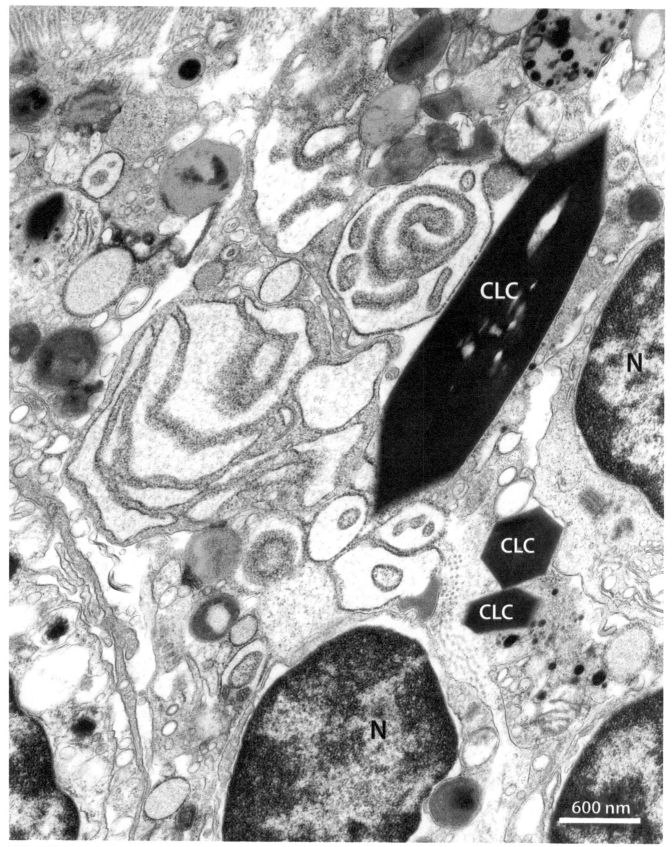

FIG. 8.58   Tissue CLCs in a human intestinal biopsy (ulcerative colitis). Hexagonal and bipyramidal CLCs of different sizes are seen in an area with eosinophil FEGs *(colored in yellow)*. The large CLC shows electron-lucent regions. Sample processed for TEM. *N*, nucleus.

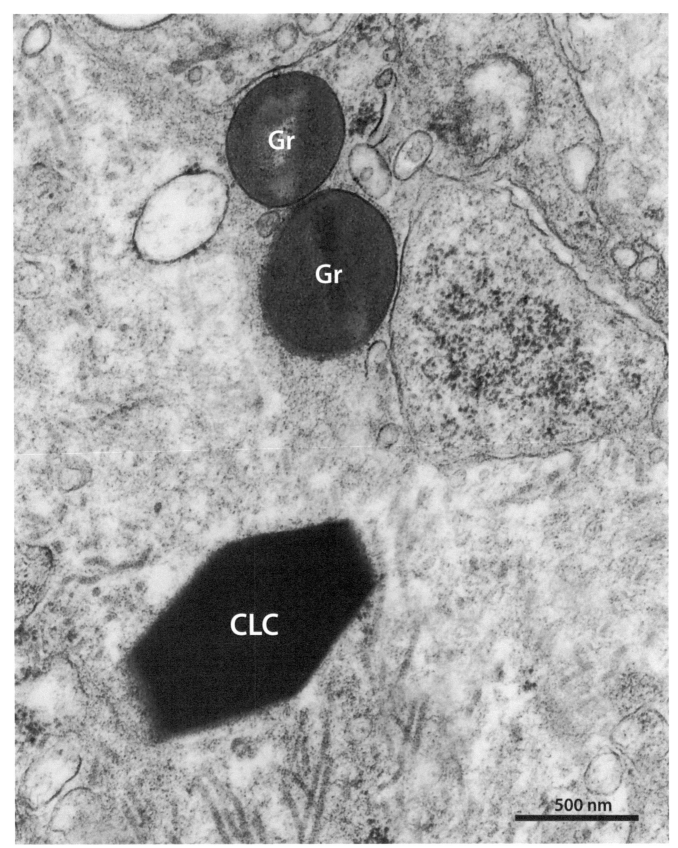

FIG. 8.59   A homogeneously electron-dense CLC seen in a human intestinal biopsy (ulcerative colitis). Eosinophil secretory granules (Gr) and EoSVs *(colored in pink)* are also observed. The ECM and surrounding cells were colored in *green*. Biopsy processed for TEM.

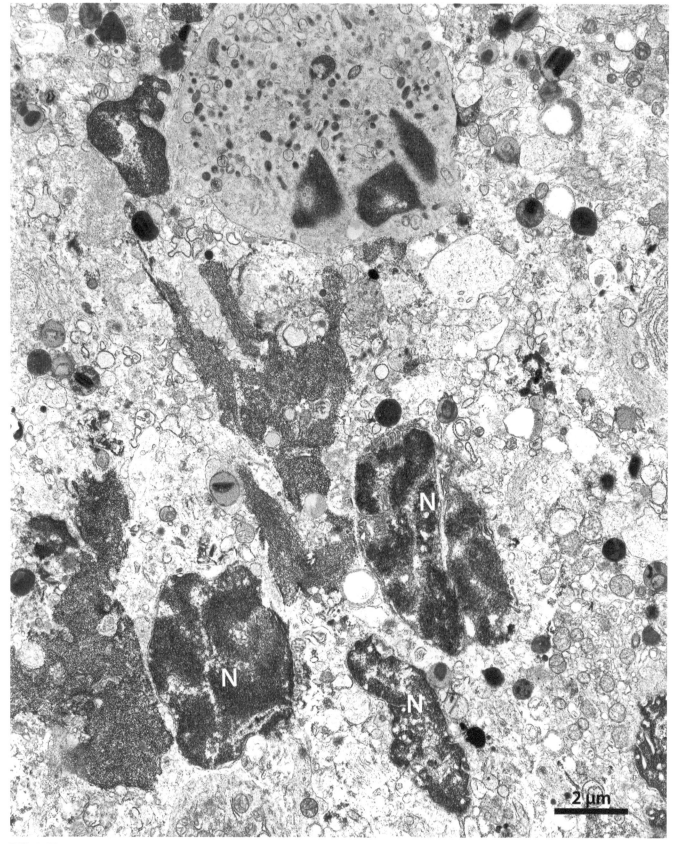

FIG. 8.60   Cytolytic tissue eosinophils (rectal biopsy, ulcerative colitis) showing ultrastructural features of ETosis. Note the highly decondensed chromatin *(colored in purple)* expanding as extracellular traps in an area with FEGs *(yellow)* and isolated nuclei (N). A single intact neutrophil *(green)* is observed. Biopsy prepared for TEM.

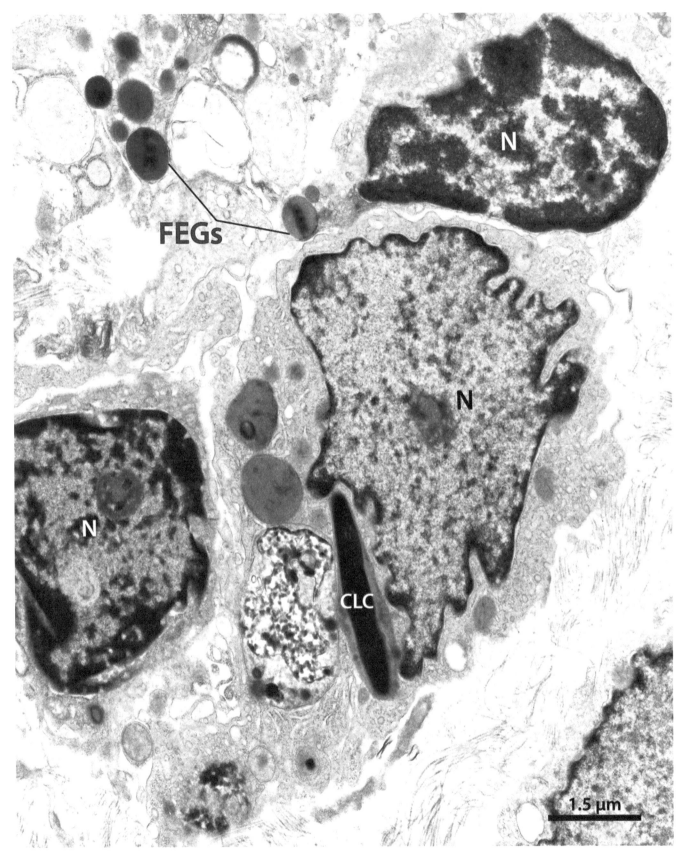

FIG. 8.61   Macrophage showing a phagocytosed CLC (intestinal biopsy, Crohn's disease). CLC is seen in a cytoplasmic membrane-bound vacuole (phagosome). Note eosinophil FEGs and the nuclei (N) of other inflammatory cells around the macrophage. Biopsy prepared for TEM.

B. Eosinophils in human diseases

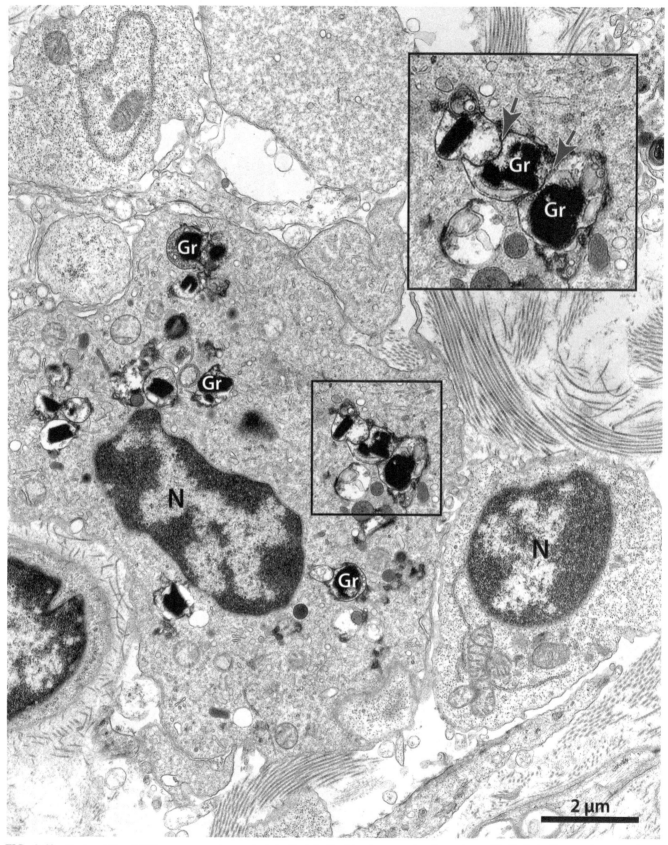

FIG. 8.62   Eosinophils exhibiting compound exocytosis in the intestinal lamina propria of a patient with Crohn's disease. Arrows indicate fused secretory granules (Gr) observed in higher magnification *(boxed area)*. A lymphocyte is seen in the eosinophil vicinity. *N*, nucleus. Biopsy processed for TEM.

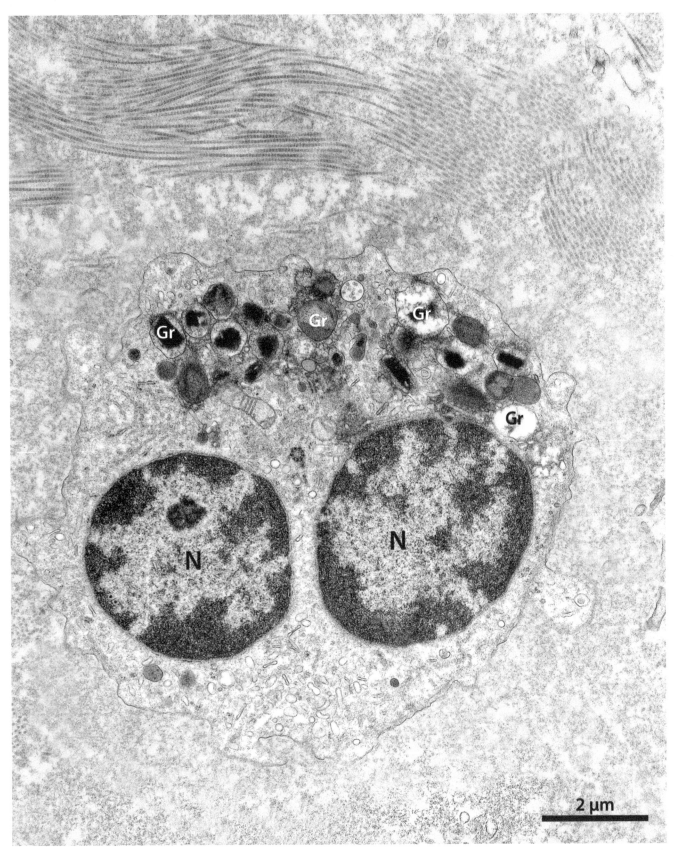

FIG. 8.63 TEM showing a mucosal eosinophil in an intestinal Crohn's disease biopsy. Secretory granules (Gr) are fused and show content losses. Note the bilobed nucleus (N) and granules accumulated in the peripheral cytoplasm in one side of the cell. ECM colored in *green*.

B. Eosinophils in human diseases

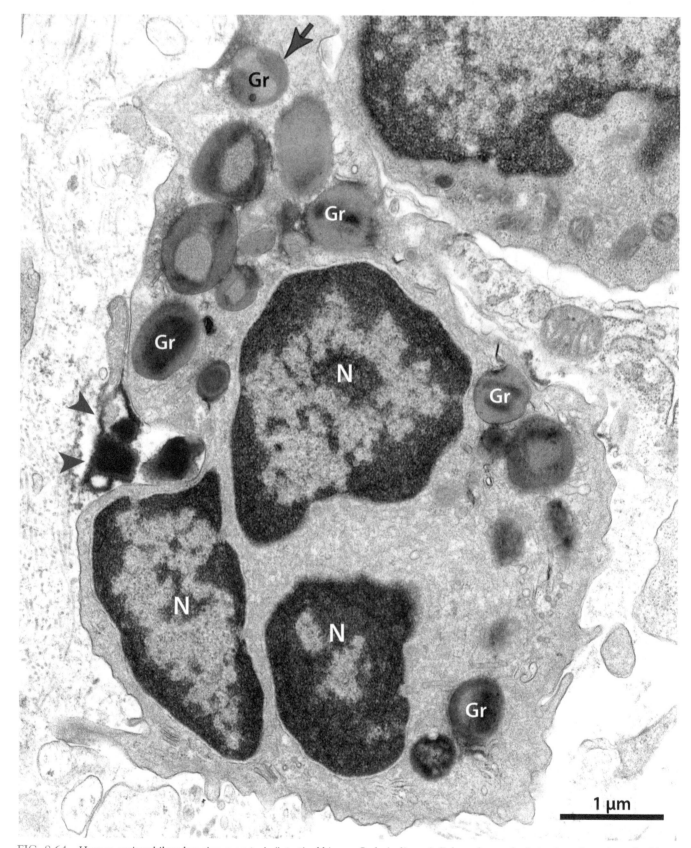

FIG. 8.64   Human eosinophil undergoing exocytosis (intestinal biopsy, Crohn's disease). Released granule-derived products *(arrowheads)* are seen in a cell surface-related pocket. Note a granule (Gr) in the process of fusion with the plasma membrane *(arrow)* and three nuclear (N) lobes. Sample prepared for TEM.

B. Eosinophils in human diseases

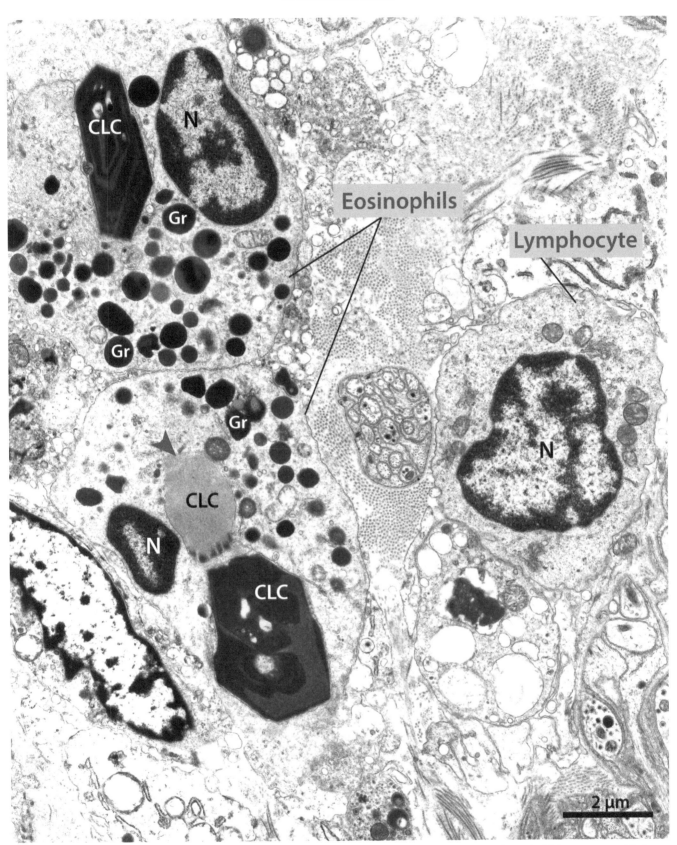

FIG. 8.65   CLCs formed within the cytoplasm of tissue human eosinophils (intestinal biopsy, Crohn's disease). CLCs are arranged in layers with different electron densities. One CLC *(arrowhead)* is mostly electron-lucent. A lymphocyte is seen in the vicinity. Sample prepared for TEM. *N*, nucleus.

B. Eosinophils in human diseases

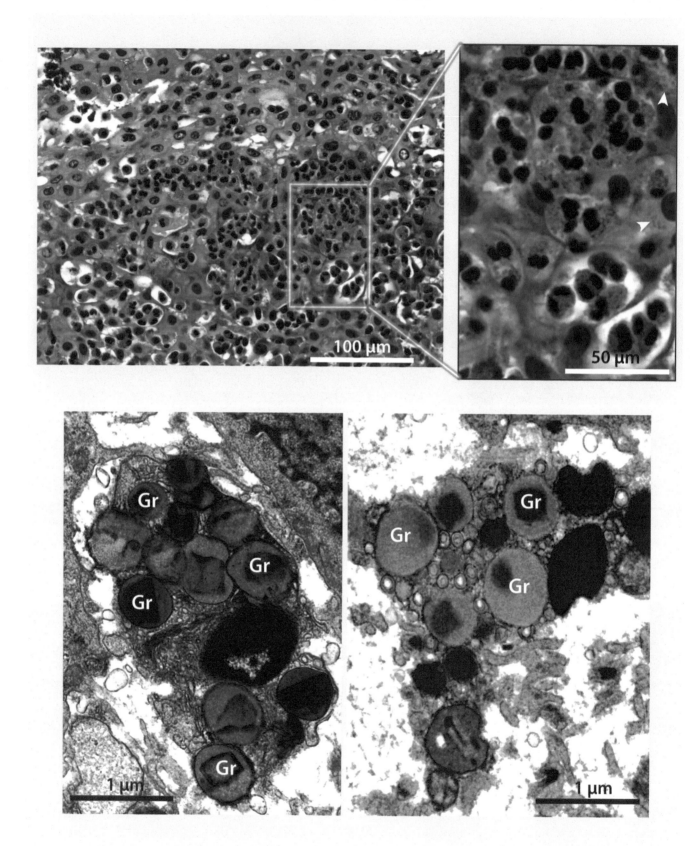

FIG. 8.66   Eosinophils infiltrated in the epithelium of a patient with eosinophilic esophagitis (EoE). In the top panels, clusters of eosinophils form abscesses and degranulate *(arrowheads)*. TEM[17] *(bottom panels)* shows released clusters of secretory granules (Gr) and EoSVs *(highlighted in pink)*. Histological image *(stained with HE) courtesy of Rhonda Yantiss.*

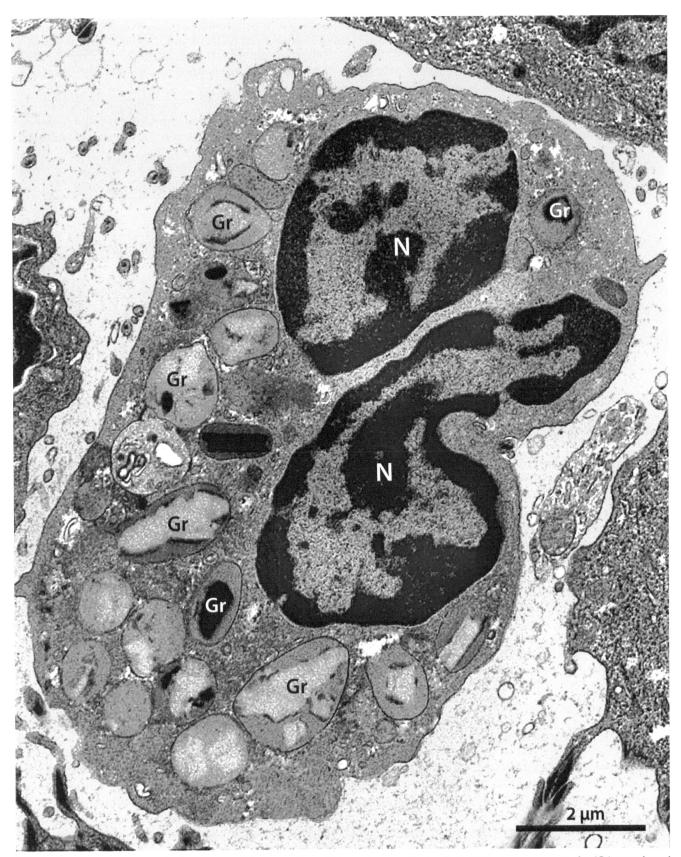

FIG. 8.67   Electron micrograph of an eosinophil in an esophageal biopsy from a patient with EoE. Most secretory granules (Gr) are enlarged and show content losses indicative of PMD. Sample prepared for TEM.[17] N, nucleus.

B. Eosinophils in human diseases

Cytolysis (Fig. 8.66) and PMD (Fig. 8.67) were recognized by TEM as the processes by which granule proteins are deposited in the disease-affected esophagus, with cytolysis being more prominent.[17,178] Quantitative analyses of 1672 electron micrographs taken from esophageal biopsies of 9 patients with EoE found that almost all of the eosinophils invading the esophagus ($n = 900$) showed changes of activation and degranulation with marked cytoplasmic vesiculation and varying degrees of membrane disruption.[17] Most eosinophils (80.6%) had evidence of cytolysis and FEGs as well as free EoSVs (Fig. 8.66) were frequently observed.[17] Considering that the normal esophageal mucosa does not contain eosinophils, the presence of eosinophils with a highly activated ultrastructural profile in the EoE esophageal epithelium indicates a pathogenic role for these cells in this disease. EoE is a Th2-associated inflammatory disease resulting from a combination of genetic, immunologic, and environmental factors, including allergic sensitization to multiple foods. The pathogenesis of EoE addressing the interplay and pathways involving activated eosinophils, Th-2 cytokines such as IL-5 and IL-13, and mast cells have been reviewed in detail elsewhere.[422]

8.7 Infectious diseases

8.7.1 Parasitic infections

Historically, eosinophils have been associated with the host's response to helminth infections and with a host-protective and helminthotoxic function. Helminths are a very diverse group of parasites comprising cestodes, trematodes, and nematodes with unique life cycles. Blood eosinophilia is a central feature of several diseases caused by these parasites and generally correlated with Th2 cell-mediated immune responses, including the production of IL-5, which enhances eosinophilopoiesis and eosinophil activation.[3,423,424] A comprehensive list of parasitic diseases leading to blood eosinophilia is summarized in other reviews.[368,425]

Earlier studies have demonstrated the ability of both human and mouse eosinophils to adhere to and kill helminth larvae in vitro. This effect was attributed, at least in part, to the release of granule contents, especially MBP onto the parasite surface. However, the views that eosinophil-mediated immune responses to parasites are beneficial to the host have been challenged. In many experimental models of helminth infection, eosinophils were not shown to have any protective effects, and data supporting a role for human eosinophils as defensive effector cells against in vivo helminth infection are scarce. Currently, the primary role of eosinophils in parasitic infections is still unclear (reviewed in[3,368,423,424,426]).

Tissue eosinophilia characterizes numerous human helminth diseases, including trichinellosis,[427] strongyloidiasis,[428] and schistosomiasis.[429,430] Following blood eosinophilia, eosinophils migrate to the site of infection where they release their products through degranulation. However, there is a lack of clarity regarding the eosinophil degranulation mechanisms in vivo in target tissues of helminth diseases and its significance in both humans and experimental models. For example, there is considerable debate on the role of eosinophils during schistosomiasis, an ancient neglected tropical disease that is still considered one of the most common parasitic diseases worldwide with substantial morbidity and mortality.[431,432] The pathology of schistosomiasis mansoni results from the parasite eggs that become trapped in host tissues, particularly in the liver and intestines. In these organs, schistosomiasis mansoni triggers an intense granulomatous inflammation characterized by the dense accumulation of eosinophils.[424,433,430] It remains uncertain if eosinophils act as major effector cells against the parasite; as immunomodulators of the immune response; as participants in tissue homeostasis and metabolism, which could favor establishment and maintenance of parasitic worms in their hosts; or merely as operators in remodeling and clearance of debris following injury.[424,434–438]

Our group has been studying the inflammatory response involved in natural (wild reservoirs), experimental (mice models addressed in Chapter 9), and human *Schistosoma mansoni* infections[352,433,439,440] with a special interest in the eosinophil immunobiology and their degranulation mechanisms during this disease.

We analyzed intestinal biopsies obtained from patients with chronic mild schistosomiasis, which is the usual form of schistosomiasis observed in endemic regions as a consequence of repeated exposures. The pathology of this disease is represented by isolated granulomas and by the presence of a variable number of macrophages and eosinophils.[441] Our ultrastructural studies detected a high number of eosinophils in the intestinal mucosa, both migrating across the epithelium (Fig. 8.68) and distributed in the lamina propria as intact (with no disrupted plasma membrane) (Figs. 8.69) or cytolytic cells (with a breakdown of the plasma membrane and regions with FEGs) (Figs. 8.70–8.74). Other cells found in the lamina propria included mast cells, macrophages, and plasma cells.

Quantitative TEM analyses showed that all eosinophils found in the *S. mansoni*-infected intestinal samples exhibited morphological signs of degranulation either through PMD (38%) (Fig. 8.69) or predominantly by cytolysis (62%) (Figs. 8.70–8.74). Eosinophils undergoing PMD showed 88% of their population of secretory granules as enlarged, nonfused granules exhibiting content losses (Fig. 8.69). These granules coexisted in the same thin section with some

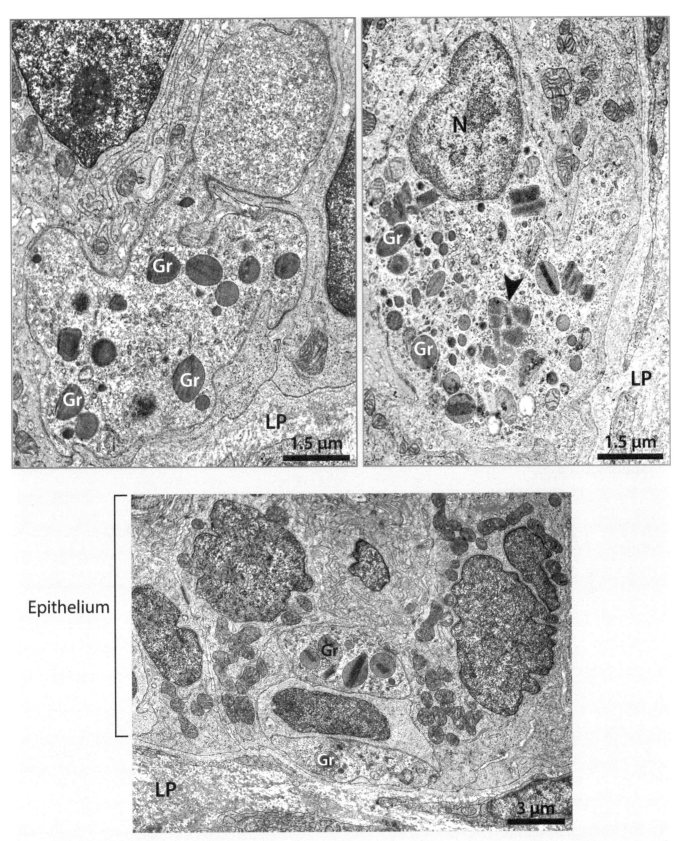

FIG. 8.68 Transmigrating eosinophils in the intestinal epithelium of a patient with chronic schistosomiasis mansoni. Note cellular processes extending between enterocytes. Secretory granules (Gr), some of them fused *(arrowhead)*, are observed. Biopsy prepared for TEM. Enterocytes were colored in *pink* with their nuclei in *purple*, RER in *light blue*, and mitochondria in *green*. LP, lamina propria; N, nucleus.

B. Eosinophils in human diseases

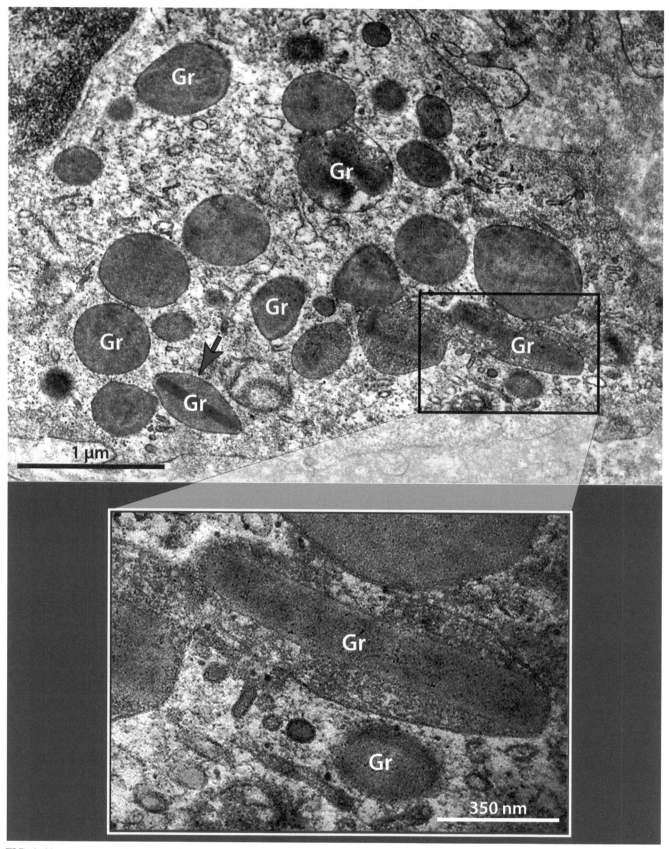

FIG. 8.69 TEM of a human eosinophil in the intestinal lamina propria (rectal biopsy, chronic schistosomiasis mansoni). Most specific granules (Gr) are enlarged and show core disarrangement and partial content losses indicative of PMD. The boxed area shows a vesiculating granule. A single intact granule with typical morphology is observed *(arrow)*. ECM was colored in *green*.

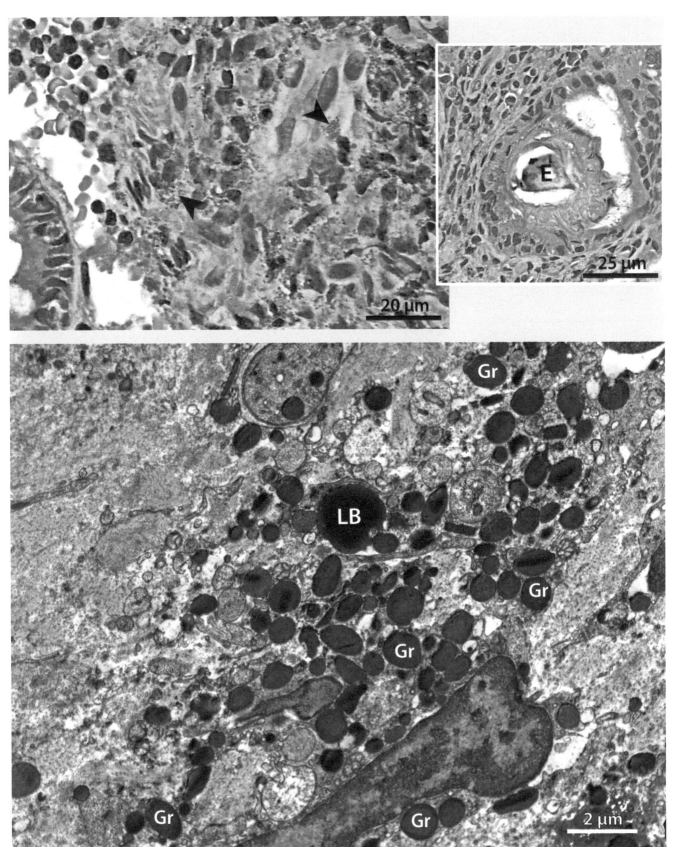

FIG. 8.70 Cytolytic human eosinophils (rectal biopsy, chronic schistosomiasis mansoni) seen by both light microscopy (HE staining, *top panels*) and TEM *(bottom panel)*. Note extensive degranulation *(arrowheads)* and released granules (Gr). A granuloma with parasite egg (E) and chromatolytic eosinophil nuclei *(purple)* are noted. *LB*, lipid body.

B. Eosinophils in human diseases

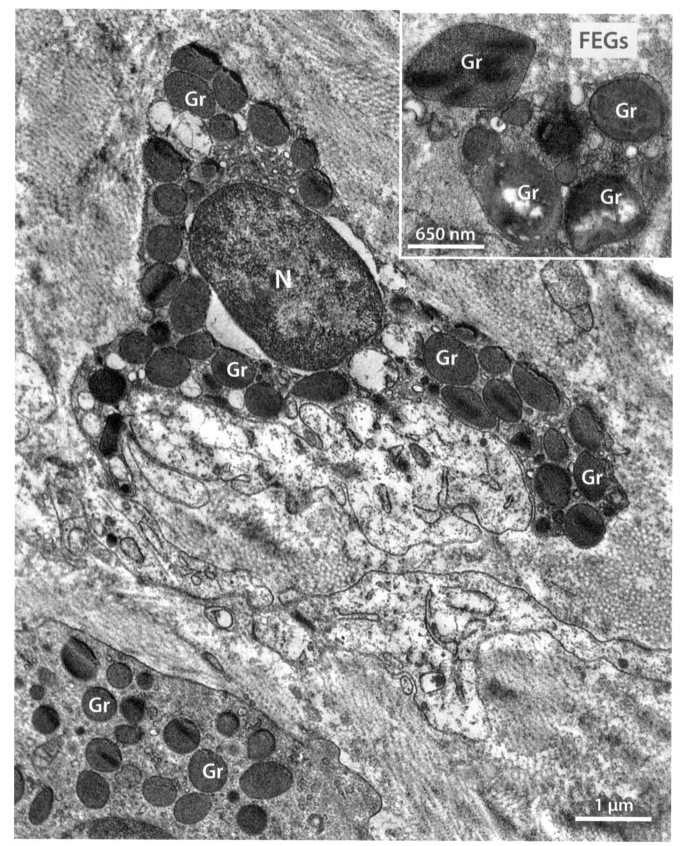

FIG. 8.71 Electron micrographs showing a cytolytic eosinophil *(center)* and FEGs in human chronic schistosomiasis mansoni. Note plasma membrane partially disrupted and damaged nuclear envelope. Released granules (Gr) show partial content losses. At the bottom, part of an intact eosinophil. Rectal biopsy prepared for TEM. *N*, nucleus.

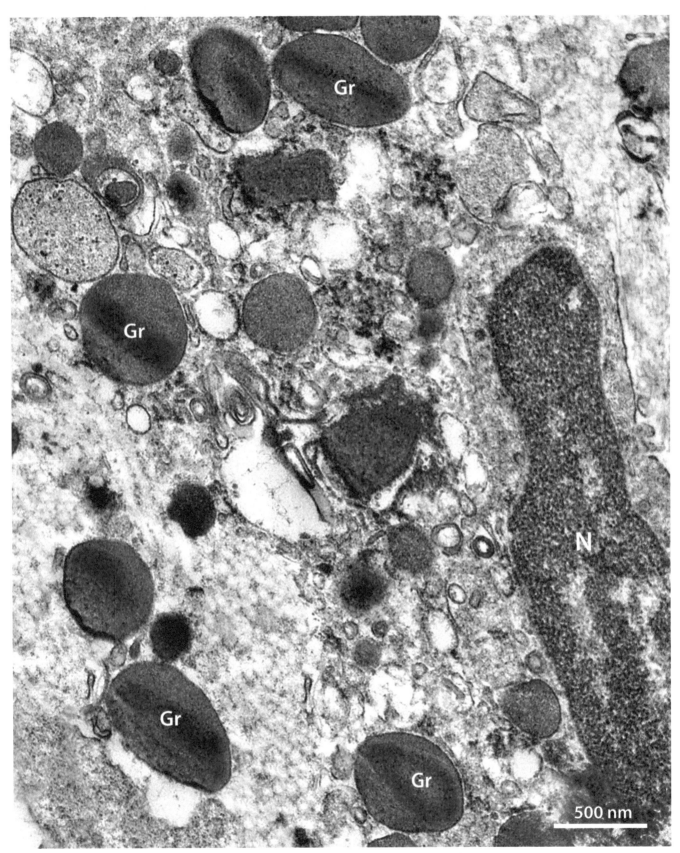

FIG. 8.72 Electron micrograph showing FEGs (Gr) and EoSVs *(colored in pink)* in the intestinal lamina propria of a patient with chronic schistosomiasis mansoni. Rectal biopsy prepared for TEM. *N*, nucleus.

B. Eosinophils in human diseases

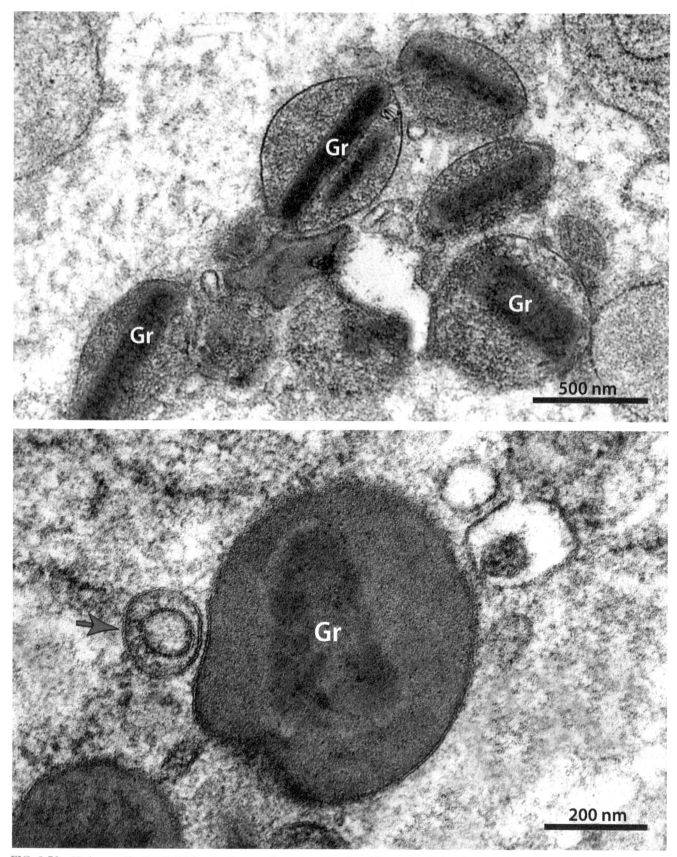

FIG. 8.73   High-magnification TEM showing free membrane-bound eosinophil granules (Gr) and an intact EoSV *(arrow)* in the human intestinal lamina propria (rectal biopsy, chronic schistosomiasis mansoni). Note the trilaminar structure of the delimiting granule and EoSV membranes.

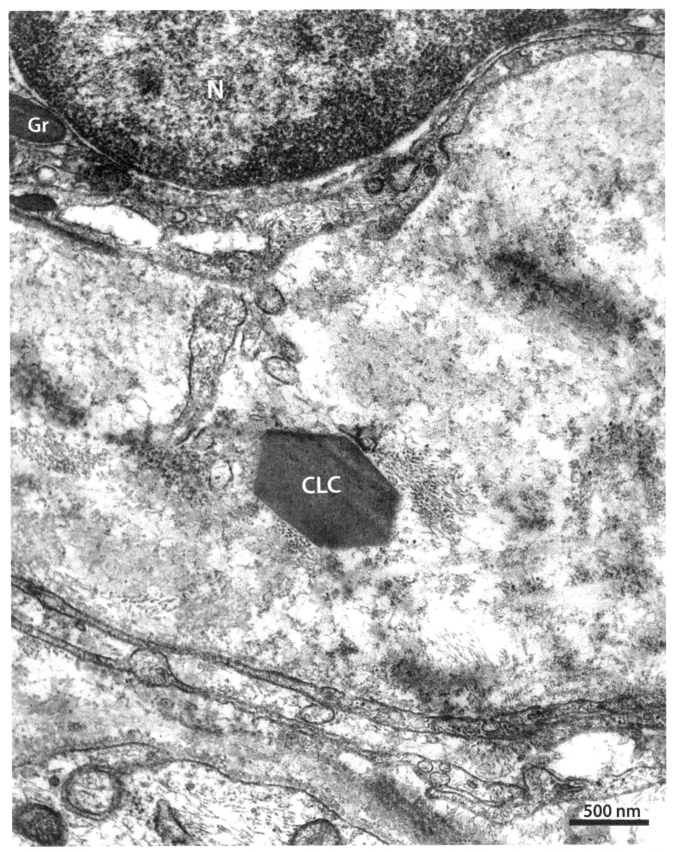

FIG. 8.74   A typical hexagonal CLC observed in the intestinal lamina propria close to an eosinophil in a patient with chronic schistosomiasis mansoni. Rectal biopsy prepared for TEM. *Gr*, specific granule; *N*, nucleus.

B. Eosinophils in human diseases

fused and resting granules, which represented 9% and 3% of the population of cytoplasmic granules, respectively (Fig. 8.69). Thus, compound exocytosis was not observed as a significant degranulation mode operating in vivo in eosinophils recruited by chronic schistosomiasis.

Eosinophil cytolysis was detected in extensive tissue areas by both histopathological (Fig. 8.70) and TEM (Figs. 8.70–8.74) analyses. Cytolysis was detected in different degrees, ranging from an initial stage with the plasma membrane partially fragmented (Fig. 8.71) up to complete disappearance of this membrane and most cell organelles and just FEGs being recognized in the extracellular matrix (Fig. 8.72). Isolated or clustered membrane-bound FEGs, as well as cell-free EoSVs, were frequently observed (Fig. 8.73), while CLCs were occasionally identified (Fig. 8.74). Our findings show that active degranulating eosinophils persist in target tissues of the chronic human schistosomiasis mansoni after different cycles of reinfection.

8.7.2 Fungal infections

Eosinophils have been implicated as important immune cells in infections caused by fungal pathogens and in allergic diseases associated with fungal sensitization/colonization (reviewed in[313]). Both blood and tissue eosinophilia have been reported in many of these diseases with a potential protective antifungal role against species such as *Alternaria alternata*, *Cryptococcus neoformans*, and *Aspergillus fumigatus*.[313]

One prototypical allergic fungal disorder with consistent involvement of eosinophils is ABPA, caused by the environmental and opportunistic fungus *A. fumigatus*. The interactions between this fungal species and human eosinophils have been studied with the use of molecular and ultrastructural approaches.[47,49] As featured in Chapter 6, EETosis was identified as a significant mechanism underlying eosinophil degranulation in response to the in vitro interaction with *A. fumigatus*. The ultrastructural signature of EETosis, represented by nuclear changes and release of both EETs and membrane-bound FEGs (Chapter 6), was detected by TEM, while SEM revealed the typical filamentous 3D aspect of extracellular traps. Both TEM and SEM revealed EETs entrapping conidia, small spores (2–3 μm) produced by *Aspergillus* species, which are critical for the fungal dispersion (see Figs. 3.36 in Chapter 3 and 6.17, 6.18, and 6.26 in Chapter 6).[47,49,313]

8.7.3 Viral infections

In general, active viral infections are associated with decreased numbers of eosinophils in the peripheral blood (reviewed in[368,426]). Eosinopenia was also correlated with the severity of the coronavirus disease-2019 (COVID-19), caused by SARS-CoV-2 (severe acute respiratory syndrome coronavirus-type 2) (reviewed in[442,443]). On the other hand, blood eosinophilia develops in association with some viral diseases, for example, HIV infection, which may increase eosinophil percentages in the blood due to secondary reasons.[368] Other viral infections such as viral myocarditis and respiratory syncytial virus (RSV) pneumonia lead to tissue eosinophilia in the absence of blood eosinophilia.[426]

Eosinophils contain granule-derived molecules with potential antiviral activity such as the eosinophil-associated ribonucleases (EARs) ECP and EDN and the cytokines IL-12 and IFN-γ, and it is accepted that these cells release their products mainly through PMD in response to respiratory viruses, for example, RSV and influenza A virus (IAV) (reviewed in[444–446]). However, most studies focused on eosinophil-virus interactions are performed in vitro and/or using mouse models of viral infections. The interactions between human eosinophils and pathogenic virus remain poorly understood, and it is not clearly defined if eosinophils are able to reduce viral infectivity in vivo.[445,447]

Human eosinophils express CD4, which serves as a receptor for HIV-1 and HIV-2,[448] and can be infected with HIV-1.[204] We demonstrated that blood-derived, HIV-infected eosinophils show in vitro marked increases in the numbers of emptying secretory granules, vesiculotubular structures (EoSVs), and lipid bodies, which also increase in size in response to the infection (see Figs. 4.27–4.30 in Chapter 4).[204] However, distinct budding of virions from the surface of eosinophils, as occurs in lymphocytes, or large collections of intracellular virions, as occurs in monocytes, were not observed. HIV-1 induced apoptosis and secondary necrosis in eosinophils (Fig. 8.75), indicating that these cells are susceptible to in vitro infection.[204]

Viruses are obligate intracellular pathogens that depend on host cell metabolism for their replication. TEM is the only imaging technique allowing the direct visualization of viruses due to its nanometer-scale resolution.[449] TEM can reveal the presence of viruses in varied biological samples such as cell suspensions, tissue sections, or mammalian cells grown in vitro in contact with clinical samples.[449] Positive-strand RNA (+ RNA) viruses, which encompass many pathogens including SARS-CoV-2, remodel cellular membranes to support replication of the viral genome.[450] The observation that viruses induce membrane alterations in infected cells was made decades ago by EM, and, currently, both conventional and 3D EM have been applied for a better understanding of how viruses assemble and replicate inside the cells.[449–451]

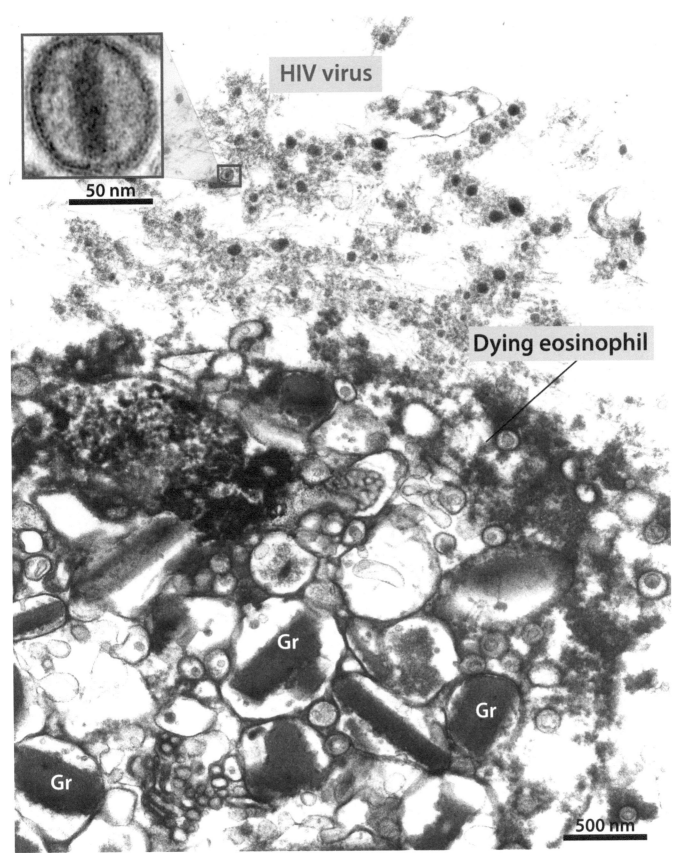

HIV virus

50 nm

Dying eosinophil

Gr

Gr

Gr

500 nm

FIG. 8.75   TEM of a human HIV-1-infected eosinophil undergoing secondary necrosis (necrosis after apoptosis). Note the dense cytoplasmic matrix with aggregated organelles (features of apoptosis) and disrupted plasma membrane (feature of necrosis). Viral particles are seen in the extracellular medium. One enveloped virus and its electron-dense core (nucleocapsid) are shown in higher magnification. *Gr*, secretory granules.

B. Eosinophils in human diseases

Thus, the ultrastructural identification of viral replication compartments, intracellular viral interactions, and virus ability to sequester host cell components is crucial to understand viral infections.[451] Nevertheless, it should be emphasized that the identification of viruses at the ultrastructural level is not always straightforward, and there are common cellular structures such as coated vesicles, MVBs, and cross-sections of the RER that apparently look similar to virus particles, although they lack the virus electron-dense core(s) (nucleocapsid), which requires high magnification (~ 90,000 ×) to be clearly identified (Fig. 8.75).[452] Therefore, several virus aspects (size, shape, internal pattern of the nucleocapsid, and surface spikes if present) should be considered to avoid mistaking cell organelles for viral particles.[453]

8.7.4 Bacterial infections

While blood eosinopenia has been associated with varied bacterial infections,[454,455] the role played by eosinophils during infections involving bacterial pathogens remains unsolved. Several lines of evidence show that eosinophils respond to pathogenic bacteria with activation and degranulation but whether these cells act as effector or immunomodulatory cells during active bacterial infections in vivo remains to be established (reviewed in[445]). Human eosinophils express diverse members of the pattern-recognition receptor (PRR) family, including all toll-like receptors (except TLR-8), nucleotide-binding oligomerization domain (NOD)-like receptors (NLR) 1 and 2, and the C-type lectin receptor Dectin-1 that recognize bacterial components upon activation and trigger immune responses.[445,456] Eosinophils, both viable and cytolytic, are also able to release extracellular DNA traps in response to the stimulation with bacteria such as *Escherichia coli* or *Staphylococcus aureus* (reviewed in[165]).

A protective role for eosinophils against bacterial infections in the gut has been indicated, notably anti-*Clostridium difficile* colitis.[457] This Gram-positive, spore-forming anaerobic bacterium infects the colon when the normal microbiota has been disrupted, and increased numbers of eosinophils seem to promote a balanced inflammatory environment that effectively combats the pathogen.[457] We analyzed by TEM intestinal biopsies (excised ileal pouch) from a patient with pouchitis, an inflammatory condition that can develop from surgical construction of ileal reservoirs, and stool positive for *C. difficile*.[42] *C. difficile* infection is the most common nosocomial pathogen responsible for severe colitis and frequently complicated with IBDs (reviewed in[458]).

Eosinophils were the most prominent inflammatory cells in the inflamed pouch, found both intraepithelially and in the lamina propria. These cells were activated and had ultrastructural profiles compatible with several types of degranulation events (PMD, cytolysis, and exocytosis to a lesser degree). Moreover, we found the extensive release of FEGs in edematous areas of the lamina propria together with coccal and bacillary bacterial forms, including bacilli with a thick capsule consistent with clostridial species (Fig. 8.76).[42] Interestingly, in the ileal lumen, we found part of an eosinophil among many bacteria (Fig. 8.77). This finding, in association with considerable numbers of eosinophils transmigrating through the epithelium in bacterial pouchitis, may be reflecting a higher activity of these cells as immune surveillance actors, a role also reported at steady state.[395] Altogether, our EM analyses indicate a role for eosinophils in the host response of the gut to bacterial inflammation.[42]

8.8 Tumors

The tumor microenvironment is a complex site in which local components, including blood and lymphatic endothelial cells, mesenchymal cells, fibroblasts, and immune cells alongside the extracellular matrix interact closely with tumor cells and can affect their growth in different ways. Thus, the survival, proliferation, invasiveness, and therapeutics responsiveness of tumor cells are influenced by their dynamic interactions with the microenvironment (reviewed in[459]). Most solid tumors are abnormally vascularized and infiltrated by diverse immune cells, which populate the stromal region and can potentially execute both protumor and antitumor functions.[459]

Eosinophils are a common component of inflammatory infiltrates of several solid and hematological tumors. This eosinophil infiltration in the tumor stroma is referred to as tumor-associated tissue eosinophilia (TATE). Tumor-infiltrating eosinophils have been described for over a century in solid tumors of the head and neck, bladder, prostate, penis, colorectal, breast, and ovary, among other locations, as well as in clonal myeloid neoplasms (chronic and acute myeloid leukemia, systemic mastocytosis, and myelodysplastic/myeloproliferative syndromes) and lymphoid neoplasms (Hodgkin lymphoma, T-cell non-Hodgkin lymphoma). However, the functional role of eosinophils in tumors, the factors involved in their recruitment, and whether the increased numbers of these cells have or do not have a beneficial influence on cancer prognosis are still poorly defined (reviewed in[460–464]).

Eosinophils have been considered to act as both effector and regulatory cells within the tumor microenvironment, with a potential prognostic/predictive role. The antitumoral cytotoxic activity of eosinophils has been

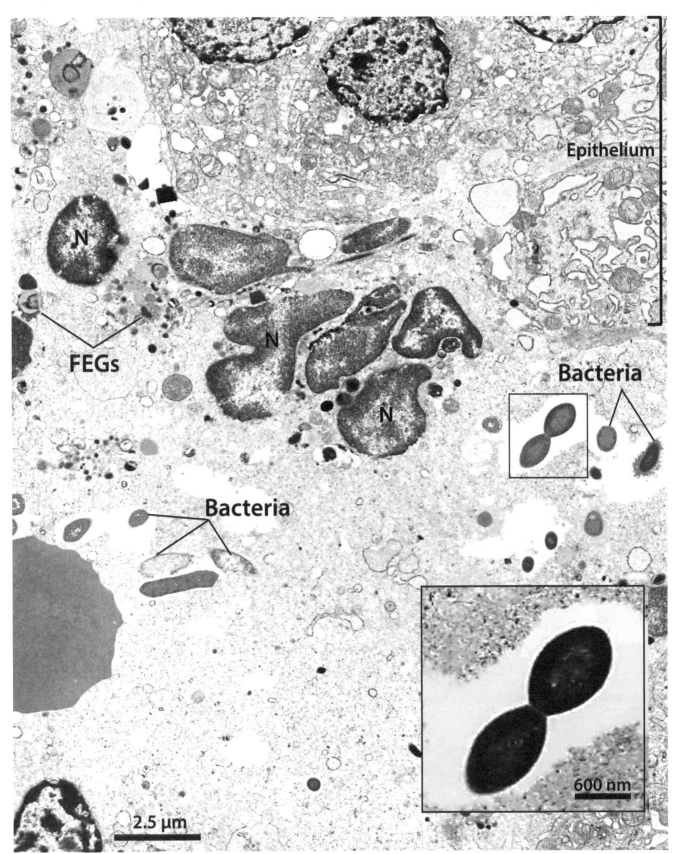

FIG. 8.76 Eosinophil FEGs in the intestinal lamina propria of a patient with bacterial pouchitis. Note released FEGs *(orange)* and nuclei (N) from cytolytic cells and bacteria *(green)* with varied morphology. Dividing bacteria are seen in higher magnification *(box)*. Epithelial cells were pseudocolored to indicate nuclei *(purple)*, RER *(light blue)*, mitochondria *(green)*, and cytoplasm *(pink)*. Ileal pouch biopsy prepared for TEM.

B. Eosinophils in human diseases

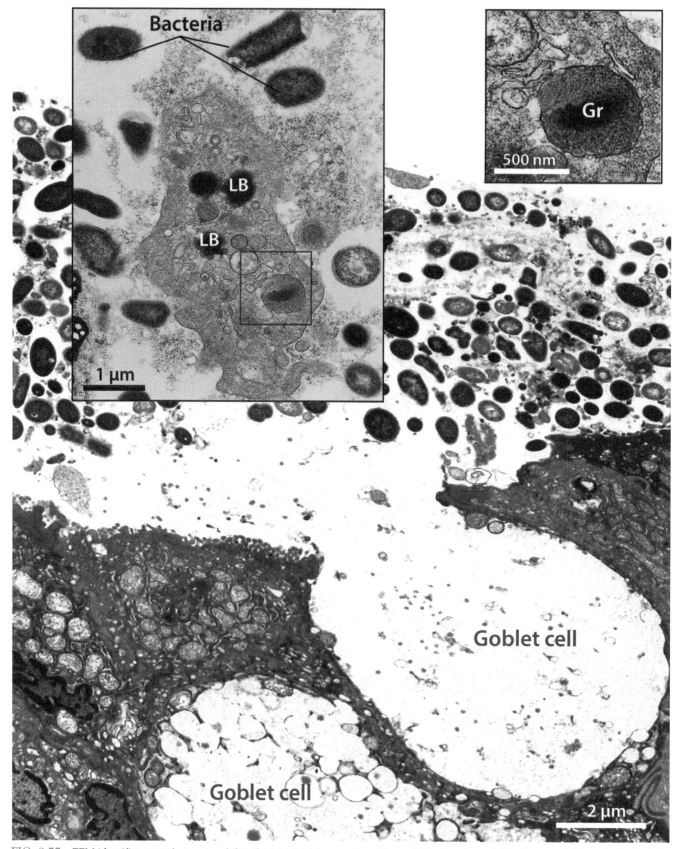

FIG. 8.77   TEM identifies part of an eosinophil in the intestinal lumen of a patient with bacterial pouchitis (ileal pouch biopsy). Osmiophilic lipid bodies (LB), EoSVs *(colored in pink)*, and a specific granule (Gr, *boxed area* shown in higher magnification) enabled unambiguous eosinophil recognition. A diversity of bacteria is seen in the lumen. Note epithelial absorptive and secretory (goblet) cells.

suggested by the observation of degranulating eosinophils and their deposited granule-derived proteins as well as by eosinophil-tumor cell interactions in the local vicinity of tumors, but the tumoricidal effects of eosinophils are not well understood. The ability of eosinophils to modulate inflammatory responses within tissue microenvironments also suggests that this function represents an important activity after their recruitment to solid tumors.[460] Hence, there is still much to learn regarding the impact of eosinophils in cancer.[463]

The clinical relevance of eosinophils in different tumor entities has been controversial.[463] In Hodgkin lymphomas and in carcinomas of the kidney, thyroid, gallbladder, pancreas, and breast, the infiltration of eosinophils is considered unfavorable. Moreover, because tumor-associated blood eosinophilia generally occurs in the disease progression phase (when the tumor has already acquired metastatic capacity), its occurrence is often associated with a poor prognosis. Oppositely, several studies have indicated an improved prognosis with eosinophil infiltration or degranulation in different types of solid tumors including nasopharyngeal carcinoma, gastric carcinoma, laryngeal carcinoma, pulmonary adenocarcinoma, colorectal tumors, bladder carcinoma, prostate cancer, and penile cancer (reviewed in[461,463]). In other types of tumors, such as oral squamous cell carcinoma, eosinophils are associated with both improved and poor prognosis (reviewed in[465,466]). Since the microenvironment is a significant factor that participates in the control of the biological properties of tumors with multifaceted interactions between inflammatory cells and stromal components as an ecosystem,[467,468] better understanding of the eosinophil role will likely depend on exploring this cell under this perspective.

TEM has played a significant role in diagnostic pathology, particularly in the field of tumor diagnosis, complementing light microscopic observations.[42–44,469–471] Although TEM is limited by the size of a specimen that can be examined, a large amount of information can be obtained from a small sample and is helpful in determining the source of metastasis from an unknown primary tumor. TEM provides diagnostic criteria with findings of typical subcellular structures, for example, the long, wavy, villous processes of mesothelial cells in mesotheliomas; desmosomes and bundles of cytokeratin of squamous cells in squamous cell carcinoma; the microvillous-lined lumina of adenocarcinomas; and Weibel-Palade bodies of endothelial cells in vascular tumors among many other features, which can be revealing of the tumor type and origin.[470] Additionally, not only the morphological features of the invading cell but also the fine details of the stromal inflammatory infiltrate and their interactions can be disclosed by ultrastructural studies.

Other groups and we have been investigating tumor-infiltrated eosinophils at the ultrastructural level.[155,247,472–474] Collectively, TEM analyses of human biopsies revealed that eosinophils at the tumor sites (i) can be found isolated or forming aggregates, (ii) interact with tumor cells and/or with other cells from the tumor microenvironment, (iii) undergo degranulation mainly through PMD and/or cytolysis, and (iv) frequently deposit FEGs and occasionally CLCs in the extracellular matrix. These ultrastructural features were fully or partially observed by us in oral squamous cell carcinoma (OSCC) (Figs. 8.78–8.82), gastric carcinoma (Figs. 8.83 and 8.84),[155] pancreatic tumors (Fig. 8.85),[247] mesotheliomas (Fig. 8.86), and intradural tumor (Figs. 8.87 and 8.88). In some situations, for example, eosinophils collected from the pleural fluid from a patient with lung adenocarcinoma, we also found the increased formation of lipid bodies within the eosinophil cytoplasm (Fig. 8.89), which clearly reflects the high activated status of these cells (Chapter 4).

OSCC is one of the most common tumors worldwide, affecting inside/around the oral cavity and comprising the tumor epithelium and the surrounding connective tissue stroma.[475] Eosinophils are consistently observed in histological sections of OSCC biopsies (Fig. 8.78).[465,476] Our analyses of OSCC biopsies detected a high number of eosinophils (isolated or aggregated) in the tumor stroma (Figs. 8.78–8.82). Eosinophil groups comprised intact cells (Figs. 8.78 and 8.79), eosinophils in the early stage of cytolysis with plasma membrane partially disrupted and spilling secretory granules into the extracellular matrix (Fig. 8.80), and clusters of FEGs between intact cells (Fig. 8.81), indicating different stages of degranulation. PMD (Figs. 8.78 and 8.82) was frequently observed in addition to cytolysis, but exocytosis was absent. Eosinophils undergoing PMD were observed in close apposition to other cells from the tumor stroma (Figs. 8.78 and 8.79). The ultrastructure of OSCC with a focus on the characteristics of the epithelial cells has been reported in other studies.[477,478]

By investigating gastric carcinomas with an emphasis on TATE (Fig. 8.83) in several ultrastructural studies, Caruso and colleagues reported different aspects of infiltrating eosinophils, which were markedly found in tubulopapillary and poorly cohesive carcinoma types.[155,367,474,479,480] By grading 460 eosinophils localized in the tumor stroma, the authors found that most of them (65%) showed different degrees of PMD characterized by ragged loss of the granule cores up to or complete losses of both core and matrix contents.[155] Eosinophils participated in the stromal reaction by interacting with tumor cells, mast cells, and each other. While eosinophils with signs of classic exocytosis were absent, deposition of FEGs was observed with occasional CLCs.[474] Accordingly, in our EM analyses, eosinophils undergoing PMD (Fig. 8.83) as well as FEGs and free EoSVs (Fig. 8.84) were detected in a gastric carcinoma biopsy.

One aspect of the eosinophil biology that is at the frontline of the eosinophil research is the ability/significance of the DNA extracellular traps released by eosinophils in EADs (reviewed in[48,165,312,318,481]). As noted, our ultrastructural studies have detected EETs in vivo from cytolytic eosinophils during EADs, including IBDs (Fig. 8.60 and

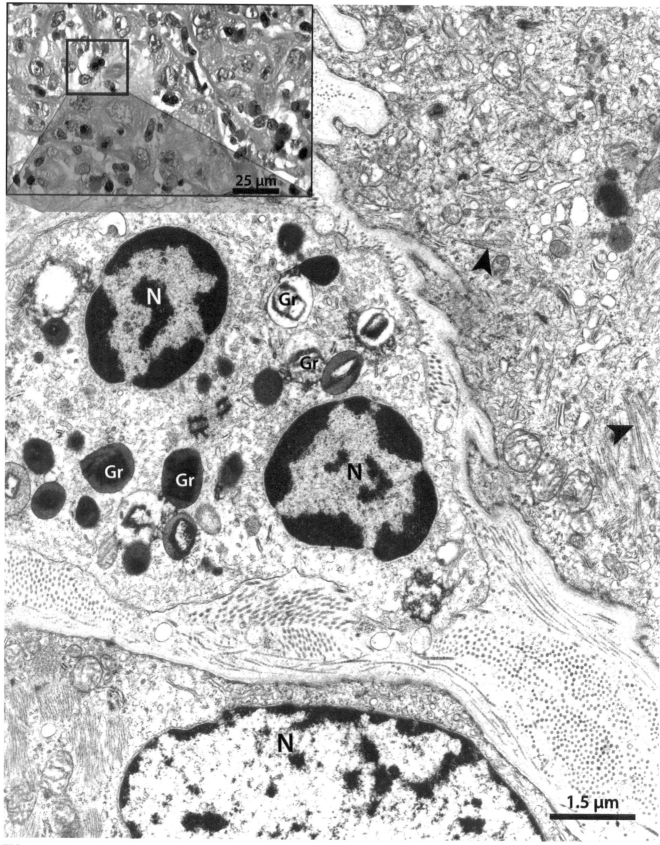

FIG. 8.78   Eosinophils infiltrated in the tumor stroma of a human oral squamous cell carcinoma (OSCC) observed by both light microscopy and TEM. Several eosinophils are seen in the histological image. Note the typical eosinophil bilobed nucleus and acidophilic cytoplasm (box). TEM shows an eosinophil located close to cells with cytoplasmic bundles of cytokeratin *(arrowheads)*. Granule (Gr) changes are associated with PMD. N, nucleus. *Histological image (stained with HE) courtesy of Tomaso Lombardi.*

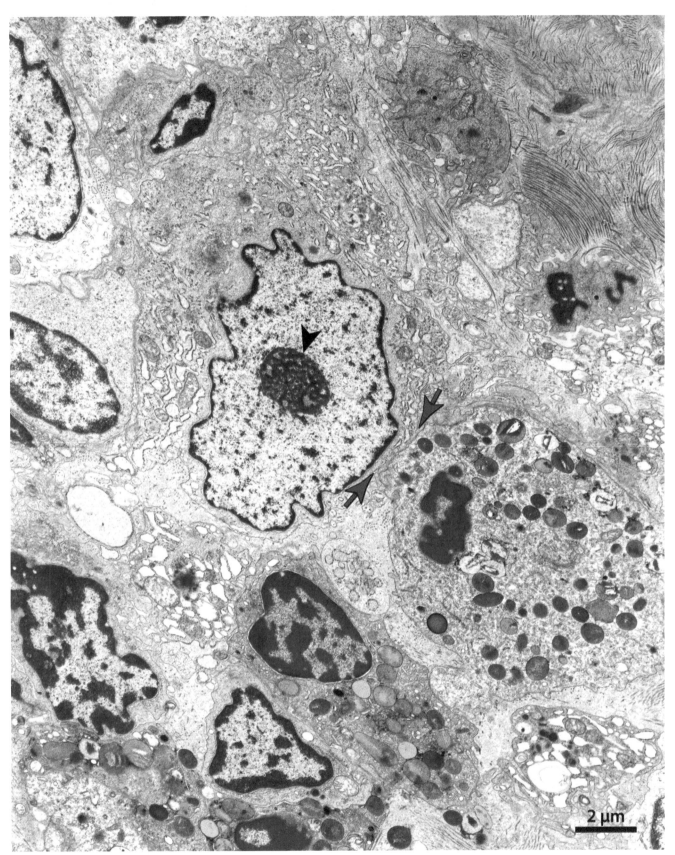

FIG. 8.79   TEM showing eosinophils infiltrated in the tumor stroma of a human OSCC. Note close interaction *(arrows)* between one eosinophil and a tumor cell depicting the irregular nucleus and well-developed nucleolus *(arrowhead)*. Eosinophils were pseudocolored to indicate nuclei *(purple)*, granules *(yellow)*, and cytoplasm *(pink)*.

B. Eosinophils in human diseases

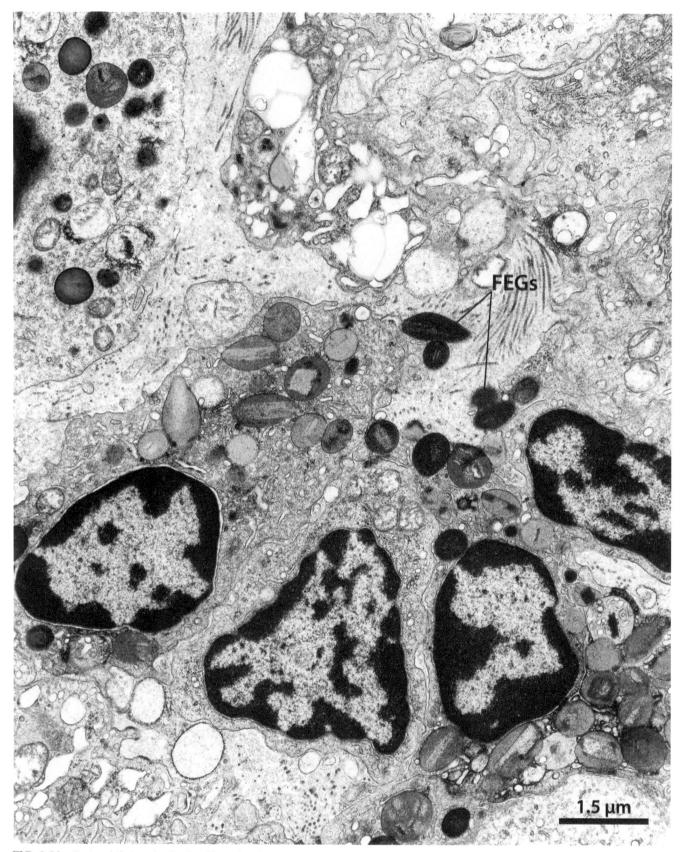

FEGs

1.5 μm

FIG. 8.80   Eosinophils *(pseudocolored)* infiltrated in the tumor stroma of a human OSCC. One eosinophil is partially disrupted, and part of their granules *(yellow)* spilled into the ECM. Biopsy (lower lip) prepared for TEM. Nuclei *(purple)*; cytoplasm *(pink)*; FEGs, free extracellular granules.

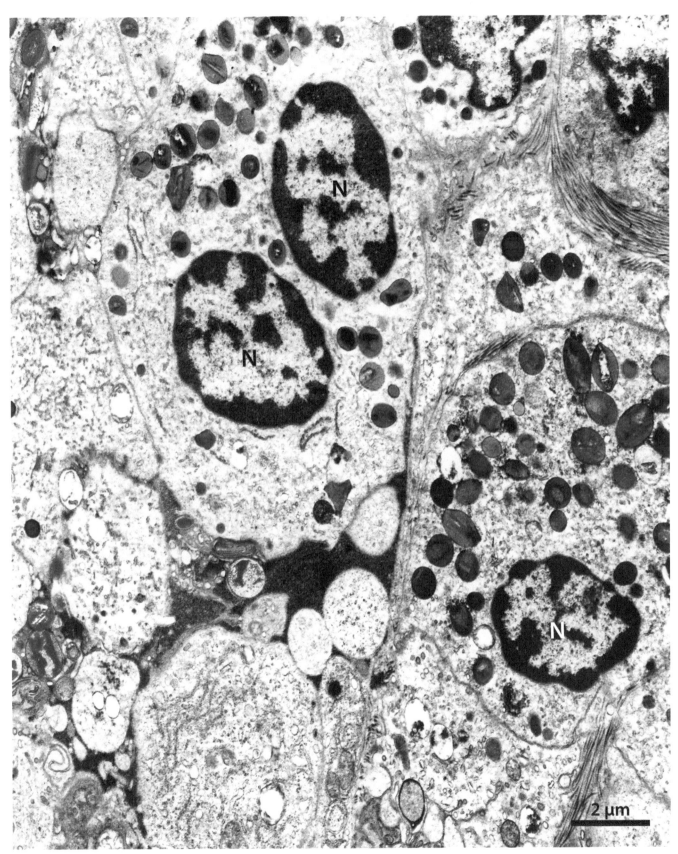

FIG. 8.81   A nest of eosinophils infiltrated in the tumor stroma of a human OSCC. Note intact eosinophils and FEGs *(colored in yellow)*. Biopsy (lower lip) prepared for TEM. Residual cytoplasm *(pink)*; *N*, nucleus.

B. Eosinophils in human diseases

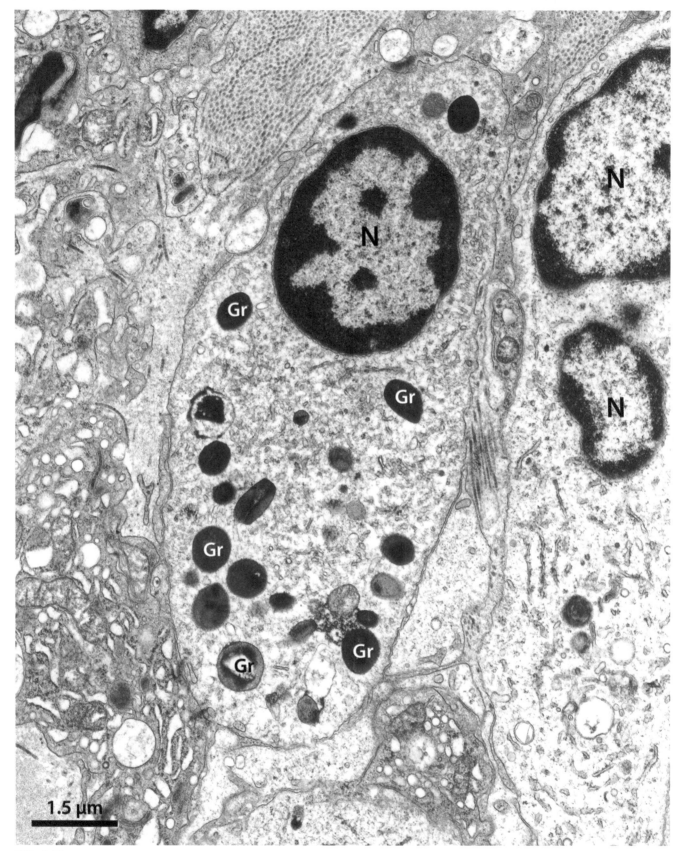

FIG. 8.82   Electron micrograph showing an eosinophil undergoing PMD in the tumor stroma of a human OSCC. Biopsy (lower lip) prepared for TEM. *Gr*, secretory granules; *N*, nucleus.

B. Eosinophils in human diseases

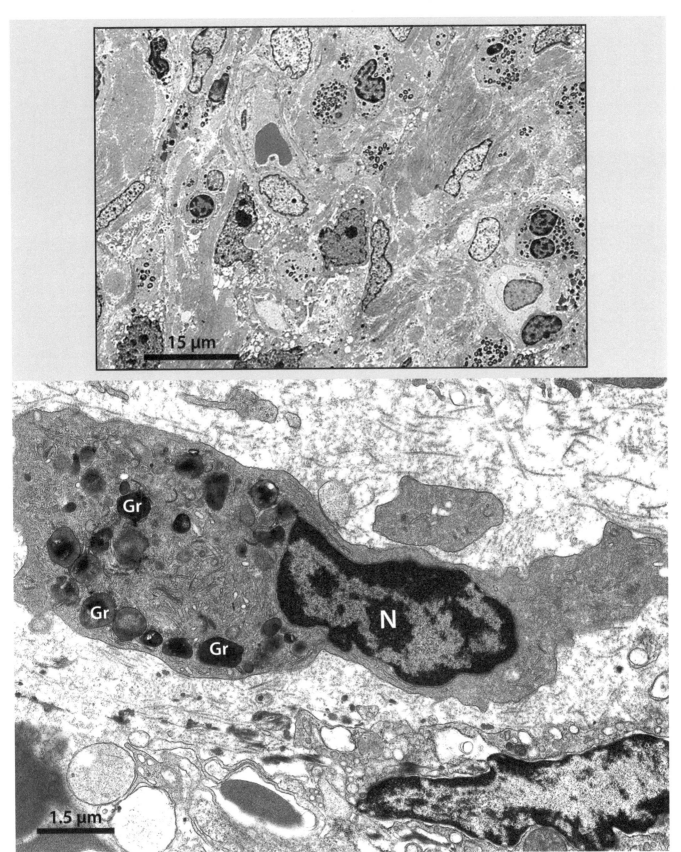

FIG. 8.83   TEM showing tumor-associated tissue eosinophilia in a human gastric carcinoma. Note in the *top panel* at low magnification the high number of infiltrated eosinophils *(pseudocolored)*. In higher magnification, an elongated eosinophil shows mobilized secretory granules (Gr) with focal losses. *N*, nucleus. *Top panel image courtesy of Rosario Caruso.*

B. Eosinophils in human diseases

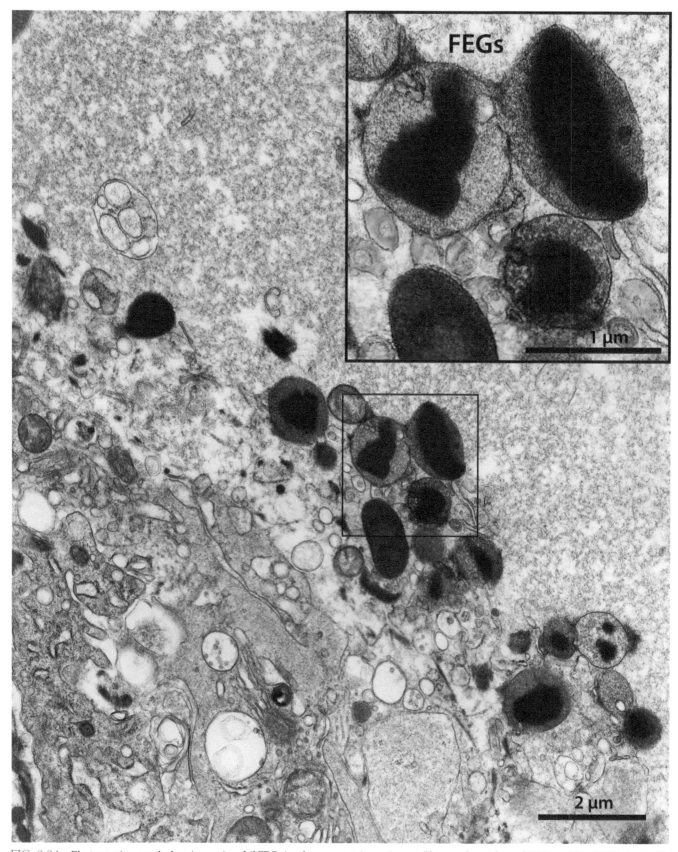

FIG. 8.84 Electron micrograph showing eosinophil FEGs in a human gastric carcinoma. The membrane-bound FEGs and free EoSVs *(colored in pink)* are shown in higher magnification *(boxed area)*. Biopsy prepared for TEM.

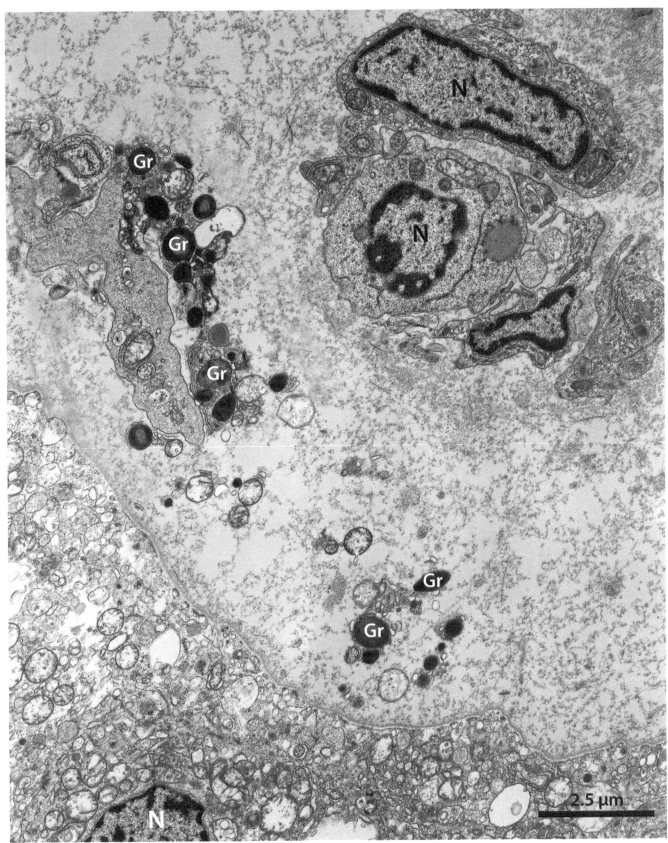

FIG. 8.85   TEM reveals a cytolytic eosinophil in the tumor stroma *(colored in green)* of a patient with neoplasm of the pancreas. Other infiltrated inflammatory cells are seen in the stromal area. Part of the cytoplasm of a tumor cell was colored in *brown* (bottom). Note released specific granules (Gr). *N*, nucleus.

B. Eosinophils in human diseases

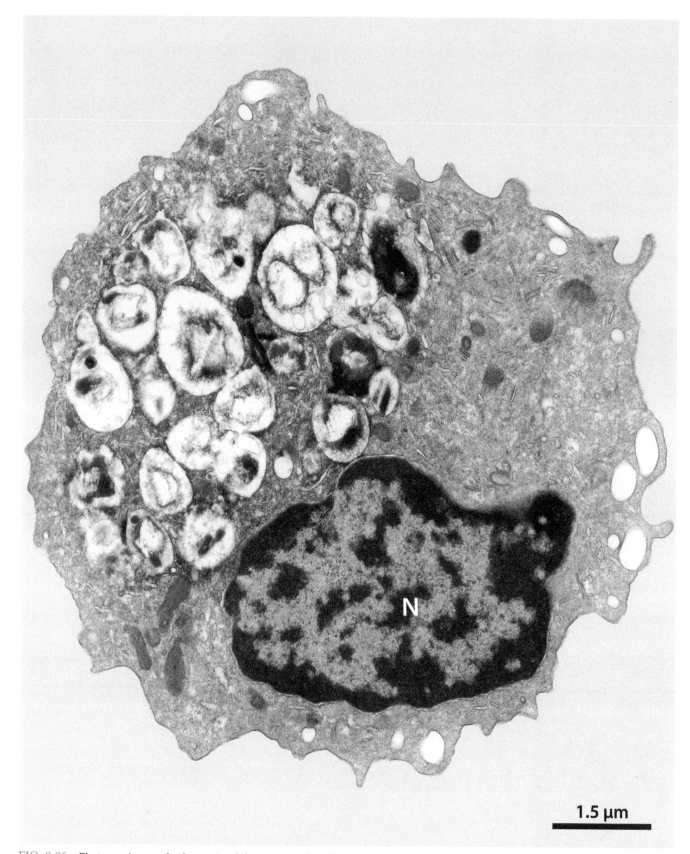

FIG. 8.86   Electron micrograph of an eosinophil seen in the pleural fluid from a patient with malignant pleural mesothelioma. Secretory granules *(colored in yellow)* are enlarged and mostly electron-lucent with extensive losses of their contents. Some granules show evidence of fusion. Sample prepared for TEM. *N*, nucleus.

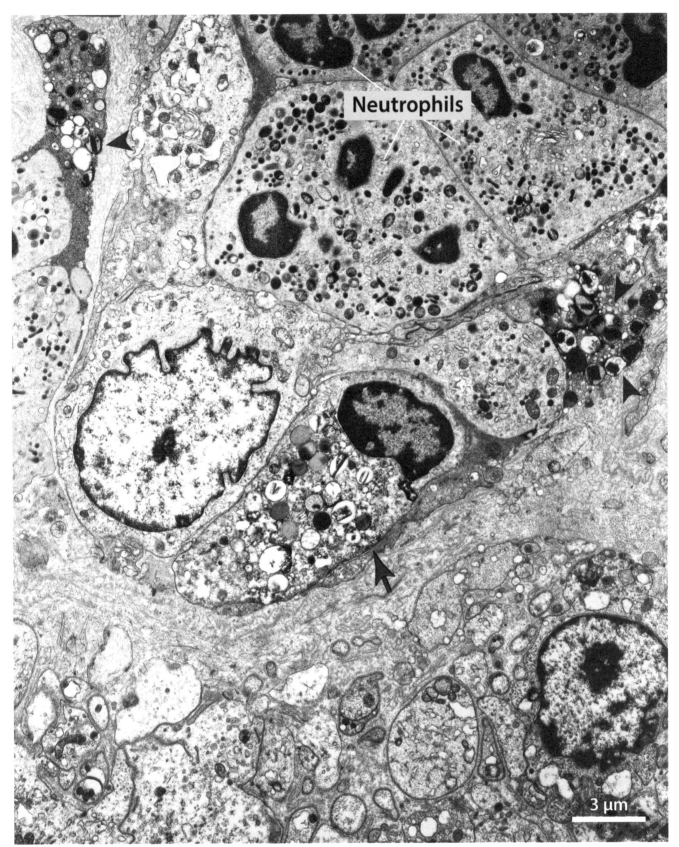

Neutrophils

FIG. 8.87 TEM showing infiltrated eosinophils *(pseudocolored)* and neutrophils in a human tumor of the central nervous system (intradural spinal tumor). Both intact *(arrow)* and cytolytic *(arrowheads)* eosinophils with greatly mobilized secretory granules *(yellow)* are observed. One eosinophil is displayed in higher magnification in Fig. 8.88.

B. Eosinophils in human diseases

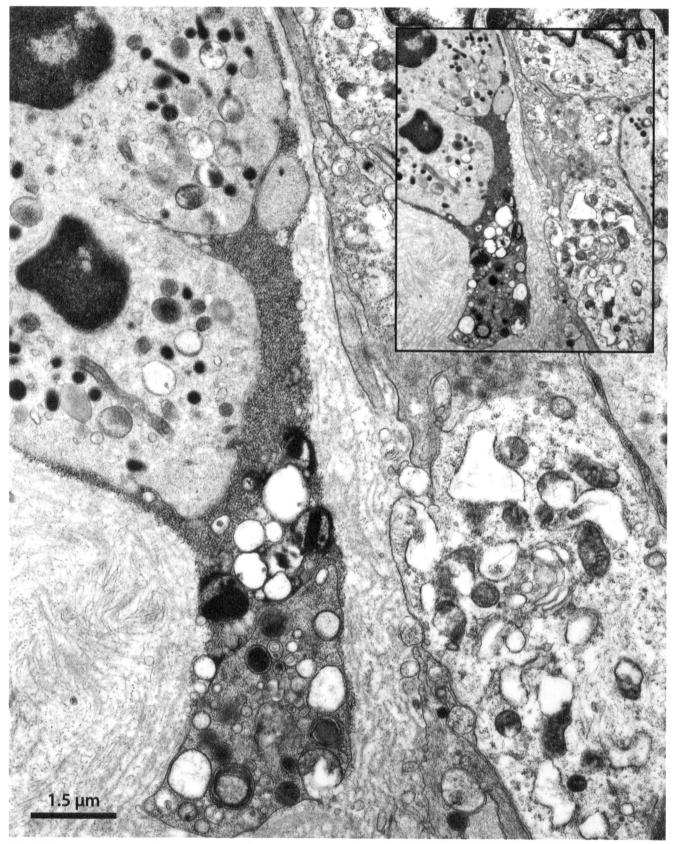

FIG. 8.88   A human eosinophil showing typical ultrastructure of ETosis in an intradural spinal tumor. Highly decondensed chromatin *(colored in purple)* appears as long extracellular traps (ETs) extending into neighboring neutrophils. Note free FEGS *(yellow)* and residual cytoplasm *(pink)*. Biopsy prepared for TEM.

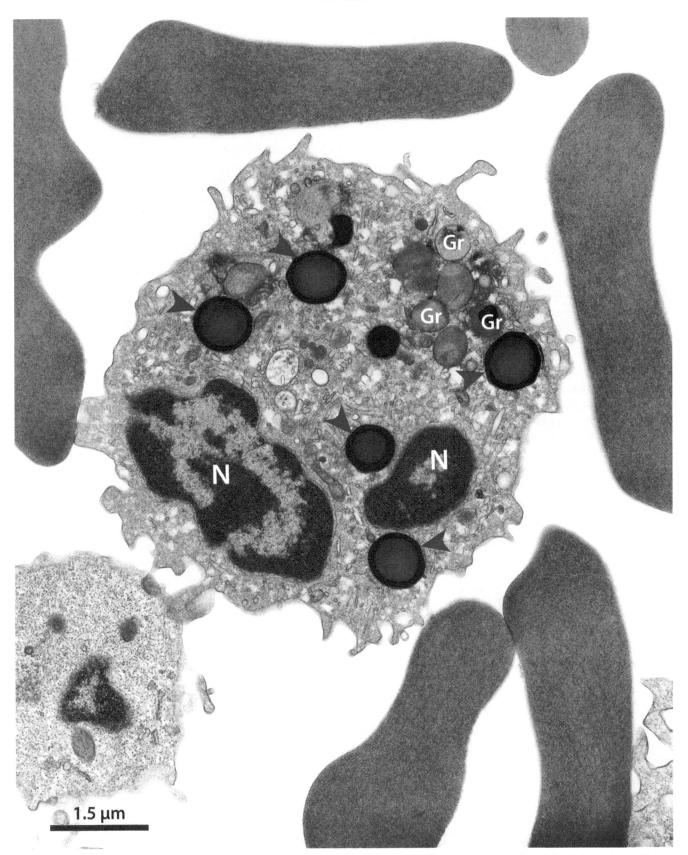

FIG. 8.89   An activated eosinophil showing many lipid bodies *(arrowheads)* in the cytoplasm is seen in the pleural fluid of a patient with lung adenocarcinoma. Note red blood cells *(colored in red)* surrounding the eosinophil. Sample prepared for TEM. *Gr*, secretory granules; *N*, nucleus.

B. Eosinophils in human diseases

Figs. 6.19, 6.20, and 6.31 in Chapter 6), HES (Figs. 6.21, 6.24, and 6.29 in Chapter 6), and ECRS (Figs. 6.22 and 6.30 in Chapter 6). In Fig. 8.88, we illustrate a representative ultrastructural signature of EETosis found in a biopsy from a patient with an intradural spinal tumor. Extracellular traps released by neutrophils have been studied in the context of cancer biology,[482] but the involvement of eosinophils remains to be investigated.

8.9 Ultrastructure of CLCs in eosinophilic diseases

The intriguing relationship between CLCs and eosinophils has been the object of great clinical and scientific interest. As noted, CLCs are frequently formed in many diseases in association with eosinophilic inflammation, and the proximity of these morphologically diverse crystals with eosinophils has been reported for more than 150 years (reviewed in[164,483]).

CLCs are composed mainly of CLC-P/Gal-10, one of the most abundant proteins within eosinophils from healthy donors (Fig. 8.90).[277,278] CLC-P/Gal-10 has been exclusively described in human eosinophils, basophils, and regulatory T cells with no evidence for the expression of this protein in the murine system.[201,484] CLC-P/Gal-10 is not a granule-stored protein and is localized predominantly in the peripheral cytoplasm of eosinophils, even after activation.[37] The subcellular sites of Gal-10 within resting and activated eosinophils are shown in Chapter 5 (Figs. 5.28–5.35).[37]

We demonstrated that substantial amounts of CLC-P/Gal-10 are released under active eosinophil cytolysis and that increased extracellular concentration of CLC-P/Gal-10 is capable of forming CLCs.[41] CLCs can be commonly found in body fluids, secretions, and tissues from patients with EADs, and both CLCs and levels/expression of Gal-10 are considered biomarkers of eosinophil involvement in inflammation.[35,36,272,279] In this atlas, the formation of CLCs in vivo at sites of eosinophilic inflammation is demonstrated in tissue biopsy specimens of patients with ECRS (Figs. 8.8–8.10, Fig. 3.34 in Chapter 3, and Fig. 6.30 in Chapter 6), HES (Figs. 8.37 and 8.38; Fig. 6.29 in Chapter 6), IBDs (Figs. 8.58, 8.59, 8.61, and 8.65; Figs. 6.14 and 6.31 in Chapter 6), and schistosomiasis mansoni (Fig. 8.74).

The ultrastructure of CLCs is intriguing. Based on TEM studies focused on EADs, we documented that these crystals in vivo (i) appear with different sizes and morphologies. Very large (several micrometers) as well as very small (lower than 1 μm) (for example, see Fig. 6.14 in Chapter 6), mostly unnoticeable by light microscopy, crystals are observed by TEM, and these structures are readily identified as bipyramidal, hexagonal, or amorphous structures. (ii) They show variable electron densities. Strongly electron-dense CLCs are often observed, but the presence of moderately electron-dense or even CLCs with electron-lucent focal areas is revealed by TEM in considerable numbers. (iii) They can exhibit a layering effect, that is, layers with distinct electron densities in the same crystal likely reflecting the dynamics of crystal formation/material deposition. (iv) They can show surface-associated amorphous material, while their internal region has a repetitive pattern typical of crystallization. The amorphous material might indicate newly deposited material during the process of crystal growth. (v) They can be seen as stacked structures, that is, piles of two or more crystals. Thus, under TEM, CLCs can be defined as heterogeneous structures in terms of sizes, morphologies, and electron densities. The diversity of CLC ultrastructure is summarized in Fig. 8.91, while Fig. 8.92 illustrates a representative large CLC in which its internal structural pattern is seen in high magnification. There is still much to learn about how CLCs grow within eosinophils and in the extracellular matrix (after Gal-10 release from eosinophils), what controls their size and shape, and whether other proteins such as lysopalmitoylphospholipases[274] are contributing to their composition/functional activity.

As noted, CLCs found in vivo are considered markers of eosinophilic inflammation. In tissue biopsies, CLCs are frequently observed scattered in the extracellular matrix in close relationship with degranulated eosinophils and FEGs (Figs. 8.9, 8.10, 8.37, 8.38, 8.58, 8.59, and 8.90).[40,274] The localization of CLCs in the proximity of FEGs and free nuclei/chromatin fibers (Figs. 8.8–8.10; see also Figs. 6.29–6.31 in Chapter 6) is particularly interesting because cytolysis with the release of extracellular traps from the nuclear origin is involved in eosinophil degranulation and CLC formation.[41,164] CLCs can also be identified occasionally within eosinophils with intact plasma membranes[40] (Figs. 8.65), thus showing the ability of CLC-P/Gal-10 protein to assemble by auto-crystallization in the cytoplasm.[41,485] In vitro, we demonstrated that CLCs form spontaneously from lysates of eosinophils[273] and are rarely seen within cultured eosinophils (see Fig. 6.35 in Chapter 6).

Phagocytosed CLCs are seen by TEM in the cytoplasm of inflammatory macrophages infiltrated in tissue sites with eosinophilic inflammation, for example, in the skin of patients with mastocytoma[486] or HES[40,291] (Fig. 8.40), chronic obstructive pulmonary disease,[487] and intestinal lamina propria of patients with Crohn's disease (Fig. 8.61). We also found accumulation of CLCs in the cytoplasm of inflammatory macrophages in bacterially infected colitis (Fig. 8.93), a disease characterized by massive formation of CLCs associated with prominent eosinophil infiltration and degranulation.

CLCs are not just inert crystals produced from activated eosinophils undergoing cytolysis but are able to trigger immune responses. CLCs induced the release of the proinflammatory cytokine IL-1β upon their phagocytosis by primary human macrophages in vitro,[488] while biosimilar Gal-10-composed crystals injected into the airways of naive mice led to an innate immune response.[280]

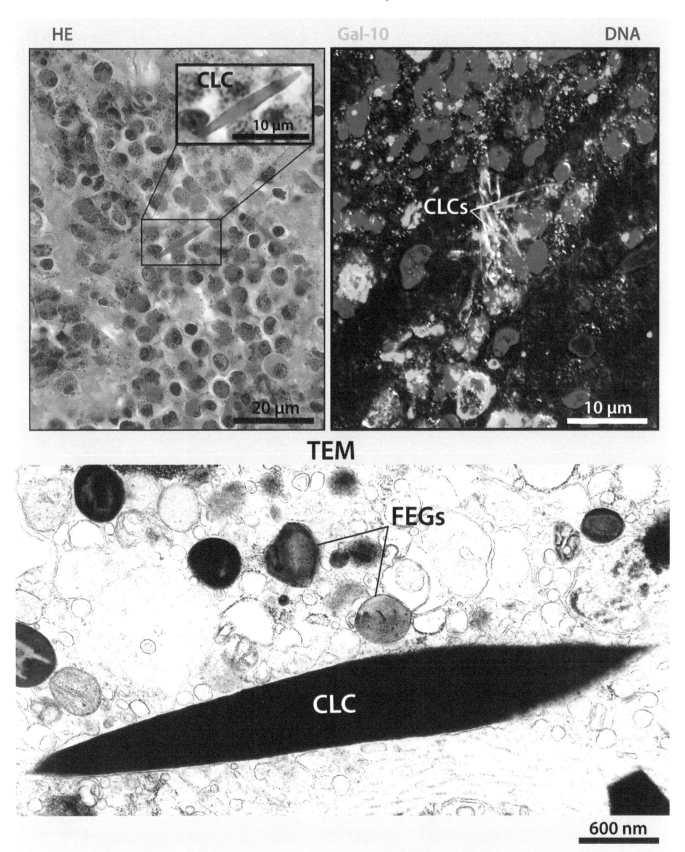

FIG. 8.90   CLCs observed by both light microscopy (LM) and TEM in biopsies (nasal polyps) of patients with severe ECRS. Top panels show CLCs under LM, stained with HE or immunolabeled for Gal-10 *(green)*. Nuclei were labeled blue with Hoechst 33342 for DNA detection. Note the high number of infiltrated eosinophils and FEGs surrounding CLCs. The bipyramidal morphology of CLCs is observed in all images. *Micrographs (top panels) courtesy of Shigeharu Ueki.*

B. Eosinophils in human diseases

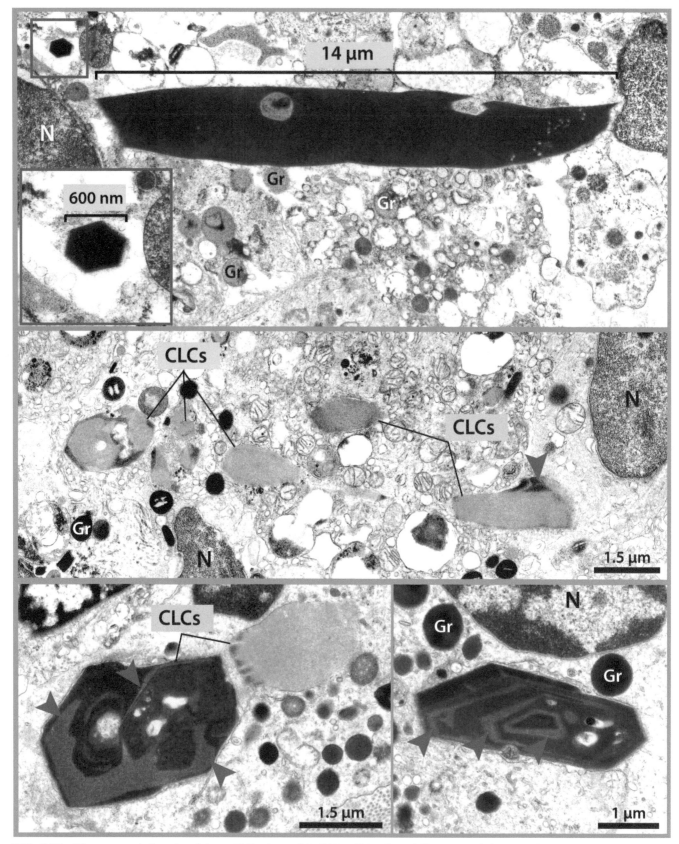

FIG. 8.91   Ultrastructural diversity of tissue CLCs observed in EADs. CLCs show different morphologies and electron densities and form layers *(arrowheads)*. *Top panel*: a tiny CLC (*boxed area* seen in higher magnification) is close to a very large one. Biopsies are from IBDs, ECRS, and HES. *Gr*, secretory granules; *N*, nucleus.

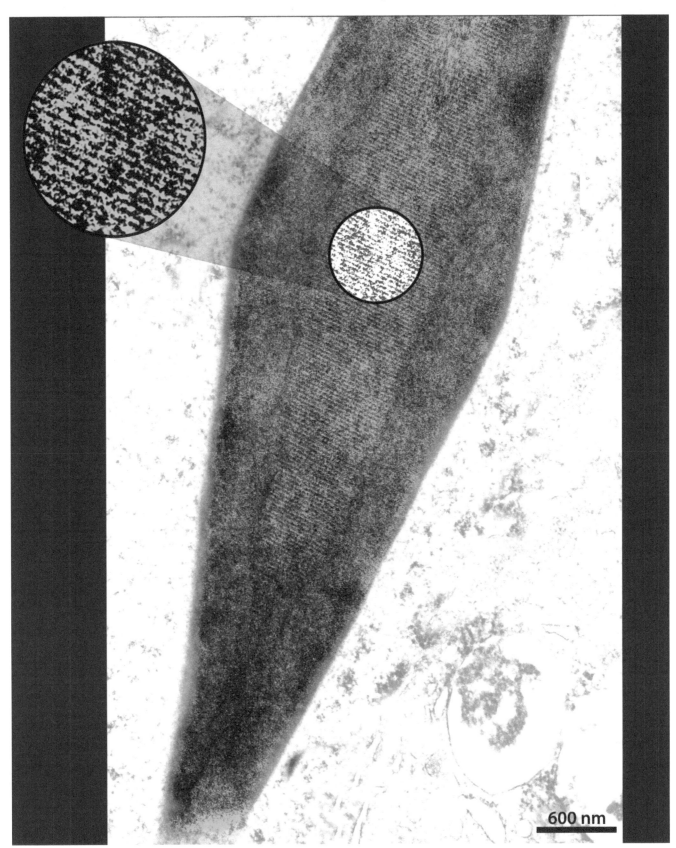

FIG. 8.92 High-magnification electron micrograph showing a large bipyramidal CLC *(pseudocolored in green)* found within a human eosinophil. CLCs are crystallized proteins. Note the regular repeating internal unit of the crystal structure while the periphery is homogeneous.

B. Eosinophils in human diseases

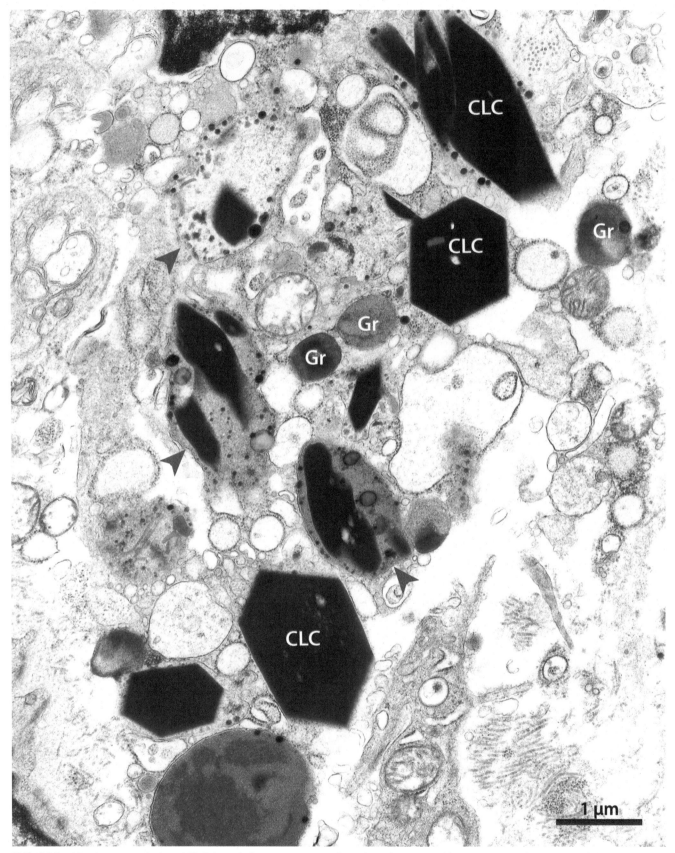

FIG. 8.93   Massive accumulation of CLCs in an inflammatory macrophage from the intestinal lamina propria of a patient with bacterial colitis. The cell is disrupted and shows CLCs in the process of digestion within phagosomes *(arrowheads)*, while others are free in the cytoplasm *(colored in yellow)*. Note eosinophil secretory granules (Gr) in the macrophage cytoplasm. The macrophage nucleus *(purple)* is partially seen. Biopsy prepared for TEM.

The cell biology of mouse eosinophils

9

Ultrastructure of mouse eosinophils

9.1 Overview

Mice models have contributed greatly to our understanding of the mechanisms underlying EADs and other human diseases. These models include not only wild-type (WT) mice but also mice with genetically manipulated numbers of eosinophils (increased or decreased) as well as mice with added, modified, or deleted specific genes involved in different aspects of the eosinophil biology (transgenic and knockout/knock-in mice).[347,359,489,490]

As expected, there are similarities and differences, both structural and functional, aside from unresolved issues, in comparing mouse and human eosinophils, which have been addressed in several reviews.[3,51,426] For example, differences are noted in the repertoire of surface receptors and granule contents. Moreover, in contrast to human eosinophils, which express large amounts of CLC/Gal-10 in their cytoplasm[37] (Chapter 5), mouse eosinophils lack a gene encoding this protein, consistent with the lack of CLC formation in mouse models of human diseases.[51] Crystals that may be found in mice are formed by Ym1, a 40 kD protein that expresses chitinase activity but is not derived from eosinophils.[491]

While eosinophilopoiesis is considered to differ in terms of the origin of their lineage-committed EoPs,[6] both mouse and human eosinophil myelocytes exhibit the definitive signature of the eosinophil lineage—the presence of crystalloid granules in their cytoplasm. However, there are open questions regarding the capacity of mouse eosinophils to store preformed cytokines in their specific granules and if signaling mechanisms of differential secretion are operating in eosinophils from this species, as documented for human eosinophils.[3] Furthermore, mouse eosinophils are able to degranulate through PMD,[492] but this ability is not reflected in their ultrastructure to the same degree as that observed for human eosinophils.

In this chapter, we detail the ultrastructural organization of mature and immature mouse eosinophils, highlighting similarities/differences between them and human eosinophils. We also discuss the ultrastructure of mouse eosinophils in response to activation conditions and in the context of representative diseases.

9.2 Mature mouse eosinophils: General morphology

The ultrastructural features of mouse eosinophils here described are based on electron microscopic studies of numerous samples obtained from different mouse strains and models in multiple organs, body fluids, and culture systems. Particularly, eosinophils from the following WT and mutant/transgenic (Tg) strains/models of mice are shown in this chapter: WT Balb/c, WT Swiss Webster, WT C57BL/6, WT C3HeB/FeJ, IL-5 Tg,[493–495] pallid [homozygous for the mutant gene pallid (pa/pa)],[496] deficient in two NFAT (nuclear factor of activated T cells) proteins—NFATp and NFAT4—(NFAT DKO),[497] IL-4 Tg,[498] and vascular permeability factor/vascular endothelial growth factor (VPF/VEGF) Tg.[499] Further details of studies with these varied mouse strains are found in the cited references.

In general, mature mouse eosinophils are slightly smaller in terms of diameter (9–12 μm) than human eosinophils (12–15 μm)[51] and show similar subcellular features but with subtle ultrastructural differences compared to the human.

Eosinophil Ultrastructure. https://doi.org/10.1016/B978-0-12-813755-0.00009-5

9.2.1 Nucleus

Mature mouse eosinophils observed in the bone marrow, peripheral blood, and tissues show ring-shaped, polymorphic nuclei with several constrictions at distinct levels of depth.[51,500] Thus, in thin sections, the nuclei of these cells appear with a diversity of morphologies due to the different section planes. These morphologies include the ring shape seen entirely through the plane section, with a small (Fig. 9.1) or large (Fig. 9.2) ring center, or partially (Figs. 9.3–9.7), and multiple (up to 5) nuclear compartments (multilobed) (Figs. 9.8–9.17). Ring-shaped nuclei are also observed in eosinophils from other rodents such as rats.[501]

Although the eosinophil nucleus of both humans and mice are polymorphic and can show several lobes, a bilobed nuclear appearance, as seen by light microscopy and TEM, is characteristically found and much more frequent in human eosinophils (see Chapter 2) than in mouse eosinophils. Therefore, it is not accurate to refer to mouse eosinophils as cells with bilobed nuclei because it is not a pattern for their nuclei.

A common ultrastructural feature of both mouse and human eosinophil nuclei is the condensed marginal (beneath the nuclear envelope) chromatin (heterochromatin) distinct from the remaining electron-lucent chromatin (euchromatin) (Fig. 9.11). However, we observed that the nuclei of murine eosinophils occupy, in general, more volume in the cytoplasm than those of human eosinophils do (~ 25% for humans and ~ 35% for mice). Nucleoli in both murine and human eosinophils are more prominent in immature cells than in mature cells and therefore not frequently seen in electron micrographs from mature eosinophils. The nuclear features of mouse eosinophils as seen in electron micrographs are summarized in Fig. 9.18.

9.2.2 Specific granules

Secretory (specific) granules of mouse eosinophils show typical ultrastructure with an electron-dense core surrounded by an electron-lucent matrix (Figs. 9.19 and 9.20). In thin sections, these granules appear as elliptical, or more rarely, circular profiles, and are limited by a well-defined unit membrane (Figs. 9.19 and 9.20).[502] Moreover, at times, the granule core appears split even in resting cells (Fig. 9.14). This appearance is distinct from that observed when granules are fused and show multiple cores within a membrane-bound large chamber.

Under the electron microscope, at very high magnifications (at least 125,000 ×), the crystalline nature of the granule core with its highly ordered structure is clearly observed in mouse eosinophils (Figs. 9.21 and 9.22). In earlier elegant work using TEM at high magnifications, Miller and colleagues described in detail the ultrastructure of the secretory granule core in eosinophils from varied locations (bone marrow, peripheral blood, peritoneal fluid, liver, and lamina propria of the intestinal mucosa) in humans and rodents (mice, rats, and guinea pigs).[502] In all species, the granule core consists of a very regular array of alternating and equidistant electron-dense and less dense bands, which usually run parallel to the long axis of the core from one end to the other. Less frequently, the granule bands can also run parallel to the short axis and occasionally form a square lattice by running in both directions.[502] In accordance with the prior study, we identified the bands of the crystalline cores paralleled to both the long (Fig. 9.21) and short (Fig. 9.22) core axis.

By comparing the specific granules of resting human and mouse eosinophils in numerous electron micrographs, we recognized that, in mice, the granules appear more elliptical/elongated and contain more uniform cores in terms of shape compared with human eosinophils (Fig. 9.23). We agree with Miller and colleagues[502] who found more varied shapes of granule cores in human eosinophils than in rodents.[502] Our ultrastructural analyses also detected at times the presence of membranous structures within granules when the granules were observed at very high magnifications (Figs. 9.21 and 9.22), thus indicating that specific granules from mouse eosinophils can internally be complex, compartmentalized organelles as seen in humans.[20] Future studies are needed to better understand the structural organization of the specific granules in mouse eosinophils. The phenomenon of reversal of the core "staining" exhibited by specific granules in human eosinophils when variable staining methods are used, as discussed in Chapter 2, can also occur in mice eosinophils (Fig. 9.24). It is common to see in the literature EM images of mouse eosinophils showing in the cytoplasm the entire population of specific granules depicting reversed density.

9.2.3 Other organelles and vesicular system

Mouse eosinophils contain lipid bodies (LBs), which show the same dark osmiophilic aspect (Figs. 9.12 and 9.15) seen in human eosinophils (Fig. 9.23). However, LBs in mice eosinophils appear in electron micrographs at less frequency and with less variable size than in humans in both resting and activated cells. Moreover, it is not established yet if mouse LBs have internal membranes, as identified in LBs from human eosinophils.[94] Endosomes, both early and late (MVBs) with characteristic ultrastructural features of human organelles (Chapter 2), are also seen in the cytoplasm of mouse eosinophils (Fig. 9.11).

Mouse Eosinophil

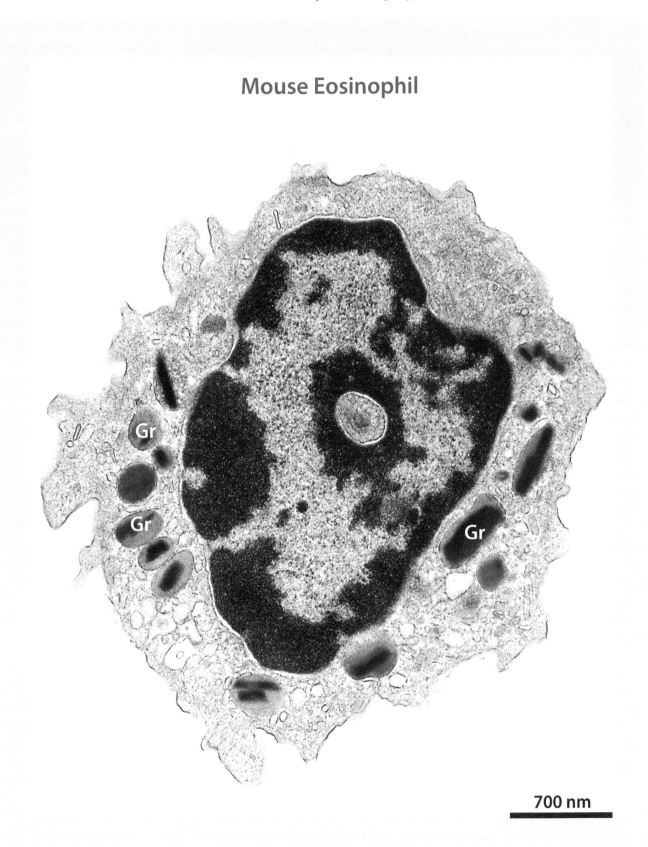

700 nm

FIG. 9.1   Ultrastructure of a mature mouse eosinophil isolated from the spleen. Note the ring-shaped morphology of the nucleus with a small ring center and secretory granules (Gr) in the cytoplasm. Eosinophils from an IL-5 transgenic mouse, which shows eosinophilia in the spleen,[493] were prepared for TEM.

C. The cell biology of mouse eosinophils

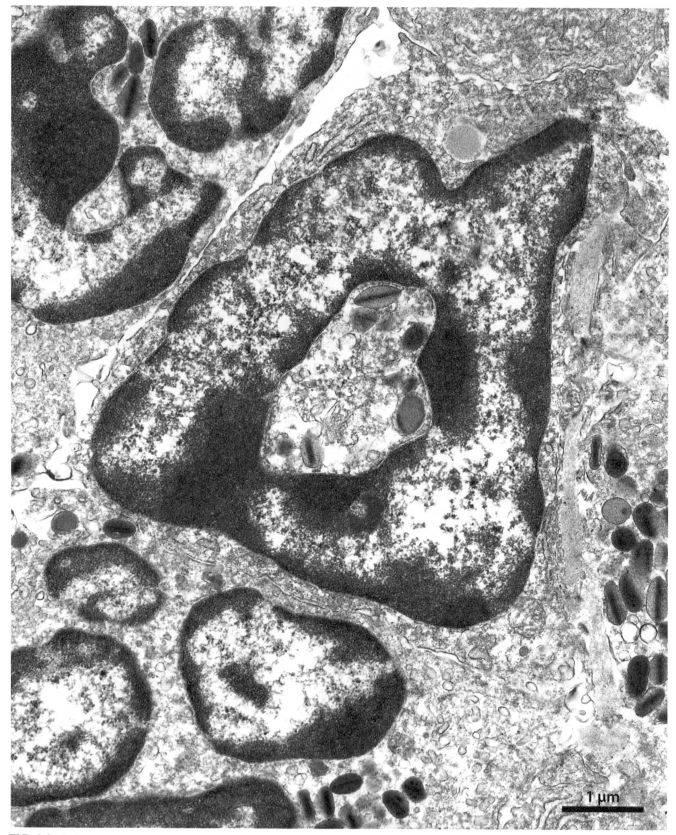

FIG. 9.2 Inflammatory eosinophils infiltrated in the liver of a mouse infected with *Schistosoma mansoni*. One eosinophil shows the ring-shaped nuclear morphology *(purple)* with a large, irregular ring center. Specific granules were colored in *yellow* and the cytoplasm in *pink*. Liver samples from WT Swiss mice, infected as described,[352] were prepared for TEM.

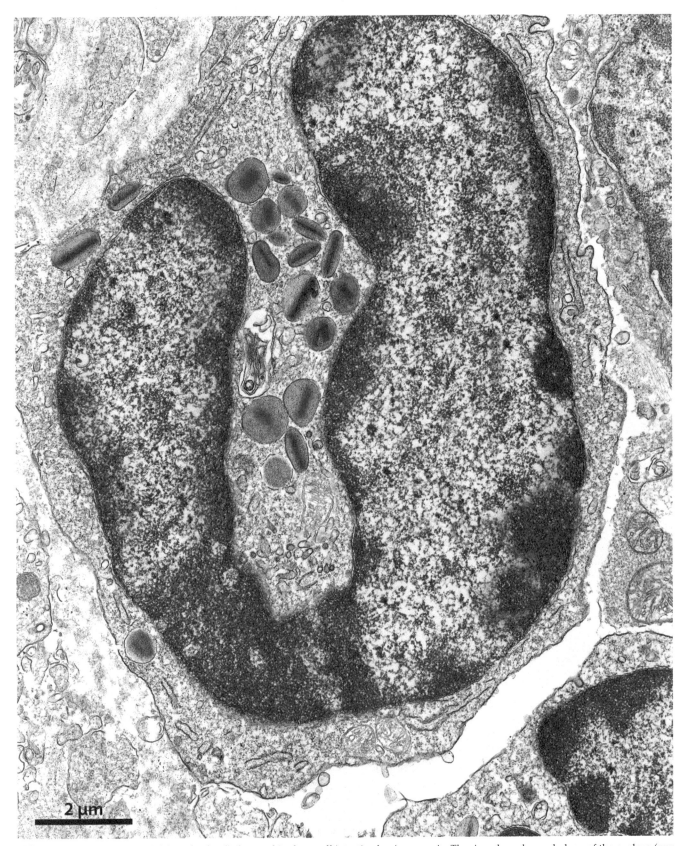

FIG. 9.3   A mouse eosinophil *(pseudocolored)* observed in the small intestine lamina propria. The ring-shaped morphology of the nucleus *(purple)* is partially observed. Note a few mitochondria *(green)*. Specific granules *(yellow)*; cytoplasm *(pink)*. Sample from a WT Swiss Webster mouse prepared for TEM.

C. The cell biology of mouse eosinophils

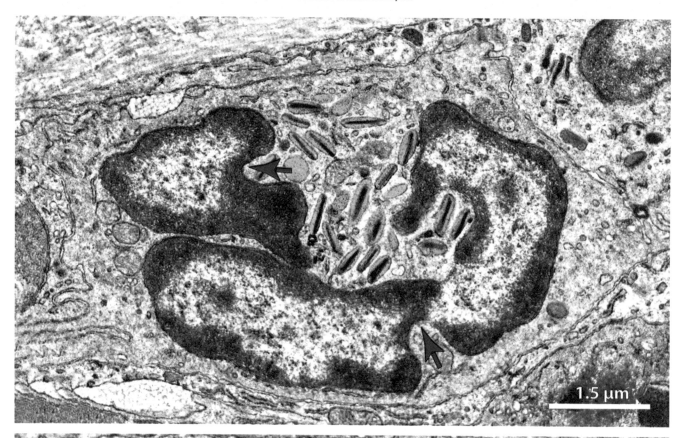

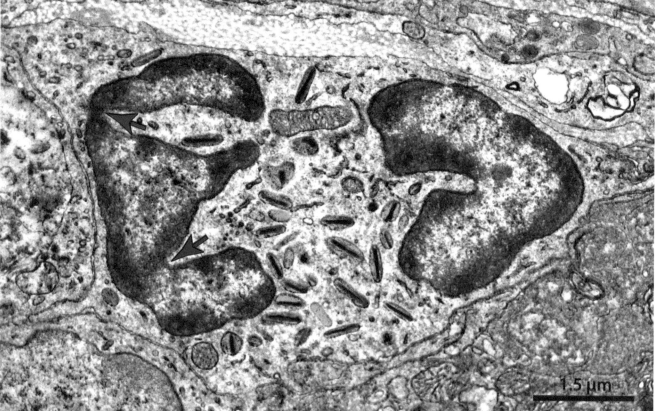

FIG. 9.4   Electron micrograph showing mouse eosinophils in the small intestine lamina propria. Note the polymorphic nuclei *(colored in purple)* with constrictions at different degrees of depth *(arrows)*. Specific granules were colored in *yellow* and the cytoplasm in *pink*. Samples from WT Swiss Webster mice were prepared for TEM.

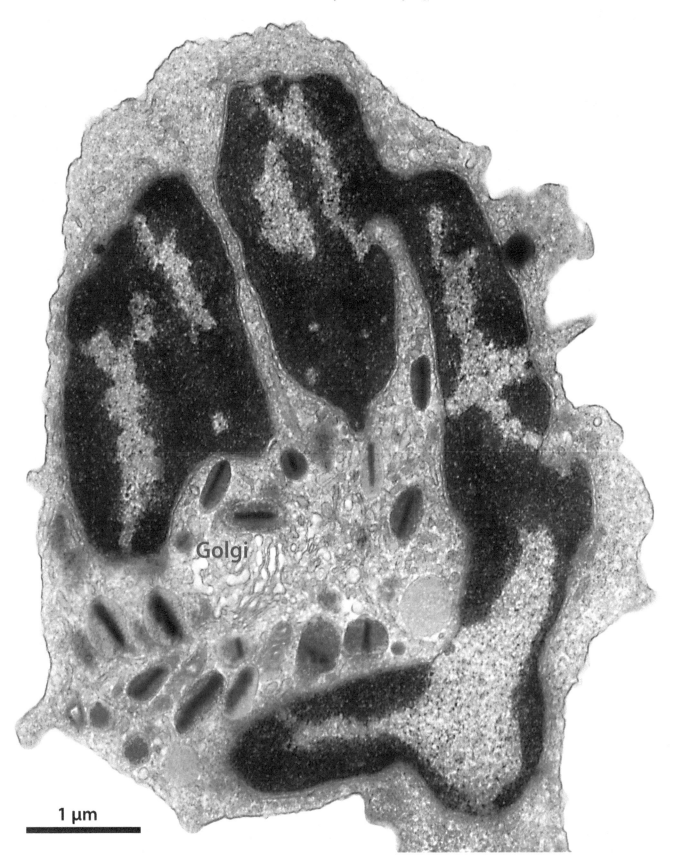

Golgi

1 μm

FIG. 9.5   A mouse eosinophil *(pseudocolored)* isolated from the spleen. The ring-shaped morphology of the nucleus *(purple)* is partially seen. Note specific granules *(yellow)* and the Golgi area. Eosinophils from an IL-5 Tg mouse, which shows eosinophilia in the spleen,[493] were prepared for TEM.

C. The cell biology of mouse eosinophils

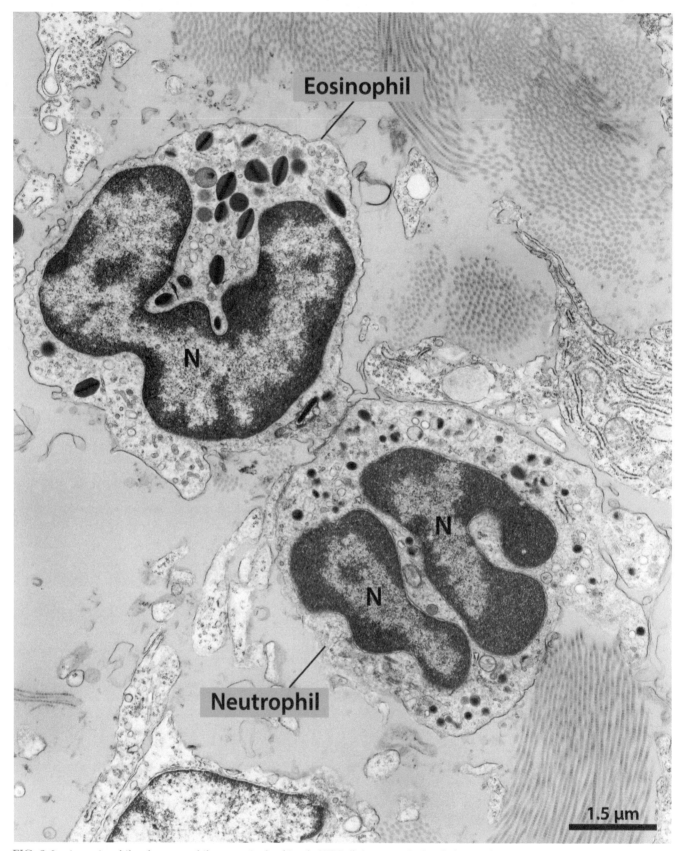

FIG. 9.6   An eosinophil and a neutrophil are seen in the skin of a WT Balb/c mouse. Both cells have polymorphic nuclei (N). Eosinophil shows the typical crystalloid granules while the neutrophil shows elongated or round small dense granules. ECM was colored in *green*. Samples prepared for TEM.

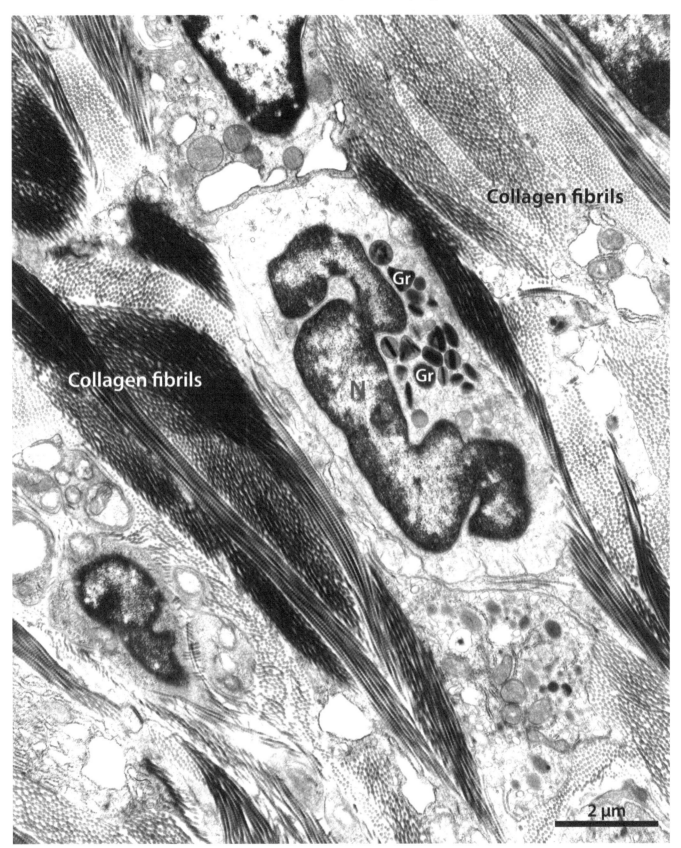

Collagen fibrils

Collagen fibrils

Gr

Gr

N

2 μm

FIG. 9.7 Ultrastructure of a dermal eosinophil in the ear of a mutant mouse showing the typical polymorphic nucleus (N) and crystalloid granules (Gr). Note the high amount of collagen fibers in the dermis. Samples from pallid (*pa/pa*) mice[496] were prepared for TEM.

C. The cell biology of mouse eosinophils

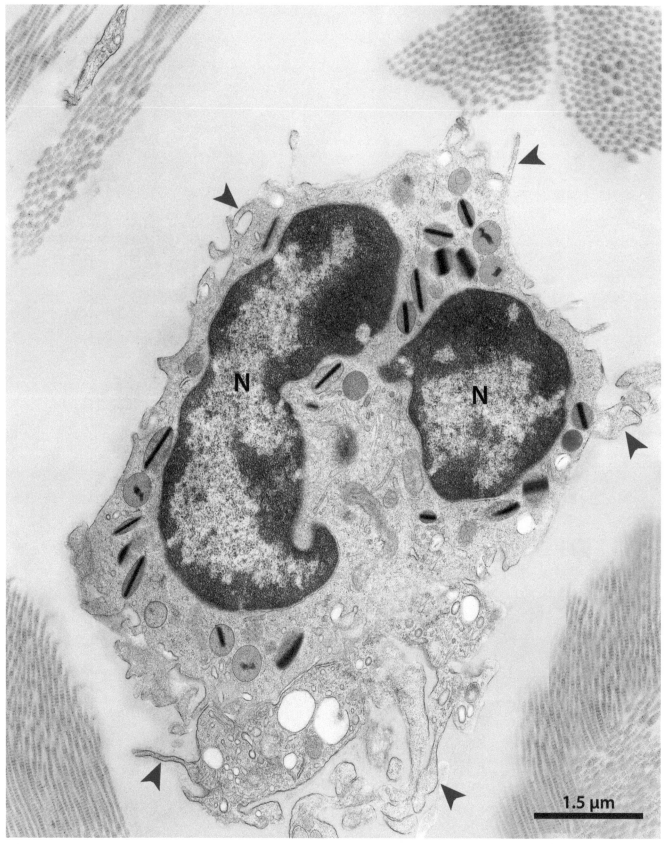

FIG. 9.8 Tissue mouse eosinophil in the skin of a WT Balb/c. Note specific granules with typical electron-dense cores, polymorphic nucleus (N), and blunt, irregular processes at the cell surface *(arrowheads)*. ECM colored in *green*. Samples prepared for TEM.

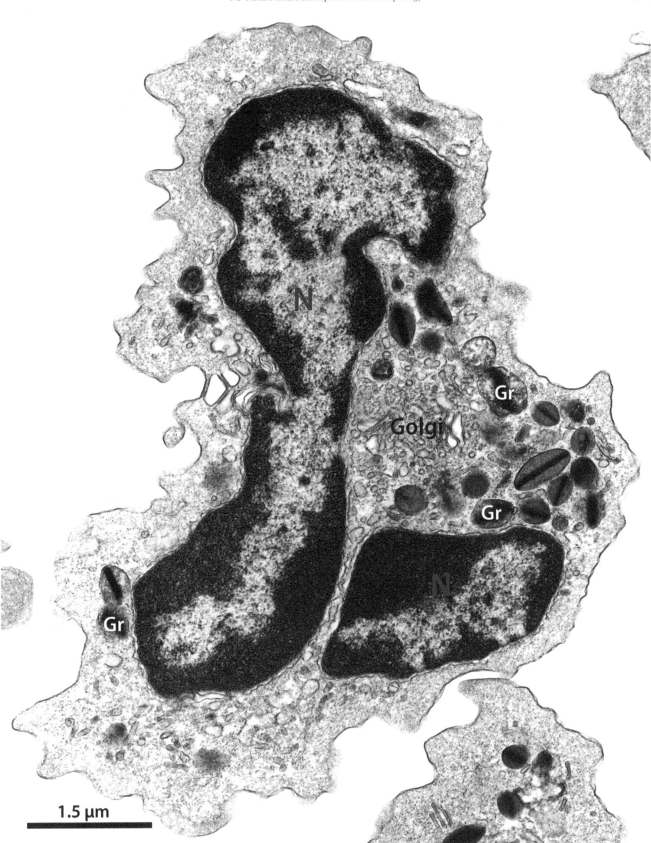

FIG. 9.9   Ultrastructure of a mouse eosinophil isolated from the lung. Secretory granules (Gr), a large polymorphic nucleus (N), and the Golgi region are observed. Eosinophils from WT Balb/c mice were prepared for TEM.

C. The cell biology of mouse eosinophils

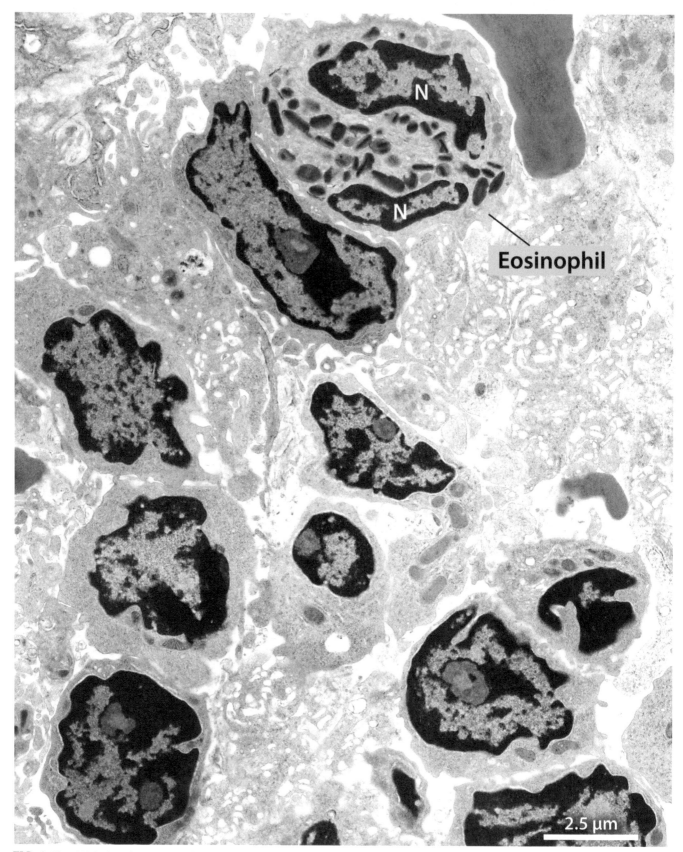

FIG. 9.10   TEM of a spleen section showing a mature eosinophil and many lymphocytes. Note the polymorphic eosinophil nucleus (N) while lymphocytes show monolobed nuclei *(colored in purple)* occupying most parts of the cytoplasm *(yellow)*. Red blood cells *(red)*. Sample from a C57BL/6 mouse.

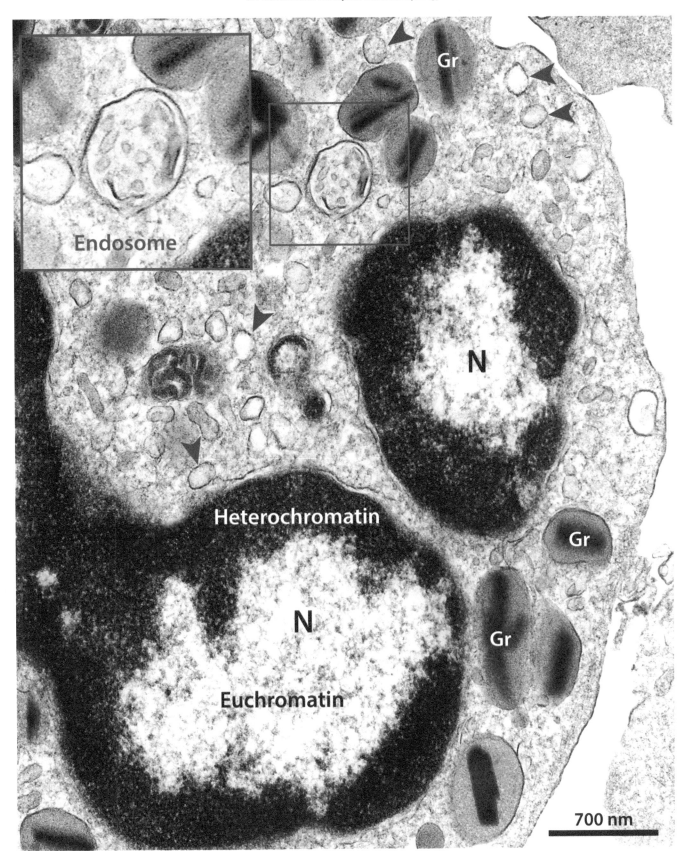

FIG. 9.11 A mature mouse eosinophil isolated from the spleen. A late endosome with several luminal vesicles, cytoplasmic vesicles *(arrowheads)*, and two nuclear (N) compartments are observed. Note the marginal heterochromatin (electron-dense) and the more central euchromatin (electron-lucent). Eosinophils from an IL-5 Tg mouse, which shows eosinophilia in the spleen,[493] were prepared for TEM. *Gr*, specific granules.

C. The cell biology of mouse eosinophils

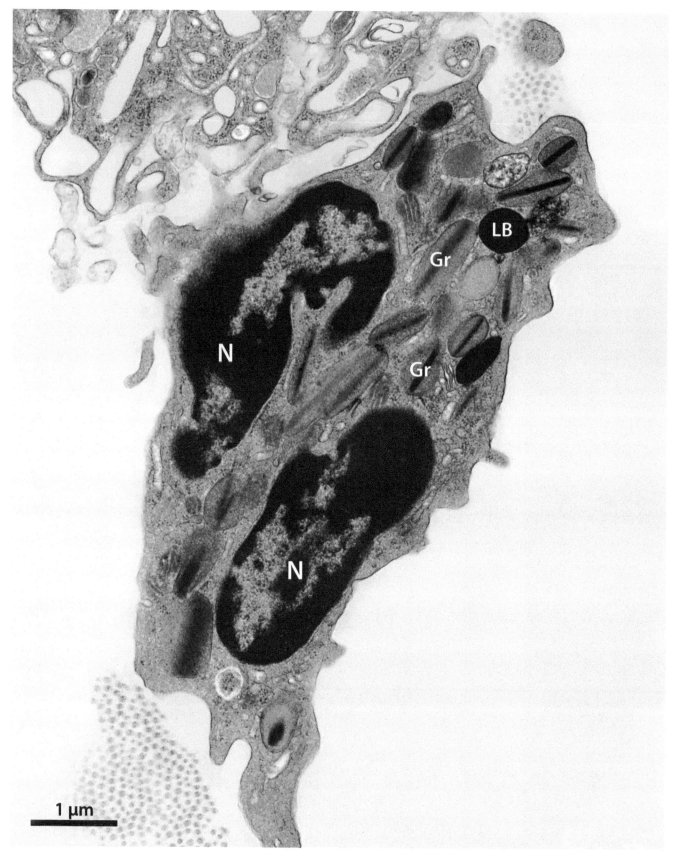

FIG. 9.12   A dermal mature eosinophil of a mutant VPF/VEGF Tg mouse. The nucleus (N) appears with two lobes while specific granules (Gr) and a LB are seen in the cytoplasm. ECM colored in *green*. Tail samples from VPF/VEGF Tg mice[499] were prepared for TEM.

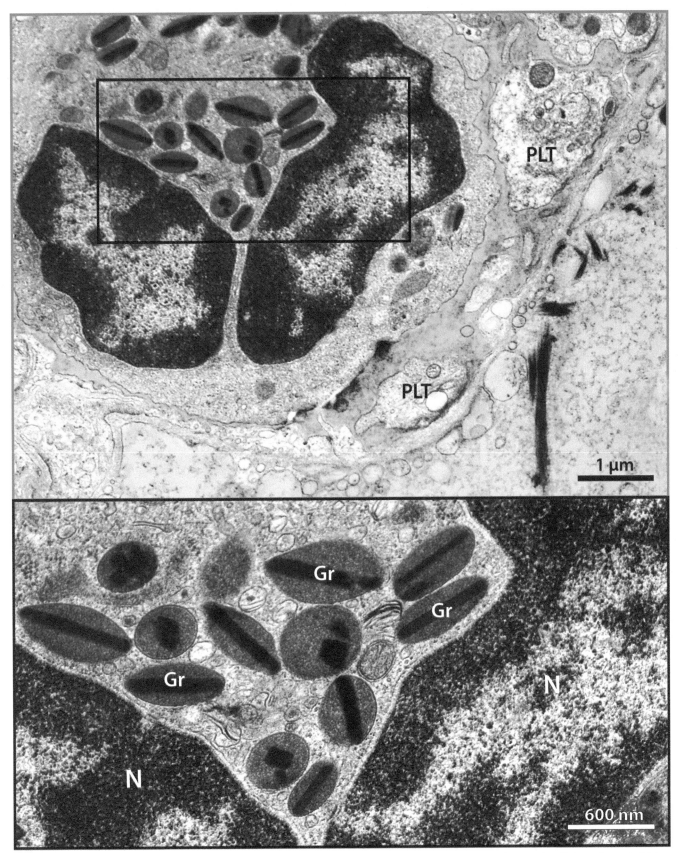

FIG. 9.13   A mature eosinophil within a small blood vessel in the lung. Two nuclear (N) lobes and specific granules (Gr) are shown in higher magnification (boxed area). Note platelets (PLT) as part of the blood. Vessel interior *(colored in pink)*; Endothelium *(yellow)*; ECM *(green)*. Samples from IL-4 Tg mice prepared for TEM.[498]

C. The cell biology of mouse eosinophils

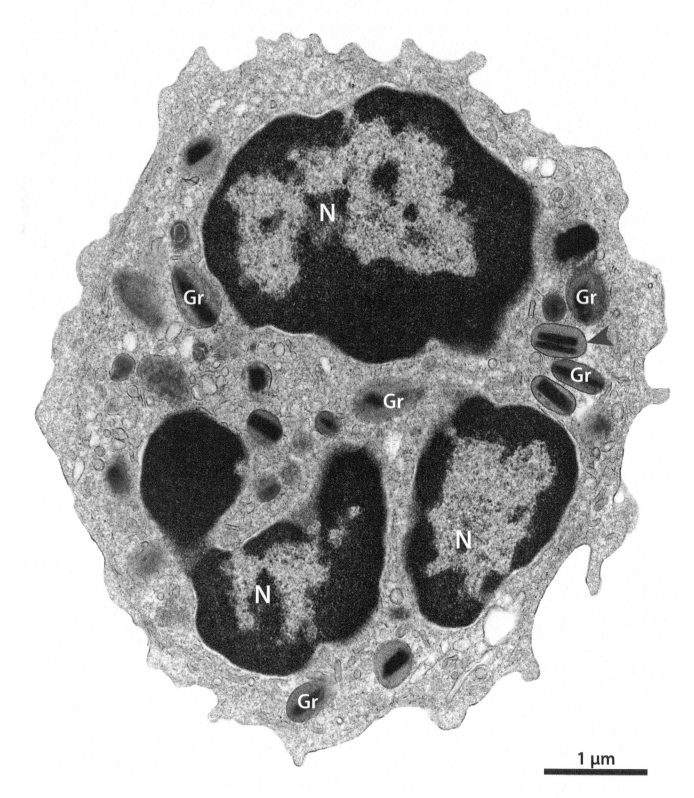

FIG. 9.14   A mature mouse eosinophil isolated from the spleen. Three nuclear lobes (N) and specific granules (Gr) are observed. One granule shows a split core *(arrowhead)*. Eosinophils from an IL-5 Tg mouse, which shows eosinophilia in the spleen,[493] were prepared for TEM.

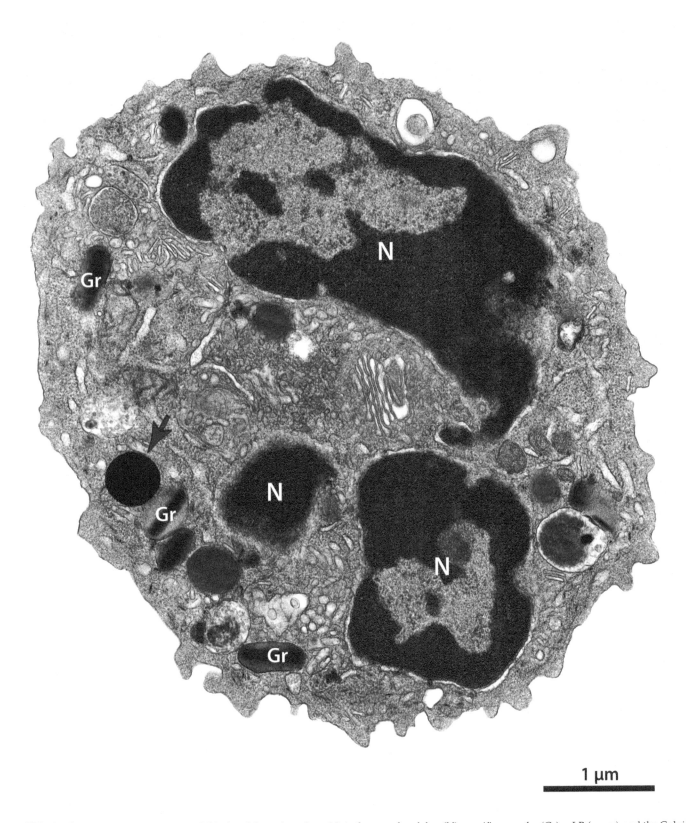

FIG. 9.15   A mature mouse eosinophil isolated from the spleen. Note three nuclear lobes (N), specific granules (Gr), a LB *(arrow)*, and the Golgi complex *(highlighted in yellow)*. Eosinophils from an IL-5 Tg mouse, which shows eosinophilia in the spleen,[493] were prepared for TEM.

C. The cell biology of mouse eosinophils

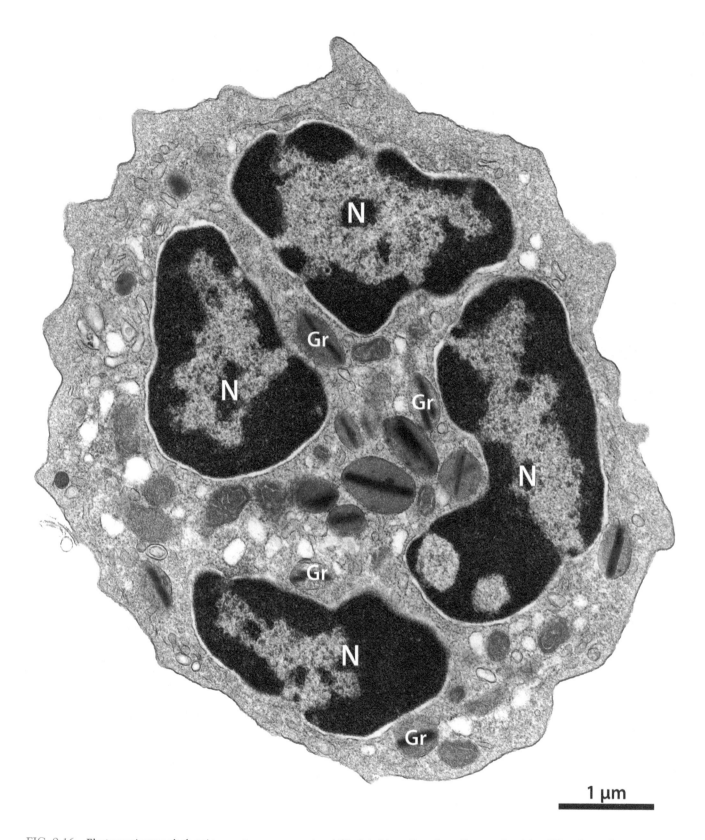

FIG. 9.16   Electron micrograph showing a mature mouse eosinophil isolated from the spleen. Four nuclear lobes (N) and specific granules (Gr) are observed. Eosinophils from an IL-5 Tg mouse, which shows eosinophilia in the spleen,[493] were prepared for TEM.

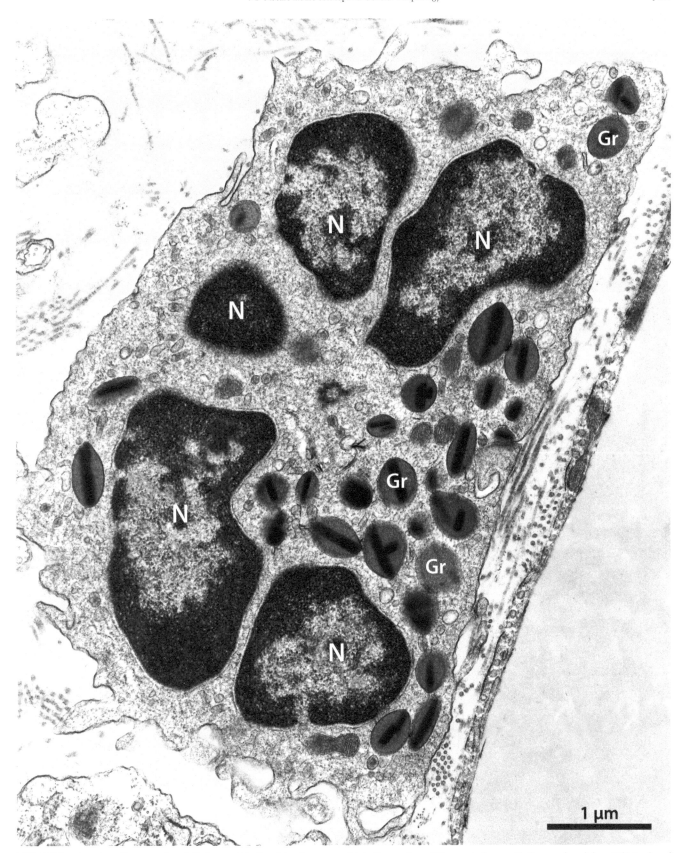

FIG. 9.17 A mature tissue eosinophil observed in the skin of a mouse. Note the nucleus (N) morphology with five nuclear lobes. ECM colored in *green*. Samples from a WT C3HeB/FeJ mouse was prepared for TEM. *Gr*, specific granules.

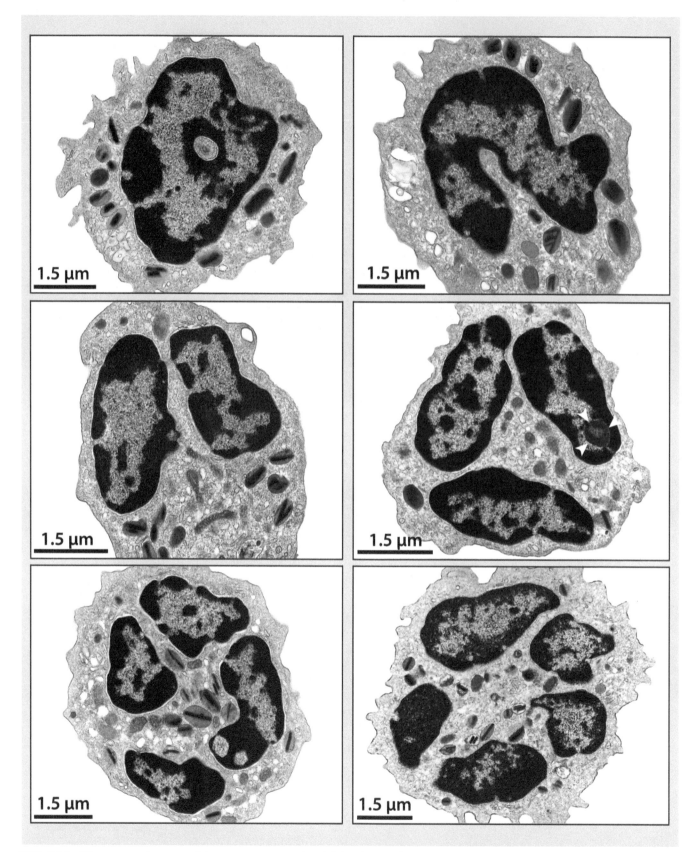

FIG. 9.18   Ultrastructural aspects of the mature mouse eosinophil. Nuclei *(purple)* are polymorphic and appear with different morphologies (from ring-shaped up to five nuclear lobes) depending on the section plane. The nucleolus *(arrowheads)* is seen in one of the nuclei. Eosinophils isolated from the spleen or lung were prepared for TEM.

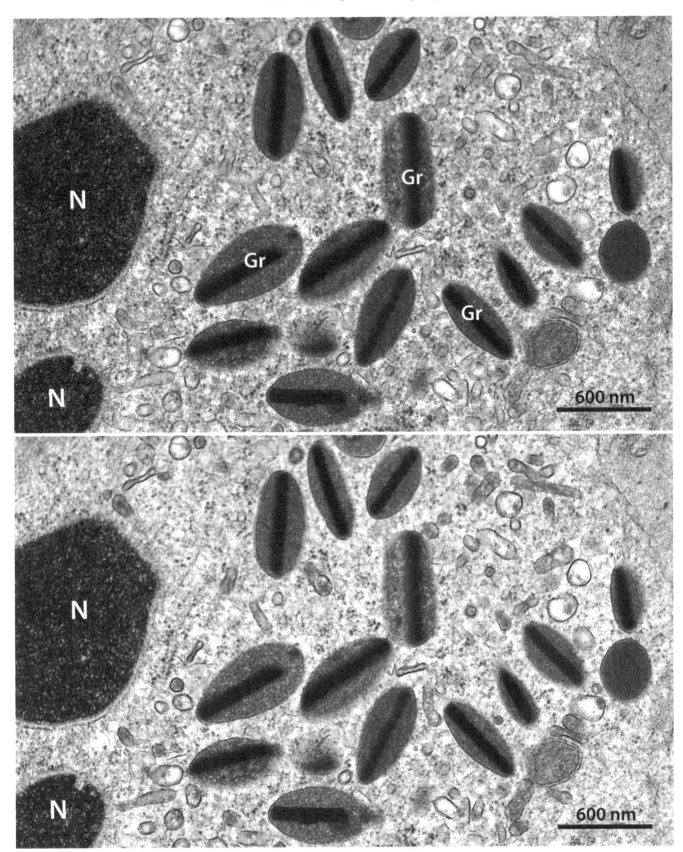

FIG. 9.19 Part of the cytoplasm of a mouse mature eosinophil showing secretory (specific) granules (Gr) and vesiculotubular structures. Vesicles were highlighted in *pink* in the *bottom panel*. A spleen fragment from a WT C57BL/6 mouse was prepared for TEM. *N*, nucleus.

C. The cell biology of mouse eosinophils

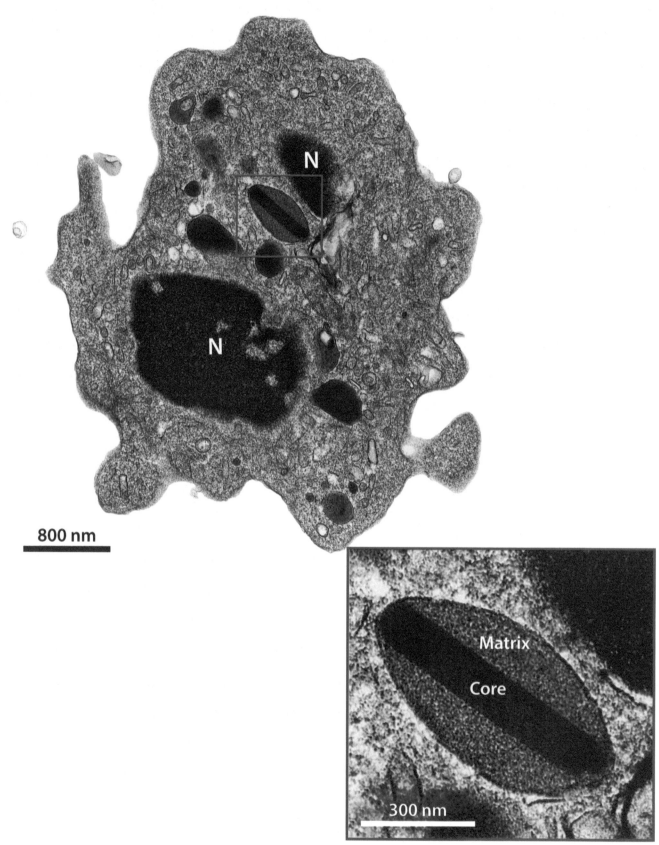

FIG. 9.20   Ultrastructure of the specific granule in the cytoplasm of a mouse tissue eosinophil. The electron-dense core appears with regular morphology surrounded by a less dense matrix, all delimited by a membrane. Eosinophils isolated from an IL-5 Tg mouse were prepared for TEM. *N*, nucleus.

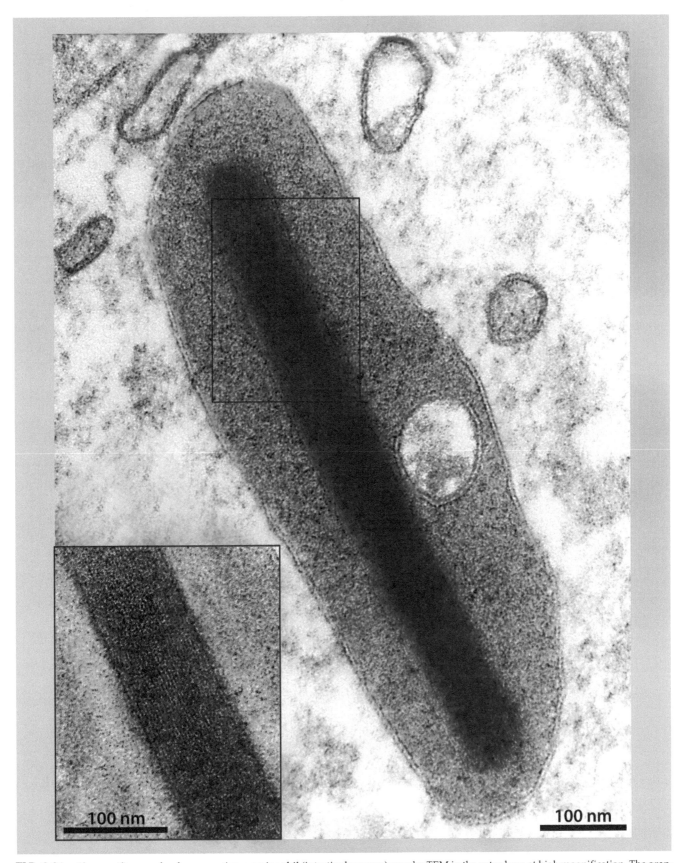

FIG. 9.21 The specific granule of a mouse tissue eosinophil (intestinal mucosa) seen by TEM in the cytoplasm at high magnification. The granule core shows a regular longitudinal array of dense bands *(boxed area)*. The same structure of the bilayer membrane delimiting the granule is seen in an intragranular compartment *(purple)* and surrounding cytoplasmic vesicles *(pink)*.

C. The cell biology of mouse eosinophils

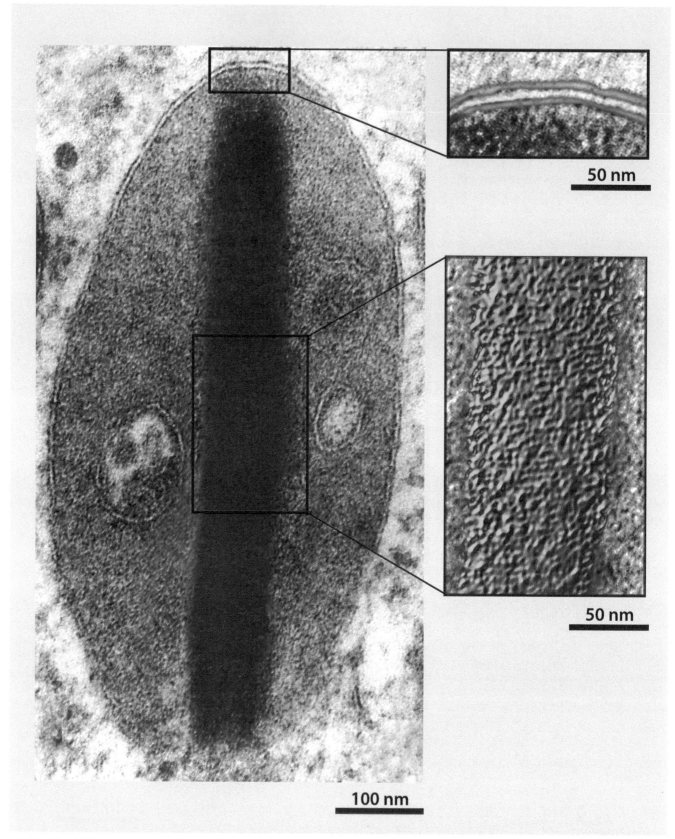

FIG. 9.22   The specific granule of a mouse tissue eosinophil seen at high magnification. The dense bands of the granule core *(highlighted in yellow)* run parallel to the short axis. Note the trilaminar aspect of the limiting granule membrane *(boxed area)* also seen in intragranular membranous compartments. Eosinophil observed in the intestinal mucosa of a Swiss mouse by TEM.

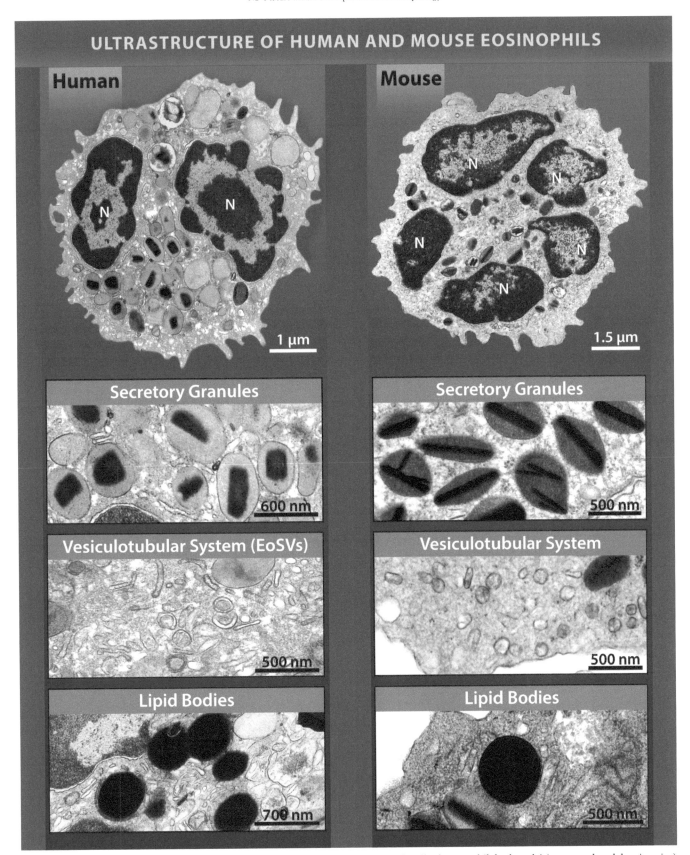

ULTRASTRUCTURE OF HUMAN AND MOUSE EOSINOPHILS

Human

Mouse

1 µm

1.5 µm

Secretory Granules

Secretory Granules

600 nm

500 nm

Vesiculotubular System (EoSVs)

Vesiculotubular System

500 nm

500 nm

Lipid Bodies

Lipid Bodies

700 nm

500 nm

FIG. 9.23 Ultrastructural aspects of human and mouse mature eosinophils. Both cells show multilobed nuclei (more nuclear lobes in mice), variable numbers of surface folds, crystalloid granules (more elongated and with more regular cores in mice), a vesiculotubular system, and osmiophilic electron-dense lipid bodies (more common in human eosinophils). Typical EoSVs are seen only in human eosinophils but an analogous system with large vesicles is also found in mice.

C. The cell biology of mouse eosinophils

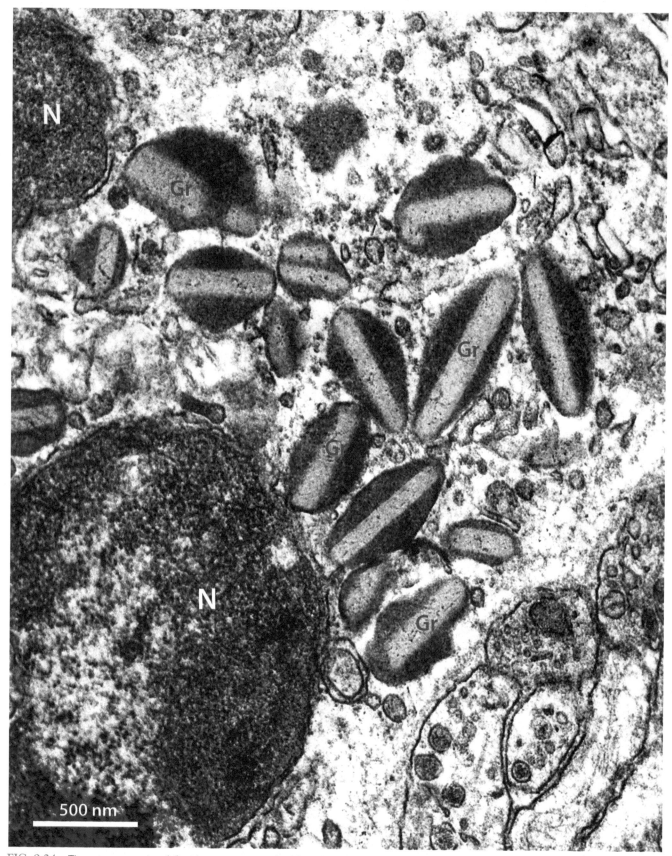

FIG. 9.24   Tissue mouse eosinophil with secretory granules (Gr) exhibiting reversed electron density. The central cores appear less dense compared with the granule matrices. Intestinal sample from a WT Swiss mouse prepared for TEM. *N*, nucleus.

Mature mouse eosinophils contain in the cytoplasm a vesiculotubular system composed of large round vesicles (70–200 nm in diameter) and tubular profiles[40] (Figs. 9.11, 9.19, and 9.21), which respond to cell activation with the amplified formation of its elements (addressed in Sections 9.4 and 9.5). This system is analogous to the human EoSVs, which are recognized markers of eosinophil activation.[18,21] However, the mouse vesiculotubular system does not clearly show the particular morphology of EoSVs and, therefore, cannot be considered as an eosinophil lineage marker as documented for humans.[21] Whether this particular system in mice can originate from the specific granules and carry cytokines, as identified for human eosinophils,[18,21] remains to be elucidated. On the other hand, a vesicular transport of MBP-1 in mouse eosinophils was already identified.[352]

Mitochondria, RER, and Golgi follow in mice the same pattern observed in human eosinophils regarding quantities and cytoplasmic distribution; i.e., mature mouse eosinophils contain low amounts of mitochondria (Fig. 9.3) and peripheral RER. The Golgi complex appears occasionally in electron micrographs (Figs. 9.5, 9.9, and 9.15). In mature mouse eosinophils, as in humans, RER is basically detected around the nucleus as part of the nuclear envelope. Moreover, other features of mature mice eosinophils are shared to some extent with mature human eosinophils. The plasma membrane has variable numbers of short folds and processes (Figs. 9.8 and 9.23) and can exhibit filopodia and uropods in response to cell activation, as shown in Section 9.4.

9.3 Immature mouse eosinophils

In the bone marrow of mice, lineage-committed EoPs are derived from granulocyte/macrophage progenitors while human eosinophils, as addressed in Chapter 7, differentiate from common myeloid progenitors.[6] During the process of eosinophil maturation, lineage-committed EoPs undergo a series of remarkable morphological changes in both humans and mice. In addition to the process of granulogenesis encompassing events of content filling, condensation, and crystallization, which culminate with the formation of mature secretory granules (specific granules), an assortment of morphological alterations is observed. These include reduction in cell size, nuclear lobulation, progressive condensation of the nuclear chromatin, nucleolar diminution, and significant reduction in the amount of the ER, Golgi, and mitochondria (Fig. 9.25).

As discussed for human immature eosinophils (Chapter 7), the sequence of morphological events that leads to a mature eosinophil has been classically divided into a number of stages. However, considering that eosinophilopoiesis is a continuum of gradual morphologic changes and that EM images represent cell sections, i.e., part of the cell cytoplasm, we consider that the terms early, middle, late, and mature eosinophils are more appropriate to indicate their stages of development when seen under the electron microscope (Fig. 9.25). These stages have also been referred to as early, middle, late, and mature eosinophilic myelocytes.[40,503] It is also important to emphasize that our understanding of eosinophil granulogenesis in both humans (Chapter 7)[18] and mice involves the formation of coreless granules (immature specific granules), which subsequently undergo crystallization, becoming cored mature granules (specific granules). Thus, as in human eosinophils, mouse eosinophils bear a single granule population.

We have been studying the ultrastructure of mouse eosinophils in cultures from bone marrow-derived cells or directly in bone marrow specimens. Fig. 9.26 shows in low magnification a representative EM view of an ex vivo culture system that generates large numbers of eosinophils at high purity from unselected mouse bone marrow progenitors.[504] Eosinophils are observed at varying stages of development with a mixture of morphologically distinct granules in the cytoplasm.

The earliest ultrastructural feature that characterizes cells of the eosinophilic lineage, thus distinguishing them from a common precursor, is the presence of large cytoplasmic granules.[360] Developing eosinophils containing predominantly large granules in the process of content accumulation and a round monolobed nucleus with prominent nucleolus are in an early stage of development (Fig. 9.27). These initial immature granules appear partially filled with variable amounts of material moderately granular and/or electron-dense accompanied or not by intragranular vesicles (Figs. 9.28–9.31). During maturation, immature specific granules, both in humans and rodents, undergo progressive condensation and crystallization.[18]

In cultures of mouse eosinophils, many cells are observed at the middle stage of development characterized by the presence of a segmented nucleus and round, electron-dense granules, full of contents in the cytoplasm (Figs. 9.28–9.31). These electron-dense granules are frequently interspaced with granules in earlier steps of maturation with heterogeneous content and also round granules showing initial events of condensation/crystallization. Thus, mouse eosinophils in the middle stage of development contain a mixture of large granules with distinct appearances, but not typically cored granules (Figs. 9.28–9.31). Of note, all granules are delimited by a true membrane and should not be misunderstood with the osmiophilic LBs that, as a rule, are not membrane-bound organelles (Chapter 2).

The presence of elliptical cored granules in the cytoplasm indicates a late stage of eosinophil development (Fig. 9.32), which progresses to mature eosinophils (Fig. 9.33). The number of eosinophils exhibiting a profile of

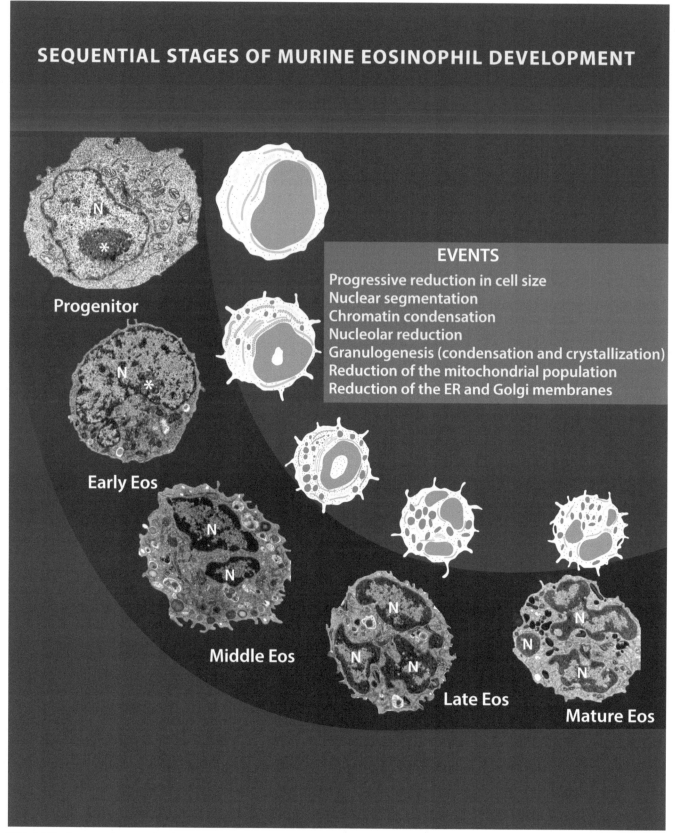

FIG. 9.25 Sequential stages of eosinophil development and morphological events in mice. In the early stage of differentiation from a progenitor cell, eosinophils are characterized by a large, eccentric, and round nucleus with a prominent nucleolus (*), which undergo progressive lobulation. Nuclear changes occur in parallel with reduction in cell size and in the amount of mitochondria, endoplasmic reticulum (ER), and Golgi membranes. Granulogenesis involves content filling, condensation, and crystallization.

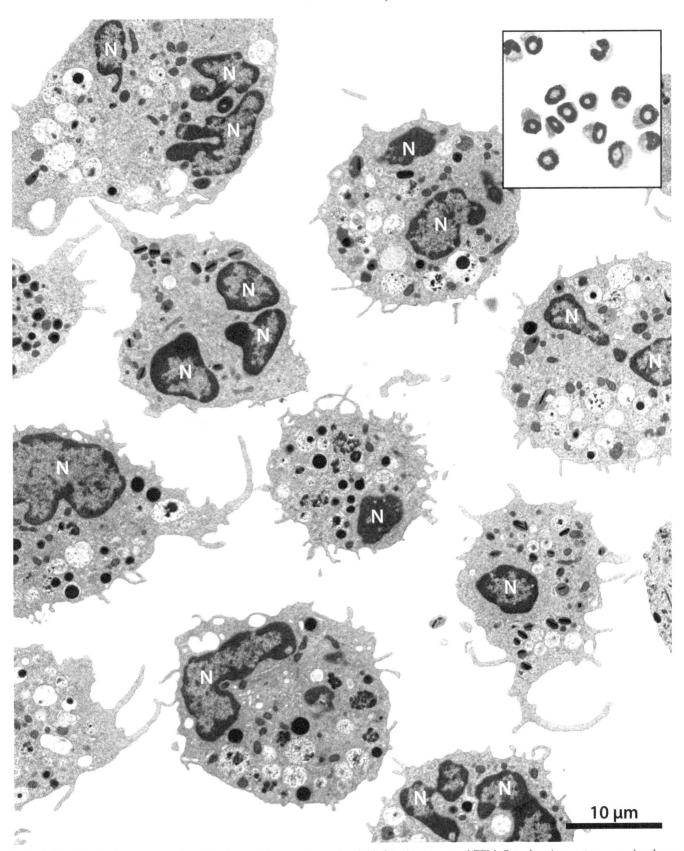

FIG. 9.26   Developing mouse eosinophils observed in a culture system by light microscopy and TEM. Cytoplasmic secretory granules show different degrees of maturation. The ring-shaped, polymorphic nuclei (N) are noted at different section planes by TEM. Cultures were prepared from normal mouse bone marrow as described.[504,524] *Electron micrograph courtesy of Amali Samarasinghe.*

C. The cell biology of mouse eosinophils

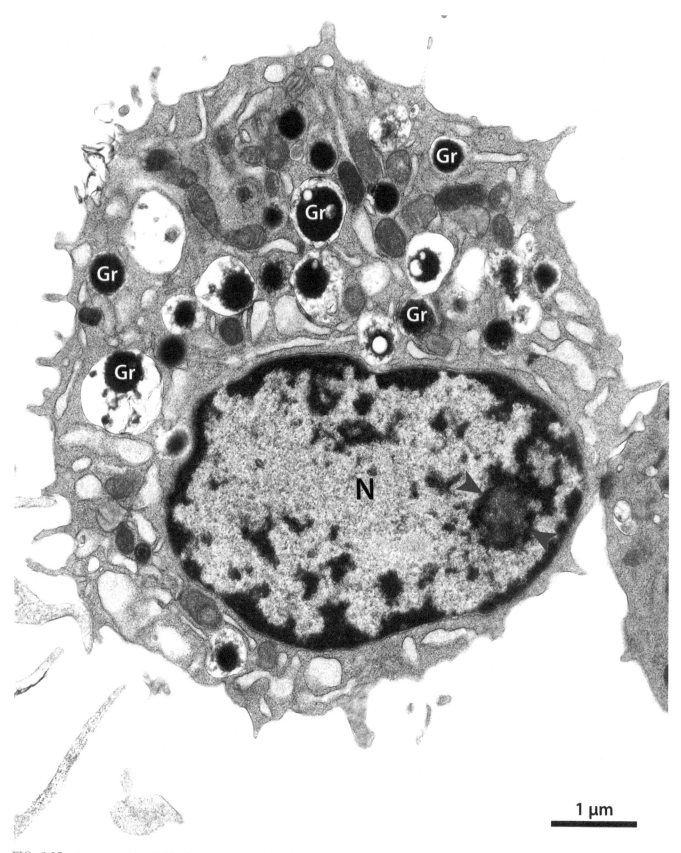

FIG. 9.27   A mouse eosinophil in the early stage of development. Heterogeneous materials are being accumulated and condensed within immature specific granules (Gr). Note the large monolobed and eccentric nucleus (N) with a prominent nucleolus *(arrowheads)*. Bone marrow-derived cells were cultured and prepared for TEM.

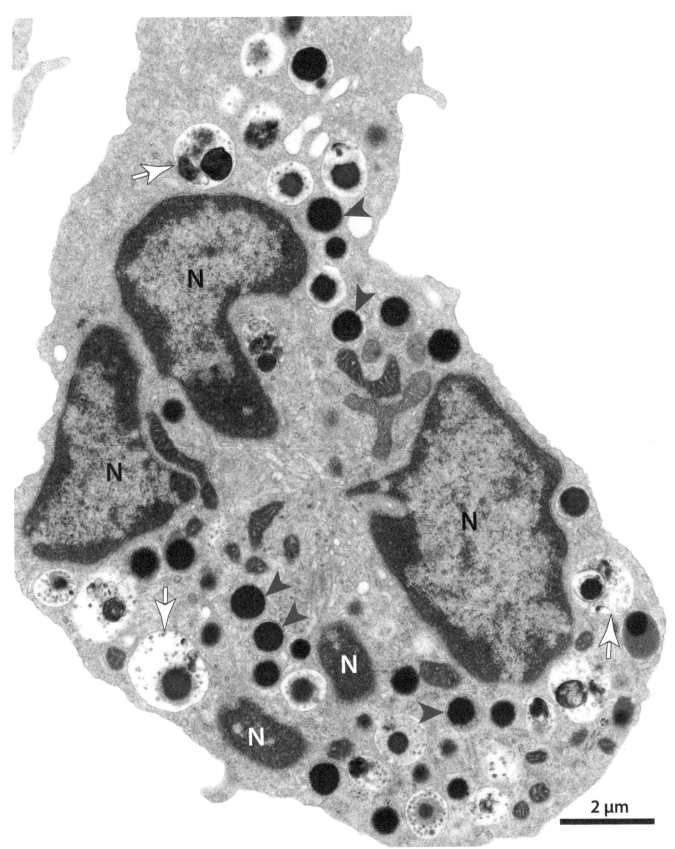

FIG. 9.28   An immature mouse eosinophil in the middle stage of development shows most granules full of electron-dense contents *(arrowheads)*. Some large granules are partially filled with tiny vesicles and heterogeneous content *(arrows)*. Note the segmented nucleus (N). Bone marrow-derived cells were cultured and prepared for TEM.[513]

C. The cell biology of mouse eosinophils

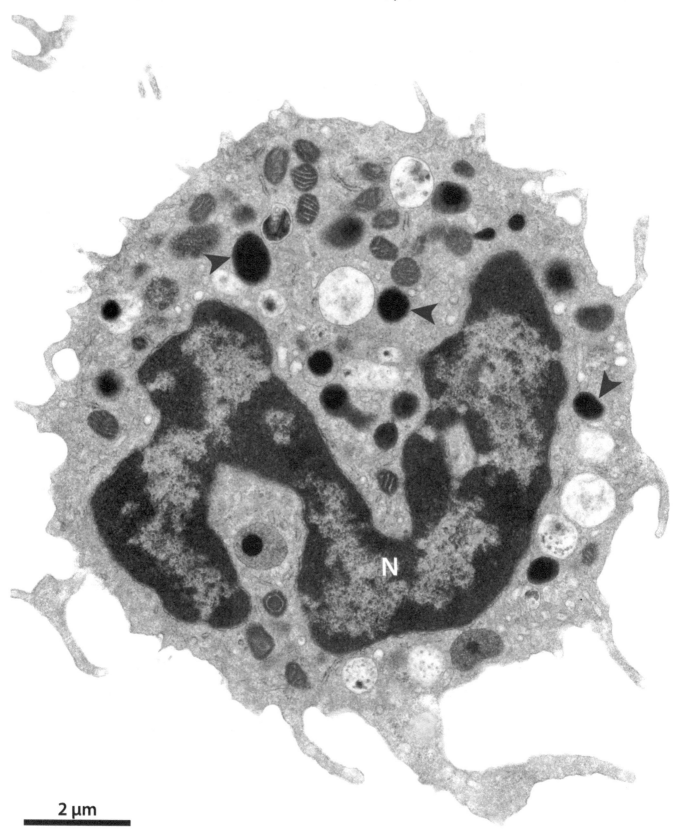

2 µm

FIG. 9.29   Ultrastructure of an immature mouse eosinophil in the middle stage of development. Electron-dense immature specific granules *(arrowheads)* and a segmented nucleus (N), typical of this stage, are observed. The cell surface shows many projections *(highlighted in yellow)*, as seen in leukocytes. Bone marrow-derived cells were cultured and prepared for TEM.[513]

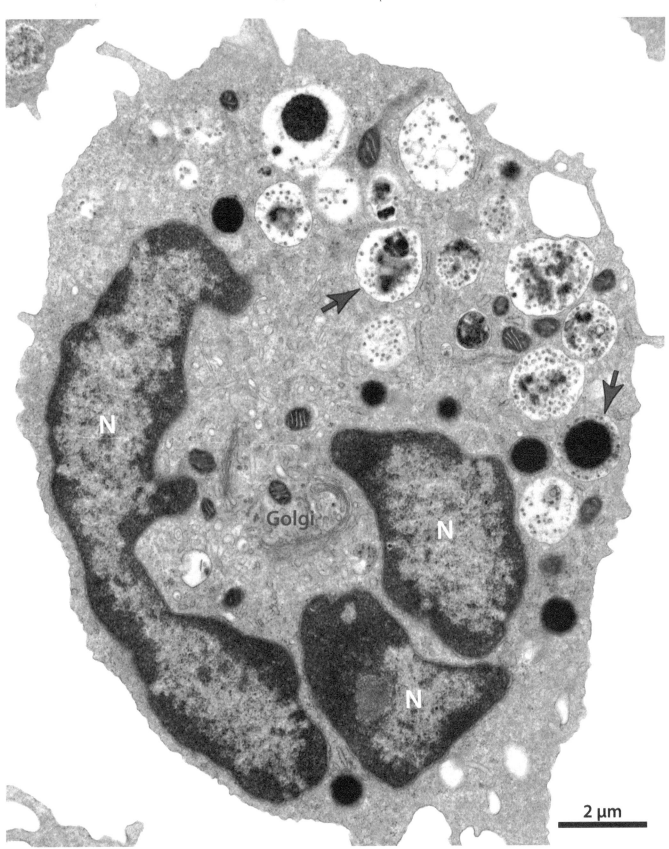

FIG. 9.30 Ultrastructure of an immature mouse eosinophil in the middle stage of development. *Arrows* indicate two granules with different degrees of vesicle deposition and content condensation, which results in spherical granules completely filled with a very electron-dense material. The Golgi complex and segmented nucleus (N) are noted. Bone marrow-derived cells were cultured and prepared for TEM.[513]

C. The cell biology of mouse eosinophils

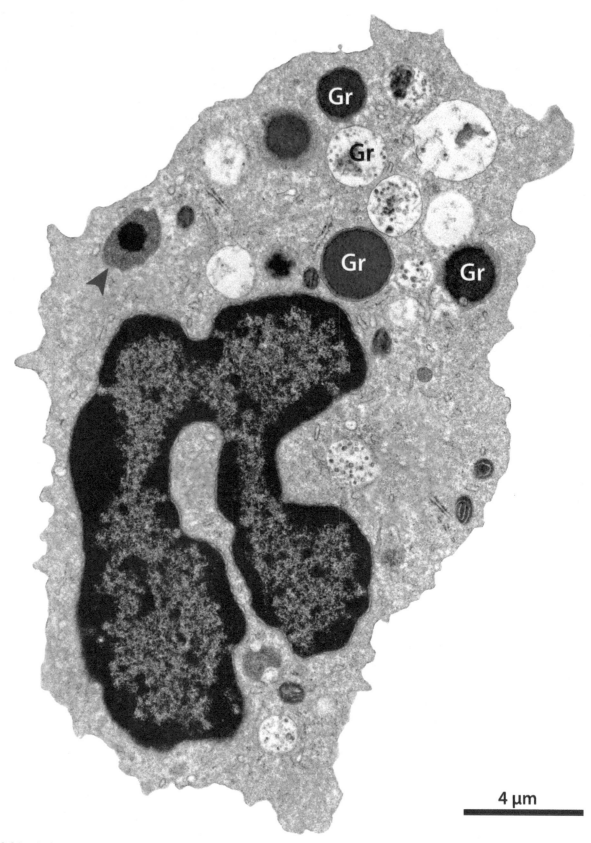

FIG. 9.31   An immature mouse eosinophil in the middle stage of development shows a mixture of immature specific granules (Gr) in different stages of development. A granule in the process of core formation is indicated *(arrowhead)*. The ring-shaped aspect of the nucleus *(pseudocolored)* is partially observed. Bone marrow-derived cells were cultured and prepared for TEM.[513]

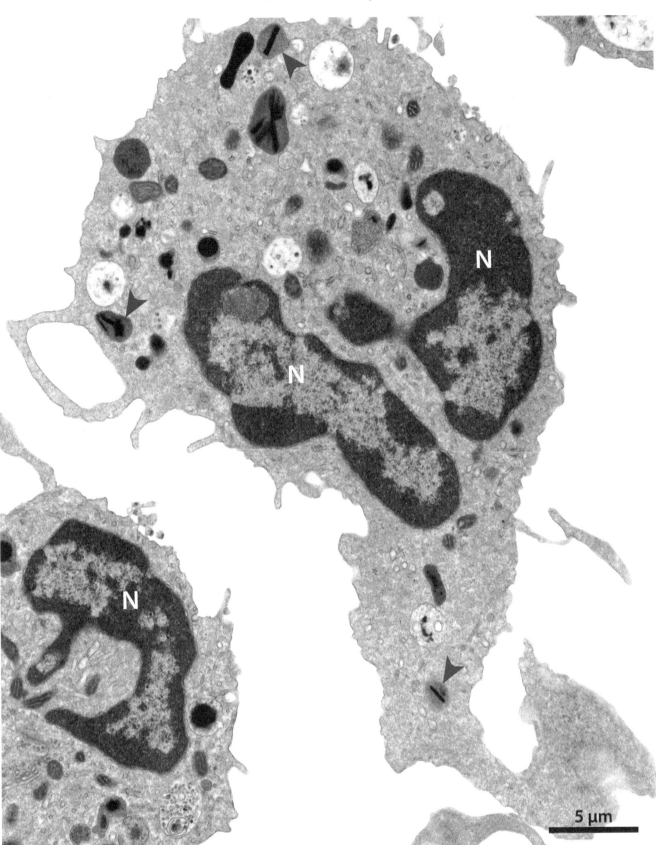

FIG. 9.32 Immature mouse eosinophil at the late stage of development. The nucleus (N) is multilobed, and the cytoplasm contains a small number of cored granules *(arrowheads)*. Bone marrow-derived cells were cultured and prepared for TEM.[513] *Image courtesy of Amali Samarasinghe.*

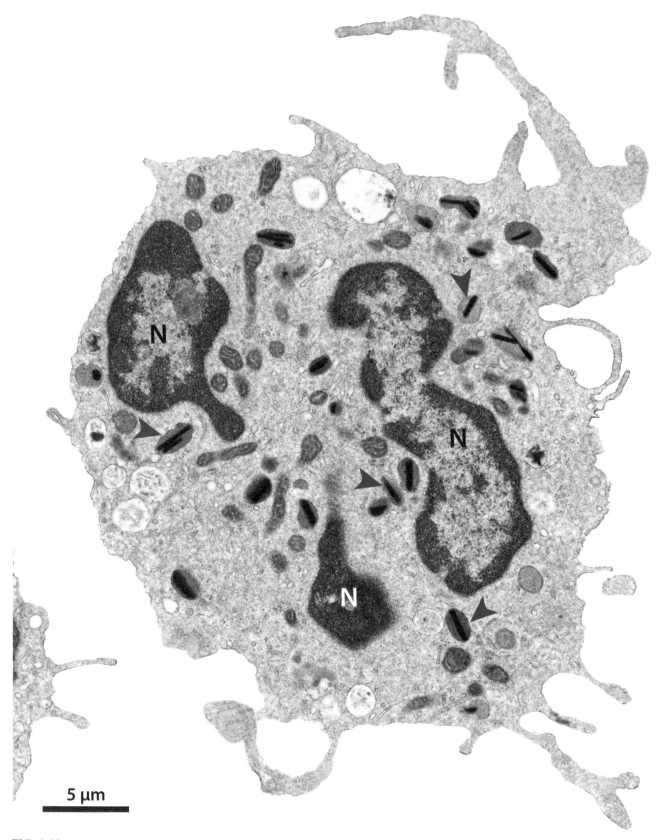

5 µm

FIG. 9.33   A mature mouse eosinophil seen in a culture from bone marrow-derived cells. The nucleus (N) is multilobed, and most secretory granules exhibit typical ultrastructure *(arrowheads)*. Sample prepared for TEM.[513]

mature cells in terms of the ultrastructure, which is considered by us as having at least 60% of their total granule population as cored granules (Fig. 9.33), is usually low in cultures of mouse eosinophils, although these cells exhibit typical eosinophil markers (IL-5-receptor α chain and Siglec F).[504]

An important feature of the eosinophil developmental morphology is related to the nucleus, which is initially round, becoming ring-shaped and segmented, with progressive condensation of the peripheral chromatin (Fig. 9.25) as a morphological pattern in mature cells (Fig. 9.11). Based on ultrastructural observations of mouse eosinophils arising in cultures, we noticed that the nuclear changes seem to occur earlier compared with humans. Human eosinophils that exhibit electron-dense, content-full granules (one feature of the intermediate stage of development), in general, show a round monolobed nucleus (for example, see Figs. 7.17–7.19 in Chapter 7), while in mice, the nuclear constrictions/lobulation process is usually already established in parallel with the presence of these granules (Figs. 9.28–9.30). Thus, it seems that nuclear constrictions start being formed in earlier stages of mouse eosinophil development compared with humans. Considering that the nuclei of mouse eosinophils are more polymorphic than those from human eosinophils (Fig. 9.23), the nuclear alterations can be occurring in different dynamics.

As indicated in Fig. 9.25, classical secretory organelles (RER and Golgi) undergo consistent reduction during mouse eosinophilopoiesis, a morphological pattern also observed in humans (Chapter 7). Eosinophil maturation also leads to a significant reduction of the mitochondrial population. Our quantitative EM analyses of immature and mature mouse eosinophils developing in cultures, which evaluated the area occupied by the mitochondria in the eosinophil cytoplasm, found that this area is reduced by ~ 90% from an initial stage (progenitor) up to a mature eosinophil (Fig. 9.25). Our ultrastructural analyses also identified that mitochondria within developing eosinophils interact with the ER (Fig. 9.34), a classical association recognized in other cell types, and undergo autophagy (mitophagy), a novel observation (Fig. 9.34), which might explain the progressive reduction of their numbers. Electron tomography applied to immature eosinophils revealed, at 3D levels, structural events consistent with mitochondrial fusion and vesiculation, indicating an active profile for these organelles in concert with other events during eosinophil development (Fig. 9.34).

In eosinophil cultures, it is common to find scattered cells in the process of apoptosis. It is recognized that eosinophils die spontaneously with an apoptotic profile briefly after differentiation in the absence of exogenous stimuli.[334,338] The canonical features of apoptosis (Chapter 6) can be observed in cultures of mouse eosinophils by TEM (Figs. 9.35 and 9.36).

The ultrastructural characteristics of immature mouse eosinophils were also studied directly in the bone marrow. In this case, samples can be obtained by aspirating the cells from the bone marrow (Fig. 9.37) or fixing in toto bone fragments, which enables the observation of the bone marrow microenvironment with excellent tissue preservation (Fig. 9.38). One mouse model used to investigate immature eosinophils was NFAT DKO. This animal is characterized by a dramatic and selective increase in Th2 cytokines such as IL-4 and IL-5 and a marked increase of eosinophils in the bone marrow (approximately 60% of bone marrow cells) compared with WT controls,[497] thus enabling the feasible findings of eosinophils in different stages of development in their microenvironment.

Bone marrow thin sections of NFAT DKO mice showed many eosinophils in close proximity with each other or with other cells, such as neutrophils (Fig. 9.38). The ultrastructural features typical of immature eosinophils could be appreciated in this model (Figs. 9.39–9.41). The secretory organelles Golgi and RER, which are very active in immature eosinophils, typically showed dilated cisternae (Figs. 9.39 and 9.40), and several mitochondrial profiles indicating high mitochondrial density were observed in eosinophil thin sections (Fig. 9.41). Eosinophils with variable proportions of coreless and cored granules or containing just cored granules in the cytoplasm were present in the bone marrow of NFAT DKO mice (Figs. 9.38–9.41), thus reflecting the gradual crystallization of the single granule population.[18]

NFAT DKO mice are also characterized by massive splenomegaly with the formation of granulomas along with eosinophilia in the spleen.[497] Our ultrastructural analyses of this organ revealed the presence of immature eosinophils exhibiting the same morphological features observed in eosinophils from the bone marrow (Figs. 9.42 and 9.43), thus indicating the occurrence of extramedullary hematopoiesis in the spleen. As noted in Chapter 7, the generation and survival of eosinophils are highly dependent on IL-5, a major cytokine that regulates the eosinophil differentiation from hematopoietic precursors and subsequent maturation.[345] It was demonstrated that constitutive overexpression of IL-5 can lead to extramedullary hematopoiesis in the spleen, as observed in IL-5 Tg compared with WT mice.[497] Accordingly, in addition to our ultrastructural observations in the spleen of NFAT DKO mice, we detected by TEM the presence of immature eosinophils in the spleen of IL-5 Tg mice (Figs. 9.44 and 9.45). Extramedullary formation of eosinophils likely occurs in other organs in response to intense recruitment of eosinophils during inflammatory diseases. Particularly, we documented the presence of immature eosinophils in the liver on a mouse model of schistosomiasis mansoni, as addressed in Section 9.5.

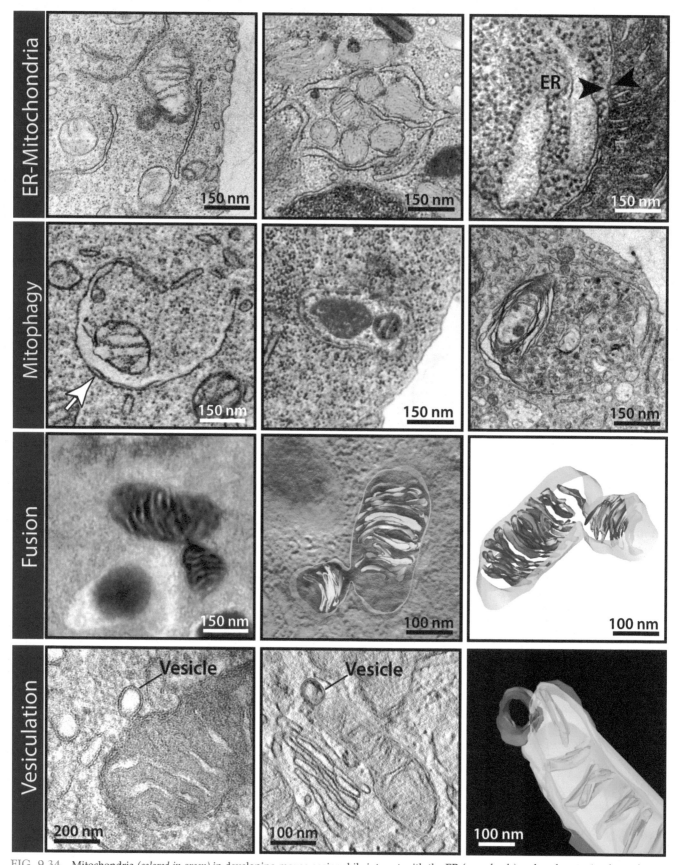

FIG. 9.34 Mitochondria *(colored in green)* in developing mouse eosinophils interact with the ER *(arrowheads)* and undergo mitophagy, fusion, and vesiculation. Note sequential stages of mitophagy [mitochondria been enclosed by a phagophore *(arrow)* and inside a double-membrane vacuole]. Bone marrow-derived cells[504] were prepared for TEM. Electron tomography shows fusion and vesiculation at 3D.

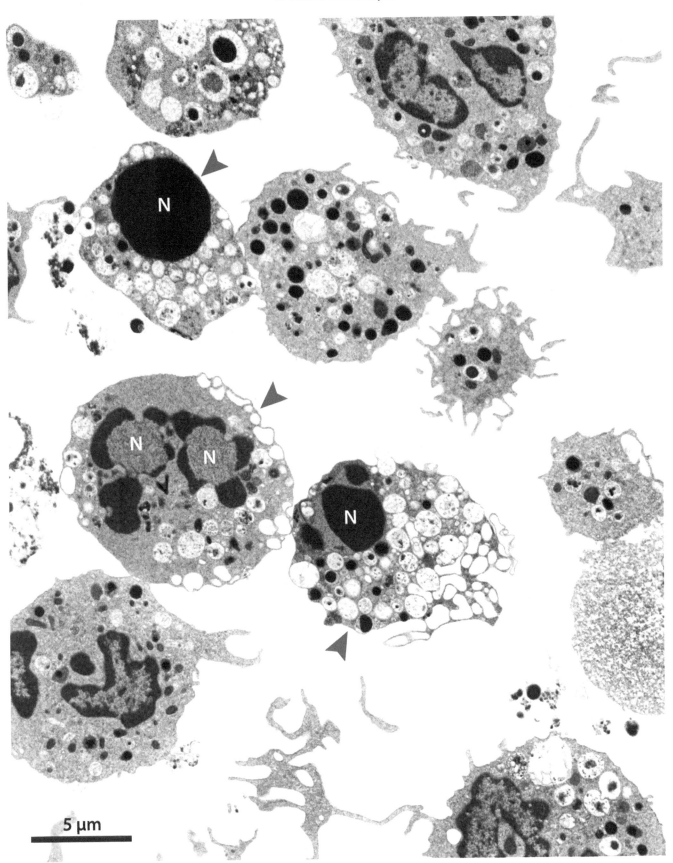

FIG. 9.35   Developing mouse eosinophils observed in a culture system by TEM. Note apoptotic eosinophils *(arrowheads)* with typical nuclear (N) changes (chromatin condensation and fragmentation with formation of cup-shaped marginal structures). Culture from bone marrow-derived cells.[513] *Image courtesy of Amali Samarasinghe.*

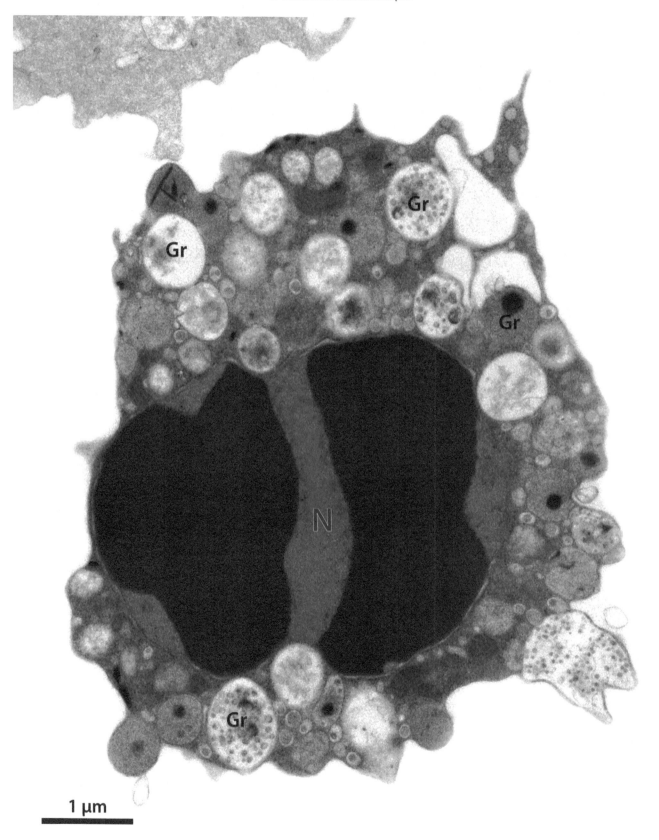

FIG. 9.36   A mouse eosinophil showing canonical ultrastructural features of apoptosis. The nucleus (N) is highly condensed and underwent fragmentation (karyorrhexis). Cell shrinkage, aggregation of cytoplasmic organelles, and electron-dense matrix are also observed. Bone marrow-derived cells were cultured and prepared for TEM.[513] *Gr*, secretory granules.

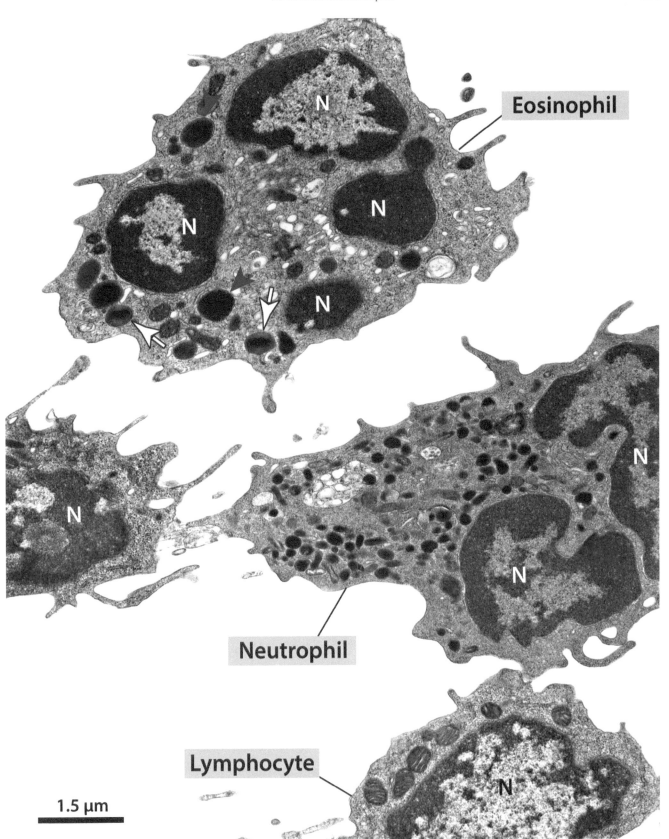

FIG. 9.37   A mouse eosinophil seen in the bone marrow with immature coreless *(arrowheads)* and mature cored *(arrows)* secretory granules in the cytoplasm. Note the multilobed eosinophil nucleus. Other leukocytes are also observed. Hematopoietic bone marrow from a WT C57BL/6 mouse was flushed and processed for TEM.[503] *N*, nucleus.

C. The cell biology of mouse eosinophils

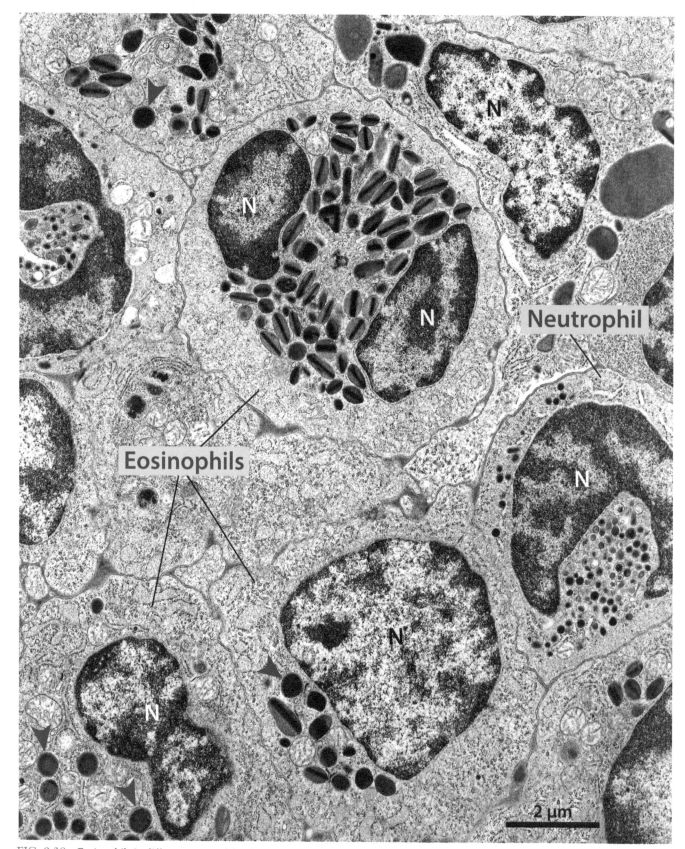

FIG. 9.38   Eosinophils in different stages of development in the bone marrow of a NFAT DKO mouse. Eosinophils show the cytoplasm full of cored granules or with a mixture of coreless *(arrowheads)* and cored granules. A neutrophil is also seen. Samples from NFAT DKO mice, which express high levels of Th2 cytokines (IL-4 and IL-5),[497] were prepared for TEM. *N*, nucleus.

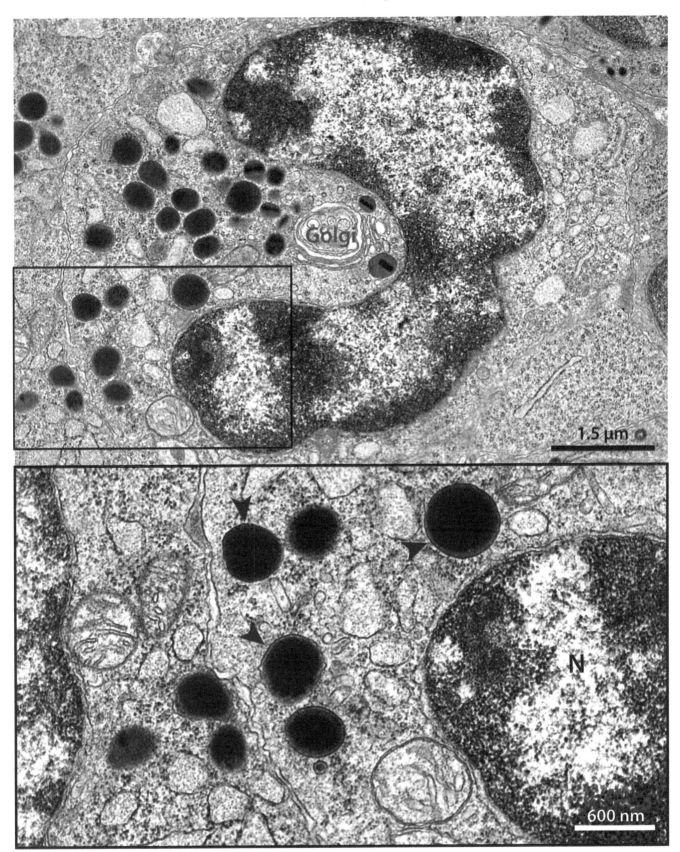

FIG. 9.39   Ultrastructure of an mouse immature eosinophil in the middle stage of development in the bone marrow. Note an incompletely seg-mented nucleus *(colored in purple, top panel)*, expanded Golgi, RER *(blue)*, and electron-dense coreless granules *(arrowheads, boxed area)*. Mitochondria *(green)*; cytoplasm *(pink)*. Samples from NFAT DKO mice, which express high levels of Th2 cytokines,[497] were prepared for TEM. *N*, nucleus.

C. The cell biology of mouse eosinophils

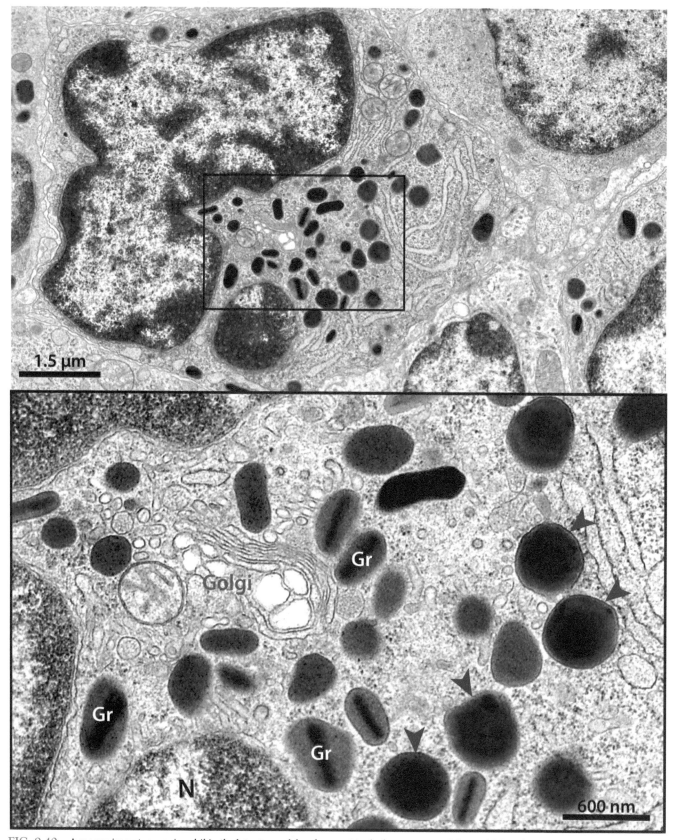

FIG. 9.40   A mouse immature eosinophil in the late stage of development in the bone marrow. Mature (Gr) and immature coreless *(arrowheads)* specific granules are shown in higher magnification *(boxed area)*. Note also an enlarged Golgi, peripheral RER *(colored in blue)*, and a segmented nucleus *(purple, top panel)*. Mitochondria *(green)*; cytoplasm *(pink)*. Samples from NFAT DKO mice[497] were prepared for TEM. *N*, nucleus.

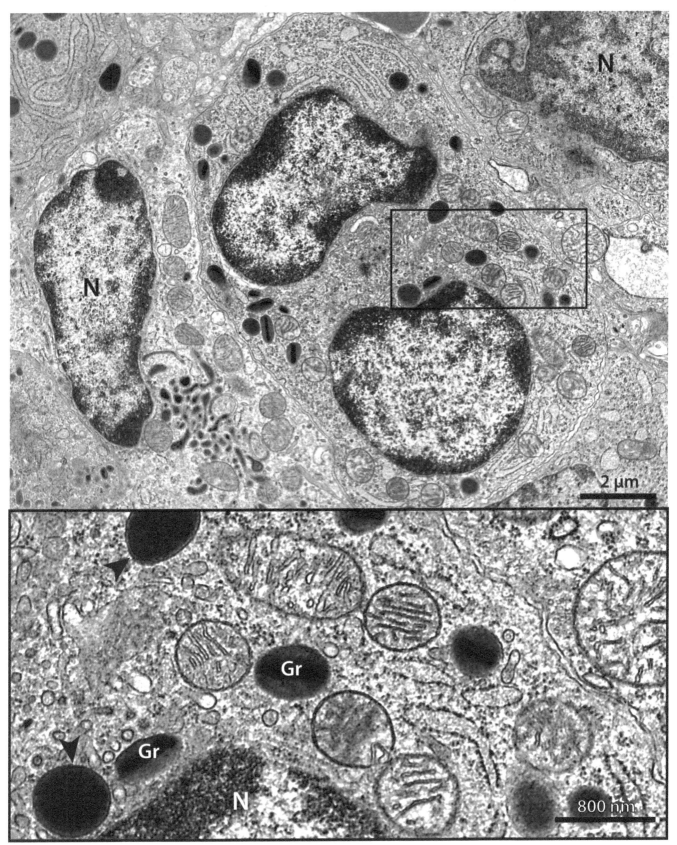

FIG. 9.41 A mouse immature eosinophil in the late stage of development in the bone marrow. Note immature *(arrowheads)* and mature (Gr) specific granules, many mitochondria *(green)*, and RER cisternae *(blue)*. Cytoplasm colored in *pink*. Samples from NFAT DKO mice[497] were prepared for TEM. *N*, nucleus.

C. The cell biology of mouse eosinophils

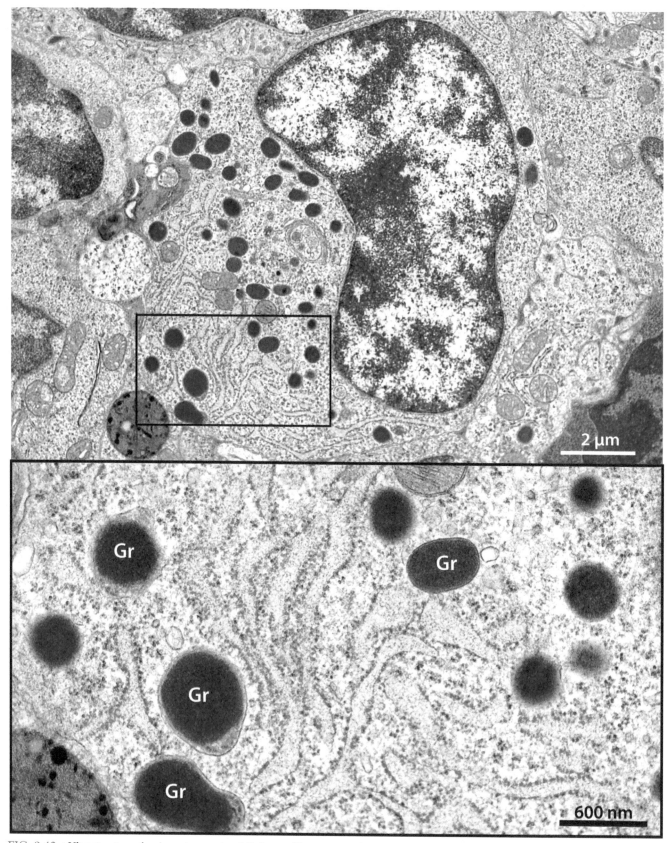

FIG. 9.42   Ultrastructure of an immature eosinophil observed in a mouse spleen. Note the large monolobed nucleus *(colored in purple, top panel)*, extensive cisternae of RER *(blue)*, and round immature specific granules (Gr). Mitochondria *(green)*; cytoplasm *(pink)*. Samples from NFAT DKO mice[497] were prepared for TEM.

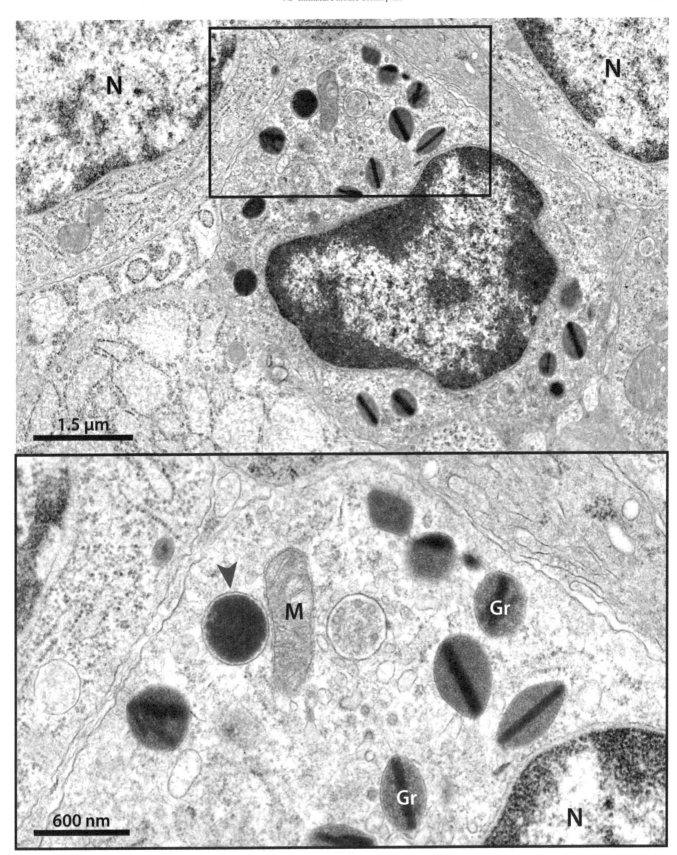

FIG. 9.43 An immature eosinophil in the spleen of a NFAT DKO mouse. An immature *(arrowhead)* and several mature (Gr) secretory granules, a mitochondrion (M), and an endosome *(orange)* are shown in higher magnification. The cytoplasm was colored in *pink*. Samples from NFAT DKO mice, which show eosinophilia in the spleen,[497] were prepared for TEM. *N*, nucleus.

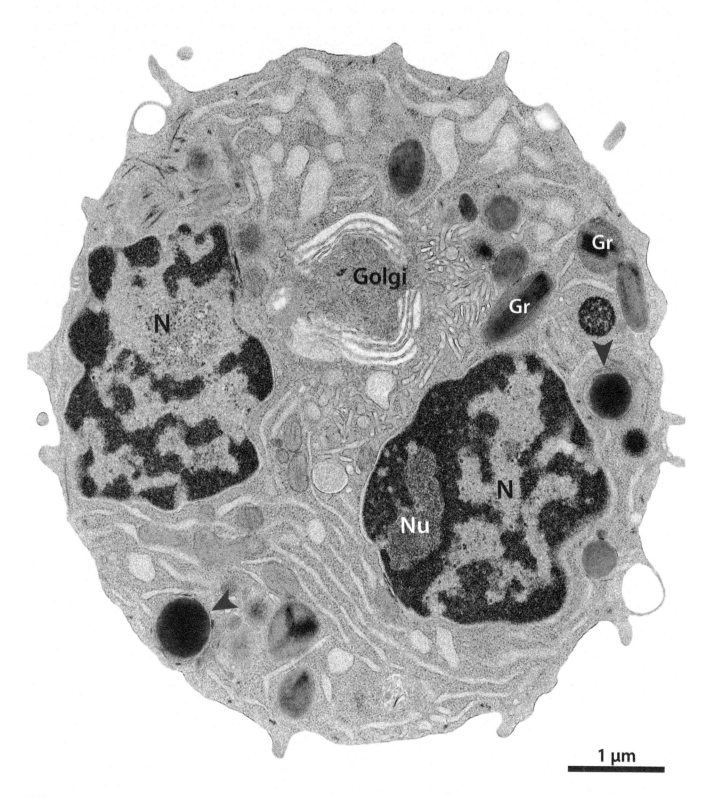

FIG. 9.44   Immature mouse eosinophil isolated from the spleen of an IL-5 Tg mouse. Expanded Golgi, developed RER *(colored in blue)*, fully electron-dense granules *(arrowheads)*, and prominent nucleolus (Nu) denote immaturity. Note mature specific granules (Gr) with typical central crystalloid. Eosinophils from IL-5 Tg mice, which show spleen eosinophilia,[493] were prepared for TEM. *N*, nucleus.

C. The cell biology of mouse eosinophils

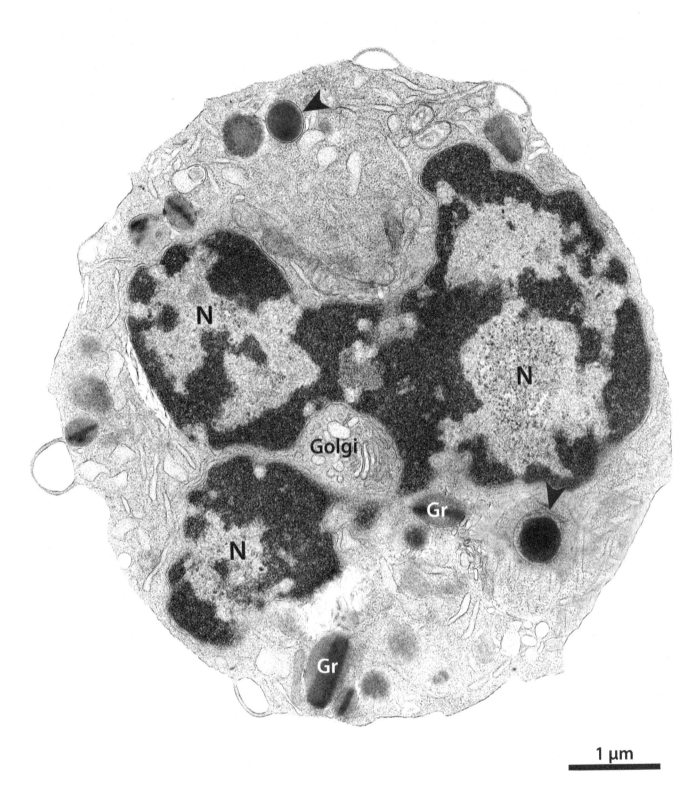

FIG. 9.45   Immature mouse eosinophil isolated from the spleen of an IL-5 Tg mouse. Note the large nucleus (N) in the process of segmentation, immature *(arrowheads)* and mature (Gr) specific granules, and multiple RER cisternae *(colored in blue)*. Eosinophils from IL-5 Tg mice, which shows eosinophilia in the spleen,[493] were prepared for TEM. *N*, nucleus.

C. The cell biology of mouse eosinophils

9.4 Activation and degranulation of mouse eosinophils

Eosinophils, both as end-stage effector cells and as collaborative interacting components of innate immune responses, rely for their activity on the secretion of varied proteins, including cationic proteins and cytokines, which are stored in their specific granules (Chapter 3). Both MBP-1 and EPX are found in specific granules of mouse eosinophils. However, human and mouse eosinophils differ in their stores of other cationic proteins. While human eosinophils contain only two EARs—EDN and ECP, mouse eosinophils enclose at least 6 EARs in their secretory granules.[51,505] Regarding cytokines, although mouse eosinophils express and are able to secrete these proteins,[51] it remains to be elucidated if the collection of cytokines stored in the specific granules of human eosinophils are also found as preformed stores in the specific granules of mouse eosinophils.[3]

The capacity of mouse eosinophils to secrete their granule contents both in vitro and in vivo has been addressed in many studies and models. In spite of concerns raised for mouse eosinophils in the past regarding the apparently higher thresholds for degranulation[506] or if they exhibit typical ultrastructural signs of PMD[507] as extensively reported for human eosinophils,[148] the current view is that mouse eosinophils are able to degranulate in response to different stimuli, both in vitro and in vivo, and in the context of diseases.[352,492,504,508]

Our in vitro ultrastructural studies combined with other approaches to detect secretion in mouse models after activation with stimuli such as GM-CSF (Fig. 9.46) or CCL11 (Fig. 9.47)[492] revealed PMD-underlying granule secretion. The spectrum of granule ultrastructural changes indicative of PMD included, as frequent events, enlargement, reduced electron density, matrix coarsening, and disassembled cores, and to a lesser extent, loss of the matrix or core contents (Figs. 9.46 and 9.47). Quantitative TEM analyses in eosinophil sections showing the entire cell profile and nucleus showed that the activation with CCL11 induced significant increases (fivefold) in the numbers of granules with these morphological signs.[492] These altered nonfused granules were always intermingled, in the same cell section, with resting, nonmobilized granules,[492] thus reflecting that not all granules respond concomitantly to stimulation, which is a common characteristic of PMD in human eosinophils.[148]

Although the ultrastructure of PMD in mouse eosinophils is comparable to that observed in human eosinophils, there are some differences and aspects that should be considered. First, in general, PMD ultrastructural changes in activated mouse eosinophils are less pronounced than those seen in human eosinophils. Second, in mice, granule changes underlying PMD seem more concentrated in the matrices than in the cores.[492] In contrast, in human eosinophils, disarrangement of both granule cores and matrices are common PMD findings in response to cell activation.[148] Third, matrix coarsening, which is often observed in swollen granules in mouse eosinophils responding to activation, is a frequent sign of PMD, both in vitro (Figs. 9.46 and 9.47) and in vivo (Fig. 9.48). Fourth, to our view, the phenomenon of granule enlargement is an important indication of PMD occurrence on eosinophils from mice. The specific granules of mouse eosinophils show a more elliptical/elongated morphology and are less prone to undergo swelling than specific granules of human eosinophils. The occurrence of this event is a sign that the granules are being responsive to the microenvironment with mobilization of their contents. Therefore, in mice, the ultrastructural features associated with PMD are subtler compared with human eosinophils, but they can be detected and quantitated under detailed examination and comparison with unstimulated/resting cells.

Other characteristics associated with cell activation are noted in mouse eosinophils (Figs. 9.48–9.50). We have been observing an increased formation of cytoplasmic vesiculotubular structures (Figs. 9.48 and 9.50; also addressed in Section 9.5), shape changes (Fig. 9.49), including uropod formation (Fig. 9.46), and release of extracellular vesicles (Fig. 9.49) in mouse eosinophils activated during different conditions/diseases compared with resting eosinophils.

Mouse eosinophils also undergo degranulation through compound exocytosis, a process less frequently found in vivo (Fig. 9.51), and cytolysis (Fig. 9.52), which together with PMD are more commonly reported. The ability of mouse eosinophils to produce extracellular DNA traps in response to activation has also been described with the use of EM, for example, when these cells are cocultured with microfilariae in vitro.[509] Interestingly, we identified ETosis in a mouse tissue eosinophil resident in the small intestine lamina propria of a normal, healthy WT mouse (Fig. 9.52), an event that may be related to the natural higher activation of these cells in the intestine, as discussed in Chapter 8.

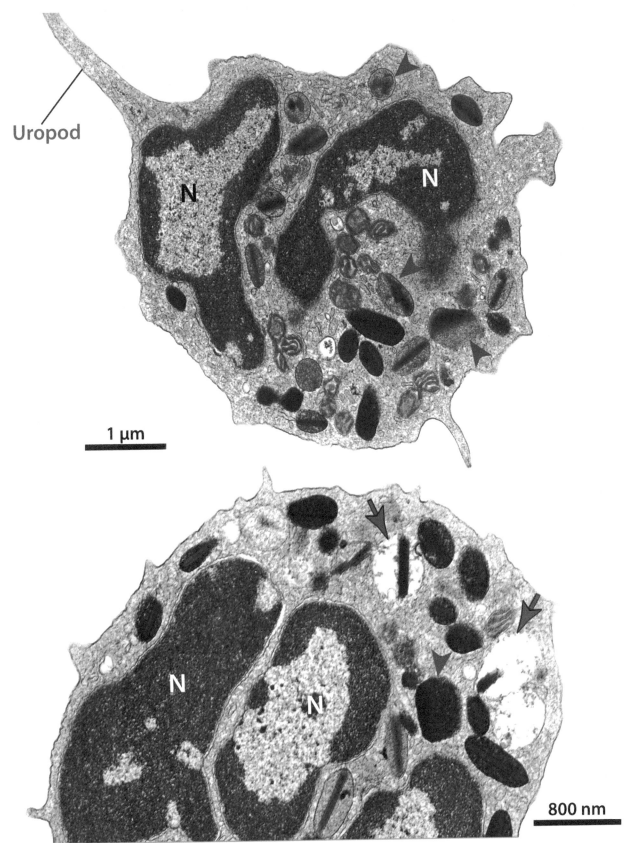

FIG. 9.46 Mouse eosinophils undergoing PMD in response to GM-CSF stimulation. Granules are swollen with structural disarrangement and matrix coarsening *(arrowheads)* and content losses *(arrows)*. Note uropod formation. Eosinophils from IL-5 Tg mice were stimulated and processed for TEM.[492] *N*, nucleus.

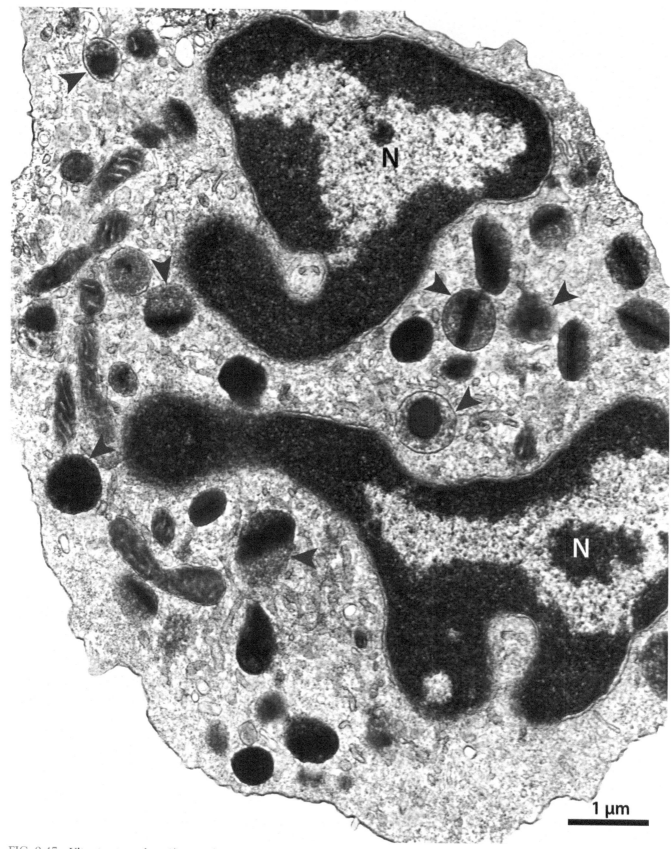

FIG. 9.47   Ultrastructure of specific granules within a CCL11-activated mouse eosinophil. Activated granules *(arrowheads)* lost their elongated morphology, becoming swollen with coarsening of the matrices and disarranged contents. Eosinophils isolated from the spleen of IL-5 Tg mice were stimulated and processed for TEM as described.[492] *N*, nucleus.

C. The cell biology of mouse eosinophils

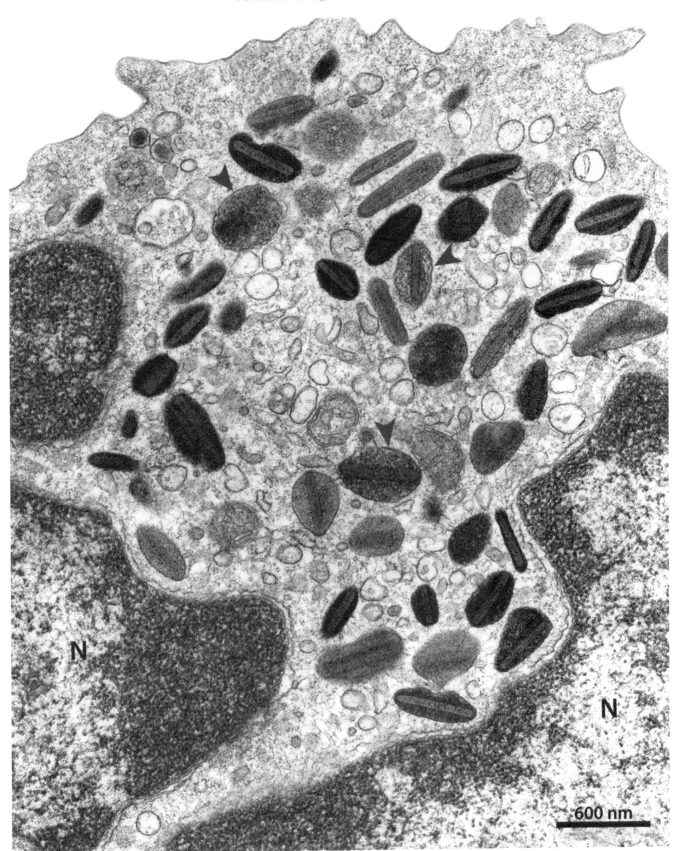

FIG. 9.48   An activated eosinophil from the pleural fluid of a mouse infected with *Mycobacterium bovis* bacillus Calmette-Guérin. Note enlarged specific granules with matrix coarsening *(arrowheads)* and a high number of cytoplasmic vesicles *(highlighted in pink)*. WT C57BL/6 mice were infected as before.[216] N, nucleus.

C. The cell biology of mouse eosinophils

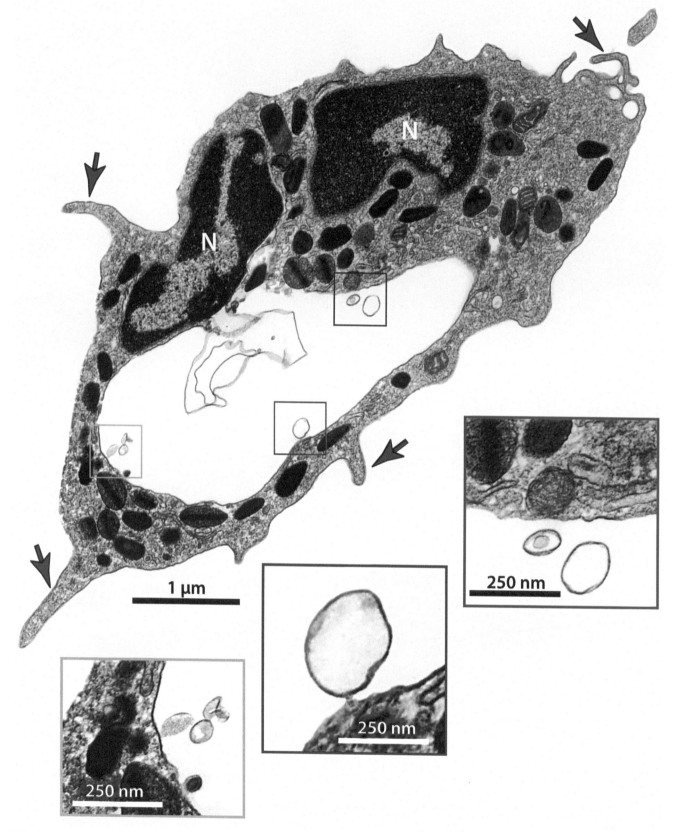

FIG. 9.49   Ultrastructure of an activated mouse eosinophil showing shape changes and production of extracellular vesicles (EVs). Note protrusive structures *(arrows)* and EVs *(boxes)* at the cell surface. Cells were isolated from the spleen of IL-5 Tg mice, stimulated with GM-CSF, and prepared for TEM. *N*, nucleus.

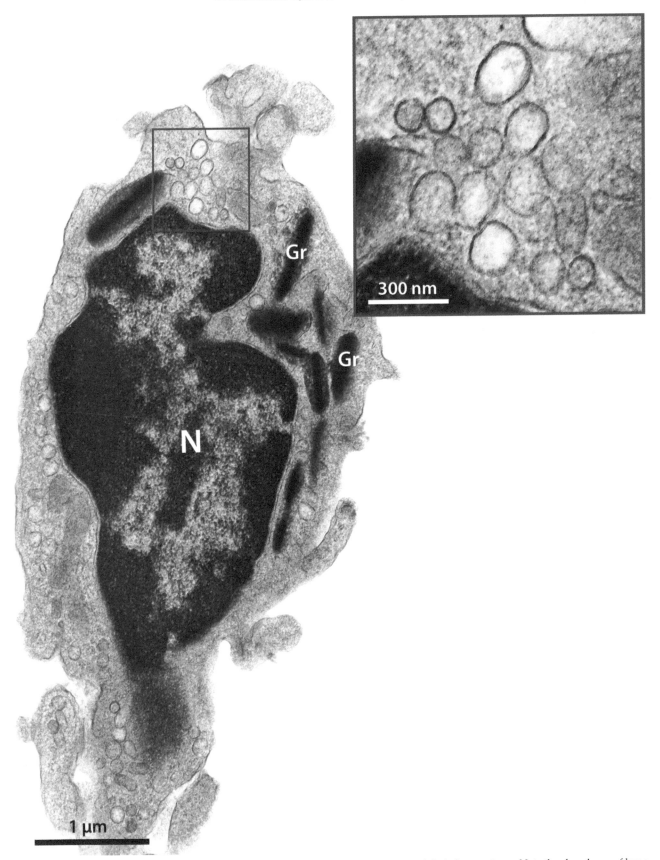

FIG. 9.50 Ultrastructure of an infiltrating eosinophil in a tumor site *(skin)* from a mouse model of plasmacytoma. Note the abundance of large vesicles distributed in the cytoplasm. Vesicles (70–200 nm in diameter) are seen in higher magnification *(box)*. Samples from WT BALB/c mice immunized with plasmacytoma cells[525] were processed for TEM. *Gr*, specific granules; *N*, nucleus.

C. The cell biology of mouse eosinophils

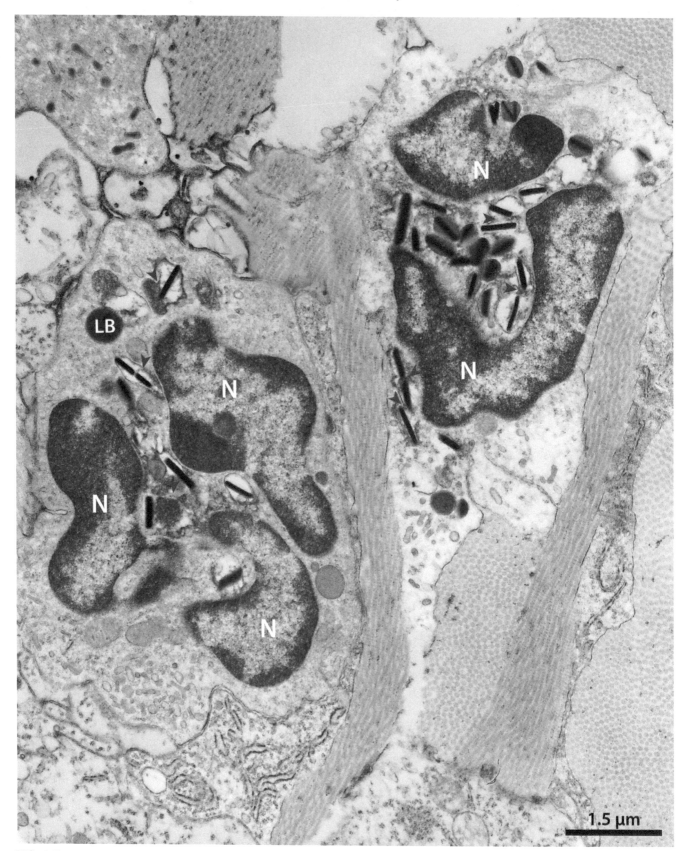

FIG. 9.51   Ultrastructure of inflammatory eosinophils in a tumor site (skin) from a mouse model of plasmacytoma. Specific granules show evidence of fusion *(arrowheads)*. ECM colored in *green*. Samples from WT BALB/c mice immunized with plasmacytoma cells[525] were processed for TEM. *LB*, lipid body; *N*, nucleus.

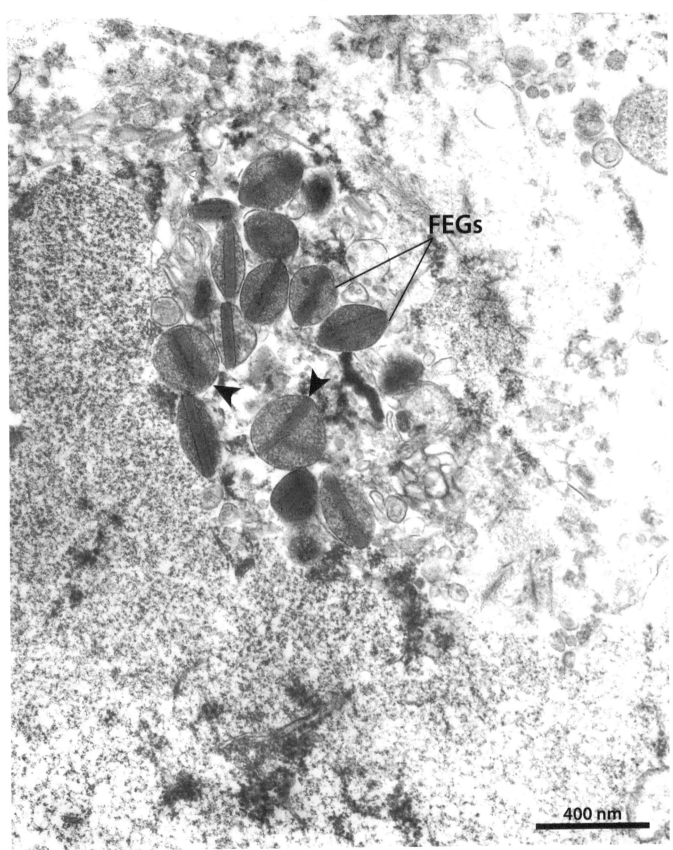

FIG. 9.52 A cytolytic mouse eosinophil showing ultrastructural features of ETosis in the small intestine lamina propria. Note the highly decondensed chromatin *(colored in purple)* spreading in an area with FEGs, some of them appearing as round, swollen organelles *(arrowheads)*. Samples from WT Swiss mice were prepared for TEM.

9.5 Mice in the context of EADs

To illustrate the ultrastructural features of tissue eosinophils in mouse models of human EADs, two representative diseases were selected: asthma and schistosomiasis mansoni.

9.5.1 Asthma

Asthma is a chronic and complex inflammatory disease of the respiratory tract with multifaceted immune responses occurring concurrently in the lung. The eosinophil is a key component of airway infiltrates in asthma phenotypes (eosinophilic asthma) with a role associated with several clinical features (mucus accumulation, airway hyperresponsiveness, and tissue remodeling) of the disease, but this role is not completely understood. The effects of drugs targeting eosinophils or IL-5 have been explored to understand the possible benefits of eosinophil depletion, for example, improving airway remodeling, in the context of asthma (reviewed in[6]).

Mouse models of allergic airway inflammation have been used for understanding the asthma pathophysiology and testing potential drug therapies (reviewed in[490,510,511]). These models have been developed with the use of different methods of sensitization to an antigen, a process that leads to physiological and immunological responses that mimic human allergic responses.[511] In mouse models, these responses can differ to some extent in terms of the mouse species and method of sensitization/antigen used.[490,510,511] Sensitized and challenged mouse models that feature eosinophil recruitment to the airways have also been used to study a potential beneficial role played by these cells during viral infections.[512,513]

We evaluated the ultrastructure of eosinophils in lung samples of a mouse model of asthma [BALB/c mice sensitized and challenged with ovalbumin (OVA)], which display some key features of the asthma airway responsiveness and bronchial inflammation along with hyperproduction of Th2 cytokines.[514] The lungs of these allergic mice were characterized by infiltration of a large number of eosinophils, seen in clusters (Fig. 9.53) or individually (Figs. 9.54–9.56) in the extracellular matrix, which was highly expanded by collagen fibrils, a central pathological feature of airway remodeling in asthma.[515]

PMD (Figs. 9.55 and 9.56) was the main mechanism observed in eosinophils infiltrated in the allergic lung, in accordance with previous ultrastructural work.[506] The morphological features of PMD as described above for mouse eosinophils stimulated in vitro were observed in vivo, but neither exocytosis nor cytolysis was observed. Inflammatory eosinophils showed expanded/elongated morphology (Figs. 9.57–9.59), with clear polarization characterized by nuclei and granule population located in opposite sites (Figs. 9.54–9.56), extensive protrusive structures at the cell surface (Figs. 9.58 and 9.60), and amplification of the vesiculotubular system (Figs. 9.57–9.60), which is analogous to the human EoSVs (Fig. 9.23). All of these characteristics are reflecting cell activation (Chapter 4).

Our quantitative analyses showed a significant increase of the cytoplasmic vesicles (70–200 nm in diameter) compared with resting eosinophils, a finding also found in eosinophils from mice infected with schistosomiasis mansoni, as discussed below. By analyzing the ultrastructure of mouse eosinophils in other models of experimental diseases, such as *Mycobacterium bovis* infection (Fig. 9.48) and plasmacytoma (Fig. 9.50), we also observed evidence for increased vesicle formation, thus indicating that the mouse vesiculotubular system, as in human eosinophils,[18,21] can be a consistent indicator of activation.

Of note, polarized eosinophils (nucleus and granule population displayed in opposite sides) were frequently observed in the allergic lung (Figs. 9.54–9.56 and 9.58). This finding drew our attention because it is also a phenomenon found in biopsies of human EADs, for example, in IBDs (Fig. 4.36 in Chapter 4), eosinophilic myocarditis (Fig. 4.37 in Chapter 4), and HES (Fig. 8.35 in Chapter 8). The relationship between such polarization and degranulation remains to be investigated.

In addition to the remarkable eosinophil infiltration, a high number of plasma cells (Figs. 9.54, 9.57, 9.58, and 9.61) were observed in the allergic lung with activated eosinophils frequently observed in proximity or physical interaction with these cells (Figs. 9.57, 9.58, and 9.61). Our ultrastructural analyses of many human biopsies identified eosinophil-plasma cell interactions as a common finding in EADs such as ECRS (Fig. 8.5 in Chapter 8) and IBDs (Figs. 8.44 and 8.46 in Chapter 8). Eosinophils are necessary to maintain plasma cells in both bone marrow and intestinal lamina propria, contributing to tissue immune homeostasis (reviewed in[413]). The mouse model of asthma is mimicking a human in vivo event that may be associated with a broader immunomodulatory activity of eosinophils in the context of diseases.

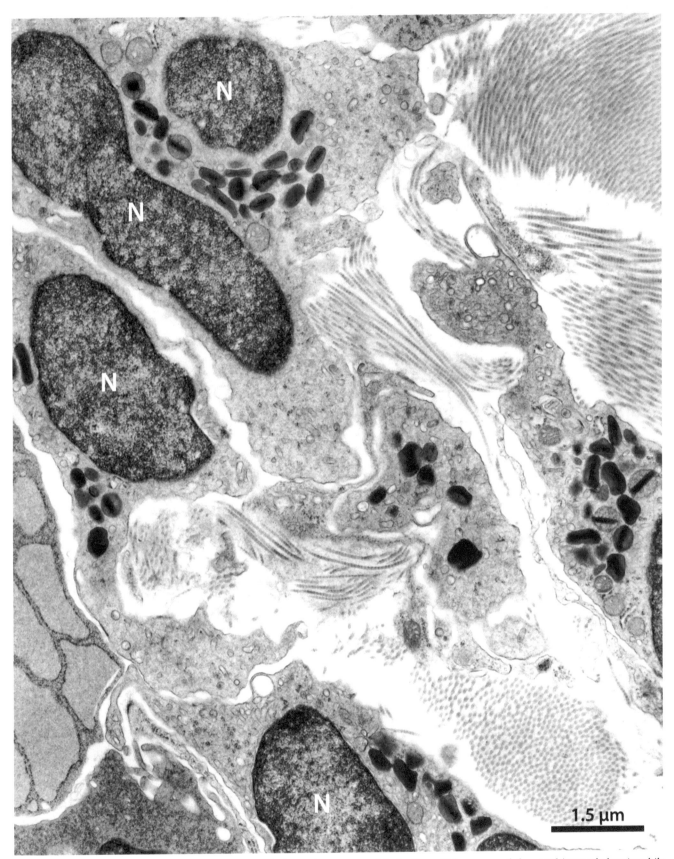

FIG. 9.53   Ultrastructure of infiltrating eosinophils in the lung of a mouse model of asthma. Note a group of elongated/expanded eosinophils with typical crystalloid granules in the cytoplasm. Samples from OVA-induced asthmatic mice (WT Balb/c)[514] were processed for TEM. *N*, nucleus.

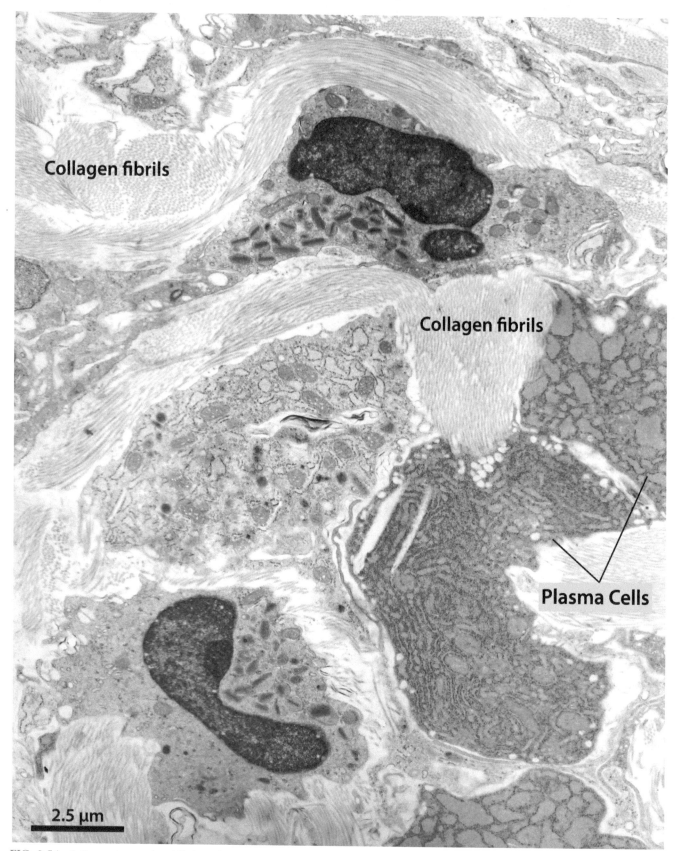

FIG. 9.54   Inflammatory eosinophils *(pseudocolored)* and plasma cells in the lung of a mouse model of asthma. Note abundant bundles of collagen fibrils, an asthma feature. Eosinophils are shown in higher magnification in Figs. 9.55 and 9.56. Samples from OVA-induced asthmatic mice (WT Balb/c)[514] were processed for TEM.

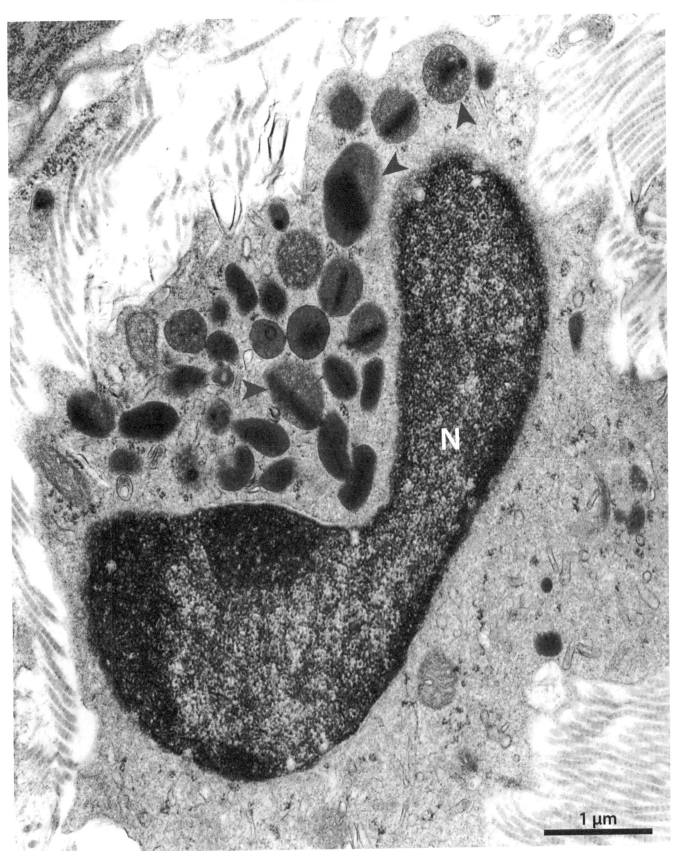

FIG. 9.55 Inflammatory eosinophil infiltrated in the lung of a mouse model of asthma. Cell shows PMD signs *(arrowheads)* and clear polarization with granules and nucleus (N) in opposite sides. Samples from OVA-induced asthmatic mice (WT Balb/c)[514] were processed for TEM. *N,* nucleus.

C. The cell biology of mouse eosinophils

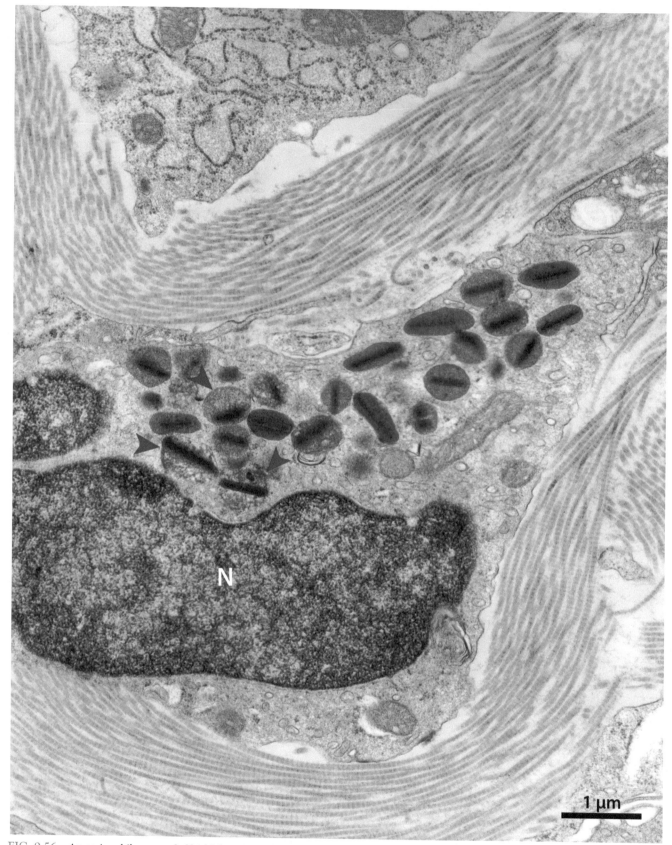

FIG. 9.56   An eosinophil surrounded by high amount of collagen fibrils *(colored in green)* in the lung of a mouse model of asthma. Note that the specific granules are localized in one side of the cell. *Arrowheads* indicate granules undergoing PMD. Samples from OVA-induced asthmatic mice (WT Balb/c)[514] were processed for TEM. *N*, nucleus.

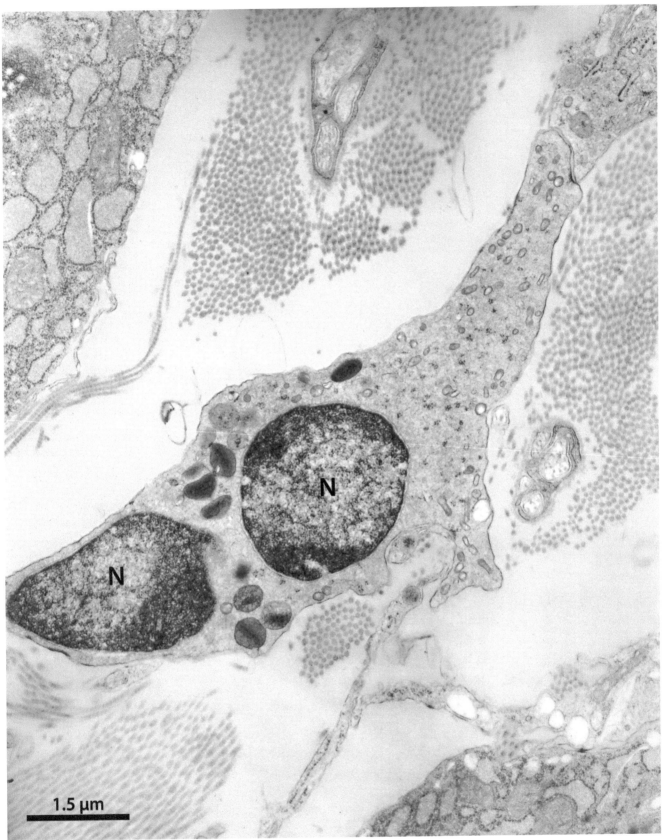

FIG. 9.57 Electron micrograph showing an activated eosinophil in the lung of a mouse model of asthma. Eosinophil is elongated and with a high number of vesiculotubular structures *(colored in pink)* in the cytoplasm. Plasma cells *(pseudocolored)* with typical dilated RER *(blue)* are partially observed. ECM colored in *green*. Samples from OVA-induced asthmatic mice (WT Balb/c)[514] were processed for TEM. *N*, nucleus.

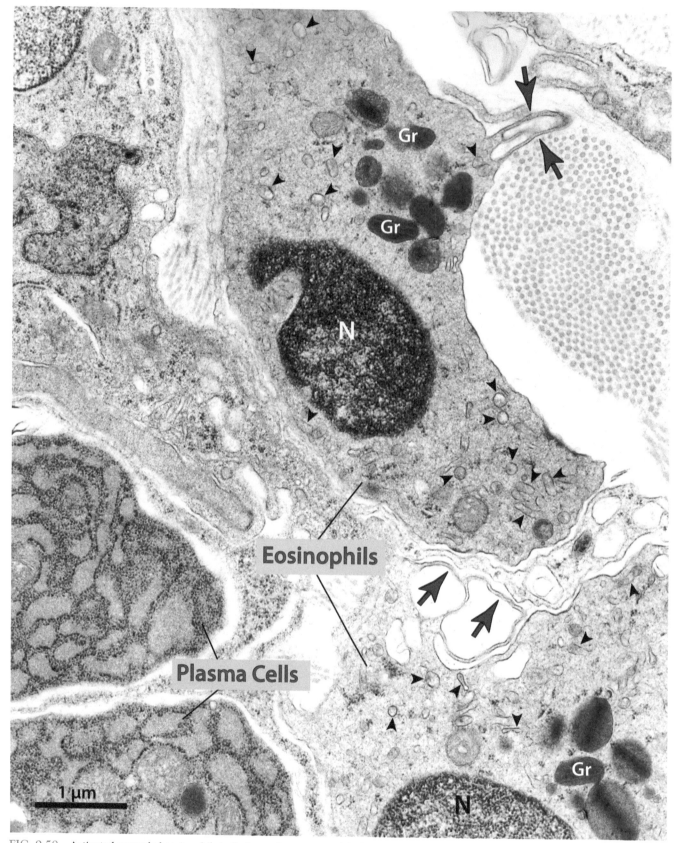

FIG. 9.58 Activated expanded eosinophils in the lung of a mouse model of asthma. Note folds and filopodia *(arrows)* at the cell surface, many vesiculotubular structures in the cytoplasm *(arrowheads)*, and the polarized morphology. Adjacent plasma cells are observed. Samples from OVA-induced asthmatic mice (WT Balb/c)[514] were processed for TEM. *Gr*, specific granules; *N*, nucleus.

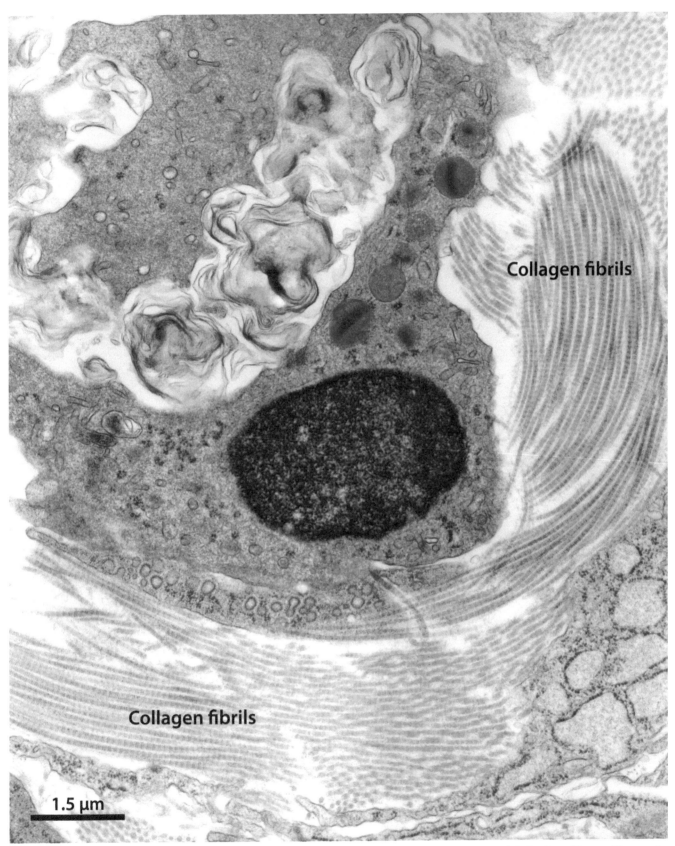

FIG. 9.59 Electron micrograph showing elongated eosinophils *(pseudocolored)* in the lung of a mouse model of asthma. The specific granules *(yellow)* are displaced to one side of the cell. Note extensive deposition of collagen fibrils in the ECM. Cytoplasm *(pink)*; nucleus *(purple)*. Samples from OVA-induced asthmatic mice (WT Balb/c)[514] were processed for TEM.

C. The cell biology of mouse eosinophils

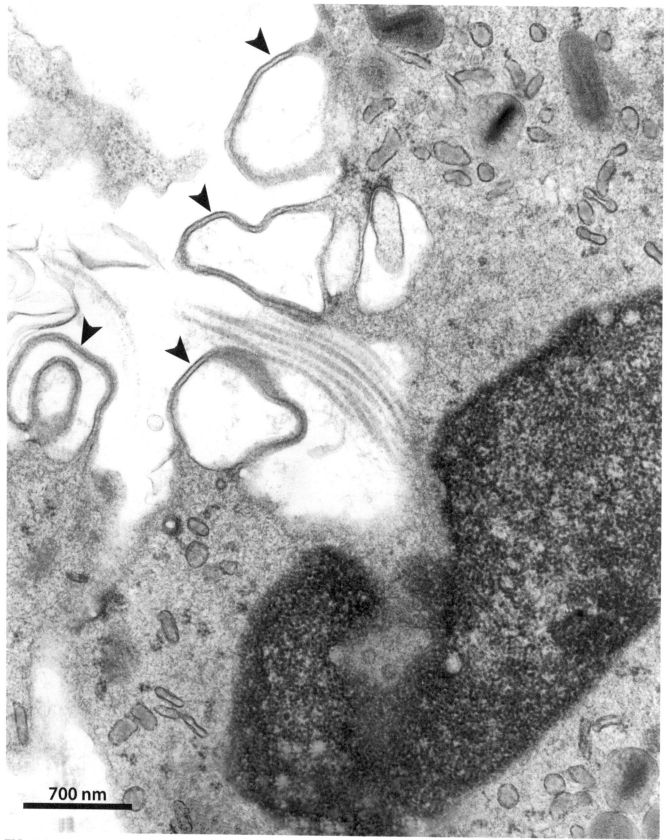

FIG. 9.60   Electron micrograph showing the surface of an pseudocolored eosinophil in a mouse allergic lung. Note membrane extensions *(arrowheads)* and numerous vesiculotubular structures *(highlighted in pink)* in the peripheral cytoplasm. A few specific granules *(yellow)* and part of the nucleus *(purple)* are observed. Samples from OVA-induced asthmatic mice (WT Balb/c)[514] were processed for TEM.

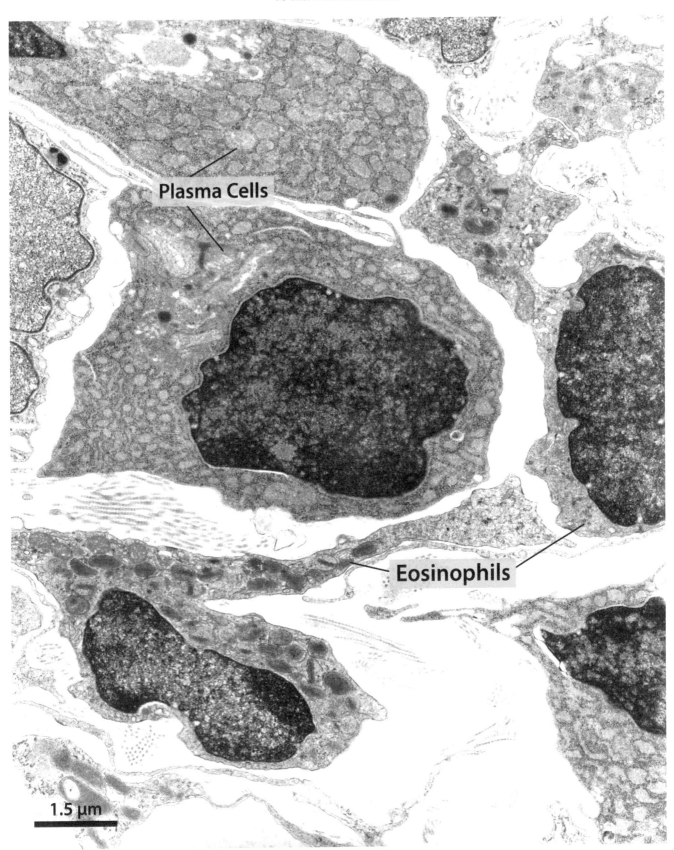

Plasma Cells

Eosinophils

1.5 μm

FIG. 9.61 Eosinophils interact with plasma cells in a mouse allergic lung. Note elongated and polarized eosinophils in close apposition to plasma cells with characteristic abundant RER *(blue)* in the cytoplasm. Cells were pseudocolored. Cytoplasm *(pink)*; nuclei *(purple)*; specific granules *(yellow)*. Samples from OVA-induced asthmatic mice (WT Balb/c)[514] were processed for TEM.

C. The cell biology of mouse eosinophils

9.5.2 Schistosomiasis mansoni

Eosinophils are extraordinarily involved in many parasitic diseases (Chapter 8), including schistosomiasis, caused by trematode worms of the genus *Schistosoma*. Here, we explore the eosinophil ultrastructure in a mouse model of acute hepatic schistosomiasis mansoni,[352,433] one of the most prevalent forms of this disease.[430] In hepatic schistosomiasis, granuloma formation (Figs. 9.62 and 9.63) arises when schistosome eggs become trapped in the presinusoidal capillary venules of the host liver, settling in the hepatic tissue after crossing the vessels.[432,516,517] Granulomas, sites containing highly organized collections of inflammatory and resident cells, occupy an extensive area of the infected liver, triggering severe hepatic fibrosis, hepatosplenomegaly, and portal hypertension, and can lead to death.[429,517,518]

A spectrum of immune cells such as lymphocytes, macrophages, neutrophils, and eosinophils colonize hepatic granulomas. These cells are sequentially recruited to the granuloma and, once at this site, arrange themselves into tightly clustered populations with varied lifespan and poorly understood functional roles.[440] Our group has been studying the composition and distribution of granulomas induced by the *Schistosoma mansoni* infection in target organs with the use of high-resolution whole slide imaging, which enables a detailed visualization and quantitation of the number, type, frequency, and areas of granulomas and inflammatory infiltrates in entire tissue sections. These morphometric histopathological evaluations in mouse models showed that eosinophils correspond to 60% of all cells within the most frequent types of hepatic granuloma in the acute infection.[352,433] Reasons for the dominance of the eosinophil population within the hepatic schistosomal granuloma remain unsolved.[440]

Ultrastructural analyses of tissue samples from *S. mansoni*-infected mice showed a high number of eosinophils with typical morphology forming tight groups of cells in the liver (Figs. 9.62–9.64). Eosinophils were in contact with each other (Figs. 9.62–9.64) and/or with other inflammatory cells, such as neutrophils (Figs. 9.62 and 9.63), plasma cells (Fig. 9.62), and lymphocytes (Fig. 9.65). Interestingly, eosinophils with morphological signs of immaturity, represented by a voluminous less-segmented nucleus with a prominent nucleolus, a small number of coreless electron-dense granules, and higher amounts of RER strands in the cytoplasm were present as a component of the inflammatory infiltrates (Figs. 9.65 and 9.66). Whether these not fully differentiated eosinophils represent cells released from the bone marrow with immaturity features or cells in the process of extramedullary differentiation remains to be established. Indeed, the hepatic schistosomal granuloma has been shown to function as an active extramedullary hematopoietic site, producing potentially all the myeloid lineages.[519–521] We consider the hepatic granuloma as an integrating and evolving ecological niche in which the functional activities of its cell populations occur in an orchestrated way in response to microenvironmental gradients.[440] Thus, local hematopoietic growth factors can be locally leading to differentiation of EoPs.

By analyzing the population of eosinophils infiltrated in the liver to identify the ultrastructural signs of degranulation, we found that while 19.1% of eosinophils were intact, most eosinophils (66.4%) exhibited predominantly features of PMD (Figs. 9.67-9.69). Characteristic features of cytolysis (Fig. 9.69) were observed in 13.0% of eosinophils, and just 1.5% showed classical or compound exocytosis (Fig. 9.69).[352]

Because MBP-1 has been associated with the immunopathogenesis of various helminthic diseases, including schistosomiasis mansoni, we also applied pre-embedding immunonanogold EM (detailed in Chapter 5) to investigate the subcellular localization of this protein in eosinophil-rich inflammatory infiltrates in the liver. We observed clear labeling for MBP-1 in the entire population of inflammatory eosinophils while other infiltrated immune cells were completely negative (Fig. 9.70).[352]

As pointed out in this chapter, our qualitative and quantitative TEM analyses show that mouse eosinophils have a cytoplasmic vesiculotubular system composed of reasonably large vesicles (70–200 nm in diameter) that can be considered analogous to the human EoSVs (Fig. 9.23) and that is amplified in response to cell activation during different conditions. Experimental *S. mansoni* infection in mice induces a significant increase (~ 500%) in the numbers of these vesicles in liver-infiltrated eosinophils (Fig. 9.67) compared to uninfected tissue eosinophils.[352] Although the origin of these vesicles remains to be investigated, the presence of vesicles in the proximity of mobilized secretory granules was noted (Fig. 9.68). We also analyzed the population of vesicles in the cytoplasm of mouse eosinophils infiltrated in the liver after ultrastructural immunolabeling for MBP-1. Our single-cell analyses at high resolution were revealing in demonstrating MBP-1-positive vesicles distributed in the cytoplasm and attached to or surrounding the surface of emptying granules (Fig. 9.70). Computational analyses showed that ~ 20% of all cytoplasmic vesicles from the same size range were carrying MBP-1,[352] thus demonstrating, for the first time, that a vesicle-mediated transport for MBP-1 release is operating in activated mouse eosinophils as already documented for human eosinophils.[22]

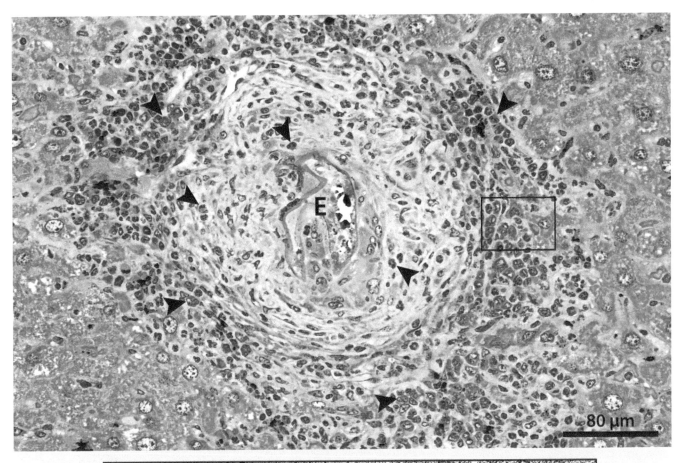

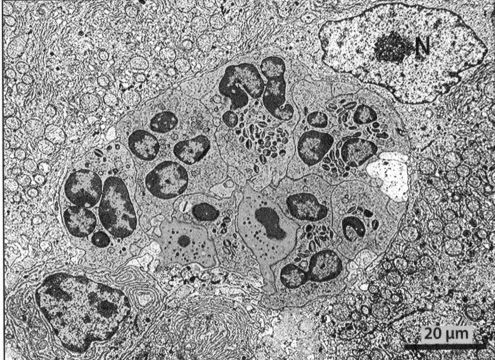

FIG. 9.62 Infiltrating eosinophils in the liver of *Schistosoma mansoni*-infected mice. Eosinophils are seen in a granuloma by both light micros-copy (*arrowheads*, HE staining, *top panel*) and TEM at low magnification (*pseudocolored in pink, bottom panel*). Eosinophils (*boxed area*) are in contact with each other, neutrophils (green), and plasma cells (*purple*) as revealed by TEM. Note the large nucleus (N) of a surrounding hepatocyte. WT Swiss mice were acutely infected.[352] *E*, parasite egg.

C. The cell biology of mouse eosinophils

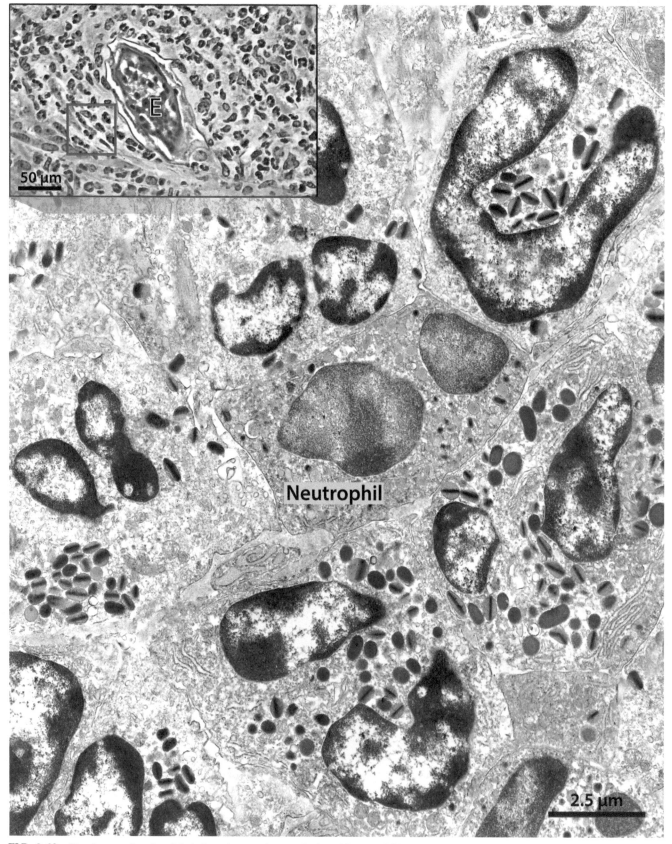

FIG. 9.63 Dominance of eosinophils in hepatic granulomas of mice with acute *Schistosoma mansoni* infection. Note the typical acidophilic eosinophil cytoplasm (box) seen by light microscopy and clustered eosinophils showing their unique ultrastructure with crystalloid granules and polymorphic nuclei by TEM. One neutrophil is seen among eosinophils. Liver samples were prepared for both histology (HE staining, *top panel*) and TEM.[352,433] *E*, parasite egg.

C. The cell biology of mouse eosinophils

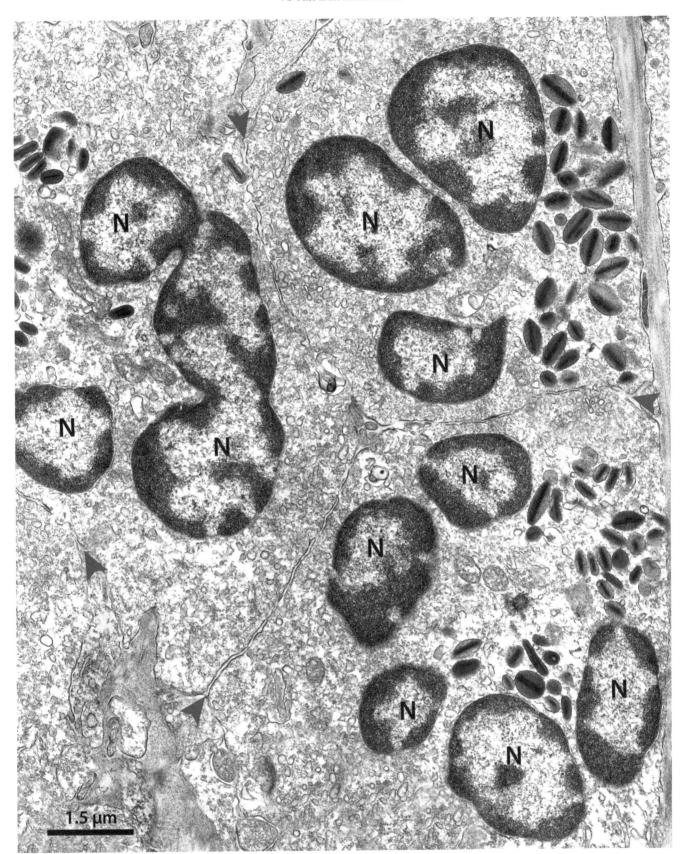

FIG. 9.64   Electron micrograph showing three eosinophils in the liver of a mouse with acute *Schistosoma mansoni* infection. Cells are in tight contact with each other *(arrowheads)*. Note the varied nuclear (N) morphologies. ECM colored in *green*. Liver sample prepared for TEM.[352]

C. The cell biology of mouse eosinophils

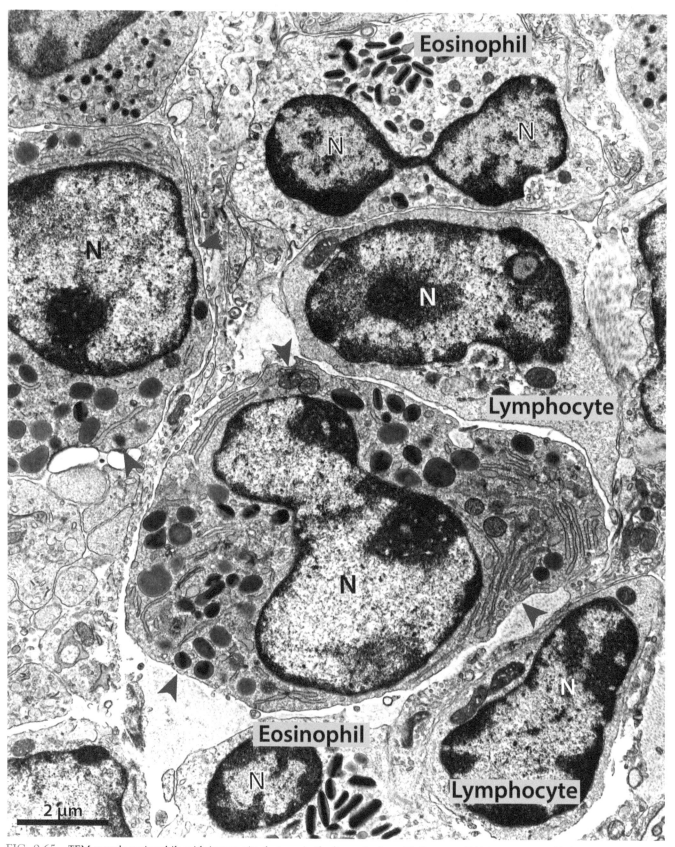

FIG. 9.65   TEM reveals eosinophils with immaturity features in the liver of mice with the acute *Schistosoma mansoni* infection. Incompletely mature eosinophils *(arrowheads)* are intermingled with mature eosinophils and other cell types in the inflammatory infiltrate. One eosinophil is shown in higher magnification in Fig. 9.66. *N*, nucleus.

C. The cell biology of mouse eosinophils

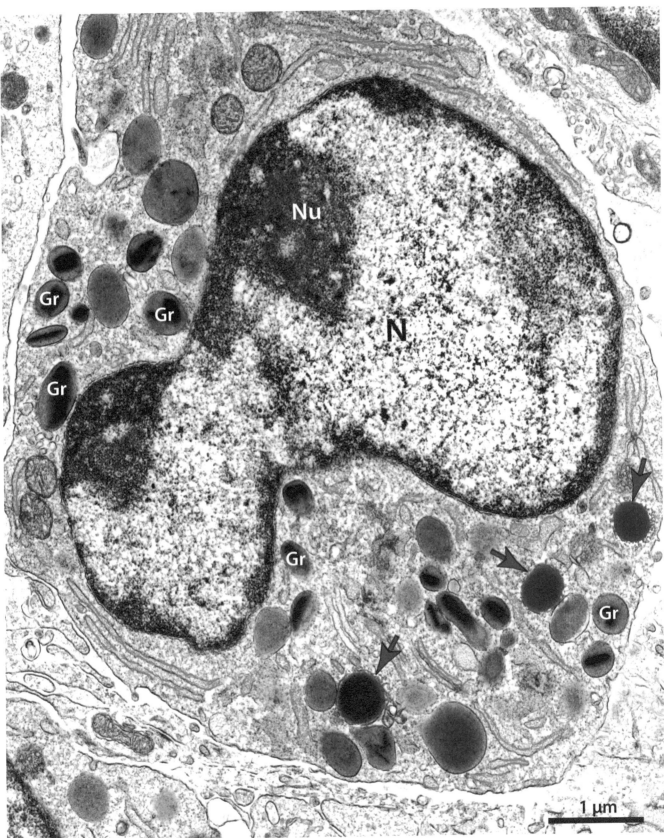

FIG. 9.66 Higher-magnification electron micrograph of an immature eosinophil in the liver of a *Schistosoma mansoni*-infected mouse. The cytoplasm contains cisternae of RER *(colored in blue)*, some coreless granules *(arrows)*, and large, predominantly euchromatic nucleus (N) with prominent nucleolus (Nu). Sample prepared for TEM. *Gr*, mature specific granules.

C. The cell biology of mouse eosinophils

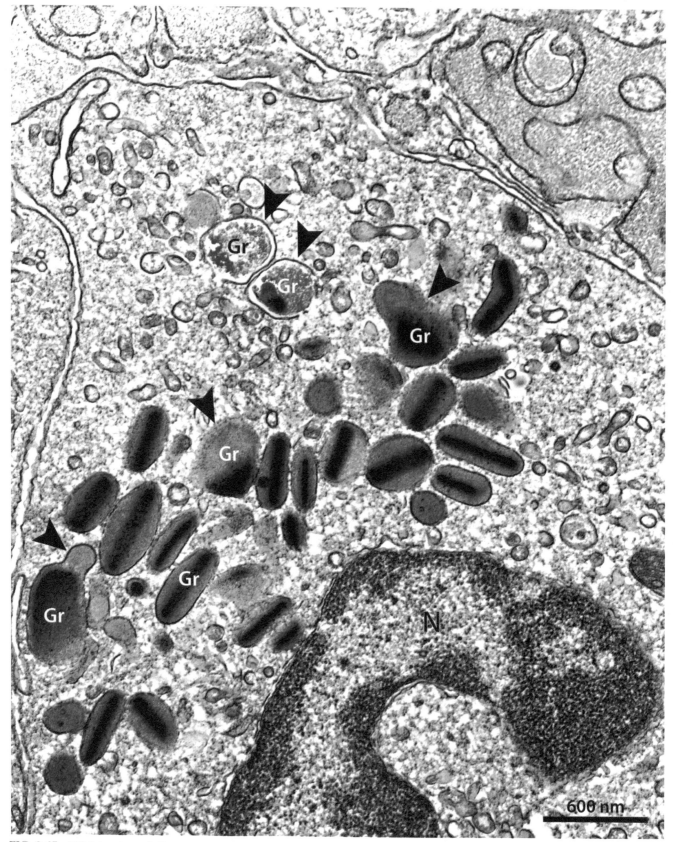

FIG. 9.67   TEM showing an inflammatory eosinophil undergoing PMD in the liver of a mouse model of acute schistosomiasis mansoni. Specific granules (Gr) show enlargement, structural disarrangement and content mobilization *(arrowheads)*. Note the abundance of vesiculotubular structures in the cytoplasm *(colored in pink)*. N, nucleus.

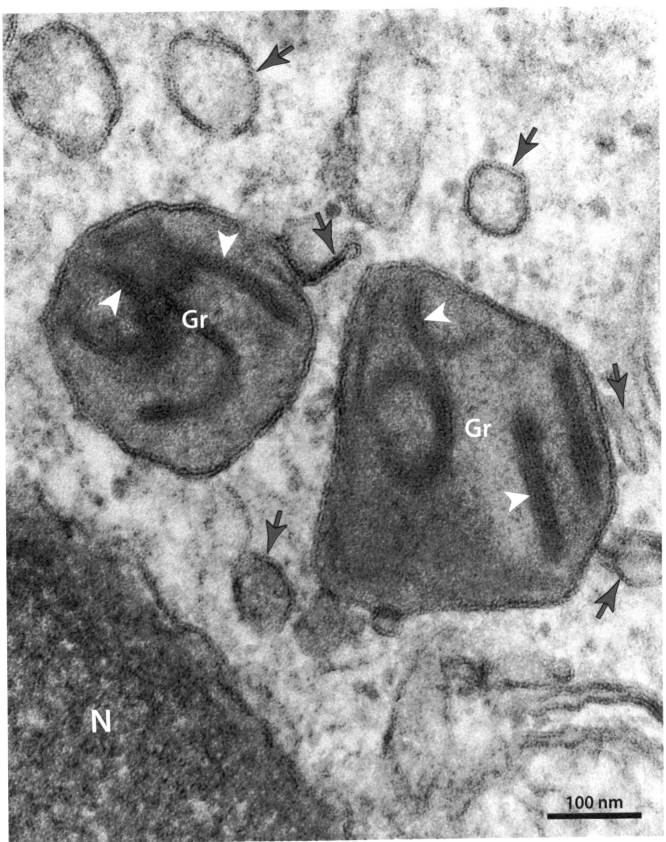

FIG. 9.68 Two specific granules (Gr) are seen at high magnification in the cytoplasm of an inflammatory tissue eosinophil elicited by murine schistosomiasis mansoni. Granules became swollen, and their disarranged crystalloids *(arrowheads)* are spread into the matrices. Note membrane-bound tubulovesicular structures *(arrows)* in contact and close to the granules, a feature of PMD. Liver sample prepared for TEM. *N*, nucleus.

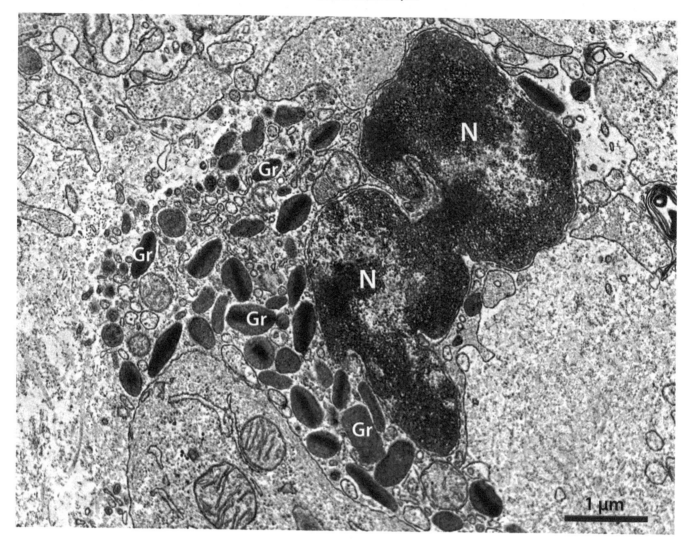

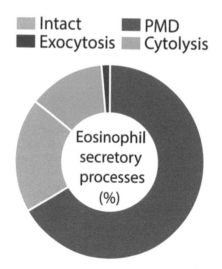

FIG. 9.69 Electron micrograph showing a cytolytic eosinophil in the liver of a *Schistosoma mansoni*-infected mouse. Quantitative analyses demonstrated that 13.1% of the inflammatory eosinophils were undergoing cytolysis while most of them (66.4%) exhibited PMD signs (*n* = 108 cells).[352] Samples prepared for TEM. *Gr*, specific granules; *N*, nucleus.

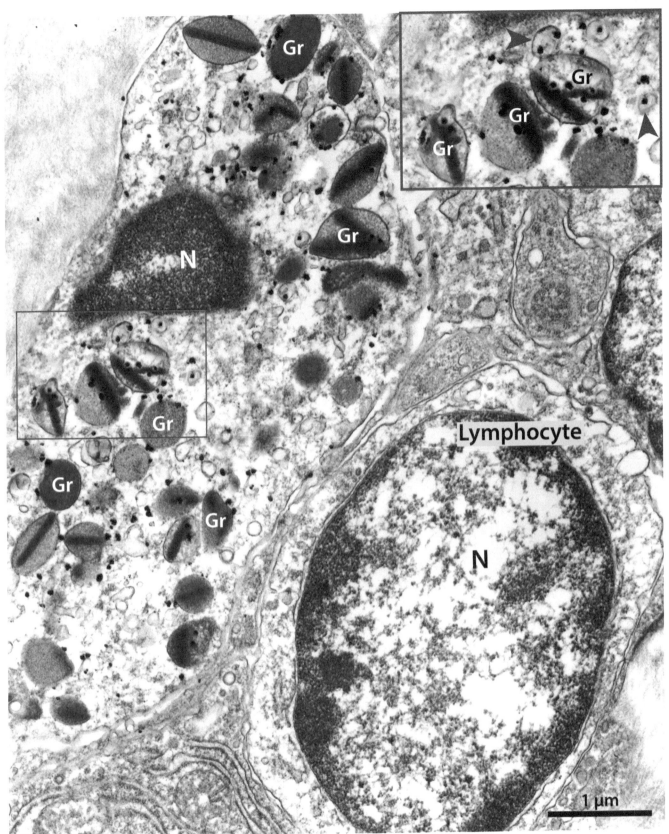

FIG. 9.70 Ultrastructural immunolocalization of MBP-1 in the liver from a *S. mansoni*-infected mouse. Positivity is associated with specific granules (Gr) and cytoplasmic vesicles *(arrowheads)* in an infiltrating eosinophil while an adjacent lymphocyte appears negative.[352] Samples were prepared for immunogold EM.[50] *N*, nucleus.

References

1. Kay AB. The early history of the eosinophil. *Clin Exp Allergy.* 2015;45(3):575–582.
2. Beiträge PE. zur Kenntnis der granulirten Bindegewbszellen und der ecosinophilen. *Leukocythen Arch Anat Physiol (Leipzig).* 1879;3:166–169.
3. Weller PF, Spencer LA. Functions of tissue-resident eosinophils. *Nat Rev Immunol.* 2017;17(12):746–760.
4. Lee JJ, Jacobsen EA, McGarry MP, Schleimer RP, Lee NA. Eosinophils in health and disease: the LIAR hypothesis. *Clin Exp Allergy.* 2010;40(4):563–575.
5. Furuta GT, Atkins FD, Lee NA, Lee JJ. Changing roles of eosinophils in health and disease. *Ann Allergy Asthma Immunol.* 2014;113(1):3–8.
6. Bochner BS. The eosinophil: for better or worse, in sickness and in health. *Ann Allergy Asthma Immunol.* 2018;121(2):150.
7. Melo RCN, Liu L, Xenakis JJ, Spencer LA. Eosinophil-derived cytokines in health and disease: unraveling novel mechanisms of selective secretion. *Allergy.* 2013;68(3):274–284.
8. Spencer LA, Bonjour K, Melo RCN, Weller PF. Eosinophil secretion of granule-derived cytokines. *Front Immunol.* 2014;5:496.
9. Melo RCN, Dvorak AM, Weller PF. Eosinophil ultrastructure. In: Lee JJ, Rosenberg HF, eds. *Eosinophils in Health and Disease.* Elsevier; 2013:20–27.
10. Rosenberg HF, Dyer KD, Foster PS. Eosinophils: changing perspectives in health and disease. *Nat Rev Immunol.* 2013;13(1):9–22.
11. Kautz J, Demarsh QB. An electron microscope study of sectioned cells of peripheral blood and bone marrow. *Blood.* 1954;9(1):24–38.
12. Pease DC. Marrow cells seen with the electron microscopy after ultrathin sectioning. *Rev Hematol.* 1955;10(2):300–313. discussion, 324–344.
13. Grey CE, Biesele JJ. Thin-section electron microscopy of circulating white blood cells. *Rev Hematol.* 1955;10(2):283–299.
14. Pease DC. An electron microscopic study of red bone marrow. *Blood.* 1956;11(6):501–526.
15. Goodman JR, Reilly EB, Moore RE. Electron microscopy of formed elements of normal human blood. *Blood.* 1957;12(5):428–442.
16. Neves JS, Perez SA, Spencer LA, et al. Eosinophil granules function extracellularly as receptor-mediated secretory organelles. *Proc Natl Acad Sci U S A.* 2008;105(47):18478–18483.
17. Saffari H, Hoffman LH, Peterson KA, et al. Electron microscopy elucidates eosinophil degranulation patterns in patients with eosinophilic esophagitis. *J Allergy Clin Immunol.* 2014;133(6):1728–1734. e1721.
18. Melo RCN, Weller PF. Contemporary understanding of the secretory granules in human eosinophils. *J Leukoc Biol.* 2018;104(1):85–93.
19. Melo RCN, Spencer LA, Perez SA, Ghiran I, Dvorak AM, Weller PF. Human eosinophils secrete preformed, granule-stored interleukin-4 through distinct vesicular compartments. *Traffic.* 2005;6(11):1047–1057.
20. Melo RCN, Perez SA, Spencer LA, Dvorak AM, Weller PF. Intragranular vesiculotubular compartments are involved in piecemeal degranulation by activated human eosinophils. *Traffic.* 2005;6(10):866–879.
21. Melo RCN, Spencer LA, Dvorak AM, Weller PF. Mechanisms of eosinophil secretion: large vesiculotubular carriers mediate transport and release of granule-derived cytokines and other proteins. *J Leukoc Biol.* 2008;83(2):229–236.
22. Melo RCN, Spencer LA, Perez SA, et al. Vesicle-mediated secretion of human eosinophil granule-derived major basic protein. *Lab Investig.* 2009;89(7):769–781.
23. Carmo LAS, Bonjour K, Ueki S, et al. CD63 is tightly associated with intracellular, secretory events chaperoning piecemeal degranulation and compound exocytosis in human eosinophils. *J Leukoc Biol.* 2016;100(2):391–401.
24. Melo RCN, Weller PF. Vesicular trafficking of immune mediators in human eosinophils revealed by immunoelectron microscopy. *Exp Cell Res.* 2016;347(2):385–390.
25. Carmo LAS, Bonjour K, Spencer LA, Weller PF, Melo RCN. Single-Cell analyses of human eosinophils at high resolution to understand compartmentalization and vesicular trafficking of interferon-gamma. *Front Immunol.* 2018;9:1542.
26. Melo RCN, Dvorak AM, Weller PF. Contributions of electron microscopy to understand secretion of immune mediators by human eosinophils. *Microsc Microanal.* 2010;16(6):653–660.
27. Frey TG, Perkins GA, Ellisman MH. Electron tomography of membrane-bound cellular organelles. *Annu Rev Biophys Biomol Struct.* 2006;35:199–224.
28. Lewis DM, Lewis JC, Loegering DA, Gleich GJ. Localization of the guinea pig eosinophil major basic protein to the core of the granule. *J Cell Biol.* 1978;77(3):702–713.
29. Egesten A, Alumets J, von Mecklenburg C, Palmegren M, Olsson I. Localization of eosinophil cationic protein, major basic protein, and eosinophil peroxidase in human eosinophils by immunoelectron microscopic technique. *J Histochem Cytochem.* 1986;34(11):1399–1403.
30. Akuthota P, Melo RCN, Spencer LA, Weller PF. MHC Class II and CD9 in human eosinophils localize to detergent-resistant membrane microdomains. *Am J Respir Cell Mol Biol.* 2012;46(2):188–195.
31. Spencer LA, Melo RCN, Perez SA, Bafford SP, Dvorak AM, Weller PF. Cytokine receptor-mediated trafficking of preformed IL-4 in eosinophils identifies an innate immune mechanism of cytokine secretion. *Proc Natl Acad Sci U S A.* 2006;103(9):3333–3338.
32. Radke AL, Reynolds LE, Melo RCN, Dvorak AM, Weller PF, Spencer LA. Mature human eosinophils express functional Notch ligands mediating eosinophil autocrine regulation. *Blood.* 2009;113(13):3092–3101.
33. Ghafouri B, Irander K, Lindbom J, Tagesson C, Lindahl M. Comparative proteomics of nasal fluid in seasonal allergic rhinitis. *J Proteome Res.* 2006;5(2):330–338.
34. Devouassoux G, Pachot A, Laforest L, et al. Galectin-10 mRNA is overexpressed in peripheral blood of aspirin-induced asthma. *Allergy.* 2008;63(1):125–131.
35. De Re V, Simula MP, Cannizzaro R, et al. Galectin-10, eosinophils, and celiac disease. *Ann N Y Acad Sci.* 2009;1173:357–364.
36. Chua JC, Douglass JA, Gillman A, O'Hehir RE, Meeusen EN. Galectin-10, a potential biomarker of eosinophilic airway inflammation. *PLoS One.* 2012;7(8), e42549.
37. Melo RCN, Wang H, Silva TP, et al. Galectin-10, the protein that forms Charcot-Leyden crystals, is not stored in granules but resides in the peripheral cytoplasm of human eosinophils. *J Leukoc Biol.* 2020;108(1):139–149.

38. Acharya KR, Ackerman SJ. Eosinophil granule proteins: form and function. *J Biol Chem*. 2014;289(25):17406–17415.

39. Flaum MA, Schooley RT, Fauci AS, Gralnick HR. A clinicopathologic correlation of the idiopathic hypereosinophilic syndrome. I. Hematologic manifestations. *Blood*. 1981;58(5):1012–1020.

40. Dvorak AM, Ackerman SJ, Weller PF. Subcellular morphology and biochemistry of eosinophils. In: Harris JR, ed. *Blood Cell Biochemistry Megakaryocytes, Platelets, Macrophages, and Eosinophils*. Vol. 2. New York: Plenus Press; 1991:237–344.

41. Ueki S, Tokunaga T, Melo RCN, et al. Charcot-Leyden crystal formation is closely associated with eosinophil extracellular trap cell death. *Blood*. 2018;132(20):2183–2187.

42. Dvorak A, Monahan-Earley R. *Diagnostic Ultrastructural Pathology I. A Text-Atlas of Case Studies Illustrating the Correlative Clinical-Ultrastructural Pathologic Approach to Diagnosis*. Vol. 1. Boca Raton: CRC; 1992. pp. 1–495.

43. Dvorak AM, Monahan-Earley R. *Diagnostic Ultrastructural Pathology II. A Text-Atlas of Case Studies Emphasizing Respiratory and Nervous Systems*. Vol. 2. Boca Raton: CRC; 1992. pp. 1–405.

44. Dvorak AM, Monahan-Earley R. *Diagnostic Ultrastructural Pathology III. A Text-Atlas of Case Studies Emphasizing Endocrine and Hematopoietic Systems*. Vol. 3. Boca Raton: CRC; 1992. pp. 1–442.

45. Karreman MA, Mercier L, Schieber NL, Shibue T, Schwab Y, Goetz JG. Correlating intravital multi-photon microscopy to 3D electron microscopy of invading tumor cells using anatomical reference points. *PLoS One*. 2014;9(12), e114448.

46. Ueki S, Melo RCN, Ghiran I, Spencer LA, Dvorak AM, Weller PF. Eosinophil extracellular DNA trap cell death mediates lytic release of free secretion-competent eosinophil granules in humans. *Blood*. 2013;121(11):2074–2083.

47. Barroso MV, Gropillo I, Detoni M, et al. Structural and signaling events driving Aspergillus fumigatus-induced human eosinophil extracellular trap release. *Front Microbiol*. 2021;12:633696.

48. Ueki S, Tokunaga T, Fujieda S, et al. Eosinophil ETosis and DNA traps: a new look at eosinophilic inflammation. *Curr Allergy Asthma Rep*. 2016;16(8):54.

49. Muniz VS, Silva JC, Braga YAV, et al. Eosinophils release extracellular DNA traps in response to *Aspergillus fumigatus*. *J Allergy Clin Immunol*. 2018;141(2):571–585. e577.

50. Melo RCN, Morgan E, Monahan-Earley R, Dvorak AM, Weller PF. Pre-embedding immunogold labeling to optimize protein localization at subcellular compartments and membrane microdomains of leukocytes. *Nat Protoc*. 2014;9(10):2382–2394.

51. Lee JJ, Jacobsen EA, Ochkur SI, et al. Human versus mouse eosinophils: "that which we call an eosinophil, by any other name would stain as red". *J Allergy Clin Immunol*. 2012;130(3):572–584.

52. Dvorak AM, Weller PF. Ultrastrutural analysis of human eosinophils. In: Marrone G, ed. *Human Eosinophils: Biological and Clinical Aspects*. Vol. 76. Basel: Karger; 2000:1–28.

53. Dvorak A. Ultrastructural studies on mechanisms of human eosinophil activation and secretion. In: Gleich GJ, Kay AB, eds. *Eosinophils in Allergy and Inflammation*. New York: Marcel Dekker, Inc.; 1994:159–209.

54. Dvorak AM. Similarities in the ultrastructural morphology and developmental and secretory mechanisms of human basophils and eosinophils. *J Allergy Clin Immunol*. 1994;94(6 Pt 2):1103–1134.

55. Monahan RA, Dvorak HF, Dvorak AM. Ultrastructural localization of nonspecific esterase activity in guinea pig and human monocytes, macrophages, and lymphocytes. *Blood*. 1981;58(6):1089–1099.

56. Dvorak AM, Ishizaka T. Human eosinophils in vitro. An ultrastructural morphology primer. *Histol Histopathol*. 1994;9(2):339–374.

57. Sparrevohn S, Wulff HR. The nuclear segmentation of eosinophils under normal and pathological conditions. *Acta Haematol*. 1967;37(2):120–125.

58. Wulfhekel U, Dullmann J, Bartels H, Hausmann K. On the ultrastructure and cytochemistry of eosinophil-myelomonocytic leukemias. *Virchows Arch A Pathol Anat Histol*. 1975;365(4):289–308.

59. Kim TH, Gu HJ, Lee WI, Lee J, Yoon HJ, Park TS. Chronic eosinophilic leukemia with FIP1L1-PDGFRA rearrangement. *Blood Res*. 2016;51(3):204–206.

60. Boyer DF. Blood and Bone Marrow Evaluation for Eosinophilia. *Arch Pathol Lab Med*. 2016;140(10):1060–1067.

61. Dvorak AM, Monahan-Earley R. *Procedural Guide to Specimen Handling for the Ultrastructural Pathology Service Laboratory. Diagnostic Ultrastructural Pathology I*. Boca Raton: CRC Press; 1992:473–480.

62. Baumann O, Walz B. Endoplasmic reticulum of animal cells and its organization into structural and functional domains. *Int Rev Cytol*. 2001;205:149–214.

63. Friedman JR, Voeltz GK. The ER in 3D: a multifunctional dynamic membrane network. *Trends Cell Biol*. 2011;21(12):709–717.

64. Hu J, Prinz WA, Rapoport TA. Weaving the web of ER tubules. *Cell*. 2011;147(6):1226–1231.

65. Shibata Y, Shemesh T, Prinz WA, Palazzo AF, Kozlov MM, Rapoport TA. Mechanisms determining the morphology of the peripheral ER. *Cell*. 2010;143(5):774–788.

66. Appenzeller-Herzog C, Ellgaard L. The human PDI family: versatility packed into a single fold. *Biochim Biophys Acta*. 2008;1783(4):535–548.

67. Laurindo FR, Pescatore LA, Fernandes DC. Protein disulfide isomerase in redox cell signaling and homeostasis. *Free Radic Biol Med*. 2012;52(9):1954–1969.

68. Turano C, Coppari S, Altieri F, Ferraro A. Proteins of the PDI family: unpredicted non-ER locations and functions. *J Cell Physiol*. 2002;193(2):154–163.

69. Dias FF, Amaral KB, Carmo LAS, et al. Human eosinophil leukocytes express protein disulfide isomerase in secretory granules and vesicles: ultrastructural studies. *J Histochem Cytochem*. 2014;62(6):450–459.

70. Bozzola JJ, Russell LD. *Electron Microscopy, Principles and Techniques for Biologists*. Boston, MA: Jones & Bartlett Publishers; 1999. pp. 1–670.

71. Lodish H, Berk A, Kaiser CA, et al. *Molecular Cell Biology*. 8th ed. New York: W.H. Freeman; 2016. pp. 1–1280.

72. Bainton DF, Farquhar MG. Segregation and packaging of granule enzymes in eosinophilic leukocytes. *J Cell Biol*. 1970;45(1):54–73.

73. Bainton DF, Farquhar MG. Origin of granules in polymorphonuclear leukocytes. Two types derived from opposite faces of the Golgi complex in developing granulocytes. *J Cell Biol*. 1966;28(2):277–301.

74. Bainton DF, Farquhar MG. Differences in enzyme content of azurophil and specific granules of polymorphonuclear leukocytes. I. Histochemical staining of bone marrow smears. *J Cell Biol*. 1968;39(2):286–298.

75. Ghadially F, Parry E. Probable significance of some morphological variations in the eosinophil granule revealed by the electron microscope. *Nature*. 1965;206(4984):632.

76. Faller A. Zur Frage von Struktur und Aufbau der eosinophilen Granula. *Z Zellforsch Mikrosk Anat*. 1966;69(1):551–565.

77. Hudson G. Eosinophil granules and uranyl acetate an electron microscope study of guinea-pig bone marrow. *Exp Cell Res.* 1967;46(1):121–128.

78. Hudson G, Heap P. Ultrastructure of the eosinophil granule-internum; the problem of reversed relative density. *Z Zellforsch Mikrosk Anat.* 1969;93(3):332–335.

79. Mitsui T, Kami K, Ochi S, Wada M, Kushida H, Fujita K. Atypical eosinophil leukocytes found in a germ-free mouse as revealed by electron microscopy. *Okajimas Folia Anat Jpn.* 1974;51(4):203–217.

80. Poole J. Electron microscopy of polymorphonuclear leucocytes. *Br J Dermatol.* 1969;81:11–18.

81. Zucker-Franklin D. Eosinophil function related to cutaneous disorders. *J Invest Dermatol.* 1978;71(1):100–105.

82. Muniz VS, Weller PF, Neves JS. Eosinophil crystalloid granules: structure, function, and beyond. *J Leukoc Biol.* 2012;92(2):281–288.

83. Pols MS, Klumperman J. Trafficking and function of the tetraspanin CD63. *Exp Cell Res.* 2009;315(9):1584–1592.

84. Mahmudi-Azer S, Downey GP, Moqbel R. Translocation of the tetraspanin CD63 in association with human eosinophil mediator release. *Blood.* 2002;99(11):4039–4047.

85. Melo RCN, D'Avila H, Wan HC, Bozza PT, Dvorak AM, Weller PF. Lipid bodies in inflammatory cells: structure, function, and current imaging techniques. *J Histochem Cytochem.* 2011;59(5):540–556.

86. Melo RCN, Weller PF. Unraveling the complexity of lipid body organelles in human eosinophils. *J Leukoc Biol.* 2014;96(5):703–712.

87. Coimbra A, Lopes-Vaz A. The presence of lipid droplets and the absence of stable sudanophilia in osmium-fixed human leukocytes. *J Histochem Cytochem.* 1971;19(9):551–557.

88. Solley GO, Maldonado JE, Gleich GJ, et al. Endomyocardiopathy with eosinophilia. *Mayo Clin Proc.* 1976;51:697–708.

89. Fujimoto T, Ohsaki Y, Suzuki M, Cheng J. Imaging lipid droplets by electron microscopy. *Methods Cell Biol.* 2013;116:227–251.

90. Melo RCN, D'Avila H, Bozza PT, Weller PF. Imaging lipid bodies within leukocytes with different light microscopy techniques. *Methods Mol Biol.* 2011;689:149–161.

91. Hayes TL, Lindgren FT, Gofman JW. A quantitative determination of the osmium tetroxide-lipoprotein interaction. *J Cell Biol.* 1963;19(1):251–255.

92. Cheng J, Fujita A, Ohsaki Y, Suzuki M, Shinohara Y, Fujimoto T. Quantitative electron microscopy shows uniform incorporation of triglycerides into existing lipid droplets. *Histochem Cell Biol.* 2009;132(3):281–291.

93. Wan HC, Melo RCN, Jin Z, Dvorak AM, Weller PF. Roles and origins of leukocyte lipid bodies: proteomic and ultrastructural studies. *FASEB J.* 2007;21(1):167–178.

94. Melo RCN, Paganoti GF, Dvorak AM, Weller PF. The internal architecture of leukocyte lipid body organelles captured by three-dimensional electron microscopy tomography. *PLoS One.* 2013;8(3), e59578.

95. Melo RCN, Dvorak AM, Weller PF. Electron tomography and immunonanogold electron microscopy for investigating intracellular trafficking and secretion in human eosinophils. *J Cell Mol Med.* 2008;12(4):1416–1419.

96. Olzmann JA, Carvalho P. Dynamics and functions of lipid droplets. *Nat Rev Mol Cell Biol.* 2019;20(3):137–155.

97. Brasaemle DL. Thematic review series: adipocyte biology. The perilipin family of structural lipid droplet proteins: stabilization of lipid droplets and control of lipolysis. *J Lipid Res.* 2007;48(12):2547–2559.

98. Tauchi-Sato K, Ozeki S, Houjou T, Taguchi R, Fujimoto T. The surface of lipid droplets is a phospholipid monolayer with a unique fatty acid composition. *J Biol Chem.* 2002;277(46):44507–44512.

99. Kimmel AR, Brasaemle DL, McAndrews-Hill M, Sztalryd C, Londos C. Adoption of PERILIPIN as a unifying nomenclature for the mammalian PAT-family of intracellular lipid storage droplet proteins. *J Lipid Res.* 2010;51(3):468–471.

100. Brasaemle DL, Dolios G, Shapiro L, Wang R. Proteomic analysis of proteins associated with lipid droplets of basal and lipolytically stimulated 3T3-L1 adipocytes. *J Biol Chem.* 2004;279(45):46835–46842.

101. Wolins NE, Rubin B, Brasaemle DL. TIP47 associates with lipid droplets. *J Biol Chem.* 2001;276(7):5101–5108.

102. Bickel PE, Tansey JT, Welte MA. PAT proteins, an ancient family of lipid droplet proteins that regulate cellular lipid stores. *Biochim Biophys Acta.* 2009;1791(6):419–440.

103. Weller PF, Dvorak AM. Arachidonic acid incorporation by cytoplasmic lipid bodies of human eosinophils. *Blood.* 1985;65(5):1269–1274.

104. Weller PF, Monahan-Earley RA, Dvorak HF, Dvorak AM. Cytoplasmic lipid bodies of human eosinophils. Subcellular isolation and analysis of arachidonate incorporation. *Am J Pathol.* 1991;138(1):141–148.

105. Melo RCN, Weller PF, Dvorak AM. Activated human eosinophils. *Int Arch Allergy Immunol.* 2005;138(4):347–349.

106. Wu H, Carvalho P, Voeltz GK. Here, there, and everywhere: The importance of ER membrane contact sites. *Science.* 2018;361(6401), eaan5835.

107. Klumperman J. Architecture of the mammalian Golgi. *Cold Spring Harb Perspect Biol.* 2011;3(7), a005181.

108. Dvorak AM, Estrella P, Ishizaka T. Vesicular transport of peroxidase in human eosinophilic myelocytes. *Clin Exp Allergy.* 1994;24(1):10–18.

109. Dvorak A, Ishizaka T, Weller P, Ackerman S. Ultrastructural contributions to the understanding of the cell biology of human eosinophils: mechanisms of growth factor-induced development, secretion, and resolution of released constituents from the microenvironment. In: *Eosinophils: Biological and Clinical Aspects.* Boca Raton, FL: CRC Press; 1993:13–32.

110. Gordon DE, Bond LM, Sahlender DA, Peden AA. A targeted siRNA screen to identify SNAREs required for constitutive secretion in mammalian cells. *Traffic.* 2010;11(9):1191–1204.

111. Muppirala M, Gupta V, Swarup G. Syntaxin 17 cycles between the ER and ERGIC and is required to maintain the architecture of ERGIC and Golgi. *Biol Cell.* 2011;103(7):333–350.

112. Muppirala M, Gupta V, Swarup G. Emerging role of tyrosine phosphatase, TCPTP, in the organelles of the early secretory pathway. *Biochim Biophys Acta.* 2013;1833(5):1125–1132.

113. Carmo LAS, Dias FF, Malta KK, et al. Expression and subcellular localization of the Qa-SNARE syntaxin17 in human eosinophils. *Exp Cell Res.* 2015;337(2):129–135.

114. Fujiwara T, Oda K, Yokota S, Takatsuki A, Ikehara Y. Brefeldin A causes disassembly of the Golgi complex and accumulation of secretory proteins in the endoplasmic reticulum. *J Biol Chem.* 1988;263(34):18545–18552.

115. Huotari J, Helenius A. Endosome maturation. *EMBO J.* 2011;30(17):3481–3500.

116. Naslavsky N, Caplan S. The enigmatic endosome – sorting the ins and outs of endocytic trafficking. *J Cell Sci.* 2018;131(13), jcs216499.

117. Klumperman J, Raposo G. The complex ultrastructure of the endolysosomal system. *Cold Spring Harb Perspect Biol.* 2014;6(10):a016857.

118. Jovic M, Sharma M, Rahajeng J, Caplan S. The early endosome: a busy sorting station for proteins at the crossroads. *Histol Histopathol.* 2010;25(1):99–112.

119. Poteryaev D, Datta S, Ackema K, Zerial M, Spang A. Identification of the switch in early-to-late endosome transition. *Cell.* 2010;141(3):497–508.

120. Marchetti A, Mercanti V, Cornillon S, Alibaud L, Charette SJ, Cosson P. Formation of multivesicular endosomes in Dictyostelium. *J Cell Sci.* 2004;117(25):6053–6059.

121. McMahon HT, Boucrot E. Molecular mechanism and physiological functions of clathrin-mediated endocytosis. *Nat Rev Mol Cell Biol.* 2011;12(8):517–533.

122. Robinson MS. Forty years of clathrin-coated vesicles. *Traffic.* 2015;16(12):1210–1238.

123. Sachse M, Urbe S, Oorschot V, Strous GJ, Klumperman J. Bilayered clathrin coats on endosomal vacuoles are involved in protein sorting toward lysosomes. *Mol Biol Cell.* 2002;13(4):1313–1328.

124. Murk JL, Posthuma G, Koster AJ, et al. Influence of aldehyde fixation on the morphology of endosomes and lysosomes: quantitative analysis and electron tomography. *J Microsc.* 2003;212(Pt 1):81–90.

125. Akuthota P, Carmo LAS, Bonjour K, et al. Extracellular microvesicle production by human eosinophils activated by "inflammatory" stimuli. *Front Cell Dev Biol.* 2016;4:117.

126. Kobayashi T, Vischer UM, Rosnoblet C, et al. The tetraspanin CD63/lamp3 cycles between endocytic and secretory compartments in human endothelial cells. *Mol Biol Cell.* 2000;11(5):1829–1843.

127. Tai PC, Spry CJ. The mechanisms which produce vacuolated and degranulated eosinophils. *Br J Haematol.* 1981;49(2):219–226.

128. Cowdry EV. The vital staining of mitochondria with Janus green and diethylsafranin in human blood cells. *Internat Monatsch f Anat u Physiol.* 1914;31:267.

129. Peachman KK, Lyles DS, Bass DA. Mitochondria in eosinophils: functional role in apoptosis but not respiration. *Proc Natl Acad Sci U S A.* 2001;98(4):1717–1722.

130. Loud AV. A quantitative stereological description of the ultrastructure of normal rat liver parenchymal cells. *J Cell Biol.* 1968;37(1):27–46.

131. Friedman JR, Nunnari J. Mitochondrial form and function. *Nature.* 2014;505(7483):335–343.

132. Zick M, Rabl R, Reichert AS. Cristae formation-linking ultrastructure and function of mitochondria. *Biochim Biophys Acta.* 2009;1793(1):5–19.

133. Scott RB, Still WJ. Glycogen in human peripheral blood leukocytes. II. The macromolecular state of leukocyte glycogen. *J Clin Invest.* 1968;47(2):353–359.

134. Robinson JM, Karnovsky ML, Karnovsky MJ. Glycogen accumulation in polymorphonuclear leukocytes, and other intracellular alterations that occur during inflammation. *J Cell Biol.* 1982;95(3):933–942.

135. Davoine F, Lacy P. Eosinophil cytokines, chemokines, and growth factors: emerging roles in immunity. *Front Immunol.* 2014;5:570.

136. Jamieson JD, Palade GE. Intracellular transport of secretory proteins in the pancreatic exocrine cell: I. Role of the peripheral elements of the Golgi complex. *J Cell Biol.* 1967;34(2):577–596.

137. Jamieson JD, Palade GE. Intracellular transport of secretory proteins in the pancreatic exocrine cell: II. Transport to condensing vacuoles and zymogen granules. *J Cell Biol.* 1967;34(2):597–615.

138. Watson P, Stephens DJ. ER-to-Golgi transport: form and formation of vesicular and tubular carriers. *Biochim Biophys Acta.* 2005;1744(3):304–315.

139. Mazzeo C, Canas JA, Zafra MP, et al. Exosome secretion by eosinophils: a possible role in asthma pathogenesis. *J Allergy Clin Immunol.* 2015;135(6):1603–1613.

140. McLaren DJ, Mackenzie CD, Ramalho-Pinto FJ. Ultrastructural observations on the in vitro interaction between rat eosinophils and some parasitic helminths (Schistosoma mansoni, Trichinella spiralis and Nippostrongylus brasiliensis). *Clin Exp Immunol.* 1977;30(1):105–118.

141. Inoue Y, Matsuwaki Y, Shin SH, Ponikau JU, Kita H. Nonpathogenic, environmental fungi induce activation and degranulation of human eosinophils. *J Immunol.* 2005;175(8):5439–5447.

142. Dvorak AM, Onderdonk AB, McLeod RS, et al. Ultrastructural identification of exocytosis of granules from human gut eosinophils in vivo. *Int Arch Allergy Immunol.* 1993;102(1):33–45.

143. Dvorak HF, Mihm Jr MC, Dvorak AM, et al. Morphology of delayed type hypersensitivity reactions in man. I. Quantitative description of the inflammatory response. *Lab Investig.* 1974;31(2):111–130.

144. Dvorak AM, Mihm Jr MC, Dvorak HF. Degranulation of basophilic leukocytes in allergic contact dermatitis reactions in man. *J Immunol.* 1976;116(3):687–695.

145. Dvorak AM. Ultrastructural evidence for release of major basic protein-containing crystalline cores of eosinophil granules in vivo: cytotoxic potential in Crohn's disease. *J Immunol.* 1980;125(1):460–462.

146. Dvorak AM, Furitsu T, Letourneau L, Ishizaka T, Ackerman SJ. Mature eosinophils stimulated to develop in human cord blood mononuclear cell cultures supplemented with recombinant human interleukin-5. Part I. Piecemeal degranulation of specific granules and distribution of Charcot-Leyden crystal protein. *Am J Pathol.* 1991;138(1):69–82.

147. Dvorak AM, Ackerman SJ, Furitsu T, Estrella P, Letourneau L, Ishizaka T. Mature eosinophils stimulated to develop in human-cord blood mononuclear cell cultures supplemented with recombinant human interleukin-5. II. Vesicular transport of specific granule matrix peroxidase, a mechanism for effecting piecemeal degranulation. *Am J Pathol.* 1992;140(4):795–807.

148. Melo RCN, Weller PF. Piecemeal degranulation in human eosinophils: a distinct secretion mechanism underlying inflammatory responses. *Histol Histopathol.* 2010;25(10):1341–1354.

149. Friesen CA, Andre L, Garola R, Hodge C, Roberts C. Activated duodenal mucosal eosinophils in children with dyspepsia: a pilot transmission electron microscopic study. *J Pediatr Gastroenterol Nutr.* 2002;35(3):329–333.

150. Karawajczyk M, Seveus L, Garcia R, et al. Piecemeal degranulation of peripheral blood eosinophils: a study of allergic subjects during and out of the pollen season. *Am J Respir Cell Mol Biol.* 2000;23(4):521–529.

151. Erjefalt JS, Greiff L, Andersson M, Adelroth E, Jeffery PK, Persson CG. Degranulation patterns of eosinophil granulocytes as determinants of eosinophil driven disease. *Thorax.* 2001;56(5):341–344.

152. Armengot M, Garin L, Carda C. Eosinophil degranulation patterns in nasal polyposis: an ultrastructural study. *Am J Rhinol Allergy.* 2009;23(5):466–470.

153. Ahlstrom-Emanuelsson CA, Greiff L, Andersson M, Persson CG, Erjefalt JS. Eosinophil degranulation status in allergic rhinitis: observations before and during seasonal allergen exposure. *Eur Respir J.* 2004;24(5):750–757.

154. Cheng JF, Ott NL, Peterson EA, et al. Dermal eosinophils in atopic dermatitis undergo cytolytic degeneration. *J Allergy Clin Immunol.* 1997;99(5):683–692.

155. Caruso RA, Ieni A, Fedele F, et al. Degranulation patterns of eosinophils in advanced gastric carcinoma: an electron microscopic study. *Ultrastruct Pathol.* 2005;29(1):29–36.

156. Vanheel H, Vicario M, Boesmans W, et al. Activation of eosinophils and mast cells in functional dyspepsia: an ultrastructural evaluation. *Sci Rep.* 2018;8(1):5383.

157. Raqib R, Moly PK, Sarker P, et al. Persistence of mucosal mast cells and eosinophils in Shigella-infected children. *Infect Immun.* 2003;71(5):2684–2692.

158. Qadri F, Bhuiyan TR, Dutta KK, et al. Acute dehydrating disease caused by Vibrio cholerae serogroups O1 and O139 induce increases in innate cells and inflammatory mediators at the mucosal surface of the gut. *Gut.* 2004;53(1):62–69.

159. Persson C, Uller L. Primary lysis of eosinophils as a major mode of activation of eosinophils in human diseased tissues. *Nat Rev Immunol.* 2013;13(12):902.

160. Persson CG, Erjefalt JS. Eosinophil lysis and free granules: an in vivo paradigm for cell activation and drug development. *Trends Pharmacol Sci.* 1997;18(4):117–123.

161. Erjefalt JS, Persson CG. New aspects of degranulation and fates of airway mucosal eosinophils. *Am J Respir Crit Care Med.* 2000;161(6):2074–2085.

162. Radonjic-Hoesli S, Wang X, de Graauw E, et al. Adhesion-induced eosinophil cytolysis requires the receptor-interacting protein kinase 3 (RIPK3)-mixed lineage kinase-like (MLKL) signaling pathway, which is counterregulated by autophagy. *J Allergy Clin Immunol.* 2017;140(6):1632–1642.

163. Ueki S, Konno Y, Takeda M, et al. Eosinophil extracellular trap cell death-derived DNA traps: Their presence in secretions and functional attributes. *J Allergy Clin Immunol.* 2016;137(1):258–267.

164. Ueki S, Miyabe Y, Yamamoto Y, et al. Charcot-leyden crystals in eosinophilic inflammation: active cytolysis leads to crystal formation. *Curr Allergy Asthma Rep.* 2019;19(8):35.

165. Mukherjee M, Lacy P, Ueki S. Eosinophil extracellular traps and inflammatory pathologies-untangling the web! *Front Immunol.* 2018;9:2763.

166. Neves JS, Perez SAC, Melo RCN, Weller PF, Spencer LA. A new paradigm for eosinophil granule-dependent secretion. *Commun Integr Biol.* 2009;2(6):1–3.

167. Neves JS, Weller PF. Functional extracellular eosinophil granules: novel implications in eosinophil immunobiology. *Curr Opin Immunol.* 2009;21(6):694–699.

168. Persson CG, Erjefalt JS. "Ultimate activation" of eosinophils in vivo: lysis and release of clusters of free eosinophil granules (Cfegs). *Thorax.* 1997;52(6):569–574.

169. Persson CG. Centennial notions of asthma as an eosinophilic, desquamative, exudative, and steroid-sensitive disease. *Lancet.* 1997;350(9083):1021–1024.

170. Erjefalt JS, Greiff L, Andersson M, et al. Allergen-induced eosinophil cytolysis is a primary mechanism for granule protein release in human upper airways. *Am J Respir Crit Care Med.* 1999;160(1):304–312.

171. Watanabe K, Misu T, Inoue S, Edamatsu H. Cytolysis of eosinophils in nasal secretions. *Ann Otol Rhinol Laryngol.* 2003;112(2):169–173.

172. Greiff L, Erjefalt JS, Andersson M, Svensson C, Persson CG. Generation of clusters of free eosinophil granules (Cfegs) in seasonal allergic rhinitis. *Allergy.* 1998;53(2):200–203.

173. Uller L, Andersson M, Greiff L, Persson CG, Erjefalt JS. Occurrence of apoptosis, secondary necrosis, and cytolysis in eosinophilic nasal polyps. *Am J Respir Crit Care Med.* 2004;170(7):742–747.

174. Erjefalt JS, Andersson M, Greiff L, et al. Cytolysis and piecemeal degranulation as distinct modes of activation of airway mucosal eosinophils. *J Allergy Clin Immunol.* 1998;102(2):286–294.

175. Gonzalez EB, Swedo JL, Rajaraman S, Daniels JC, Grant JA. Ultrastructural and immunohistochemical evidence for release of eosinophilic granules in vivo: cytotoxic potential in chronic eosinophilic pneumonia. *J Allergy Clin Immunol.* 1987;79(5):755–762.

176. Fox B, Seed WA. Chronic eosinophilic pneumonia. *Thorax.* 1980;35(8):570–580.

177. Grantham JG, Meadows 3rd JA, Gleich GJ. Chronic eosinophilic pneumonia. Evidence for eosinophil degranulation and release of major basic protein. *Am J Med.* 1986;80(1):89–94.

178. Capocelli KE, Fernando SD, Menard-Katcher C, Furuta GT, Masterson JC, Wartchow EP. Ultrastructural features of eosinophilic oesophagitis: impact of treatment on desmosomes. *J Clin Pathol.* 2015;68(1):51–56.

179. Buzas EI, György B, Nagy G, Falus A, Gay S. Emerging role of extracellular vesicles in inflammatory diseases. *Nat Rev Rheumatol.* 2014;10(6):356–364.

180. Groot Kormelink T, Mol S, de Jong EC, Wauben MHM. The role of extracellular vesicles when innate meets adaptive. *Semin Immunopathol.* 2018;40(5):439–452.

181. Twu O, Johnson PJ. Parasite extracellular vesicles: mediators of intercellular communication. *PLoS Pathog.* 2014;10(8), e1004289.

182. Lawson C, Vicencio JM, Yellon DM, Davidson SM. Microvesicles and exosomes: new players in metabolic and cardiovascular disease. *J Endocrinol.* 2016;228(2):R57–R71.

183. van der Pol E, Böing AN, Harrison P, Sturk A, Nieuwland R. Classification, functions, and clinical relevance of extracellular vesicles. *Pharmacol Rev.* 2012;64(3):676–705.

184. Canas JA, Sastre B, Mazzeo C, et al. Exosomes from eosinophils autoregulate and promote eosinophil functions. *J Leukoc Biol.* 2017;101(5):1191–1199.

185. Lötvall J, Hill AF, Hochberg F, et al. Minimal experimental requirements for definition of extracellular vesicles and their functions: a position statement from the International Society for Extracellular Vesicles. *J Extracell Vesicles.* 2014;3, 26913.

186. Andreu Z, Yáñez-Mó M. Tetraspanins in extracellular vesicle formation and function. *Front Immunol.* 2014;5:442.

187. Kernen P, Wymann MP, von Tscharner V, et al. Shape changes, exocytosis, and cytosolic free calcium changes in stimulated human eosinophils. *J Clin Invest.* 1991;87(6):2012–2017.

188. Johansson MW. Eosinophil activation status in separate compartments and association with asthma. *Front Med.* 2017;4:75.

189. Kroegel C, Dewar A, Yukawa T, Venge P, Barnes PJ, Chung KF. Ultrastructural characterization of platelet-activating factor-stimulated human eosinophils from patients with asthma. *Clin Sci (Lond).* 1993;84(4):391–399.

190. Malm-Erjefalt M, Greiff L, Ankerst J, et al. Circulating eosinophils in asthma, allergic rhinitis, and atopic dermatitis lack morphological signs of degranulation. *Clin Exp Allergy.* 2005;35(10):1334–1340.

191. Reber L, Da Silva CA, Frossard N. Stem cell factor and its receptor c-Kit as targets for inflammatory diseases. *Eur J Pharmacol.* 2006;533(1-3):327–340.

192. Liu LY, Bates ME, Jarjour NN, Busse WW, Bertics PJ, Kelly EA. Generation of Th1 and Th2 chemokines by human eosinophils: evidence for a critical role of TNF-alpha. *J Immunol.* 2007;179(7):4840–4848.

193. Spencer LA, Szela CT, Perez SA, et al. Human eosinophils constitutively express multiple Th1, Th2, and immunoregulatory cytokines that are secreted rapidly and differentially. *J Leukoc Biol.* 2009;85(1):117–123.

194. Dvorak AM, Costa JJ, Monahan-Earley RA, Fox P, Galli SJ. Ultrastructural analysis of human skin biopsy specimens from patients receiving recombinant human stem cell factor: subcutaneous injection of rhSCF induces dermal mast cell degranulation and granulocyte recruitment at the injection site. *J Allergy Clin Immunol.* 1998;101(6):793–806.

195. Okuda M, Takenaka T, Kawabori S, Ogami Y. Ultrastructural study of the specific granule of the human eosinophil. *J Submicrosc Cytol.* 1981;13(3):465–471.

196. Parmley RT, Spicer SS. Altered tissue eosinophils in Hodgkin's disease. *Exp Mol Pathol.* 1975;23(1):70–82.

197. Dvorak AM, Monahan RA, Osage JE, Dickersin GR. Crohn's disease: transmission electron microscopic studies. II. Immunologic inflammatory response. Alterations of mast cells, basophils, eosinophils, and the microvasculature. *Hum Pathol.* 1980;11(6):606–619.

198. Torpier G, Colombel JF, Mathieu-Chandelier C, et al. Eosinophilic gastroenteritis: ultrastructural evidence for a selective release of eosinophil major basic protein. *Clin Exp Immunol.* 1988;74(3):404–408.

199. Stirling JW. Ultrastructural localization of lysozyme in human colon eosinophils using the protein A-gold technique: effects of processing on probe distribution. *J Histochem Cytochem.* 1989;37(5):709–714.

200. Henderson WR, Chi EY. Ultrastructural characterization and morphometric analysis of human eosinophil degranulation. *J Cell Sci.* 1985;73(1):33–48.

201. El-Hashimi W. Charcot-Leyden crystals. Formation from primate and lack of formation from nonprimate eosinophils. *Am J Pathol.* 1971;65(2):311–324.

202. Dinter A, Berger EG. Golgi-disturbing agents. *Histochem Cell Biol.* 1998;109(5-6):571–590.

203. Zeck-Kapp G, Kroegel C, Riede UN, Kapp A. Mechanisms of human eosinophil activation by complement protein C5a and platelet-activating factor: similar functional responses are accompanied by different morphologic alterations. *Allergy.* 1995;50(1):34–47.

204. Weller PF, Marshall WL, Lucey DR, Rand TH, Dvorak AM, Finberg RW. Infection, apoptosis, and killing of mature human eosinophils by human immunodeficiency virus-1. *Am J Respir Cell Mol Biol.* 1995;13(5):610–620.

205. Bozza PT, Melo RCN, Bandeira-Melo C. Leukocyte lipid bodies regulation and function: contribution to allergy and host defense. *Pharmacol Ther.* 2007;113(1):30–49.

206. Melo RCN, Weller PF. Lipid droplets in leukocytes: Organelles linked to inflammatory responses. *Exp Cell Res.* 2016;340(2):193–197.

207. Dias FF, Zarantonello VC, Parreira GG, Chiarini-Garcia H, Melo RCN. The intriguing ultrastructure of lipid body organelles within activated macrophages. *Microsc Microanal.* 2014;20(3):869–878.

208. Weller PF. Leukocyte lipid bodies – structure and function as "eicosasomes". *Trans Am Clin Climatol Assoc.* 2016;127:328–340.

209. Ishida T, Matsumura Y, Miyake A, Amitani R. Ultrastructural observation of eosinophils in bronchoalveolar lavage fluid in eosinophilic pneumonia. *Nihon Kyobu Shikkan Gakkai Zasshi.* 1992;30(11):1951–1956.

210. Weller PF, Bubley GJ. The idiopathic hypereosinophilic syndrome. *Blood.* 1994;83(10):2759–2779.

211. Melo RCN, Sabban A, Weller PF. Leukocyte lipid bodies: inflammation-related organelles are rapidly detected by wet scanning electron microscopy. *J Lipid Res.* 2006;47(11):2589–2594.

212. Weller PF, Lee CW, Foster DW, Corey EJ, Austen KF, Lewis RA. Generation and metabolism of 5-lipoxygenase pathway leukotrienes by human eosinophils: predominant production of leukotriene C4. *Proc Natl Acad Sci U S A.* 1983;80(24):7626–7630.

213. Henderson WR, Harley JB, Fauci AS. Arachidonic acid metabolism in normal and hypereosinophilic syndrome human eosinophils: generation of leukotrienes B4, C4, D4 and 15-lipoxygenase products. *Immunology.* 1984;51(4):679–686.

214. Bozza PT, Yu W, Cassara J, Weller PF. Pathways for eosinophil lipid body induction: differing signal transduction in cells from normal and hypereosinophilic subjects. *J Leukoc Biol.* 1998;64(4):563–569.

215. Melo RCN, Fabrino DL, Dias FF, Parreira GG. Lipid bodies: Structural markers of inflammatory macrophages in innate immunity. *Inflamm Res.* 2006;55(8):342–348.

216. D'Avila H, Melo RCN, Parreira GG, Werneck-Barroso E, Castro-Faria-Neto HC, Bozza PT. *Mycobacterium bovis* bacillus Calmette-Guerin induces TLR2-mediated formation of lipid bodies: intracellular domains for eicosanoid synthesis in vivo. *J Immunol.* 2006;176(5):3087–3097.

217. Hayes TL, Hawkins JN. Osmium tetroxide-triglyceride interaction as a function of degree of unsaturation. An X-ray fluorescence study. Ucrl-11184. *UCRL US At Energy Comm.* 1963;72:110–112.

218. Adams CW, Abdulla YH, Bayliss OB. Osmium tetroxide as a histochemical and histological reagent. *For Hist.* 1967;9(1):68–77.

219. Beil WJ, Weller PF, Tzizik DM, Galli SJ, Dvorak AM. Ultrastructural immunogold localization of tumor necrosis factor-alpha to the matrix compartment of eosinophil secondary granules in patients with idiopathic hypereosinophilic syndrome. *J Histochem Cytochem.* 1993;41(11):1611–1615.

220. Beil WJ, Weller PF, Peppercorn MA, Galli SJ, Dvorak AM. Ultrastructural immunogold localization of subcellular sites of TNF-alpha in colonic Crohn's disease. *J Leukoc Biol.* 1995;58(3):284–298.

221. Singer SJ, Kupfer A. The directed migration of eukaryotic cells. *Annu Rev Cell Biol.* 1986;2:337–365.

222. Gilbert SH, Perry K, Fay FS. Mediation of chemoattractant-induced changes in [Ca2+]i and cell shape, polarity, and locomotion by InsP3, DAG, and protein kinase C in newt eosinophils. *J Cell Biol.* 1994;127(2):489–503.

223. Han ST, Mosher DF. IL-5 induces suspended eosinophils to undergo unique global reorganization associated with priming. *Am J Respir Cell Mol Biol.* 2014;50(3):654–664.

224. Choi EN, Choi MK, Park CS, Chung IY. A parallel signal-transduction pathway for eotaxin- and interleukin-5-induced eosinophil shape change. *Immunology.* 2003;108(2):245–256.

225. Ip WK, Wong CK, Wang CB, Tian YP, Lam CWK. Interleukin-3,-5, and granulocyte macrophage colony-stimulating factor induce adhesion and chemotaxis of human eosinophils via p38 mitogen-activated protein kinase and nuclear factor κB. *Immunopharmacol Immunotoxicol.* 2005;27(3):371–393.

226. Bandeira-Melo C, Phoofolo M, Weller PF. Extranuclear lipid bodies, elicited by CCR3-mediated signaling pathways, are the sites of chemokine-enhanced leukotriene C4 production in eosinophils and basophils. *J Biol Chem.* 2001;276(25):22779–22787.

227. Mattila PK, Lappalainen P. Filopodia: molecular architecture and cellular functions. *Nat Rev Mol Cell Biol.* 2008;9(6):446–454.

228. Innocenti M. New insights into the formation and the function of lamellipodia and ruffles in mesenchymal cell migration. *Cell Adhes Migr.* 2018;12(5):401–416.

229. Fais S, Malorni W. Leukocyte uropod formation and membrane/cytoskeleton linkage in immune interactions. *J Leukoc Biol.* 2003;73(5):556–563.

230. Lampinen M, Carlson M, Hakansson LD, Venge P. Cytokine-regulated accumulation of eosinophils in inflammatory disease. *Allergy.* 2004;59(8):793–805.

231. Koster AJ, Klumperman J. Electron microscopy in cell biology: integrating structure and function. *Nat Rev Mol Cell Biol.* 2003;9(Suppl):6–10.

232. Griffiths G, Lucocq JM. Antibodies for immunolabeling by light and electron microscopy: not for the faint hearted. *Histochem Cell Biol.* 2014;142(4):347–360.

233. Schikorski T. Pre-embedding immunogold localization of antigens in mammalian brain slices. *Methods Mol Biol.* 2010;657:133–144.

234. Baschong W, Stierhof YD. Preparation, use, and enlargement of ultrasmall gold particles in immunoelectron microscopy. *Microsc Res Tech.* 1998;42(1):66–79.

235. Roth J, Bendayan M, Orci L. Ultrastructural localization of intracellular antigens by the use of protein A-gold complex. *J Histochem Cytochem.* 1978;26(12):1074–1081.

236. Feng D, Nagy JA, Brekken RA, et al. Ultrastructural localization of the vascular permeability factor/vascular endothelial growth factor (VPF/VEGF) receptor-2 (FLK-1, KDR) in normal mouse kidney and in the hyperpermeable vessels induced by VPF/VEGF-expressing tumors and adenoviral vectors. *J Histochem Cytochem.* 2000;48(4):545–556.

237. Yamaguchi H, Nakagawa I, Yamamoto A, Amano A, Noda T, Yoshimori T. An initial step of GAS-containing autophagosome-like vacuoles formation requires Rab7. *PLoS Pathog.* 2009;5(11), e1000670.

238. Yamazaki H, Oda M, Takahashi Y, et al. Relation between ultrastructural localization, changes in caveolin-1, and capillarization of liver sinusoidal endothelial cells in human hepatitis C-related cirrhotic liver. *J Histochem Cytochem.* 2013;61(2):169–176.

239. Yamamoto A, Masaki R. Pre-embedding nanogold silver and gold intensification. *Methods Mol Biol.* 2010;657:225–235.

240. Dvorak AM, Morgan ES, Tzizik DM, Weller PF. Prostaglandin endoperoxide synthase (cyclooxygenase): ultrastructural localization to nonmembrane-bound cytoplasmic lipid bodies in human eosinophils and 3T3 fibroblasts. *Int Arch Allergy Immunol.* 1994;105(3):245–250.

241. Gleich GJ. Mechanisms of eosinophil-associated inflammation. *J Allergy Clin Immunol.* 2000;105(4):651–663.

242. Trivedi SG, Lloyd CM. Eosinophils in the pathogenesis of allergic airways disease. *Cell Mol Life Sci.* 2007;64(10):1269–1289.

243. Gleich GJ, Loegering DA, Kueppers F, Bajaj SP, Mann KG. Physiochemical and biological properties of the major basic protein from guinea pig eosinophil granules. *J Exp Med.* 1974;140(2):313–332.

244. Lewis DM, Loegering DA, Gleich GJ. Isolation and partial characterization of a major basic protein from rat eosinophil granules. *Proc Soc Exp Biol Med.* 1976;152(4):512–515.

245. Gleich GJ, Loegering DA, Mann KG, Maldonado JE. Comparative properties of the Charcot-Leyden crystal protein and the major basic protein from human eosinophils. *J Clin Invest.* 1976;57(3):633–640.

246. Dvorak AM, Furitsu T, Estrella P, Letourneau L, Ishizaka T, Ackerman SJ. Ultrastructural localization of major basic protein in the human eosinophil lineage in vitro. *J Histochem Cytochem.* 1994;42(11):1443–1451.

247. Dvorak AM, Letourneau L, Weller PF, Ackerman SJ. Ultrastructural localization of Charcot-Leyden crystal protein (lysophospholipase) to intracytoplasmic crystals in tumor cells of primary solid and papillary epithelial neoplasm of the pancreas. *Lab Investig.* 1990;62(5):608–615.

248. Adamko DJ, Odemuyiwa SO, Vethanayagam D, Moqbel R. The rise of the phoenix: the expanding role of the eosinophil in health and disease. *Allergy.* 2005;60(1):13–22.

249. Jacobsen EA, Taranova AG, Lee NA, Lee JJ. Eosinophils: singularly destructive effector cells or purveyors of immunoregulation? *J Allergy Clin Immunol.* 2007;119(6):1313–1320.

250. Akuthota P, Wang HB, Spencer LA, Weller PF. Immunoregulatory roles of eosinophils: a new look at a familiar cell. *Clin Exp Allergy.* 2008;38(8):1254–1263.

251. Specht S, Saeftel M, Arndt M, et al. Lack of eosinophil peroxidase or major basic protein impairs defense against murine filarial infection. *Infect Immun.* 2006;74(9):5236–5243.

252. Moqbel R, Levi-Schaffer F, Kay AB. Cytokine generation by eosinophils. *J Allergy Clin Immunol.* 1994;94(6 Pt 2):1183–1188.

253. Egesten A, Calafat J, Knol EF, Janssen H, Walz TM. Subcellular localization of transforming growth factor-alpha in human eosinophil granulocytes. *Blood.* 1996;87(9):3910–3918.

254. Moqbel R, Ying S, Barkans J, et al. Identification of messenger RNA for IL-4 in human eosinophils with granule localization and release of the translated product. *J Immunol.* 1995;155(10):4939–4947.

255. Moller GM, de Jong TA, Overbeek SE, van der Kwast TH, Postma DS, Hoogsteden HC. Ultrastructural immunogold localization of interleukin 5 to the crystalloid core compartment of eosinophil secondary granules in patients with atopic asthma. *J Histochem Cytochem.* 1996;44(1):67–69.

256. Woerly G, Lacy P, Younes AB, et al. Human eosinophils express and release IL-13 following CD28-dependent activation. *J Leukoc Biol.* 2002;72(4):769–779.

257. Persson-Dajotoy T, Andersson P, Bjartell A, Calafat J, Egesten A. Expression and production of the CXC chemokine growth-related oncogene-alpha by human eosinophils. *J Immunol.* 2003;170(10):5309–5316.

258. Bandeira-Melo C, Perez SA, Melo RCN, Ghiran I, Weller PF. EliCell assay for the detection of released cytokines from eosinophils. *J Immunol Methods.* 2003;276(1-2):227–237.

259. Bandeira-Melo C, Weller PF. Mechanisms of eosinophil cytokine release. *Mem Inst Oswaldo Cruz.* 2005;100(Suppl 1):73–81.

260. Liu LY, Wang H, Xenakis JJ, Spencer LA. Notch signaling mediates granulocyte-macrophage colony-stimulating factor priming-induced transendothelial migration of human eosinophils. *Allergy.* 2015;70(7):805–812.

261. Stow JL, Manderson AP, Murray RZ. SNAREing immunity: the role of SNAREs in the immune system. *Nat Rev Immunol.* 2006;6(12):919–929.

262. Stow JL, Murray RZ. Intracellular trafficking and secretion of inflammatory cytokines. *Cytokine Growth Factor Rev.* 2013;24(3):227–239.

263. Hoffmann HJ, Bjerke T, Karawajczyk M, Dahl R, Knepper MA, Nielsen S. SNARE proteins are critical for regulated exocytosis of ECP from human eosinophils. *Biochem Biophys Res Commun.* 2001;282(1):194–199.

264. Lacy P, Logan MR, Bablitz B, Moqbel R. Fusion protein vesicle-associated membrane protein 2 is implicated in IFN-gamma-induced piecemeal degranulation in human eosinophils from atopic individuals. *J Allergy Clin Immunol.* 2001;107(4):671–678.

265. Feng D, Flaumenhaft R, Bandeira-Melo C, Weller P, Dvorak A. Ultrastructural localization of vesicle-associated membrane protein(s) to specialized membrane structures in human pericytes, vascular smooth muscle cells, endothelial cells, neutrophils, and eosinophils. *J Histochem Cytochem.* 2001;49(3):293–304.

266. Logan MR, Lacy P, Bablitz B, Moqbel R. Expression of eosinophil target SNAREs as potential cognate receptors for vesicle-associated membrane protein-2 in exocytosis. *J Allergy Clin Immunol.* 2002;109(2):299–306.

267. Logan MR, Lacy P, Odemuyiwa SO, et al. A critical role for vesicle-associated membrane protein-7 in exocytosis from human eosinophils and neutrophils. *Allergy.* 2006;61(6):777–784.

268. Hemler ME. Tetraspanin functions and associated microdomains. *Nat Rev Mol Cell Biol.* 2005;6(10):801–811.

269. Charrin S, Jouannet S, Boucheix C, Rubinstein E. Tetraspanins at a glance. *J Cell Sci.* 2014;127(Pt 17):3641–3648.

270. Pak V, Budikhina A, Pashenkov M, Pinegin B. Neutrophil activity in chronic granulomatous disease. *Adv Exp Med Biol.* 2007;601:69–74.

271. MacGlashan Jr D. Expression of CD203c and CD63 in human basophils: relationship to differential regulation of piecemeal and anaphylactic degranulation processes. *Clin Exp Allergy.* 2010;40(9):1365–1377.

272. Furuta GT, Kagalwalla AF, Lee JJ, et al. The oesophageal string test: a novel, minimally invasive method measures mucosal inflammation in eosinophilic oesophagitis. *Gut.* 2013;62(10):1395–1405.

273. Weller PF, Goetzl EJ, Austen KF. Identification of human eosinophil lysophospholipase as the constituent of Charcot-Leyden crystals. *Proc Natl Acad Sci U S A.* 1980;77(12):7440–7443.

274. Weller PF, Wang H, Melo RCN. The Charcot-Leyden crystal protein revisited-A lysopalmitoylphospholipase and more. *J Leukoc Biol.* 2020;108(1):105–112.

275. Ackerman SJ, Corrette SE, Rosenberg HF, et al. Molecular cloning and characterization of human eosinophil Charcot-Leyden crystal protein (lysophospholipase). Similarities to IgE binding proteins and the S-type animal lectin superfamily. *J Immunol.* 1993;150(2):456–468.

276. Swaminathan GJ, Leonidas DD, Savage MP, Ackerman SJ, Acharya KR. Selective recognition of mannose by the human eosinophil Charcot-Leyden crystal protein (galectin-10): a crystallographic study at 1.8 A resolution. *Biochemistry.* 1999;38(46):15406.

277. Straub C, Pazdrak K, Young TW, et al. Toward the proteome of the human peripheral blood Eosinophil. *Proteomics Clin Appl.* 2009;3(10):1151–1173.

278. Wilkerson EM, Johansson MW, Hebert AS, et al. The peripheral blood eosinophil proteome. *J Proteome Res.* 2016;15(5):1524–1533.

279. Fukuchi M, Kamide Y, Ueki S, et al. Eosinophil ETosis-mediated release of galectin-10 in eosinophilic granulomatosis with polyangiitis. *Arthritis Reumatol.* 2021;73(9):1683–1693.

280. Persson EK, Verstraete K, Heyndrickx I, et al. Protein crystallization promotes type 2 immunity and is reversible by antibody treatment. *Science.* 2019;364(6442), eaaw4295.

281. Popa SJ, Stewart SE, Moreau K. Unconventional secretion of annexins and galectins. *Semin Cell Dev Biol.* 2018;83:42–50.

282. Furtak V, Hatcher F, Ochieng J. Galectin-3 mediates the endocytosis of beta-1 integrins by breast carcinoma cells. *Biochem Biophys Res Commun.* 2001;289(4):845–850.

283. Stechly L, Morelle W, Dessein AF, et al. Galectin-4-regulated delivery of glycoproteins to the brush border membrane of enterocyte-like cells. *Traffic.* 2009;10(4):438–450.

284. Straube T, von Mach T, Honig E, Greb C, Schneider D, Jacob R. pH-dependent recycling of galectin-3 at the apical membrane of epithelial cells. *Traffic.* 2013;14(9):1014–1027.

285. Johannes L, Jacob R, Leffler H. Galectins at a glance. *J Cell Sci.* 2018;131(9), jcs208884.

286. Repine JE, White JG, Clawson CC, Holmes BM. Effects of phorbol myristate acetate on the metabolism and ultrastructure of neutrophils in chronic granulomatous disease. *J Clin Invest.* 1974;54(1):83–90.

287. Hoidal JR, Repine JE, Beall GD, Rasp Jr FL, White JG. The effect of phorbol myristate acetate on the metabolism and ultrastructure of human alveolar macrophages. *Am J Pathol.* 1978;91(3):469–482.

288. Phaire-Washington L, Wang E, Silverstein SC. Phorbol myristate acetate stimulates pinocytosis and membrane spreading in mouse peritoneal macrophages. *J Cell Biol.* 1980;86(2):634–640.

289. Karlsson A, Nixon JB, McPhail LC. Phorbol myristate acetate induces neutrophil NADPH-oxidase activity by two separate signal transduction pathways: dependent or independent of phosphatidylinositol 3-kinase. *J Leukoc Biol.* 2000;67(3):396–404.

290. Dvorak AM, Letourneau L, Login GR, Weller PF, Ackerman SJ. Ultrastructural localization of the Charcot-Leyden crystal protein (lysophospholipase) to a distinct crystalloid-free granule population in mature human eosinophils. *Blood.* 1988;72(1):150–158.

291. Dvorak AM, Weller PF, Monahan-Earley RA, Letourneau L, Ackerman SJ. Ultrastructural localization of Charcot-Leyden crystal protein (lysophospholipase) and peroxidase in macrophages, eosinophils, and extracellular matrix of the skin in the hypereosinophilic syndrome. *Lab Investig.* 1990;62(5):590–607.

292. Calafat J, Janssen H, Knol EF, Weller PF, Egesten A. Ultrastructural localization of Charcot-Leyden crystal protein in human eosinophils and basophils. *Eur J Haematol.* 1997;58(1):56–66.

293. Bozza PT, Yu W, Penrose JF, Morgan ES, Dvorak AM, Weller PF. Eosinophil lipid bodies: specific, inducible intracellular sites for enhanced eicosanoid formation. *J Exp Med.* 1997;186(6):909–920.

294. Sun J, Dahlen B, Agerberth B, Haeggstrom JZ. The antimicrobial peptide LL-37 induces synthesis and release of cysteinyl leukotrienes from human eosinophils – implications for asthma. *Allergy.* 2013;68(3):304–311.

295. Galluzzi L, Vitale I, Aaronson SA, et al. Molecular mechanisms of cell death: recommendations of the Nomenclature Committee on Cell Death 2018. *Cell Death Differ.* 2018;25(3):486–541.

296. Tang D, Kang R, Berghe TV, Vandenabeele P, Kroemer G. The molecular machinery of regulated cell death. *Cell Res.* 2019;29(5):347–364.

297. Galluzzi L, Bravo-San Pedro JM, Vitale I, et al. Essential versus accessory aspects of cell death: recommendations of the NCCD 2015. *Cell Death Differ.* 2015;22(1):58–73.

298. Jorgensen I, Rayamajhi M, Miao EA. Programmed cell death as a defence against infection. *Nat Rev Immunol.* 2017;17(3):151–164.

299. Kroemer G, El-Deiry WS, Golstein P, et al. Classification of cell death: recommendations of the Nomenclature Committee on Cell Death. *Cell Death Differ.* 2005;12(Suppl 2):1463–1467.

300. Kroemer G, Galluzzi L, Vandenabeele P, et al. Classification of cell death: recommendations of the Nomenclature Committee on Cell Death 2009. *Cell Death Differ.* 2009;16(1):3–11.

301. Galluzzi L, Vitale I, Abrams JM, et al. Molecular definitions of cell death subroutines: recommendations of the Nomenclature Committee on Cell Death 2012. *Cell Death Differ.* 2012;19(1):107–120.

302. Loos B, Engelbrecht AM. Cell death: a dynamic response concept. *Autophagy.* 2009;5(5):590–603.

303. Brinkmann V, Reichard U, Goosmann C, et al. Neutrophil extracellular traps kill bacteria. *Science.* 2004;303(5663):1532–1535.

304. Vanden Berghe T, Linkermann A, Jouan-Lanhouet S, Walczak H, Vandenabeele P. Regulated necrosis: the expanding network of non-apoptotic cell death pathways. *Nat Rev Mol Cell Biol.* 2014;15(2):135–147.

305. Jeffery PK, Godfrey RW, Adelroth E, Nelson F, Rogers A, Johansson SA. Effects of treatment on airway inflammation and thickening of basement membrane reticular collagen in asthma. A quantitative light and electron microscopic study. *Am Rev Respir Dis.* 1992;145(4 Pt 1):890–899.

306. Yamashima T, Kubota T, Yamamoto S. Eosinophil degranulation in the capsule of chronic subdural hematomas. *J Neurosurg.* 1985;62(2):257–260.

307. Dunai Z, Bauer PI, Mihalik R. Necroptosis: biochemical, physiological and pathological aspects. *Pathol Oncol Res.* 2011;17(4):791–800.

308. Falcieri E, Gobbi P, Zamai L, Vitale M. Ultrastructural features of apoptosis. *Scanning Microsc.* 1994;8(3):653–665.

309. Steinberg BE, Grinstein S. Unconventional roles of the NADPH oxidase: signaling, ion homeostasis, and cell death. *Sci STKE.* 2007;2007(379), pe11.

310. Wartha F, Henriques-Normark B. ETosis: a novel cell death pathway. *Sci Signal.* 2008;1(21). pe25.

311. Fuchs TA, Abed U, Goosmann C, et al. Novel cell death program leads to neutrophil extracellular traps. *J Cell Biol.* 2007;176(2):231–241.

312. Fukuchi M, Miyabe Y, Furutani C, et al. How to detect eosinophil ETosis (EETosis) and extracellular traps. *Allergol Int.* 2021;70(1):19–29.

313. Silva JC, Thompson-Souza GA, Barroso MV, Neves JS, Figueiredo RT. Neutrophil and eosinophil DNA extracellular trap formation: lessons from pathogenic fungi. *Front Microbiol.* 2021;12:318.

314. Amulic B, Knackstedt SL, Abu Abed U, et al. Cell-cycle proteins control production of neutrophil extracellular traps. *Dev Cell.* 2017;43(4):449–462. e445.

315. Sollberger G, Tilley DO, Zychlinsky A. Neutrophil extracellular traps: the biology of chromatin externalization. *Dev Cell.* 2018;44(5):542–553.

316. Krautgartner WD, Klappacher M, Hannig M, et al. Fibrin mimics neutrophil extracellular traps in SEM. *Ultrastruct Pathol.* 2010;34(4):226–231.

317. Echevarría, LU, Leimgruber C, Garcia Gonzalez J, et al. Evidence of eosinophil extracellular trap cell death in COPD: does it represent the trigger that switches on the disease? *Int J Chron Obstruct Pulmon Dis.* 2017;12:885–896.

318. Ueki S, Hebisawa A, Kitani M, Asano K, Neves JS. Allergic bronchopulmonary aspergillosis–a luminal hypereosinophilic disease with extracellular trap cell death. *Front Immunol.* 2018;9:2346.

319. Omokawa A, Ueki S, Kikuchi Y, et al. Mucus plugging in allergic bronchopulmonary aspergillosis: implication of the eosinophil DNA traps. *Allergol Int.* 2018;67(2):280–282.

320. Yousefi S, Gold JA, Andina N, et al. Catapult-like release of mitochondrial DNA by eosinophils contributes to antibacterial defense. *Nat Med.* 2008;14(9):949–953.

321. Kerr JF, Wyllie AH, Currie AR. Apoptosis: a basic biological phenomenon with wide-ranging implications in tissue kinetics. *Br J Cancer.* 1972;26(4):239–257.

322. Taatjes DJ, Sobel BE, Budd RC. Morphological and cytochemical determination of cell death by apoptosis. *Histochem Cell Biol.* 2008;129(1):33–43.

323. Gotzos V, Cappelli-Gotzos B, Conti G. Cells affected by apoptosis in effusions of tumor origin in man. *Ann Pathol.* 1984;4(3):217–221.

324. Adachi T, Motojima S, Hirata A, et al. Eosinophil apoptosis caused by theophylline, glucocorticoids, and macrolides after stimulation with IL-5. *J Allergy Clin Immunol.* 1996;98(6 Pt 2):S207–215.

325. Taylor RC, Cullen SP, Martin SJ. Apoptosis: controlled demolition at the cellular level. *Nat Rev Mol Cell Biol.* 2008;9(3):231–241.

326. Battistelli M, Falcieri E. Apoptotic bodies: particular extracellular vesicles involved in intercellular communication. *Biology (Basel).* 2020;9(1):21.

327. Burattini S, Ferri P, Battistelli M, et al. Apoptotic DNA fragmentation can be revealed in situ: an ultrastructural approach. *Microsc Res Tech.* 2009;72(12):913–923.

328. Caruso S, Poon IKH. Apoptotic cell-derived extracellular vesicles: more than just debris. *Front Immunol.* 2018;9:1486.

329. Elmore S. Apoptosis: a review of programmed cell death. *Toxicol Pathol.* 2007;35(4):495–516.

330. Felton JM, Lucas CD, Rossi AG, Dransfield I. Eosinophils in the lung – modulating apoptosis and efferocytosis in airway inflammation. *Front Immunol.* 2014;5:302.

331. Poon IKH, Parkes MAF, Jiang L, et al. Moving beyond size and phosphatidylserine exposure: evidence for a diversity of apoptotic cell-derived extracellular vesicles in vitro. *J Extracell Vesicles.* 2019;8(1):1608786.

332. Sexton DW, Blaylock MG, Walsh GM. Human alveolar epithelial cells engulf apoptotic eosinophils by means of integrin- and phosphatidylserine receptor-dependent mechanisms: a process upregulated by dexamethasone. *J Allergy Clin Immunol.* 2001;108(6):962–969.

333. Silva MT. Secondary necrosis: the natural outcome of the complete apoptotic program. *FEBS Lett.* 2010;584(22):4491–4499.

334. Walsh GM. Eosinophil apoptosis and clearance in asthma. *J Cell Death.* 2013;6:17–25.

335. Vanden Berghe T, Vanlangenakker N, Parthoens E, et al. Necroptosis, necrosis and secondary necrosis converge on similar cellular disintegration features. *Cell Death Differ.* 2010;17(6):922–930.

336. Uller L, Persson CG, Erjefalt JS. Resolution of airway disease: removal of inflammatory cells through apoptosis, egression or both? *Trends Pharmacol Sci.* 2006;27(9):461–466.

337. Bickham UR, Malter JS. Apoptotic and survival signaling in eosinophils. In: Lee JJ, Rosenberg HF, eds. *Eosinophils in Health and Disease.* Elsevier; 2013:189–197.

338. Park YM, Bochner BS. Eosinophil survival and apoptosis in health and disease. *Allergy Asthma Immunol Res.* 2010;2(2):87–101.

339. Simon HU, Yousefi S, Schranz C, Schapowal A, Bachert C, Blaser K. Direct demonstration of delayed eosinophil apoptosis as a mechanism causing tissue eosinophilia. *J Immunol.* 1997;158(8):3902–3908.

340. Ilmarinen P, Moilanen E, Kankaanranta H. Regulation of spontaneous eosinophil apoptosis-a neglected area of importance. *J Cell Death.* 2014;7:1–9.

341. Shen ZJ, Malter JS. Determinants of eosinophil survival and apoptotic cell death. *Apoptosis*. 2015;20(2):224–234.

342. Matsumoto K, Schleimer RP, Saito H, Iikura Y, Bochner BS. Induction of apoptosis in human eosinophils by anti-Fas antibody treatment in vitro. *Blood*. 1995;86(4):1437–1443.

343. Druilhe A, Cai Z, Haile S, Chouaib S, Pretolani M. Fas-mediated apoptosis in cultured human eosinophils. *Blood*. 1996;87(7):2822–2830.

344. Kita H. Eosinophils: multifaceted biological properties and roles in health and disease. *Immunol Rev*. 2011;242(1):161–177.

345. Maltby S, McNagny KM, Ackerman SJ, et al. Eosinophilopoiesis. In: Lee JJ, Rosenberg HF, eds. *Eosinophils in Health and Disease*. Elsevier; 2013:73–119.

346. Abdala-Valencia H, Coden ME, Chiarella SE, et al. Shaping eosinophil identity in the tissue contexts of development, homeostasis, and disease. *J Leukoc Biol*. 2018;104(1):95–108.

347. Mack EA, Stein SJ, Rome KS, et al. Trib1 regulates eosinophil lineage commitment and identity by restraining the neutrophil program. *Blood*. 2019;133(22):2413–2426.

348. Kusano S, Kukimoto-Niino M, Hino N, et al. Structural basis of interleukin-5 dimer recognition by its α receptor. *Protein Sci*. 2012;21(6):850–864.

349. Mori Y, Iwasaki H, Kohno K, et al. Identification of the human eosinophil lineage-committed progenitor: revision of phenotypic definition of the human common myeloid progenitor. *J Exp Med*. 2009;206(1):183–193.

350. Hui CCK, McNagny KM, Denburg JA, Siracusa MC. In situ hematopoiesis: a regulator of TH 2 cytokine-mediated immunity and inflammation at mucosal surfaces. *Mucosal Immunol*. 2015;8(4):701–711.

351. Johnston LK, Bryce PJ. Understanding interleukin 33 and its roles in eosinophil development. *Front Med*. 2017;4:51.

352. Dias FF, Amaral KB, Malta KK, et al. Identification of piecemeal degranulation and vesicular transport of MBP-1 in liver-Infiltrating mouse eosinophils during acute experimental *Schistosoma mansoni* infection. *Front Immunol*. 2018;9:3019.

353. Dvorak AM, Saito H, Estrella P, Kissell S, Arai N, Ishizaka T. Ultrastructure of eosinophils and basophils stimulated to develop in human cord blood mononuclear cell cultures containing recombinant human interleukin-5 or interleukin-3. *Lab Investig*. 1989;61(1):116–132.

354. Saito H, Hatake K, Dvorak AM, et al. Selective differentiation and proliferation of hematopoietic cells induced by recombinant human interleukins. *Proc Natl Acad Sci U S A*. 1988;85(7):2288–2292.

355. Barker RL, Gleich GJ, Pease LR. Acidic precursor revealed in human eosinophil granule major basic protein cDNA. *J Exp Med*. 1988;168(4):1493–1498.

356. Popken-Harris P, Checkel J, Loegering D, et al. Regulation and processing of a precursor form of eosinophil granule major basic protein (ProMBP) in differentiating eosinophils. *Blood*. 1998;92(2):623–631.

357. Egesten A, Calafat J, Weller PF, et al. Localization of granule proteins in human eosinophil bone marrow progenitors. *Int Arch Allergy Immunol*. 1997;114(2):130–138.

358. Doyle AD, Jacobsen EA, Ochkur SI, et al. Homologous recombination into the eosinophil peroxidase locus generates a strain of mice expressing Cre recombinase exclusively in eosinophils. *J Leukoc Biol*. 2013;94(1):17–24.

359. Bettigole SE, Lis R, Adoro S, et al. The transcription factor XBP1 is selectively required for eosinophil differentiation. *Nat Immunol*. 2015;16(8):829–837.

360. Scott RE, Horn RG. Fine structural features of eosinophile granulocyte development in human bone marrow. Evidence for granule secretion. *J Ultrastruct Res*. 1970;33(1):16–28.

361. Wickramasinghe S, Porwit A, Erber W. Normal bone marrow cells: development and cytology. In: Porwit A, McCullough J, Erber W, eds. *Blood and Bone Marrow Pathology*. 2nd ed. Edinburgh: Churchil Livingstone; 2011:19–44.

362. Paul CC, Ackerman SJ, Mahrer S, Tolbert M, Dvorak AM, Baumann MA. Cytokine induction of granule protein synthesis in an eosinophil-inducible human myeloid cell line, AML14. *J Leukoc Biol*. 1994;56(1):74–79.

363. Yantiss RK. Eosinophils in the GI tract: how many is too many and what do they mean? *Mod Pathol*. 2015;28(1):S7–S21.

364. Valent P, Klion AD, Horny H-P, et al. Contemporary consensus proposal on criteria and classification of eosinophilic disorders and related syndromes. *J Allergy Clin Immun*. 2012;130(3):607–612. e609.

365. Khoury P, Akuthota P, Ackerman SJ, et al. Revisiting the NIH taskforce on the research needs of eosinophil-associated diseases (RE-TREAD). *J Leukoc Biol*. 2018;104(1):69–83.

366. Minai-Fleminger Y, Elishmereni M, Vita F, et al. Ultrastructural evidence for human mast cell-eosinophil interactions in vitro. *Cell Tissue Res*. 2010;341(3):405–415.

367. Caruso RA, Bersiga A, Rigoli L, Inferrera C. Eosinophil-tumor cell interaction in advanced gastric carcinoma: an electron microscopic approach. *Anticancer Res*. 2002;22(6C):3833–3836.

368. Kovalszki A, Weller PF. Eosinophils and eosinophilia. In: *Clinical Immunology*. Elsevier; 2019:349–361. e341.

369. Shah SA, Ishinaga H, Takeuchi K. Pathogenesis of eosinophilic chronic rhinosinusitis. *J Inflamm*. 2016;13(1):1–9.

370. Simon D, Wardlaw A, Rothenberg ME. Organ-specific eosinophilic disorders of the skin, lung, and gastrointestinal tract. *J Allergy Clin Immunol*. 2010;126(1):3–13.

371. Long H, Zhang G, Wang L, Lu Q. Eosinophilic skin diseases: a comprehensive review. *Clin Rev Allergy Immunol*. 2016;50(2):189–213.

372. Leiferman KM, Peters MS. Eosinophil-related disease and the skin. *J Allergy Clin Immunol Pract*. 2018;6(5):1462–1482. e1466.

373. Marzano AV, Genovese G. Eosinophilic dermatoses: recognition and management. *Am J Clin Dermatol*. 2020;21(4):525–539.

374. Kiehl P, Falkenberg K, Vogelbruch M, Kapp A. Tissue eosinophilia in acute and chronic atopic dermatitis: a morphometric approach using quantitative image analysis of immunostaining. *Br J Dermatol*. 2001;145(5):720–729.

375. Amber KT, Valdebran M, Kridin K, Grando SA. The role of eosinophils in bullous pemphigoid: a developing model of eosinophil pathogenicity in mucocutaneous disease. *Front Med*. 2018;5:201.

376. BinJadeed HF, Alyousef AM, Alsaif FM, Alhumidi AA, Alotaibi HO. Histologic characterization of cellular infiltration in autoimmune subepidermal bullous diseases in a tertiary hospital in Saudi Arabia. *Clin Cosmet Invest Dermatol*. 2018;11:187.

377. Miyamoto D, Santi CG, Aoki V, Maruta CW. Bullous pemphigoid. *An Bras Dermatol*. 2019;94(2):133–146.

378. Genovese G, Di Zenzo G, Cozzani E, Berti E, Cugno M, Marzano AV. New insights into the pathogenesis of bullous pemphigoid: 2019 update. *Front Immunol*. 2019;10:1506.

379. Dvorak AM, Mihm Jr MC, Osage JE, Kwan TH, Austen KF, Wintroub BU. Bullous pemphigoid, and ultrastructural study of the inflammatory response: eosinophil, basophil and mast cell granule changes in multiple biopsies from one patient. *J Invest Dermatol*. 1982;78(2):91–101.

380. Chusid MJ, Dale DC, West BC, Wolff SM. The hypereosinophilic syndrome: analysis of fourteen cases with review of the literature. *Medicine.* 1975;54(1):1–27.

381. Fauci AS, Harley JB, Robert WC, Ferrans VJ, Gralnick HR, Bjornson BH. The idiopathic hypereosinophilic syndrome: clinical, pathophysiologic, and therapeutic considerations. *Ann Intern Med.* 1982;97(1):78–92.

382. Valent P, Klion AD, Horny HP, et al. Contemporary consensus proposal on criteria and classification of eosinophilic disorders and related syndromes. *J Allergy Clin Immunol.* 2012;130(3):607–612. e609.

383. Simon HU, Rothenberg ME, Bochner BS, et al. Refining the definition of hypereosinophilic syndrome. *J Allergy Clin Immunol.* 2010;126(1):45–49.

384. Lefèvre G, Copin MC, Staumont-Sallé D, et al. The lymphoid variant of hypereosinophilic syndrome: study of 21 patients with CD3-CD4+ aberrant T-cell phenotype. *Medicine.* 2014;93(17):255–266.

385. Legrand F, Renneville A, MacIntyre E, et al. The spectrum of FIP1L1-PDGFRA-associated chronic eosinophilic leukemia: new insights based on a survey of 44 cases. *Medicine.* 2013;92(5), e1.

386. Yamamoto H, Ninomiya H, Yoshimatsu K, et al. Serum levels of major basic protein in patients with or without eosinophilia: measurement by enzyme-linked immunosorbent assay. *Br J Haematol.* 1994;86(3):490–495.

387. Khoury P, Makiya M, Klion AD. Clinical and biological markers in hypereosinophilic syndromes. *Front Med.* 2017;4:240.

388. Takeda M, Ueki S, Yamamoto Y, et al. Hypereosinophilic syndrome with abundant Charcot-Leyden crystals in spleen and lymph nodes. *Asia Pac Allergy.* 2020;10(3):e24.

389. Khoury P, Grayson PC, Klion AD. Eosinophils in vasculitis: characteristics and roles in pathogenesis. *Nat Rev Rheumatol.* 2014;10(8):474.

390. Wu EY, Hernandez ML, Jennette JC, Falk RJ. Eosinophilic granulomatosis with polyangiitis: clinical pathology conference and review. *J Allergy Clin Immunol Pract.* 2018;6(5):1496–1504.

391. Furuta S, Iwamoto T, Nakajima H. Update on eosinophilic granulomatosis with polyangiitis. *Allergol Int.* 2019;68(4):430–436.

392. Villanueva KLV, Espinoza LR. Eosinophilic vasculitis. *Curr Rheumatol Rep.* 2020;22(1):1–10.

393. Nagase H, Ueki S, Fujieda S. The roles of IL-5 and anti-IL-5 treatment in eosinophilic diseases: asthma, eosinophilic granulomatosis with polyangiitis, and eosinophilic chronic rhinosinusitis. *Allergol Int.* 2020;69(2):178–186.

394. Nishi R, Koike H, Ohyama K, et al. Differential clinicopathologic features of EGPA-associated neuropathy with and without ANCA. *Neurology.* 2020;94(16):e1726–e1737.

395. Masterson JC, Menard-Katcher C, Larsen LD, Furuta GT, Spencer LA. Heterogeneity of intestinal tissue eosinophils: potential considerations for next-generation eosinophil-targeting strategies. *Cell.* 2021;10(2):426.

396. Kato M, Kephart GM, Talley NJ, et al. Eosinophil infiltration and degranulation in normal human tissue. *Anat Rec.* 1998;252(3):418–425.

397. Akuthota P, Weller PF. Spectrum of eosinophilic end-organ manifestations. *Immunol Allergy Clin N Am.* 2015;35(3):403–411.

398. Straumann A, Kristl J, Conus S, et al. Cytokine expression in healthy and inflamed mucosa: probing the role of eosinophils in the digestive tract. *Inflamm Bowel Dis.* 2005;11(8):720–726.

399. Powell N, Walker MM, Talley NJ. Gastrointestinal eosinophils in health, disease and functional disorders. *Nat Rev Gastroenterol Hepatol.* 2010;7(3):146.

400. Loktionov A. Eosinophils in the gastrointestinal tract and their role in the pathogenesis of major colorectal disorders. *World J Gastroenterol.* 2019;25(27):3503.

401. Rothenberg ME. Eosinophilic gastrointestinal disorders (EGID). *J Allergy Clin Immunol.* 2004;113(1):11–28.

402. Baumgarten T, Sperling S, Seifert J, et al. Membrane vesicle formation as a multiple-stress response mechanism enhances Pseudomonas putida DOT-T1E cell surface hydrophobicity and biofilm formation. *Appl Environ Microbiol.* 2012;78(17):6217–6224.

403. Gajendran M, Loganathan P, Catinella AP, Hashash JG. A comprehensive review and update on Crohn's disease. *Dis Mon.* 2018;64(2):20–57.

404. Torres J, Mehandru S, Colombel J-F, Peyrin-Biroulet L. Crohn's disease. *Lancet.* 2017;389(10080):1741–1755.

405. Baumgart DC, Sandborn WJ. Crohn's disease. *Lancet.* 2012;380(9853):1590–1605.

406. Ungaro R, Mehandru S, Allen PB, Peyrin-Biroulet L, Colombel JF. Ulcerative colitis. *Lancet.* 2017;389:1756–1770.

407. Dvorak AM, Dickersin GR. Crohn's disease: transmission electron microscopic studies. I. Barrier function. Possible changes related to alterations of cell coat, mucous coat, epithelial cells, and Paneth cells. *Hum Pathol.* 1980;11(5 Suppl):561–571.

408. Dvorak AM, Osage JE, Monahan RA, Dickersin GR. Crohn's disease: transmission electron microscopic studies. III. Target tissues. Proliferation of and injury to smooth muscle and the autonomic nervous system. *Hum Pathol.* 1980;11(6):620–634.

409. Dvorak AM. Electron microscopy of Paneth cells in Crohn's disease. *Arch Pathol Lab Med.* 1980;104(7):393–394.

410. Dvorak A, Silen W. Differentiation between Crohn's disease and other inflammatory conditions by electron microscopy. *Ann Surg.* 1985;201(1):53.

411. Dvorak AM, Onderdonk AB, McLeod RS, et al. Axonal necrosis of enteric autonomic nerves in continent ileal pouches. Possible implications for pathogenesis of Crohn's disease. *Ann Surg.* 1993;217(3):260.

412. Lueschow SR, McElroy SJ. The Paneth cell: the curator and defender of the immature small intestine. *Front Immunol.* 2020;11:587.

413. Berek C. Eosinophils: important players in humoral immunity. *Clin Exp Immunol.* 2016;183(1):57–64.

414. Blanchard C, Wang N, Rothenberg ME. Eosinophilic esophagitis: pathogenesis, genetics, and therapy. *J Allergy Clin Immunol.* 2006;118(5):1054–1059.

415. Furuta GT, Katzka DA. Eosinophilic esophagitis. *N Engl J Med.* 2015;373(17):1640–1648.

416. Reed CC, Dellon ES. Eosinophilic esophagitis. *Med Clin North Am.* 2019;103(1):29–42.

417. Collins MH. Histopathology of eosinophilic esophagitis. *Dig Dis.* 2014;32(1-2):68–73.

418. Mueller S, Aigner T, Neureiter D, Stolte M. Eosinophil infiltration and degranulation in oesophageal mucosa from adult patients with eosinophilic oesophagitis: a retrospective and comparative study on pathological biopsy. *J Clin Pathol.* 2006;59(11):1175–1180.

419. Peterson KA, Gleich GJ, Limaye NS, et al. Eosinophil granule major basic protein 1 deposition in eosinophilic esophagitis correlates with symptoms independent of eosinophil counts. *Dis Esophagus.* 2019;32(11), doz055.

420. Protheroe C, Woodruff SA, De Petris G, et al. A novel histologic scoring system to evaluate mucosal biopsies from patients with eosinophilic esophagitis. *Clin Gastroenterol Hepatol.* 2009;7(7):749–755. e711.

421. Kephart GM, Alexander JA, Arora AS, et al. Marked deposition of eosinophil-derived neurotoxin in adult patients with eosinophilic esophagitis. *Am J Gastroenterol.* 2010;105(2):298.

422. O'Shea KM, Aceves SS, Dellon ES, et al. Pathophysiology of eosinophilic esophagitis. *Gastroenterology*. 2018;154(2):333–345.

423. Lee JJ, Rosenberg HF, eds. *Eosinophils in health and disease*. Elsevier; 2013. pp. 1–654.

424. Huang L, Appleton JA. Eosinophils in helminth infection: defenders and dupes. *Trends Parasitol*. 2016;32(10):798–807.

425. O'Connell EM, Nutman TB. Eosinophilia in infectious diseases. *Immunol Allergy Clin N Am*. 2015;35(3):493.

426. Klion AD, Ackerman SJ, Bochner BS. Contributions of eosinophils to human health and disease. *Annu Rev Pathol*. 2020;15:179–209.

427. Bruschi F, Korenaga M, Watanabe N. Eosinophils and Trichinella infection: toxic for the parasite and the host? *Trends Parasitol*. 2008;24(10):462–467.

428. Kassalik M, Mönkemüller K. Strongyloides stercoralis hyperinfection syndrome and disseminated disease. *Gastroenterol Hepatol*. 2011;7(11):766.

429. Colley DG, Bustinduy AL, Secor WE, King CH. Human schistosomiasis. *Lancet*. 2014;383(9936):2253–2264.

430. McManus DP, Dunne DW, Sacko M, Utzinger J, Vennervald BJ, Zhou XN. Schistosomiasis. *Nat Rev Dis Primers*. 2018;4(1):13.

431. Di Bella S, Riccardi N, Giacobbe DR, Luzzati R. History of schistosomiasis (bilharziasis) in humans: from Egyptian medical papyri to molecular biology on mummies. *Pathog Glob Health*. 2018;112(5):268–273.

432. Schwartz C, Fallon PG. Schistosoma "eggs-iting" the host: granuloma formation and egg excretion. *Front Immunol*. 2018;9:2492.

433. Amaral KB, Silva TP, Dias FF, et al. Histological assessment of granulomas in natural and experimental *Schistosoma mansoni* infections using whole slide imaging. *PLoS One*. 2017;12(9), e0184696.

434. Lenzi HL, Pacheco RG, Pelajo-Machado M, Panasco MS, Romanha WS, Lenzi JA. Immunological system and Schistosoma mansoni: co-evolutionary immunobiology. What is the eosinophil role in parasite-host relationship? *Mem Inst Oswaldo Cruz*. 1997;92(1):19–32.

435. Meeusen EN, Balic A. Do eosinophils have a role in the killing of helminth parasites? *Parasitol Today*. 2000;16(3):95–101.

436. Swartz JM, Dyer KD, Cheever AW, et al. Schistosoma mansoni infection in eosinophil lineage-ablated mice. *Blood*. 2006;108(7):2420–2427.

437. Cadman ET, Lawrence RA. Granulocytes: effector cells or immunomodulators in the immune response to helminth infection? *Parasite Immunol*. 2010;32(1):1–19.

438. Tweyongyere R, Namanya H, Naniima P, et al. Human eosinophils modulate peripheral blood mononuclear cell response to Schistosoma mansoni adult worm antigen in vitro. *Parasite Immunol*. 2016;38(8):516–522.

439. Amaral KB, Silva TP, Malta KK, et al. Natural Schistosoma mansoni infection in the wild reservoir Nectomys squamipes leads to excessive lipid droplet accumulation in hepatocytes in the absence of liver functional impairment. *PLoS One*. 2016;11(11), e0166979.

440. Malta KK, Silva TP, Palazzi C, et al. Changing our view of the Schistosoma granuloma to an ecological standpoint. *Biol Rev Camb Philos Soc*. 2021;96(4):1404–1420.

441. Andrade ZA. Pathology of human schistosomiasis. *Mem Inst Oswaldo Cruz*. 1987;82(1):17–23.

442. Lindsley AW, Schwartz JT, Rothenberg ME. Eosinophil responses during COVID-19 infections and coronavirus vaccination. *J Allergy Clin Immunol*. 2020;146(1):1–7.

443. Rosenberg HF, Foster PS. Eosinophils and COVID-19: diagnosis, prognosis, and vaccination strategies. *Semin Immunopathol*. 2021;1–10.

444. Flores-Torres AS, Salinas-Carmona MC, Salinas E, Rosas-Taraco AG. Eosinophils and respiratory viruses. *Viral Immunol*. 2019;32(5):198–207.

445. LeMessurier KS, Samarasinghe AE. Eosinophils: nemeses of pulmonary pathogens? *Curr Allergy Asthma Rep*. 2019;19(8):1–10.

446. Rosenberg HF, Dyer KD, Domachowske JB. Eosinophils and their interactions with respiratory virus pathogens. *Immunol Res*. 2009;43(1-3):128–137.

447. Rosenberg HF, Dyer KD, Domachowske JB. Respiratory viruses and eosinophils: exploring the connections. *Antivir Res*. 2009;83(1):1–9.

448. Lucey DR, Dorsky DI, Nicholson-Weller A, Weller PF. Human eosinophils express CD4 protein and bind human immunodeficiency virus 1 gp120. *J Exp Med*. 1989;169(1):327–332.

449. Roingeard P. Viral detection by electron microscopy: past, present and future. *Biol Cell*. 2008;100(8):491–501.

450. Miller S, Krijnse-Locker J. Modification of intracellular membrane structures for virus replication. *Nat Rev Microbiol*. 2008;6(5):363–374.

451. Strating JRPM, van Kuppeveld FJM. Viral rewiring of cellular lipid metabolism to create membranous replication compartments. *Curr Opin Cell Biol*. 2017;47:24–33.

452. Hopfer H, Herzig MC, Gosert R, et al. Hunting coronavirus by transmission electron microscopy – a guide to SARS-CoV-2-associated ultrastructural pathology in COVID-19 tissues. *Histopathology*. 2021;78(3):358–370.

453. Miller SE, Goldsmith, CS. Caution in identifying coronaviruses by electron microscopy. *JASN* 2020;31(9):2223–2224.

454. Lavoignet C-E, Le Borgne P, Chabrier S, et al. White blood cell count and eosinopenia as valuable tools for the diagnosis of bacterial infections in the ED. *Eur J Clin Microbiol Infect Dis*. 2019;38(8):1523–1532.

455. Karakonstantis S, Gryllou N, Papazoglou G, Lydakis C. Eosinophil count (EC) as a diagnostic and prognostic marker for infection in the internal medicine department setting. *Rom J Intern Med*. 2019;57(2):166–174.

456. Kvarnhammar AM, Cardell LO. Pattern-recognition receptors in human eosinophils. *Immunology*. 2012;136(1):11–20.

457. Buonomo EL, Cowardin CA, Wilson MG, Saleh MM, Pramoonjago P, Petri Jr WA. Microbiota-regulated IL-25 increases eosinophil number to provide protection during Clostridium difficile infection. *Cell Rep*. 2016;16(2):432–443.

458. Akiyama S, Rai V, Rubin DT. Pouchitis in inflammatory bowel disease: a review of diagnosis, prognosis, and treatment. *Intest Res*. 2021;19(1):1–11.

459. Turley SJ, Cremasco V, Astarita JL. Immunological hallmarks of stromal cells in the tumour microenvironment. *Nat Rev Immunol*. 2015;15(11):669–682.

460. Lotfi R, Lee JJ, Lotze MT. Eosinophilic granulocytes and damage-associated molecular pattern molecules (DAMPs): role in the inflammatory response within tumors. *J Immunother*. 2007;30(1):16–28.

461. Davis BP, Rothenberg ME. Eosinophils and cancer. *Cancer Immunol Res*. 2014;2(1):1–8.

462. Sakkal S, Miller S, Apostolopoulos V, Nurgali K. Eosinophils in cancer: favourable or unfavourable? *Curr Med Chem*. 2016;23(7):650–666.

463. Simon SC, Utikal J, Umansky V. Opposing roles of eosinophils in cancer. *Cancer Immunol Immunother*. 2019;68(5):823–833.

464. Grisaru-Tal S, Itan M, Klion AD, Munitz A. A new dawn for eosinophils in the tumour microenvironment. *Nat Rev Cancer*. 2020;20(10):594–607.

465. Martinelli-Kläy CP, Mendis BRRN, Lombardi T. Eosinophils and oral squamous cell carcinoma: a short review. *J Oncology*. 2009;2009.

466. Tadbir AA, Ashraf MJ, Sardari Y. Prognostic significance of stromal eosinophilic infiltration in oral squamous cell carcinoma. *J Craniofac Surg*. 2009;20(2):287–289.

467. Ariztia EV, Lee CJ, Gogoi R, Fishman DA. The tumor microenvironment: key to early detection. *Crit Rev Clin Lab Sci*. 2006;43(5-6):393–425.

468. Lu P, Weaver VM, Werb Z. The extracellular matrix: a dynamic niche in cancer progression. *J Cell Biol*. 2012;196(4):395–406.

469. Akhtar M, Bakry M. Role of electron microscopy in tumor diagnosis: a review. *Ann Saudi Med.* 1982;2(4):243–260.

470. Graham L, Orenstein JM. Processing tissue and cells for transmission electron microscopy in diagnostic pathology and research. *Nat Protoc.* 2007;2(10):2439.

471. Ghadially FN. *Diagnostic Electron Microscopy of Tumours.* 2nd ed. Butterworth-Heinemann; 1985. pp. 1–510.

472. Ono Y, Ozawa M, Tamura Y, et al. Tumor-associated tissue eosinophilia of penile cancer. *Int J Urol.* 2002;9(2):82–87.

473. Ionescu MA, Rivet J, Daneshpouy M, Briere J, Morel P, Janin A. In situ eosinophil activation in 26 primary cutaneous T-cell lymphomas with blood eosinophilia. *J Am Acad Dermatol.* 2005;52(1):32–39.

474. Caruso RA, Fedele F, Parisi A, et al. Chronic allergic-like inflammation in the tumor stroma of human gastric carcinomas: an ultrastructural study. *Ultrastruct Pathol.* 2012;36(3):139–144.

475. Speight PM, Farthing PM. The pathology of oral cancer. *Br Dent J.* 2018;225(9):841–847.

476. Deepthi G, Kulkarni PG, Nandan SRK. Eosinophils: an imperative histopathological prognostic indicator for oral squamous cell carcinoma. *J Oral Maxillofac Pathol.* 2019;23(2):307.

477. Kannan S, Kartha CC, Chandran GJ, et al. Ultrastructure of oral squamous cell carcinoma: a comparative analysis of different histological types. *Eur J Cancer B Oral Oncol.* 1994;30(1):32–42.

478. Tanaka N, Sugihara K, Odajima T, Mimura M, Kimijima Y, Ichinose S. Oral squamous cell carcinoma: electron microscopic and immunohistochemical characteristics. *Med Electron Microsc.* 2002;35(3):127–138.

479. Caruso RA, Fedele F, Zuccala V, Fracassi MG, Venuti A. Mast cell and eosinophil interaction in gastric carcinomas: ultrastructural observations. *Anticancer Res.* 2007;27(1A):391–394.

480. Caruso RA, Parisi A, Quattrocchi E, et al. Ultrastructural descriptions of heterotypic aggregation between eosinophils and tumor cells in human gastric carcinomas. *Ultrastruct Pathol.* 2011;35(4):145–149.

481. Yousefi S, Simon D, Stojkov D, Karsonova A, Karaulov A, Simon H-U. In vivo evidence for extracellular DNA trap formation. *Cell Death Dis.* 2020;11(4):1–15.

482. Wang W, Zhang J, Zheng N, Li L, Wang X, Zeng Y. The role of neutrophil extracellular traps in cancer metastasis. *Clin Transl Med.* 2020;10(6):e126.

483. Su J. A brief history of Charcot-Leyden crystal protein/galectin-10 research. *Molecules.* 2018;23(11):2931.

484. Kubach J, Lutter P, Bopp T, et al. Human CD4+CD25+ regulatory T cells: proteome analysis identifies galectin-10 as a novel marker essential for their anergy and suppressive function. *Blood.* 2007;110(5):1550–1558.

485. Schönherr R, Rudolph JM, Redecke L. Protein crystallization in living cells. *Biol Chem.* 2018;399(7):751–772.

486. Lao L-M, Kumakiri M, Nakagawa K, et al. The ultrastructural findings of Charcot-Leyden crystals in stroma of mastocytoma. *J Dermatol Sci.* 1998;17(3):198–204.

487. Carson HJ, Buschmann RJ, Weisz-Carrington P, Choi YS. Identification of Charcot-Leyden crystals by electron microscopy. *Ultrastruct Pathol.* 1992;16(4):403–411.

488. Rodríguez-Alcázar JF, Ataide MA, Engels G, et al. Charcot–Leyden crystals activate the NLRP3 inflammasome and cause IL-1β inflammation in human macrophages. *J Immunol.* 2019;202(2):550–558.

489. Lee N. Mouse models manipulating eosinophilopoiesis. In: Lee, JJ, Rosenberg, HF, eds. *Eosinophils in Health and Disease.* Elsevier; 2013:111–120.

490. Kumar RK, Foster PS. Modeling allergic asthma in mice: pitfalls and opportunities. *Am J Respir Cell Mol Biol.* 2002;27(3):267–272.

491. Guo L, Johnson RS, Schuh JCL. Biochemical characterization of endogenously formed eosinophilic crystals in the lungs of mice. *J Biol Chem.* 2000;275(11):8032–8037.

492. Shamri R, Melo RCN, Young KM, et al. CCL11 elicits secretion of RNases from mouse eosinophils and their cell-free granules. *FASEB J.* 2012;26(5):2084–2093.

493. Dent LA, Strath M, Mellor AL, Sanderson CJ. Eosinophilia in transgenic mice expressing interleukin 5. *J Exp Med.* 1990;172(5):1425–1431.

494. Wang HB, Ghiran I, Matthaei K, Weller PF. Airway eosinophils: allergic inflammation recruited professional antigen-presenting cells. *J Immunol.* 2007;179(11):7585–7592.

495. Wang HB, Weller PF. Pivotal advance: eosinophils mediate early alum adjuvant-elicited B cell priming and IgM production. *J Leukoc Biol.* 2008;83(4):817–821.

496. McGarry MP, Borchers M, Novak EK, et al. Pulmonary pathologies in pallid mice result from nonhematopoietic defects. *Exp Mol Pathol.* 2002;72(3):213–220.

497. Ranger AM, Oukka M, Rengarajan J, Glimcher LH. Inhibitory function of two NFAT family members in lymphoid homeostasis and Th2 development. *Immunity.* 1998;9(5):627–635.

498. Dvorak AM, Tepper RI, Weller PF, et al. Piecemeal degranulation of mast cells in the inflammatory eyelid lesions of interleukin-4 transgenic mice. Evidence of mast cell histamine release in vivo by diamine oxidase-gold enzyme-affinity ultrastructural cytochemistry. *Blood.* 1994;83(12):3600–3612.

499. Pettersson A, Nagy JA, Brown LF, et al. Heterogeneity of the angiogenic response induced in different normal adult tissues by vascular permeability factor/vascular endothelial growth factor. *Lab Investig.* 2000;80(1):99–115.

500. Biermann H, Pietz B, Dreier R, Schmid KW, Sorg C, Sunderkötter C. Murine leukocytes with ring-shaped nuclei include granulocytes, monocytes, and their precursors. *J Leukoc Biol.* 1999;65(2):217–231.

501. Arai M, Mantani Y, Nakanishi S, et al. Morphological and phenotypical diversity of eosinophils in the rat ileum. *Cell Tissue Res.* 2020;381:439–450.

502. Miller F, de Harven E, Palade GE. The structure of eosinophil leukocyte granules in rodents and in man. *J Cell Biol.* 1966;31(2):349–362.

503. Dvorak AM, Nabel G, Pyne K, Cantor H, Dvorak HF, Galli SJ. Ultrastructural identification of the mouse basophil. *Blood.* 1982;59(6):1279–1285.

504. Dyer KD, Moser JM, Czapiga M, Siegel SJ, Percopo CM, Rosenberg HF. Functionally competent eosinophils differentiated ex vivo in high purity from normal mouse bone marrow. *J Immunol.* 2008;181(6):4004–4009.

505. Lee JJ, Lee NA. Eosinophil degranulation: an evolutionary vestige or a universally destructive effector function? *Clin Exp Allergy.* 2005;35(8):986–994.

506. Clark K, Simson L, Newcombe N, et al. Eosinophil degranulation in the allergic lung of mice primarily occurs in the airway lumen. *J Leukoc Biol.* 2004;75(6):1001–1009.

507. Malm-Erjefalt M, Persson CG, Erjefalt JS. Degranulation status of airway tissue eosinophils in mouse models of allergic airway inflammation. *Am J Respir Cell Mol Biol*. 2001;24(3):352–359.

508. Dyer KD, Percopo CM, Fischer ER, Gabryszewski SJ, Rosenberg HF. Pneumoviruses infect eosinophils and elicit MyD88-dependent release of chemoattractant cytokines and interleukin-6. *Blood*. 2009;114(13):2649–2656.

509. Ehrens A, Lenz B, Neumann AL, et al. Microfilariae trigger eosinophil extracellular DNA traps in a dectin-1-dependent manner. *Cell Rep*. 2021;34(2):108621.

510. Gueders MM, Paulissen G, Crahay C, et al. Mouse models of asthma: a comparison between C57BL/6 and BALB/c strains regarding bronchial responsiveness, inflammation, and cytokine production. *Inflamm Res*. 2009;58(12):845–854.

511. Kianmeher M, Ghorani V, Boskabady MH. Animal model of asthma, various methods and measured parameters: a methodological review. *Iran J Allergy Asthma Immunol*. 2016;15(6):445–465.

512. Percopo CM, Dyer KD, Ochkur SI, et al. Activated mouse eosinophils protect against lethal respiratory virus infection. *Blood*. 2014;123(5):743–752.

513. Samarasinghe AE, Melo RCN, Duan S, et al. Eosinophils promote antiviral immunity in mice infected with influenza A virus. *J Immunol*. 2017;198(8):3214–3226.

514. Serra MF, Anjos-Valotta EA, Olsen PC, et al. Nebulized lidocaine prevents airway inflammation, peribronchial fibrosis, and mucus production in a murine model of asthma. *Anesthesiology*. 2012;117(3):580–591.

515. Royce SG, Cheng V, Samuel CS, Tang MLK. The regulation of fibrosis in airway remodeling in asthma. *Mol Cell Endocrinol*. 2012;351(2):167–175.

516. Lenzi HL, Kimmel E, Schechtman H, et al. Histoarchitecture of schistosomal granuloma development and involution: morphogenetic and biomechanical approaches. *Mem Inst Oswaldo Cruz*. 1998;93(1):141–151.

517. Chuah C, Jones MK, Burke ML, McManus DP, Gobert GN. Cellular and chemokine-mediated regulation in schistosome-induced hepatic pathology. *Trends Parasitol*. 2014;30(3):141–150.

518. Andrade ZA. Schistosomiasis and liver fibrosis. *Parasite Immunol*. 2009;31(11):656–663.

519. Dutra HS, Rossi MID, Azevedo SP, El-Cheikh MC, Borojevic R. Haematopoietic capacity of colony-forming cells mobilized in hepatic inflammatory reactions as compared to that of normal bone marrow cells. *Res Immunol*. 1997;148(7):437–444.

520. Rossi MID, Dutra HS, El-Cheikh MC, Bonomo A, Borojevic R. Extramedullar B lymphopoiesis in liver schistosomal granulomas: presence of the early stages and inhibition of the full B cell differentiation. *Int Immunol*. 1999;11(4):509–518.

521. Lenzi HL, Romanha Wde S, Santos RM, et al. Four whole-istic aspects of schistosome granuloma biology: fractal arrangement, internal regulation, autopoietic component and closure. *Mem Inst Oswaldo Cruz*. 2006;101(Suppl 1):219–231.

522. Dvorak AM. Images in clinical medicine. An apoptotic eosinophil. *N Engl J Med*. 1999;340(6):437.

523. Edwards G, Diercks GFH, Seelen MAJ, Horvath B, Van Doorn M, Damman J. Complement activation in autoimmune bullous dermatoses: a comprehensive review. *Front Immunol*. 2019;10:1477.

524. Dyer KD, Garcia-Crespo KE, Percopo CM, Sturm EM, Rosenberg HF. Protocols for identifying, enumerating, and assessing mouse eosinophils. *Methods Mol Biol*. 2013;1032:59–77.

525. Tepper RI, Coffman RL, Leder P. An eosinophil-dependent mechanism for the antitumor effect of interleukin-4. *Science*. 1992;257(5069):548–551.

Figure credits

Chapter 1

Fig. 1.1
(Top panel)

Spencer LA, Bonjour K, Melo RCN, Weller PF. Eosinophil secretion of granule-derived cytokines. *Front Immunol*. 2014;5:496. Licensed under the Creative Commons Attribution 4.0 International (CC BY 4.0). Link to the license: https://creativecommons.org/licenses/by/4.0/.

Chapter 2

Fig. 2.1

Akuthota P, Carmo LAS, Bonjour K, Murphy RO, Silva TP, Gamalier JP, ..., Melo RCN. Extracellular microvesicle production by human eosinophils activated by "inflammatory" stimuli. *Front Cell Dev Biol*. 2016;4:117. Licensed under the Creative Commons Attribution 4.0 International (CC BY 4.0). Link to the license: https://creativecommons.org/licenses/by/4.0/.

Fig. 2.3
(Top panel)

Melo RCN. *Células & Microscopia: Princípios e Práticas*. Barueri: Manole; 2018. pp. 1–300. Reprinted with permission of Manole.

Fig. 2.20

Melo RCN, D'Avila H, Wan HC, Bozza PT, Dvorak AM, Weller PF. Lipid bodies in inflammatory cells: structure, function, and current imaging techniques. *J Histochem Cytochem*. 2011;59(5):540–556. Reprinted with permission of SAGE under the STM Permissions Guidelines.

Fig. 2.21

Melo RCN, Paganoti GF, Dvorak AM, Weller PF. The internal architecture of leukocyte lipid body organelles captured by three-dimensional electron microscopy tomography. *PLoS One*. 2013;8(3):e59578. Licensed under the Creative Commons Attribution 4.0 International (CC BY 4.0). Link to the license: https://creativecommons.org/licenses/by/4.0/.

Fig. 2.24

Melo RCN, Paganoti GF, Dvorak AM, Weller PF. The internal architecture of leukocyte lipid body organelles captured by three-dimensional electron microscopy tomography. *PLoS One*. 2013;8(3):e59578. Licensed under the Creative Commons Attribution 4.0 International (CC BY 4.0). Link to the license: https://creativecommons.org/licenses/by/4.0/.

Fig. 2.25

Wan HC, Melo RCN, Jin Z, Dvorak AM, Weller PF. Roles and origins of leukocyte lipid bodies: proteomic and ultrastructural studies. *FASEB J*. 2007;21(1):167–178. Reprinted with permission of John Wiley and Sons.

Fig. 2.26

Melo RCN, Paganoti GF, Dvorak AM, Weller PF. The internal architecture of leukocyte lipid body organelles captured by three-dimensional electron microscopy tomography. *PLoS One*. 2013;8(3):e59578. Licensed under the Creative Commons Attribution 4.0 International (CC BY 4.0). Link to the license: https://creativecommons.org/licenses/by/4.0/.

Fig. 2.28

Wan HC, Melo RCN, Jin Z, Dvorak AM, Weller PF. Roles and origins of leukocyte lipid bodies: proteomic and ultrastructural studies. *FASEB J*. 2007;21(1):167–178. Reprinted with permission of John Wiley and Sons.

Fig. 2.30

Melo RCN, Dvorak AM. Lipid body–phagosome interaction in macrophages during infectious diseases: host defense or pathogen survival strategy?. *PLoS Pathog*. 2012;8(7):e1002729. Licensed under the Creative Commons Attribution 4.0 International (CC BY 4.0). Link to the license: https://creativecommons.org/licenses/by/4.0/.

Fig. 2.33

Melo RCN, Weller PF. Vesicular trafficking of immune mediators in human eosinophils revealed by immunoelectron microscopy. *Exp Cell Res*. 2016;347(2):385–390. Reprinted with permission of Elsevier.

Fig. 2.34

Melo RCN, Spencer LA, Perez SA, Ghiran I, Dvorak AM, Weller PF. Human eosinophils secrete preformed, granule-stored interleukin-4 through distinct vesicular compartments. *Traffic*. 2005;6(11):1047–1057. Reprinted with permission of John Wiley and Sons.

Fig. 2.39

Melo RCN. *Células & Microscopia: Princípios e Práticas*. Barueri: Manole; 2018. pp. 1–300. Reprinted with permission of Manole.

Chapter 3

Fig. 3.1	Spencer LA, Bonjour K, Melo RCN, Weller PF. Eosinophil secretion of granule-derived cytokines. *Front Immunol.* 2014;5:496. Licensed under the Creative Commons Attribution 4.0 International (CC BY 4.0). Link to the license: https://creativecommons.org/licenses/by/4.0/.
Fig. 3.2 (Partially)	Dvorak AM. Human eosinophil granule. *Int Arch Allergy Immunol.* 1996;111(1):35–35. Reprinted with permission of S. Karger AG, Basel.
Fig. 3.4	Melo RCN, Weller PF. Contemporary understanding of the secretory granules in human eosinophils. *J Leukoc Biol.* 2018;104(1):85–93. Reprinted with permission of John Wiley and Sons.
Fig. 3.5	Carmo LAS, Bonjour K, Ueki S, Neves JS, Liu L, Spencer LA, …, Melo RCN. CD63 is tightly associated with intracellular, secretory events chaperoning piecemeal degranulation and compound exocytosis in human eosinophils. *J Leukoc Biol.* 2016;100(2):391–401. Reprinted with permission of John Wiley and Sons.
Fig. 3.6	Carmo LAS, Bonjour K, Ueki S, Neves JS, Liu L, Spencer LA, …, Melo RCN. CD63 is tightly associated with intracellular, secretory events chaperoning piecemeal degranulation and compound exocytosis in human eosinophils. *J Leukoc Biol.* 2016;100(2):391–401. Reprinted with permission of John Wiley and Sons.
Fig. 3.17	Spencer LA, Bonjour K, Melo RCN, Weller PF. Eosinophil secretion of granule-derived cytokines. *Front Immunol.* 2014;5:496. Licensed under the Creative Commons Attribution 4.0 International (CC BY 4.0). Link to the license: https://creativecommons.org/licenses/by/4.0/.
Fig. 3.21	Melo RCN, Perez SA, Spencer LA, Dvorak AM, Weller PF. Intragranular vesiculotubular compartments are involved in piecemeal degranulation by activated human eosinophils. *Traffic.* 2005;6(10):866–879. Reprinted with permission of John Wiley and Sons.
Fig. 3.22	Melo RCN, Spencer LA, Perez SA, Ghiran I, Dvorak AM, Weller PF. Human eosinophils secrete preformed, granule-stored interleukin-4 through distinct vesicular compartments. *Traffic.* 2005;6(11):1047–1057. Reprinted with permission of John Wiley and Sons.
Fig. 3.23 (Drawing)	Melo RCN. *Células & Microscopia: Princípios e Práticas.* Barueri: Manole; 2018. pp. 1–300. Reprinted with permission of Manole.
Fig. 3.24 (Top panels)	Melo RCN, Spencer LA, Perez SA, Ghiran I, Dvorak AM, Weller PF. Human eosinophils secrete preformed, granule-stored interleukin-4 through distinct vesicular compartments. *Traffic.* 2005;6(11):1047–1057. Reprinted with permission of John Wiley and Sons.
Fig. 3.24 (Bottom panel)	Melo RCN, Dvorak AM, Weller PF. Contributions of electron microscopy to understand secretion of immune mediators by human eosinophils. *Microsc Microanal.* 2010;16(6):653–660. Reprinted with permission of Cambridge University Press.
Fig. 3.25	Melo RCN, Spencer LA, Perez SA, Ghiran I, Dvorak AM, Weller PF. Human eosinophils secrete preformed, granule-stored interleukin-4 through distinct vesicular compartments. *Traffic.* 2005;6(11):1047–1057. Reprinted with permission of John Wiley and Sons.
Fig. 3.28 (Partially)	Dvorak AM, Ackerman SJ, Furitsu T, Estrella P, Letourneau L, Ishizaka T. Mature eosinophils stimulated to develop in human-cord blood mononuclear cell cultures supplemented with recombinant human interleukin-5. II. Vesicular transport of specific granule matrix peroxidase, a mechanism for effecting piecemeal degranulation. *Am J Pathol.* 1992;140(4):795–807.
Fig. 3.31	Melo RCN, Dvorak AM, Weller PF. Eosinophil ultrastructure. In: Lee J, Rosenberg H, eds. *Eosinophils in Health and Disease*; vol. 1. New York: Elsevier; 2012:20–27.
Fig. 3.34	Melo RCN, Wang H, Silva TP, Imoto Y, Fujieda S, Fukuchi M, …, Weller PF. Galectin-10, the protein that forms Charcot-Leyden crystals, is not stored in granules but resides in the peripheral cytoplasm of human eosinophils. *J Leukoc Biol.* 2020;108(1):139–149. Reprinted with permission of John Wiley and Sons.
Fig. 3.36	Barroso MV, Gropillo I, Detoni MA, Thompson-Souza GA, Muniz VS, Vasconcelos CRI, …, Neves JS. Structural and signaling events driving Aspergillus fumigatus-induced human eosinophil extracellular trap release. *Front Microbiol.* 2021;12:274. Licensed under the Creative Commons Attribution 4.0 International (CC BY 4.0). Link to the license: https://creativecommons.org/licenses/by/4.0/.
Fig. 3.37	Akuthota P, Carmo LAS, Bonjour K, Murphy RO, Silva TP, Gamalier JP, …, Melo RCN. Extracellular microvesicle production by human eosinophils activated by "inflammatory" stimuli. *Front Cell Dev Biol.* 2016;4:117. Licensed under the Creative Commons Attribution 4.0 International (CC BY 4.0). Link to the license: https://creativecommons.org/licenses/by/4.0/.
Fig. 3.38	Akuthota P, Carmo LAS, Bonjour K, Murphy RO, Silva TP, Gamalier JP, …, Melo RCN. Extracellular microvesicle production by human eosinophils activated by "inflammatory" stimuli. *Front Cell Dev Biol.* 2016;4:117. Licensed under the Creative Commons Attribution 4.0 International (CC BY 4.0). Link to the license: https://creativecommons.org/licenses/by/4.0/.

Chapter 4

Fig. 4.1
(Top panel
marked "resting")

Carmo LAS, Bonjour K, Spencer LA, Weller PF, Melo RCN. Single-cell analyses of human eosinophils at high resolution to understand compartmentalization and vesicular trafficking of interferon-gamma. *Front Immunol*. 2018;9:1542. Licensed under the Creative Commons Attribution 4.0 International (CC BY 4.0). Link to the license: https://creativecommons.org/licenses/by/4.0/.

Fig. 4.4
(Top panels)

Carmo LAS, Bonjour K, Ueki S, Neves JS, Liu L, Spencer LA, …, Melo RCN. CD63 is tightly associated with intracellular, secretory events chaperoning piecemeal degranulation and compound exocytosis in human eosinophils. *J Leukoc Biol*. 2016;100(2):391–401. Reprinted with permission of John Wiley and Sons.

Fig. 4.6

Neves VH, Bonjour K, Palazzi C, Neves JS, Dvorak AM, Weller PF, and Melo RCN. Dermal-infiltrated and blood human eosinophils respond to Stem Cell Factor treatment with degranulation and increased production of sombrero vesicles. *Frontiers in Allergy* (in press). Licensed under the Creative Commons Attribution 4.0 International (CC BY 4.0). Link to the license: https://creativecommons.org/licenses/by/4.0/.

Fig. 4.7

Neves VH, Bonjour K, Palazzi C, Neves JS, Dvorak AM, Weller PF, and Melo RCN. Dermal-infiltrated and blood human eosinophils respond to Stem Cell Factor treatment with degranulation and increased production of sombrero vesicles. *Frontiers in Allergy* (in press). Licensed under the Creative Commons Attribution 4.0 International (CC BY 4.0). Link to the license: https://creativecommons.org/licenses/by/4.0/.

Fig. 4.8

Melo RCN, Perez SA, Spencer LA, Dvorak AM, Weller PF. Intragranular vesiculotubular compartments are involved in piecemeal degranulation by activated human eosinophils. *Traffic*. 2005;6(10):866–879. Reprinted with permission of John Wiley and Sons.

Fig. 4.10

Melo RCN, Perez SA, Spencer LA, Dvorak AM, Weller PF. Intragranular vesiculotubular compartments are involved in piecemeal degranulation by activated human eosinophils. *Traffic*. 2005;6(10):866–879. Reprinted with permission of John Wiley and Sons.

Fig. 4.11

Carmo LAS, Bonjour K, Spencer LA, Weller PF, Melo RCN. Single-cell analyses of human eosinophils at high resolution to understand compartmentalization and vesicular trafficking of interferon-gamma. *Front Immunol*. 2018;9:1542. Licensed under the Creative Commons Attribution 4.0 International (CC BY 4.0). Link to the license: https://creativecommons.org/licenses/by/4.0/.

Fig. 4.12

Carmo LAS, Bonjour K, Ueki S, Neves JS, Liu L, Spencer LA, …, Melo RCN. CD63 is tightly associated with intracellular, secretory events chaperoning piecemeal degranulation and compound exocytosis in human eosinophils. *J Leukoc Biol*. 2016;100(2):391–401. Reprinted with permission of John Wiley and Sons.

Fig. 4.14

Neves VH, Bonjour K, Palazzi C, Neves JS, Dvorak AM, Weller PF, and Melo RCN. Dermal-infiltrated and blood human eosinophils respond to Stem Cell Factor treatment with degranulation and increased production of sombrero vesicles. *Frontiers in Allergy* (in press). Licensed under the Creative Commons Attribution 4.0 International (CC BY 4.0). Link to the license: https://creativecommons.org/licenses/by/4.0/.

Fig. 4.15

Neves VH, Bonjour K, Palazzi C, Neves JS, Dvorak AM, Weller PF, and Melo RCN. Dermal-infiltrated and blood human eosinophils respond to Stem Cell Factor treatment with degranulation and increased production of sombrero vesicles. *Frontiers in Allergy* (in press). Licensed under the Creative Commons Attribution 4.0 International (CC BY 4.0). Link to the license: https://creativecommons.org/licenses/by/4.0/.

Fig. 4.19

Carmo LAS, Weller PF, Melo RCN. TNF-alpha-induced lipid droplets interact with secretory granules and are sites for IFN-gamma in human eosinophils. *Front Cell Dev Biol*. (in press). Licensed under the Creative Commons Attribution 4.0 International (CC BY 4.0). Link to the license: https://creativecommons.org/licenses/by/4.0/.

Fig. 4.20

Melo RCN, Paganoti GF, Dvorak AM, Weller PF. The internal architecture of leukocyte lipid body organelles captured by three-dimensional electron microscopy tomography. *PloS One*. 2013;8(3):e59578. Licensed under the Creative Commons Attribution 4.0 International (CC BY 4.0). Link to the license: https://creativecommons.org/licenses/by/4.0/.

Fig. 4.31

Melo RCN, Sabban A, Weller PF. Leukocyte lipid bodies: inflammation-related organelles are rapidly detected by wet scanning electron microscopy. *J Lipid Res*. 2006;47(11):2589–2594. Licensed under the Creative Commons Attribution 4.0 International (CC BY 4.0). Link to the license: https://creativecommons.org/licenses/by/4.0/.

Fig. 4.33

Neves VH, Palazzi C, Bonjour K, Ueki S, Weller PF, and Melo RCN (2022). In vivo ETosis of human eosinophils: the ultrastructural signature captured by TEM in eosinophilic diseases. *Front Immunol* (in press). Licensed under the Creative Commons Attribution 4.0 International (CC BY 4.0). Link to the license: https://creativecommons.org/licenses/by/4.0/.

Fig. 4.45   Akuthota P, Carmo LAS, Bonjour K, Murphy RO, Silva TP, Gamalier JP, …, Melo RCN. Extracellular microvesicle production by human eosinophils activated by "inflammatory" stimuli. *Front Cell Dev Biol.* 2016;4:117. Licensed under the Creative Commons Attribution 4.0 International (CC BY 4.0). Link to the license: https://creativecommons.org/licenses/by/4.0/.

Chapter 5

Fig. 5.1   Melo RCN, Morgan E, Monahan-Earley R, Dvorak AM, Weller PF. Pre-embedding immunogold labeling to optimize protein localization at subcellular compartments and membrane microdomains of leukocytes. *Nat Protoc.* 2014;9(10):2382–2394.

Fig. 5.2   Carmo LAS, Bonjour K, Ueki S, Neves JS, Liu L, Spencer LA, …, Melo RCN. CD63 is tightly associated with
(Bottom panel)   intracellular, secretory events chaperoning piecemeal degranulation and compound exocytosis in human eosinophils. *J Leukoc Biol.* 2016;100(2):391–401. Reprinted with permission of John Wiley and Sons.

Fig. 5.3   Weller PF, Spencer LA. Functions of tissue-resident eosinophils. *Nat Rev Immunol.* 2017;17(12):746–760.
(Top panels)   Reprinted with permission of Springer Nature.

Fig. 5.3   Melo RCN, Spencer LA, Perez SA, Neves JS, Bafford SP, Morgan ES, …, Weller PF. Vesicle-mediated secretion
(Bottom panel/left) of human eosinophil granule-derived major basic protein. *Lab Invest.* 2009;89(7):769–781.

Fig. 5.3   Melo RCN, Spencer LA, Perez SA, Ghiran I, Dvorak AM, Weller PF. Human eosinophils secrete preformed,
(Bottom panel/right) granule-stored interleukin-4 through distinct vesicular compartments. *Traffic.* 2005;6(11):1047–1057. Reprinted with permission of John Wiley and Sons.

Fig. 5.4   Melo RCN, Perez SA, Spencer LA, Dvorak AM, Weller PF. Intragranular vesiculotubular compartments are involved in piecemeal degranulation by activated human eosinophils. *Traffic.* 2005;6(10):866–879. Reprinted with permission of John Wiley and Sons.

Fig. 5.5   Melo RCN, Spencer LA, Perez SA, Neves JS, Bafford SP, Morgan ES, …, Weller PF. Vesicle-mediated secretion of human eosinophil granule-derived major basic protein. *Lab Invest.* 2009;89(7):769–781.

Fig. 5.6   Weller PF, Spencer LA. Functions of tissue-resident eosinophils. *Nat Rev Immunol.* 2017;17(12):746–760.
(Top panels and   Reprinted with permission of Springer Nature. Reprinted with permission of Springer Nature.
bottom panel/right)

Fig. 5.6   Melo RCN, Spencer LA, Perez SA, Ghiran I, Dvorak AM, Weller PF. Human eosinophils secrete preformed,
(Bottom panel/left) granule-stored interleukin-4 through distinct vesicular compartments. *Traffic.* 2005;6(11):1047–1057. Reprinted with permission of John Wiley and Sons.

Fig. 5.7   Spencer LA, Bonjour K, Melo RCN, Weller PF. Eosinophil secretion of granule-derived cytokines. *Front*
(Top panels marked *Immunol.* 2014;5:496. Licensed under the Creative Commons Attribution 4.0 International (CC BY 4.0). Link to
"conventional TEM" and the license: Link to the license: https://creativecommons.org/licenses/by/4.0/.
"MBP-immunolabeling")

Fig. 5.7   Melo RCN, Dvorak AM, Weller PF. Electron tomography and immunonanogold electron microscopy for
(Bottom panel/left) investigating intracellular trafficking and secretion in human eosinophils. *J Cell Mol Med.* 2008;12(4):1416–1419. Reprinted with permission of John Wiley and Sons.

Fig. 5.7   Spencer LA, Melo RCN, Perez SA, Bafford SP, Dvorak AM, Weller PF. Cytokine receptor-mediated trafficking
(Bottom panel/right) of preformed IL-4 in eosinophils identifies an innate immune mechanism of cytokine secretion. *Proc Natl Acad Sci.* 2006;103(9):3333–3338. Reprinted with permission of National Academy of Sciences. Copyright (2006) National Academy of Sciences, USA.

Fig. 5.8   Melo RCN, Spencer LA, Perez SA, Ghiran I, Dvorak AM, Weller PF. Human eosinophils secrete preformed, granule-stored interleukin-4 through distinct vesicular compartments. *Traffic.* 2005;6(11):1047–1057. Reprinted with permission of John Wiley and Sons.

Fig. 5.9   Carmo LAS, Bonjour K, Spencer LA, Weller PF, Melo RCN. Single-cell analyses of human eosinophils at high resolution to understand compartmentalization and vesicular trafficking of interferon-gamma. *Front Immunol.* 2018;9:1542. Licensed under the Creative Commons Attribution 4.0 International (CC BY 4.0). Link to the license: https://creativecommons.org/licenses/by/4.0/.

Fig. 5.10   Carmo LAS, Bonjour K, Spencer LA, Weller PF, Melo RCN. Single-cell analyses of human eosinophils at high resolution to understand compartmentalization and vesicular trafficking of interferon-gamma. *Front Immunol.* 2018;9:1542. Licensed under the Creative Commons Attribution 4.0 International (CC BY 4.0). Link to the license: https://creativecommons.org/licenses/by/4.0/.

Fig. 5.12   Carmo LAS, Weller PF, Melo RCN. TNF-alpha-induced lipid droplets interact with secretory granules and are sites for IFN-gamma in human eosinophils. *Front Cell Dev Biol.* (in press). Licensed under the Creative Commons Attribution 4.0 International (CC BY 4.0). Link to the license: https://creativecommons.org/licenses/by/4.0/.

Fig. 5.13 (Drawing)	Melo RCN. *Células & Microscopia: Princípios e Práticas*. Barueri: Manole; 2018. pp. 1–300. Reprinted with permission of Manole.
Fig. 5.14	Carmo LAS, Dias FF, Malta KK, Amaral KB, Shamri R, Weller PF, Melo RCN. Expression and subcellular localization of the Qa-SNARE syntaxin17 in human eosinophils. *Exp Cell Res*. 2015;337(2):129–135. Reprinted with permission of Elsevier.
Fig. 5.15	Carmo LAS, Dias FF, Malta KK, Amaral KB, Shamri R, Weller PF, Melo RCN. Expression and subcellular localization of the Qa-SNARE syntaxin17 in human eosinophils. *Exp Cell Res*. 2015;337(2):129–135. Reprinted with permission of Elsevier.
Fig. 5.20	Melo RCN, Morgan E, Monahan-Earley R, Dvorak AM, Weller PF. Pre-embedding immunogold labeling to optimize protein localization at subcellular compartments and membrane microdomains of leukocytes. *Nat Protoc*. 2014;9(10):2382–2394.
Fig. 5.21	Carmo LAS, Bonjour K, Ueki S, Neves JS, Liu L, Spencer LA, …, Melo RCN. CD63 is tightly associated with intracellular, secretory events chaperoning piecemeal degranulation and compound exocytosis in human eosinophils. *J Leukoc Biol*. 2016;100(2):391–401. Reprinted with permission of John Wiley and Sons.
Fig. 5.22	Carmo LAS, Bonjour K, Ueki S, Neves JS, Liu L, Spencer LA, …, Melo RCN. CD63 is tightly associated with intracellular, secretory events chaperoning piecemeal degranulation and compound exocytosis in human eosinophils. *J Leukoc Biol*. 2016;100(2):391–401. Reprinted with permission of John Wiley and Sons.
Fig. 5.25	Carmo LAS, Bonjour K, Ueki S, Neves JS, Liu L, Spencer LA, …, Melo RCN. CD63 is tightly associated with intracellular, secretory events chaperoning piecemeal degranulation and compound exocytosis in human eosinophils. *J Leukoc Biol*. 2016;100(2):391–401. Reprinted with permission of John Wiley and Sons.
Fig. 5.26	Carmo LAS, Bonjour K, Ueki S, Neves JS, Liu L, Spencer LA, …, Melo RCN. CD63 is tightly associated with intracellular, secretory events chaperoning piecemeal degranulation and compound exocytosis in human eosinophils. *J Leukoc Biol*. 2016;100(2):391–401. Reprinted with permission of John Wiley and Sons.
Fig. 5.27	Akuthota P, Carmo LAS, Bonjour K, Murphy RO, Silva TP, Gamalier JP, …, Melo RCN. Extracellular microvesicle production by human eosinophils activated by "inflammatory" stimuli. *Front Cell Dev Biol*. 2016;4:117. Licensed under the Creative Commons Attribution 4.0 International (CC BY 4.0). Link to the license: https://creativecommons.org/licenses/by/4.0/.
Fig. 5.28	Melo RCN, Wang H, Silva TP, Imoto Y, Fujieda S, Fukuchi M, …, Weller PF. Galectin-10, the protein that forms Charcot-Leyden crystals, is not stored in granules but resides in the peripheral cytoplasm of human eosinophils. *J Leukoc Biol*. 2020;108(1):139–149. Reprinted with permission of John Wiley and Sons.
Fig. 5.29	Melo RCN, Wang H, Silva TP, Imoto Y, Fujieda S, Fukuchi M, …, Weller PF. Galectin-10, the protein that forms Charcot-Leyden crystals, is not stored in granules but resides in the peripheral cytoplasm of human eosinophils. *J Leukoc Biol*. 2020;108(1):139–149. Reprinted with permission of John Wiley and Sons.
Fig. 5.30	Melo RCN, Wang H, Silva TP, Imoto Y, Fujieda S, Fukuchi M, …, Weller PF. Galectin-10, the protein that forms Charcot-Leyden crystals, is not stored in granules but resides in the peripheral cytoplasm of human eosinophils. *J Leukoc Biol*. 2020;108(1):139–149. Reprinted with permission of John Wiley and Sons.
Fig. 5.31	Melo RCN, Wang H, Silva TP, Imoto Y, Fujieda S, Fukuchi M, …, Weller PF. Galectin-10, the protein that forms Charcot-Leyden crystals, is not stored in granules but resides in the peripheral cytoplasm of human eosinophils. *J Leukoc Biol*. 2020;108(1):139–149. Reprinted with permission of John Wiley and Sons.
Fig. 5.33	Melo RCN, Wang H, Silva TP, Imoto Y, Fujieda S, Fukuchi M, …, Weller PF. Galectin-10, the protein that forms Charcot-Leyden crystals, is not stored in granules but resides in the peripheral cytoplasm of human eosinophils. *J Leukoc Biol*. 2020;108(1):139–149. Reprinted with permission of John Wiley and Sons.
Fig. 5.34	Melo RCN, Wang H, Silva TP, Imoto Y, Fujieda S, Fukuchi M, …, Weller PF. Galectin-10, the protein that forms Charcot-Leyden crystals, is not stored in granules but resides in the peripheral cytoplasm of human eosinophils. *J Leukoc Biol*. 2020;108(1):139–149. Reprinted with permission of John Wiley and Sons.
Fig. 5.38	Melo RCN, Weller PF. Lipid droplets in leukocytes: organelles linked to inflammatory responses. *Exp Cell Res*. 2016;340(2):193–197. Reprinted with permission of Elsevier.
Fig. 5.40 (Bottom panel)	Melo RCN, D'Avila H, Wan HC, Bozza PT, Dvorak AM, Weller PF. Lipid bodies in inflammatory cells: structure, function, and current imaging techniques. *J Histochem Cytochem*. 2011;59(5):540–556. Reprinted with permission of SAGE under the STM Permissions Guidelines.

Chapter 6

Fig. 6.1 (Partially)	Spencer LA, Bonjour K, Melo RCN, Weller PF. Eosinophil secretion of granule-derived cytokines. *Front Immunol*. 2014;5:496. Licensed under the Creative Commons Attribution 4.0 International (CC BY 4.0). Link to the license: https://creativecommons.org/licenses/by/4.0/.

Fig. 6.9 Ueki S, Melo RCN, Ghiran I, Spencer LA, Dvorak AM, Weller PF. Eosinophil extracellular DNA trap cell death mediates lytic release of free secretion-competent eosinophil granules in humans. *Blood*. 2013;121(11):2074–2083. Reprinted with permission of Elsevier.

Fig. 6.11 Melo RCN. *Células & Microscopia: Princípios e Práticas*. Barueri: Manole; 2018. pp. 1–300. Reprinted with permission of Manole.

Fig. 6.17
(Top panels) Barroso MV, Gropillo I, Detoni MA, Thompson-Souza GA, Muniz VS, Vasconcelos CRI, …, Neves JS. Structural and signaling events driving Aspergillus fumigatus-induced human eosinophil extracellular trap release. *Front Microbiol*. 2021;12:274. Licensed under the Creative Commons Attribution 4.0 International (CC BY 4.0). Link to the license: https://creativecommons.org/licenses/by/4.0/.

Fig. 6.19 Neves VH, Palazzi C, Bonjour K, Ueki S, Weller PF, and Melo RCN. In vivo ETosis of human eosinophils: the ultrastructural signature captured by TEM in eosinophilic diseases. *Front Immunol* (in press). Licensed under the Creative Commons Attribution 4.0 International (CC BY 4.0). Link to the license: https://creativecommons.org/licenses/by/4.0/.

Fig. 6.20 Spencer LA, Bonjour K, Melo RCN, Weller PF. Eosinophil secretion of granule-derived cytokines. *Front Immunol*. 2014;5:496. Licensed under the Creative Commons Attribution 4.0 International (CC BY 4.0). Link to the license: https://creativecommons.org/licenses/by/4.0/.

Fig. 6.21 Neves VH, Palazzi C, Bonjour K, Ueki S, Weller PF, and Melo RCN. In vivo ETosis of human eosinophils: the ultrastructural signature captured by TEM in eosinophilic diseases. *Front Immunol* (in press). Licensed under the Creative Commons Attribution 4.0 International (CC BY 4.0). Link to the license: https://creativecommons.org/licenses/by/4.0/.

Fig. 6.23 Neves VH, Palazzi C, Bonjour K, Ueki S, Weller PF, and Melo RCN. In vivo ETosis of human eosinophils: the ultrastructural signature captured by TEM in eosinophilic diseases. *Front Immunol* (in press). Licensed under the Creative Commons Attribution 4.0 International (CC BY 4.0). Link to the license: https://creativecommons.org/licenses/by/4.0/.

Fig. 6.24 Ueki S, Melo RCN, Ghiran I, Spencer LA, Dvorak AM, Weller PF. Eosinophil extracellular DNA trap cell death mediates lytic release of free secretion-competent eosinophil granules in humans. *Blood*. 2013;121(11):2074–2083. Reprinted with permission of Elsevier.

Fig. 6.26 Muniz VS, Silva JC, Braga YA, Melo RCN, Ueki S, Takeda M, …, Neves JS. Eosinophils release extracellular DNA traps in response to Aspergillus fumigatus. *J Allergy Clin Immunol*. 2018;141(2):571–585. Reprinted with permission of Elsevier.

Fig. 6.29 Ueki S, Tokunaga T, Melo RCN, Saito H, Honda K, Fukuchi M, …, Weller PF. Charcot-Leyden crystal formation is closely associated with eosinophil extracellular trap cell death. *Blood*. 2018;132(20):2183–2187. Reprinted with permission of Elsevier.

Fig. 6.30 Ueki S, Tokunaga T, Melo RCN, Saito H, Honda K, Fukuchi M, …, Weller PF. Charcot-Leyden crystal formation is closely associated with eosinophil extracellular trap cell death. *Blood*. 2018;132(20):2183–2187. Reprinted with permission of Elsevier.

Fig. 6.31 Neves VH, Palazzi C, Bonjour K, Ueki S, Weller PF, and Melo RCN. In vivo ETosis of human eosinophils: the ultrastructural signature captured by TEM in eosinophilic diseases. *Front Immunol* (in press). Licensed under the Creative Commons Attribution 4.0 International (CC BY 4.0). Link to the license: https://creativecommons.org/licenses/by/4.0/.

Fig. 6.36 Dvorak, AM. An apoptotic eosinophil. *N Engl J Med*. 1999;340(6):437–437.

Fig. 6.38 Weller PF, Marshall WL, Lucey DR, Rand TH, Dvorak, AM, Finberg RW. Infection, apoptosis, and killing of mature human eosinophils by human immunodeficiency virus-1. *AM J Resp Cell Mol*. 1995;13(5):610–620. Reprinted with permission of the American Thoracic Society. Copyright © 2021 American Thoracic Society. All rights reserved. The American Journal of Respiratory Cell and Molecular Biology is an official journal of the American Thoracic Society.

Chapter 7

Fig. 7.1 Melo RCN, Weller PF. Contemporary understanding of the secretory granules in human eosinophils. *J Leukoc Biol*. 2018;104(1):85–93. Reprinted with permission of John Wiley and Sons.

Fig. 7.14 Dvorak AM, Saito H, Estrella P, Kissell S, Arai N, Ishizaka T. Ultrastructure of eosinophils and basophils stimulated to develop in human cord blood mononuclear cell cultures containing recombinant human interleukin-5 or interleukin-3. *Lab Invest*. 1989;61(1):116–132.

Fig. 7.30
(Top panel—left
micrograph/partially) Dvorak AM, Saito H, Estrella P, Kissell S, Arai N, Ishizaka T. Ultrastructure of eosinophils and basophils stimulated to develop in human cord blood mononuclear cell cultures containing recombinant human interleukin-5 or interleukin-3. *Lab Invest*. 1989;61(1):116–132.

Chapter 8

Fig. 8.3	Neves VH, Palazzi C, Bonjour K, Ueki S, Weller PF, and Melo RCN. In vivo ETosis of human eosinophils: the ultrastructural signature captured by TEM in eosinophilic diseases. *Front Immunol* (in press). Licensed under the Creative Commons Attribution 4.0 International (CC BY 4.0). Link to the license: https://creativecommons.org/licenses/by/4.0/.
Fig. 8.6	Neves VH, Palazzi C, Bonjour K, Ueki S, Weller PF, and Melo RCN. In vivo ETosis of human eosinophils: the ultrastructural signature captured by TEM in eosinophilic diseases. *Front Immunol* (in press). Licensed under the Creative Commons Attribution 4.0 International (CC BY 4.0). Link to the license: https://creativecommons.org/licenses/by/4.0/.
Fig. 8.8	Ueki S, Tokunaga T, Melo RCN, Saito H, Honda K, Fukuchi M, …, Weller PF. Charcot-Leyden crystal formation is closely associated with eosinophil extracellular trap cell death. *Blood.* 2018;132(20):2183–2187. Reprinted with permission of Elsevier.
Fig. 8.9	Neves VH, Palazzi C, Bonjour K, Ueki S, Weller PF, and Melo RCN. In vivo ETosis of human eosinophils: the ultrastructural signature captured by TEM in eosinophilic diseases. *Front Immunol* (in press). Licensed under the Creative Commons Attribution 4.0 International (CC BY 4.0). Link to the license: https://creativecommons.org/licenses/by/4.0/.
Fig. 8.10	Ueki S, Tokunaga T, Melo RCN, Saito H, Honda K, Fukuchi M, …, Weller PF. Charcot-Leyden crystal formation is closely associated with eosinophil extracellular trap cell death. *Blood.* 2018;132(20):2183–2187. Reprinted with permission of Elsevier.
Fig. 8.11	Cheng JF, Ott NL, Peterson EA, George TJ, Hukee MJ, Gleich GJ, Leiferman KM. Dermal eosinophils in atopic dermatitis undergo cytolytic degeneration. *J Allergy Clin Immunol.* 1997;99(5):683–692. Reprinted with permission of Elsevier.
Fig. 8.12 (Top panel/left)	Edwards G, Diercks GF, Seelen MA, Horvath B, Van Doorn M, Damman J. Complement activation in autoimmune bullous dermatoses: a comprehensive review. *Front Immunol.* 2019;10:1477. Licensed under the Creative Commons Attribution 4.0 International (CC BY 4.0). Link to the license: https://creativecommons.org/licenses/by/4.0/.
Fig. 8.12 (Top panel/right)	SarahKayb/Wikimedia Commons, licensed under the Creative Commons Attribution ShareAlike 4.0 International (CC BY-SA 4.0). Link to the license https://creativecommons.org/licenses/by-sa/4.0/.
Fig. 8.12 (Bottom panel)	Dvorak AM, Mihm Jr MC, Osage JE, Kwan TH, Austen KF, Wintroub BU. Bullous pemphigoid, and ultrastructural study of the inflammatory response: eosinophil, basophil and mast cell granule changes in multiple biopsies from one patient. *J Investig Dermatol.* 1982;78(2):91–101.
Fig. 8.13	Dvorak AM, Mihm Jr MC, Osage JE, Kwan TH, Austen KF, Wintroub BU. Bullous pemphigoid, and ultrastructural study of the inflammatory response: eosinophil, basophil and mast cell granule changes in multiple biopsies from one patient. *J Investig Dermatol.* 1982;78(2):91–101.
Fig. 8.15	Dvorak AM, Mihm Jr MC, Osage JE, Kwan TH, Austen KF, Wintroub BU. Bullous pemphigoid, and ultrastructural study of the inflammatory response: eosinophil, basophil and mast cell granule changes in multiple biopsies from one patient. *J Investig Dermatol.* 1982;78(2):91–101.
Fig. 8.16	Dvorak AM, Mihm Jr MC, Osage JE, Kwan TH, Austen KF, Wintroub BU. Bullous pemphigoid, and ultrastructural study of the inflammatory response: eosinophil, basophil and mast cell granule changes in multiple biopsies from one patient. *J Investig Dermatol.* 1982;78(2):91–101.
Fig. 8.34	Ueki S, Tokunaga T, Melo RCN, Saito H, Honda K, Fukuchi M, …, Weller PF. Charcot-Leyden crystal formation is closely associated with eosinophil extracellular trap cell death. *Blood.* 2018;132(20):2183–2187. Reprinted with permission of Elsevier.
Fig. 8.40	Dvorak AM, Weller PF, Monahan-Earley RA, Letourneau L, Ackerman SJ. Ultrastructural localization of Charcot-Leyden crystal protein (lysophospholipase) and peroxidase in macrophages, eosinophils, and extracellular matrix of the skin in the hypereosinophilic syndrome. *Lab Invest.* 1990;62(5):590–607.
Fig. 8.57	Neves VH, Palazzi C, Bonjour K, Ueki S, Weller PF, and Melo RCN. In vivo ETosis of human eosinophils: the ultrastructural signature captured by TEM in eosinophilic diseases. *Front Immunol* (in press). Licensed under the Creative Commons Attribution 4.0 International (CC BY 4.0). Link to the license: https://creativecommons.org/licenses/by/4.0/.
Fig. 8.60	Neves VH, Palazzi C, Bonjour K, Ueki S, Weller PF, and Melo RCN. In vivo ETosis of human eosinophils: the ultrastructural signature captured by TEM in eosinophilic diseases. *Front Immunol* (in press). Licensed under the Creative Commons Attribution 4.0 International (CC BY 4.0). Link to the license: https://creativecommons.org/licenses/by/4.0/.

Chapter 9

Index

Note: Page numbers followed by *f* indicate figures.